NUREG-1821

Final Safety Evaluation Report
on the Construction Authorization
Request for the Mixed Oxide Fuel
Fabrication Facility at the Savannah
River Site, South Carolina

Docket No. 70-3098

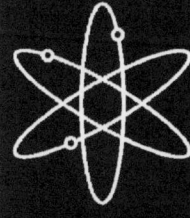

Duke Cogema Stone & Webster, L.L.C.

U.S. Nuclear Regulatory Commission
Office of Nuclear Material Safety and Safeguards
Washington, DC 20555-0001

AVAILABILITY OF REFERENCE MATERIALS
IN NRC PUBLICATIONS

NUREG-1821

Final Safety Evaluation Report
on the Construction Authorization
Request for the Mixed Oxide Fuel
Fabrication Facility at the Savannah
River Site, South Carolina

Docket No. 70-3098

Duke Cogema Stone & Webster, L.L.C.

Manuscript Completed: March 2005
Date Published: March 2005

Division of Fuel Cycle Safety and Safeguards
Office of Nuclear Material Safety and Safeguards
U.S. Nuclear Regulatory Commission
Washington, DC 20555-0001

This page intentionally left blank.

ABSTRACT

On February 28, 2001, Duke COGEMA Stone & Webster (DCS or the applicant), a limited liability company, submitted a request to the U.S. Nuclear Regulatory Commission (NRC) to construct a Mixed Oxide (MOX) Fuel Fabrication Facility (MFFF or the facility) on the U.S. Department of Energy's (DOE) Savannah River Site near Aiken, South Carolina. DCS is a DOE contractor. If its construction authorization request (CAR), as revised, is approved, DCS is expected to then submit a license application for authority to possess and use licensed material at the MFFF. Following an acceptance review of such an application, a notice of opportunity for hearing would be published in the *Federal Register*. In support of its revised CAR, DCS has submitted several items to the NRC, including a quality assurance plan and an environmental report.

Under the applicable requirements of Title 10, Part 70, "Domestic Licensing of Special Nuclear Material," of the *Code of Federal Regulations* (10 CFR Part 70), before a possession and use license may be issued, the NRC must first authorize construction of the facility. This final safety evaluation report (FSER) documents the NRC staff's review of the revised CAR and supplemental supporting information provided by the applicant. This FSER pertains only to approval of construction and does not address operational aspects of the MFFF.

If the MFFF is eventually authorized to operate, it would be a key asset of the DOE Surplus Plutonium Disposition Program (SPDP), which is being implemented as a result of a bilateral agreement with the Russian Federation. Pursuant to this agreement, the United States and the Russian Federation would each convert 34 metric tons (37.5 tons) of weapons-grade plutonium that has been declared excess to national security needs into forms less usable in nuclear weapons. As part of the SPDP, surplus U.S. weapons-grade plutonium would be converted into MOX fuel and irradiated in commercial reactors to produce electricity. Following irradiation, the resulting spent fuel would contain plutonium in a form less usable in nuclear weapons.

This page intentionally left blank.

TABLE OF CONTENTS

FIGURES

TABLES

ABBREVIATIONS

ac	alternating current
ACI	American Concrete Institute
AEC	active engineered control
AEGL	acute exposure guideline level
AFS	alternate feedstock
AHJ	authorities having jurisdiction
AISC	American Institute of Steel Construction
ALARA	as low as reasonably achievable
ALI	allowable limit on intake
ALOHA	areal locations of hazardous atmospheres
A-MIMAS	advanced micronized master blend
ANS	American Nuclear Society
ANSI	American National Standards Institute
AOA	area of applicability
AP	aqueous polishing
ARF	atmospheric release fraction
ASHRAE	American Society of Heating, Refrigeration, and Air-Conditioning Engineers
ASME	American Society of Mechanical Engineers
ASTM	American Society for Testing and Materials
AWS	American Welding Society
BA	Bachelor of Arts
BDC	baseline design criterion/criteria
BMF	fuel fabrication building
BP&V	Boiler and Pressure Vessel Code (ASME)
B.R.	breathing rate
BS	Bachelor of Science
CAAS	criticality accident alarm system
CAM	continuous air monitor
CAR	construction authorization request
CCU	criticality control unit
CEDE	committed effective dose equivalent
CFM	cubic feet per minute
CFR	*Code of Federal Regulations*
CGA	Compressed Gas Association
CM	configuration management
CRT	cargo restraint transporter
CS	conventional seismic
CSE	criticality safety evaluation
CWTS	clean water treatment system
dc	direct current
DCF	dose conversion factor
DCP	double-contingency principle
DCS	Duke Cogema Stone & Webster

DOE	U.S. Department of Energy
DPSG	Duke Project Services Group, Inc.
DR	damage ratio
DSER	draft safety evaluation report
DU	depleted uranium
DUO$_2$	depleted uranium dioxide
EALF	energy of average lethargy causing fission
ECR	emergency control room
ECRACS	emergency control room air conditioning system
EDG	emergency diesel generator
EDGFOS	emergency diesel generator fuel oil system
EDMS	electronic data management system
EIS	environmental impact statement
EMMH	external manmade hazard
ENDF	evaluated nuclear data file
EPA	Environmental Protection Agency
EPRI	Electric Power Research Institute
EQ	environmental qualification
ER	environmental report
ERDA	U.S. Energy Research and Development Administration
ERPG	emergency response planning guidelines
ETF	effluent treatment facility
FHA	fire hazards analysis
FM	Factory Mutual
FNMC	fundamental nuclear material control
FOCI	foreign ownership, control, or influence
FSER	final safety evaluation report
FTS	fluid transport system
HA	hazards analysis
HAAW	high alpha activity waste stream (containing uranium, americium, and other radioactive decay products)
HAN	hydroxylamine nitrate
HAZOP	hazard and operability (analysis)
HD	high depressurization
HEPA	high-efficiency particulate air
HEU	high-enriched uranium
HFE	human factors engineering
HPT	hydrogenated propylene tetramer
HSI	human-system interface
HVAC	heating, ventilation, and air conditioning
I&C	instrumentation and control
ICN	immediate control network
ICRP	International Commission on Radiation Protection
ICSBEP	International Criticality Safety Benchmark Experiments

IDLH	immediate danger to life and health
IEC	International Electrotechnical Commission
IEEE	Institute of Electrical and Electronics Engineers
IOC	individual (or individuals) outside the controlled area
IROFS	items relied on for safety
ISA	integrated safety analysis
JSHU	jar storage and handling unit
LANL	Los Alamos National Laboratory
LFL	lower flammability limit
LIN	local industrial network
LLNL	Lawrence Livermore National Laboratory
LLW	low-level waste
LPF	leak path factor
LWR	light-water reactor
MAPE	mean annual probability of exceedance
MAR	material at risk
MC&A	material control and accounting
MCC	motor control center
MCNP	Monte Carlo Neutron Photon
MDE	medium depressurization exhaust
MFFF	mixed oxide fuel fabrication facility
MFFP	mixed oxide fresh fuel package
MIMAS	micronized master blend
MMI	Mercalli intensity
MMIS	manufacturing management and information system
MOU	memorandum of understanding
MOX	mixed oxide
MP	mixed oxide process
MPQAP	MOX Project Quality Assurance Plan
MSDS	material safety data sheet
MTHM	metric tons heavy metal
NCS	nuclear criticality control
NCSE	nuclear criticality safety evaluation
NFPA	National Fire Protection Association
NNSA	National Nuclear Security Administration
NRC	U.S. Nuclear Regulatory Commission
NPH	natural phenomena hazards
ORNL	Oak Ridge National Laboratory
OSHA	Occupational Safety and Health Administration
PAA	preliminary accident analysis
PAG	protective action guide
PBX	public branch exchange

PC	performance categories
PDCF	pit disassembly and conversion facility
PEC	passive engineered control
PEP	personnel and equipment protection
PFOD	probability of failure on demand
PHA	preliminary hazard analysis
PIP	plutonium immobilization pit
PLC	programmable logic controller
PSCS	process safety control system
PSHA	probabilistic seismic hazard analysis
PSSC	principal structure, system, and component
PTFE	polytetrafluorethylene
Pu	plutonium
PuO_2	plutonium dioxide
PUREX	plutonium uranium reduction extraction

| QA | quality assurance |
| QL | quality level |

RACB	restricted area boundary concentration
RAI	request for additional information
RF	respirable fraction
RG	Regulatory Guide

SA	safety analysis
SAR	safety analysis report
SC	seismic category
SCALE	Standardized Computer Analyses Licensing Evaluation
SER	safety evaluation report
SFPE	Society of Fire Protection Engineers
SNM	special nuclear material
SPDP	Surplus Plutonium Disposition Program
SQ	seismic qualification
SRP	Standard Review Plan (NUREG-1718)
SRS	Savannah River Site
SSC	structure, system, and component
SST	safe secure transport
STEL	short-term exposure limit
S/U	sensitive/uncertainty

TBP	tributyl phosphate
TEDE	total effective dose equivalent
TEEL	temporary emergency exposure limit
TLV	threshold limit value
TQ	threshold quantities
TRU	transuranic

| UBC | Uniform Building Code |

UL	Underwriters Laboratories
UO_2	uranium dioxide
USL	upper subcritical limit
UPS	uninterruptible power supply
WAC	waste acceptance criteria
WG	water gauge
WTA	work task agreement
XTN	X-terminal network

LIST OF ACRONYMS FOR MFFF BUILDING AND SYSTEM DESIGNATIONS

Buildings

BAD	administration building
BAP	aqueous polishing area
BEG	emergency diesel generator bldg
BMF	MOX fuel fabrication bldg
BMP	MOX fuel fabrication area (MOX processing area)
BRP	reagent processing building
BSG	standby diesel generator bldg
BSH	safe haven buildings
BSR	shipping and receiving area
BSW	secured warehouse building
BTS	technical support building

Systems

BAS	breathing air system	KDD	dechlorination and dissolution unit
CHH	HVAC chilled water system	KDM	milling unit
CHP	process chilled water system	KDR	recanning unit
DCE	PuO$_2$ buffer storage unit	KPA	purification cycle
DCM	PuO$_2$ 3013 storage unit	KPB	solvent recovery cycle
DCP	PuO$_2$ receiving unit	KPC	acid recovery unit
DCS	decontamination system	KPF	silver recovery unit
DDP	UO$_2$ drum emptying unit	KWD	liquid waste reception unit
DMW	demineralized water system	KWG	offgas treatment unit
DRS	UO$_2$ receiving and storage unit	KWS	waste organic solvent unit
EGF	emergency diesel generator fuel oil system	MDE	medium depressurization exhaust system
GAH	argon/hydrogen system	NBX	primary blend ball milling unit
GDE	rod decladding unit	NBY	scrap milling unit
GHE	helium system	NCR	scrap processing unit
GME, GMF	rod cladding & decontamination units	NDD	PuO$_2$ container opening & handling unit
		NDP	primary dosing unit
GMK	rod tray loading unit	NDS	final dosing unit
GNO	nitrogen oxide system	NPE, NPF	homogenization & pelletizing unit
GNS	nitrogen system	NTM	jar storage & handling unit
GOX	oxygen system	NXR	powder auxiliary unit
HDE	high depressurization exhaust system	PAD	pellet repackaging unit
		PAR	scrap box loading unit
HWS	process hot water system	PFE, PFF	sintering units
IAS	instrument air system	PML	pellet handling unit
KCA	oxalic precipitation & oxidation unit	POE	process cell exhaust system
		PQE	quality control and manual sorting units
KCB	homogenization unit	PRE, PRF	grinding units
KCC	canning unit	PSE	green pellet storage unit
KCD	oxalic mother liquor recovery unit	PSF	sintered pellet storage unit
		PSI	scrap pellet storage unit
KDA	decanning unit	PSJ	ground and sorted pellet storage unit
KDB	dissolution unit	PTE	pellet inspection and sorting units

PWS	plant water system
RDO	diluent system
RHN	hydroxylamine nitrate system
RHP	hydrogen peroxide system
RHZ	hydrazine system
RMN	manganese nitrate system
RNA	nitric acid system
ROA	oxalic acid system
RSC	sodium carbonate system
RSH	sodium hydroxide system
RSN	silver nitrate system
RTP	tributyl phosphate system
SAS	service air system
SCE	rod scanning unit
SDK	rod inspection and sorting unit
SEK	helium leak test unit
SGF	standby diesel generator fuel oil system
SPS, SPC	process steam and process condensate systems
STK	rod storage unit
SXE, SXF	x-ray inspection units
TAS	assembly handling and storage unit
TCK	assembly dry cleaning unit
TCL	assembly final inspection unit
TCP	assembly dimensional inspection unit
TGM	assembly mockup loading unit
TGV	assembling mounting unit
TXE	assembly packaging unit
VHD	very high depressurization exhaust system
VRM	radiation monitoring vacuum system
WSB	waste solidification building
WVA	vehicle access portal

This page intentionally left blank

INTRODUCTION

On February 28, 2001, Duke Cogema Stone & Webster (DCS or the applicant) submitted a construction authorization request (CAR) to the U.S. Nuclear Regulatory Commission (NRC) seeking authority to construct a mixed oxide (MOX) fuel fabrication facility (MFFF or the facility) on the U.S. Department of Energy's (DOE) Savannah River Site in South Carolina.[1] DCS is under contract to the National Nuclear Security Administration (NNSA), a semi-autonomous agency within DOE charged with designing, building, and operating the proposed MFFF. Building and operating the MFFF would implement a September 2000 agreement between the United States and the Russian Federation, in which each nation agreed to the disposition of 34 metric tons (37.5 tons) of surplus weapons-grade plutonium. The MOX fuel manufactured at the proposed MFFF would be irradiated in U.S. commercial nuclear power plants, thus converting the surplus plutonium to a form that cannot be readily used to make nuclear weapons.

The NNSA will own the MFFF. In connection with its operation, the NNSA plans to construct nearby a pit disassembly and conversion facility (PDCF). The PDCF, which would not be under the NRC's jurisdiction, would be used to disassemble nuclear weapon components and convert them to an unclassified plutonium dioxide form. The plutonium dioxide from the PDCF would be the largest source of plutonium feedstock for the MFFF.

In 2001, NNSA was planning to build and operate a third surplus plutonium disposition facility, the Plutonium Immobilization Plant (PIP). As proposed, the PIP would have immobilized approximately 8.5 metric tons (9.4 tons) of surplus plutonium into a waste form suitable for direct disposal in a high-level waste repository, leaving 25.5 metric tons (28.1 tons) of surplus plutonium for conversion into MOX fuel at the MFFF. But in February 2002, NNSA announced that it was cancelling the PIP. As a result, NNSA directed DCS to redesign the MFFF to enable it to convert an additional 8.5 metric tons (9.4 tons) that was no longer slated for conversion at the PIP. This material, referred to as "alternate feedstock," represents the balance of the 34 metric tons (37.5 tons) of surplus plutonium that will not come from the proposed PDCF.

On October 31, 2002, DCS submitted a revised CAR, which replaced, in its entirety, the original CAR. The revised CAR reflects the changes necessary to process plutonium feed materials from DOE sources other than the PDCF, incorporates information previously provided in the DCS response to the NRC staff requests for additional information, and provides additional information addressing open items identified in the NRC draft safety evaluation report (DSER) dated April 30, 2002. DCS made further changes to the revised CAR, as reflected in CAR change pages dated December 12, 2002, February 18, 2003, April 1, 2003, and April 8, 2003.

The NRC staff issued another DSER on April 30, 2003, leading to the CAR change pages dated July 28, 2003. Further CAR revisions were made necessary by the November 2003

---

[1]Earlier, on December 19, 2000, DCS submitted an environmental report (ER). The staff's review of the ER (as revised July 12, 2001, October 29, 2002, December 10, 2002, January 15, 2003, June 20, 2003, August 13, 2003, and June 10, 2004), is reflected in NUREG-1767, "Environmental Impact Statement on the Construction and Operation of a Mixed Oxide Fuel Fabrication Facility at the Savannah River Site, South Carolina," published in January 2005.

change to the MFFF controlled area boundary, as reflected in the CAR change pages dated June 10, 2004. Additional CAR revisions were made on January 27, 2005, and February 9, 2005, which were needed to incorporate commitments made previously by DCS in letters to the NRC that were the bases for closing open items in the DSER. This final safety evaluation report (FSER) contains the staff's findings and conclusions on the CAR, revised as described above.

MFFF Description

The MFFF would consist of three major functional areas, (1) shipping and receiving, (2) aqueous polishing where the plutonium dioxide would be purified, and (3) MOX production where the fuel pellets, fuel rods, and fuel assemblies would be fabricated. The facility would receive depleted uranium dioxide and plutonium dioxide, purify the plutonium dioxide to remove impurities such as gallium and americium, fabricate MOX fuel containing uranium and plutonium dioxides, assemble fuel rods, and fabricate fuel assemblies. The completed fuel assemblies would be subsequently irradiated in commercial nuclear power plants authorized by the NRC to use MOX fuel.

Applicable NRC Regulations

Under the applicable requirements of Title 10, Part 70, "Domestic Licensing of Special Nuclear Material," of the *Code of Federal Regulations* (10 CFR Part 70), before a license to possess and use licensed material at the MFFF may be issued, the NRC must first authorize construction of the facility. This FSER documents the NRC staff's review of the revised CAR and supplemental supporting information provided by the applicant. This FSER only addresses regulatory requirements for approval of construction and does not address operational aspects of the proposed facility.

The following discussion summarizes the regulatory requirements applicable to the NRC staff's review of the revised CAR. In this regard, 10 CFR 70.23(b) states that the NRC will approve construction of a plutonium processing and fuel fabrication facility if it finds that the design bases of the principal structures, systems, and components (PSSCs) and the Quality Assurance (QA) Program, provide reasonable assurance of protection against natural phenomena and the consequences of potential accidents. The FSER discusses the applicant's selection of design basis functions and values and how the applicant determined that the design bases will provide reasonable assurance of protection against natural phenomena and the consequences of potential accidents. Each FSER chapter contains one or more of the NRC staff's 10 CFR 70.23(b) safety findings on the applicable PSSCs and design bases being evaluated.

Regarding the term "design bases," the NRC stated in a letter to DCS dated October 26, 1999, that the following definition of design basis, which appears in 10 CFR 50.2, "Definitions," will be applied to the proposed facility:

> Design basis means that information which identifies the specific functions to be performed by a structure, system, or component of a facility, and the specific values or range of values chosen for controlling parameters as reference bounds for design. These values may be (1) restraints derived from generally accepted

"state of the art" practices for achieving functional goals, or (2) requirements derived from analysis (based on calculation and/or experiments) of the effects of a postulated accident for which a structure, system, or component must meet its functional goals.

Other 10 CFR Part 70 requirements applicable to authorizing construction of the facility are discussed below. Baseline design criteria (BDC), listed in 10 CFR 70.64(a), cover 10 design issues, (1) quality standards and records, (2) natural phenomena hazards, (3) fire protection, (4) internal environmental conditions and dynamic effects, (5) chemical protection, (6) emergency capability, (7) utility services, (8) inspection, testing, and maintenance, (9) criticality control, and (10) instrumentation and controls. The staff has evaluated the MFFF preliminary design against these BDC, and the FSER describes how these BDC requirements are satisfied by the facility's design bases. The BDC specify design features that are required and acceptable under the conditions specified in 10 CFR 70.64(a).

Under 10 CFR 70.64(b), the facility design and layout must be based on defense-in-depth practices. As used in 10 CFR 70.64, "Requirements for New Facilities or Processes at Existing Facilities," defense-in-depth practices at new facilities mean a design philosophy, applied from the outset and through completion of the design, that is based on providing successive levels of protection such that health and safety will not be wholly dependent upon any single element of the facility design. The net effect of incorporating defense-in-depth practices is a conservatively designed facility that will exhibit greater tolerance to failures and external challenges. Chapter 5 of the FSER documents the NRC staff's findings that the facility's preliminary design is properly based on defense-in-depth practices.

As further required by 10 CFR 70.64(b), to the extent practicable, the facility design must incorporate (1) a preference for engineered controls over administrative controls to increase overall system reliability, and (2) features that will enhance safety by reducing challenges to items which will be relied upon for safety. Chapter 5 of the FSER documents the NRC staff's findings that the facility's preliminary design adequately incorporates a preference for engineered controls over administrative controls and that its design features will adequately enhance safety.

As discussed below, the set of requirements contained in 10 CFR 70.61(b–d), and (f) are also applicable to authorizing construction of the facility. Pursuant to the 10 CFR 70.61(b–d) performance requirements, the risk of accidents at the proposed facility must be limited. To properly determine the nature of such risks, and whether the proposed facility design effectively controls such risks, requires an evaluation of the safety assessment of the design bases submitted by DCS for construction approval. In FSER Chapter 5, the staff reviews the methodology used by DCS in performing the safety assessment of the facility design bases, and finds that the safety assessment adequately considered all appropriate natural phenomena, external manmade events, and internal process hazards. Identifying, and later implementing, the required controls to meet the 10 CFR 70.61(b–d) performance requirements ensures that an acceptable level of risk will be maintained during any facility operation. DCS may satisfy these performance requirements through a combination of limiting the chance that accidents at the facility would occur (prevention) and reducing the consequences of such events (mitigation). In approving the revised CAR, the staff finds that the DCS safety assessment describes an adequate strategy which, if effectively applied, will ensure that the 10 CFR 70.61 performance requirements will be met, should the facility later be authorized to operate.

The performance requirements for "credible high-consequence event[s]" (i.e., potential accidents involving high levels of radiation or hazardous chemicals) are set forth in 10 CFR 70.61(b). These provisions pertain to the protection of both workers and individuals outside the controlled area. In particular, 10 CFR 70.61(b) requires the use of controls sufficient to either make the occurrence of such accidents "highly unlikely," or make the consequences of such accidents less severe than (1) an acute 1 Sv (100 rem) total effective dose equivalent (TEDE) to a "worker," as defined in 10 CFR 70.4, (2) an acute 0.25 Sv (25 rem) TEDE to an individual outside the "controlled area," as defined in 10 CFR 20.1003, which DCS is required to designate for the proposed facility pursuant to 10 CFR 70.61(f), (3) an intake of 30 mg (6.6 × 10^{-5} lb) of soluble uranium to an individual outside the controlled area, or (4) either an acute chemical exposure that could endanger the life of a worker or an acute chemical exposure that could lead to irreversible or other serious long-lasting health effects to an individual outside the controlled area. (See 10 CFR 70.61(b)(1–4)).

The performance requirements for "credible intermediate-consequence event[s]" (i.e., potential accidents involving lower levels of radiation or hazardous chemicals than referenced in 10 CFR 70.61(b)), are set forth in 10 CFR 70.61(c). These provisions pertain to environmental protection, as well as to protecting the health of workers and individuals outside the controlled area. In particular, 10 CFR 70.61(c) requires the use of controls sufficient to either make the occurrence of such accidents "unlikely," or make the consequences of such accidents less severe than (1) an acute 0.25 Sv (25 rem) TEDE to a worker, (2) an acute 0.05 Sv (5 rem) TEDE to an individual outside the controlled area, (3) 24-hr average release of radioactive material outside the "restricted area," as defined in 10 CFR 20.1003, "Definitions, into the environment in concentrations exceeding 5000 times the values in Table 2 of Appendix B to 10 CFR Part 20, "Standards for Protection Against Radiation," or (4) either a chemical exposure that could lead to irreversible or other serious long-lasting health effects to a worker or a chemical exposure that could cause mild transient health effects to an individual outside the controlled area. (See 10 CFR 70.61(c)(1–4)).

The performance requirements applicable to nuclear criticality accidents, which are discussed in FSER Chapter 6, are set forth in 10 CFR 70.61(d). Under 10 CFR 70.61(d), the risk of such accidents is limited by requiring the use of an approved margin of subcriticality for safety. Additionally, primary reliance must be placed on the use of preventative, rather than mitigative, controls and measures to limit the risk of nuclear criticality accidents.

Two sections of 10 CFR 70.61 are not yet applicable to DCS, as they reference 10 CFR 70.62 safety program requirements which apply to the evaluation of applications for 10 CFR Part 70 licenses to possess and use licensed material. The integrated safety analysis (ISA) requirements referenced in 10 CFR 70.61(a) are set forth in 10 CFR 70.62(c); under these provisions, DCS must later show that it would be in compliance with the above-described performance requirements during any facility operations. Similarly, 10 CFR 70.61(e) also references the 10 CFR 70.62 safety program and requires that each control identified by DCS to meet the performance requirements be designated as an item relied on for safety (IROFS). Each IROFS must be available and reliable to perform its intended function when needed. Pursuant to 10 CFR 70.65, "Additional Content of Applications," DCS must list all IROFS in its

ISA summary, which is submitted as part of its application for a 10 CFR Part 70 license to possess and use licensed material.

In Section 5.5.5 of the CAR, the applicant described its general design philosophy and the defense-in-depth practices used in formulating the preliminary design of the facility. Pursuant to 10 CFR 70.64(b), in order to ensure that engineered controls are relied upon over administrative controls, to the extent practicable, DCS has established a hierarchy of controls. In further adherence to 10 CFR 70.64(b), DCS stated that it has incorporated defense-in-depth practices in its preliminary facility design. In Section 5.5.5.2 of the CAR, DCS stated that, it has incorporated defense-in-depth practices through use of the double-contingency principle for protection against criticality events and the use of the single-failure criterion. Under the latter criterion, DCS stated that for the PSSCs it has identified, each is required to be capable of carrying out its function in the event any single active component fails, whether such failure occurs within the applicable system or in an associated system that supports the component's operation. In Section 5.5.5.4 of the CAR, DCS stated that the BDC are incorporated into the facility design. The NRC staff evaluates these DCS statements in the FSER.

Potential accidents evaluated by the applicant include loss of confinement of licensed nuclear material, fire, load-handling events, explosions, nuclear criticality, natural phenomena events, external manmade events, external exposure, and events related to chemical interactions. The set of natural phenomena hazards identified by the applicant and evaluated by the staff include earthquakes, high wind, tornadoes and tornado-generated missiles, extreme temperatures, rain, snow, ice, lightning, and fires external to the MFFF. For most of the postulated hazards, the applicant has chosen a mitigation strategy, but for the explosion and nuclear criticality hazards, the applicant has chosen a prevention strategy. In FSER Chapter 5, the staff reviewed (1) the methodology used by the applicant in performing the safety assessment of the facility design bases, (2) the applicant's hazard assessments, (3) the applicant's formulation of a safety strategy, and (4) the applicant's identification of PSSCs to meet the 10 CFR 70.61 performance requirements. The staff finds that the applicant's safety assessment adequately considered all appropriate natural phenomena, external manmade events, and internal process hazards.

In the revised CAR, the applicant evaluated PSSCs primarily at the systems level, and the identified PSSCs are primarily design features/administrative controls that are to be implemented in the final design as IROFS, pursuant to 10 CFR 70.61(e). NUREG-1718, "Standard Review Plan for the Review of an Application for a Mixed Oxide (MOX) Fuel Fabrication Facility," defines PSSCs as "safety controls that are identified in the design bases as providing protection against the consequences of accidents or natural phenomena." NUREG-1718 further states that designating a control as a PSSC "is effectively synonymous with designating that control as an IROFS." The staff used NUREG-1718 as guidance in performing its review of the CAR and its supporting information. The definition of IROFS in 10 CFR 70.4, "Definitions," states, in relevant part, that IROFS are "structures, systems, equipment, components, and activities of personnel that are relied on to prevent potential accidents at a facility that could exceed the performance requirements in [10 CFR] 70.61 or to mitigate their potential consequences." As stated in the rulemaking associated with 10 CFR Part 70, Subpart H, "Additional Requirements for Certain Licensees Authorized to Possess a Critical Mass of Special Nuclear Materials," (NRC, 2001), IROFS may be described at the systems level, provided that there is enough detail to understand the function of the system in relation to the performance requirements. Accordingly, as discussed in subsequent FSER chapters, the staff finds it acceptable to identify PSSCs at the systems level. Table 5.1-3 of the FSER summarizes each PSSC identified by DCS, the safety functions of each PSSC, and the design bases associated with the PSSCs. As indicated above, the IROFS identified in any subsequent DCS application for a license to possess and use licensed material would be expected to contain more component-level controls as part of the facility's final design.

Management measures, as defined in 10 CFR 70.4, will be applied to IROFS to assure that they are available and reliable to perform their functions when needed. The ISA summary, the IROFS identified therein, and management measures will be evaluated when the staff reviews any subsequent DCS application for a license to possess and use licensed material. However, in determining, pursuant to 10 CFR 70.23(b), that the applicant's QA program will provide reasonable assurance of protection against natural phenomena and the consequences of potential accidents, an evaluation of the management measures described in CAR Chapter 15 is set forth in FSER Chapter 15.

With respect to QA issues, on June 22, 2000, DCS submitted a MOX project quality assurance plan (MPQAP). The staff's findings and conclusions on the DCS MPQAP, including Revision 2 dated January 29, 2001, are documented in a safety evaluation report (SER) dated October 1, 2001. On March 26, 2002, DCS submitted to NRC Revision 3 of the MPQAP. Revision 3 of the MPQAP incorporated all of the DCS commitments noted in the October 1, 2001, SER. By letter dated January 10, 2003, the NRC approved MPQAP, Revision 3, for the MFFF construction activities, including design, procurement, and fabrication.

Reference

(NRC, 2001) Nuclear Regulatory Commission (U.S.), Washington, D.C. "Domestic Licensing of Special Nuclear Material; Possession of a Critical Mass of Special Nuclear Material." *Federal Register.* Vol. 65, No. 181, pp. 56211-56231. September 18, 2001.

1. GENERAL INFORMATION

1.1 Facility and Process Overview

1.1.1 Conduct of Review

This chapter of the final safety evaluation report (FSER) discusses general information contained in Chapter 1 of the revised mixed oxide (MOX) fuel fabrication facility (MFFF or the facility) construction authorization request (CAR) (DCS, 2002). Chapter 1 of the MFFF revised CAR provides general information about the facility processes and the site. It consists of a general facility description, material flow, and process overview. The objective of FSER Chapter 1 is to familiarize the reader with the pertinent features of the proposed facility and the site.

1.1.1.1 General Facility Description

The proposed facility will be a "plutonium processing and fuel fabrication plant," as defined in Title 10, Section 70.4, "Definitions," of the *Code of Federal Regulations* (10 CFR 70.4). The facility will be designed to produce fuel assemblies for commercial nuclear power plants. The assemblies are composed of fuel rods which contain fuel pellets consisting of a blend of uranium and plutonium dioxides (i.e., mixed oxides). The plutonium dioxide to be used would be obtained from weapons-grade plutonium inventories held by the U.S. Department of Energy (DOE), which are declared surplus to national security needs.

The proposed MFFF would be located in the F-Area of the DOE Savannah River Site (SRS) near Aiken, South Carolina. The site encompasses approximately 0.17 km² (41 a), of which approximately 6.9×10^{-2} km²(17 a) will be developed with roads, facilities, or buildings. No roads, railroads, or waterways traverse the facility site. The nearest public transportation route is South Carolina Route 125, approximately 6.4 km (4 mi) to the west.

1.1.1.1.1 Controlled Area Boundary

With respect to the controlled area boundary, 10 CFR 70.61(f) requires that the applicant establish a controlled area, as defined in 10 CFR 20.1003, "Definitions." Section 20.1003 of 10 CFR defines the controlled area as an area outside of the restricted area but inside the site boundary, access to which can be restricted by the licensee for any reason. Section 20.1003 of 10 CFR defines the term "restricted area" as an area the "access to which is limited by the licensee for the purpose of protecting individuals against undue risks from exposure to radiation." Additionally, 10 CFR 20.1003 defines "site boundary" as the "line beyond which the land or property is not owned, leased, or otherwise controlled by the licensee." As discussed further below, the site boundary line for the proposed facility will correspond with the MFFF site perimeter, which encompasses 0.17 km² (41 a). The restricted area and controlled area are shown in Figure 1.1-2 of the CAR. Pursuant to 10 CFR Section 70.61(f), Duke Cogema Stone & Webster (DCS or the applicant) will retain authority to exclude or remove personnel and property from the controlled area. Accordingly, the staff finds that the applicant's proposed location of the controlled area boundary meets the requirements in 10 CFR 70.61(f) and 10 CFR Part 20, "Standards for Protection Against Radiation."

Individuals inside the proposed MFFF controlled area will be employed by the applicant and other DOE contractors to conduct activities licensed by the U.S. Nuclear Regulatory Commission (NRC) related to the manufacture of MOX fuel assemblies. As a result, the duties of individuals inside the proposed MFFF controlled area will involve potential exposures to radiation or to radioactive material, and, therefore, these individuals could receive an occupational dose. The staff concludes, therefore, that the applicant may consider these individuals to be workers for the purposes of demonstrating that the risks of postulated accidents and natural phenomema hazards are acceptably low.

Individuals outside the controlled area (IOC) are subject to DOE regulations, directives, and orders issued pursuant to DOE jurisdiction and authority. As such, the applicant committed to meet the performance requirements for "individuals outside the controlled area," as that phrase is used in 10 CFR 70.61, "Performance Requirements." The staff concludes that this commitment is consistent with the applicable regulatory requirement and is, therefore, acceptable.

In this FSER, the staff has adopted the applicant's usage for names of individuals for whom accident risks must be limited. That is, "facility workers," are workers in the restricted area inside a room of the MFFF near a potential accident release point. "Site workers" are workers considered to be outside the MFFF located 100 m (328 ft) from the ventilation exhaust stack. IOC refers to individuals at or beyond the proposed MFFF controlled area boundary. A fourth receptor protected under the provisions of 10 CFR 70.61 is the environment, which is understood by both the applicant and NRC staff to be all areas outside the restricted area.

The regulations in 10 CFR Part 20 define a "member of the public" as any individual, except when that individual is receiving an occupational dose. SRS employees located outside the MFFF controlled area (i.e., IOC) include both those who receive an occupational dose and those who do not. Therefore, the population outside the MFFF controlled area, and on the SRS, includes both workers and members of the public. As a result, the regulatory terms "member of the public" (from 10 CFR Part 20) and "IOC" (from 10 CFR Part 70, "Domestic Licensing of Special Nuclear Material") may apply to the same individual, depending on the regulatory context. In this FSER, when staff evaluates the risk of accidents and natural phenomena to an individual outside the MFFF controlled area, the term "individual outside the controlled area," or "IOC," is used, to be consistent with the use of that phrase in 10 CFR Part 70.61. Alternatively, when the staff evaluates the DCS proposal for protection of the individuals outside the MFFF controlled area from radiation hazards in accordance with 10 CFR Part 20 (see Chapter 10 of this FSER), the term "member of the public" is used.

1.1.1.1.2 Facility Buildings and Structures

Facility buildings consist of the mixed oxide (MOX) fuel fabrication building (the main building on the site), the emergency diesel generator building, the standby diesel generator building, the secured warehouse building, the administration building, the technical support building, and the reagents processing building. Miscellaneous site structures consist of a gas storage pad; heating, ventilation, air-conditioning, and process chiller pads; diesel fuel filling stations; electrical transformers; and other minor structures.

The main building would be the MOX fuel fabrication building. This building would contain all of the plutonium dioxide handling, fuel processing, and fuel fabrication operations of the facility. It would be a reinforced concrete building having a footprint of approximately 91.5 m (300 ft) by 137 m (450 ft) by approximately 22.3 m (73 ft) above grade. The building would be composed of three major functional areas, (1) the MOX processing area, (2) the aqueous polishing (AP) area, and (3) the shipping and receiving area. In the AP area, plutonium dioxide feedstock received from either the pit disassembly and conversion facility (PDCF), or from alternate feedstock, would be purified to remove impurities, such as gallium and americium. The purified plutonium dioxide would then be blended with depleted uranium dioxide powder and processed into MOX fuel, and ultimately fuel assemblies, in the MOX processing area. In the shipping and receiving area, plutonium and uranium dioxides would be received along with other materials necessary to produce fuel assemblies. Completed fuel assemblies would be shipped to commercial nuclear power plants.

Most reagents (e.g., nitric acid, hydrogen peroxide, hydroxylamine nitrate (HAN), hydrazine, oxalic acid, sodium carbonate, diluent (HPT), and tributyl phosphate (TBP)) would be stored and solutions would be prepared in the reagent processing building (BRP) for use in the AP area of the facility. The building would be divided into discrete areas to segregate chemicals and the associated equipment and vessels to prevent inadvertent chemical interaction. It would have a below-grade collection tank room that would receive waste chemicals from the building. A loading dock at one end of the building would be used for unloading and transfer of chemical containers and drums. Liquid chemical containers would be located inside curbed areas to contain accidental spills. The applicant does not intend to store, process, or commingle radioactive materials or radiochemicals in this building. Chemicals would be transferred to the AP area from the BRP via piping located in a concrete, below-grade trench between the two buildings.

1.1.1.2 Material Flow

The facility would receive plutonium dioxide from the PDCF, located on the SRS near the facility, as well as other DOE sources (i.e., alternate feedstock). The material would be transported to the shipping and receiving area of the facility in approved shipping containers. The material would be unloaded and inspected according to the Material Control and Accounting (MC&A) and Radiation Protection Programs. The material would then be moved to the AP or MOX Processing Area. The facility also would receive depleted uranium dioxide at the material receipt area of the secured warehouse building, where it would be inspected according to the MC&A and Radiation Protection Programs. The depleted uranium dioxide would be trucked to the shipping and receiving area of the facility. Fresh MOX fuel assemblies would be stored in the assembly storage vault in the facility before shipping offsite. For shipping to commercial power plants, the assemblies would be moved to the shipping and receiving area of the facility where they would be loaded into a MOX fresh fuel transportation package that had been approved by the NRC, and then loaded onto a secure transport vehicle for transport to commercial power plants for irradiation.

Airborne effluents from the MOX fuel fabrication building would be treated, passed through a final two-stage high-efficiency particulate air filter to remove radioactive particles, and then discharged through a continuously monitored stack. The exhaust streams would come from building ventilation systems; gloveboxes; process vents of tanks, vessels and other equipment; and the sintering furnaces.

Liquid waste streams containing radioactive materials would be sampled, characterized, and transferred to the SRS Waste Management Program for final processing and disposal. No radioactive liquid waste streams would be released from the proposed MFFF to the environment. Liquid waste streams include high alpha activity waste containing americium, gallium, and silver from the dissolution process; uranium solutions containing low-enriched uranium; alkaline solutions; liquid low-level waste (e.g., acid recovery condensate; room heating, ventilation, and air conditioning (HVAC) condensate; laboratory rinsing; and sanitary washing); and solvent streams.

Solid radioactive wastes would be typically placed in 55 gal (208.2 L) drums, assayed, and transferred to SRS for processing and disposal under the SRS Waste Management Program. The wastes would be compacted to reduce volume to the extent possible. These wastes include transuranic and low-level wastes which include uranium and/or plutonium contamination.

1.1.1.3 Process Overview

The facility would have two main process operations, (1) an AP process that serves to remove impurities, such as americium and gallium (i.e., polishing), and (2) the MOX fuel fabrication, or MOX process (MP), which processes the plutonium and depleted uranium dioxides into fuel pellets, fuel rods, and fuel assemblies. A summary of the major processes in the facility is provided below. A more detailed discussion of process chemistry and chemical safety is provided in Chapter 8 and Sections 11.2 and 11.3 of this FSER.

1.1.1.3.1 AP Process Overview

All feedstock, both from the PDCF and from other DOE sources, will be received as plutonium dioxide. The plutonium dioxide received at the MFFF would contain small amounts of impurities that must be removed for use of the MOX fuel in reactors. Feedstock from the PDCF will contain impurities such as gallium, americium, and highly enriched uranium. The diversity of impurities and the level of impurities will be higher in alternate feedstock. Some of this alternate feedstock may have higher than normal salt contaminants (other than chlorides), some will contain chloride contaminants, and some will contain small amounts of uranium. The AP process is used to remove these impurities. The AP process consists of three major steps, (1) dissolution, (2) purification, and (3) conversion.

In the dissolution step, the plutonium dioxide powder received from the PDCF and other DOE sources would be placed into solution by electrolytic dissolution with silver in nitric acid.

The purification step involves purification of the plutonium solution in pulsed columns by solvent extraction. The solvent mixture would be TPB dissolved in HPT solvent. Nitrate impurities, such as americium, gallium, and silver, remain in the aqueous phase and would be routed to an acid recovery unit after solvent washing. The plutonium and uranium solvent stream would be mixed in a pulsed column with a stripping solution of nitric acid, HAN, and hydrazine, which would reduce plutonium to a trivalent state and allow its transfer from the solvent phase back to the aqueous phase stripping solution. The organic solvent, stripped of plutonium, would be mixed with an additional stripping solution in a plutonium barrier that assures that nearly all plutonium is removed before the organic solvent is transferred to the uranium stripping process. In the uranium stripping step, the uranium would be removed from the solvent phase using dilute nitric acid. The uranium stream would be diluted with depleted uranium before being transferred to SRS as waste. The remaining solvent stream, stripped of plutonium and uranium, would be routed to solvent recovery mixer-settlers to be recycled. In the purified plutonium stream, the plutonium valence would be adjusted back to plutonium (IV) by driving nitrous dioxide fumes through the plutonium solution in an oxidation column. The offgas would be routed through an offgas treatment system and discharged to the atmosphere.

In the conversion step, the plutonium (IV) would be converted to a PuO_2 powder using a continuous oxalate conversion process. In this step, the plutonium (IV) would react with excess oxalic acid to precipitate plutonium oxalate. The plutonium oxalate would be collected on a filter, dried in a screw calciner to produce purified PuO_2 powder, which is then blended and stored in sealed cans. The spent oxalic mother liquors would be concentrated, reacted with manganese to destroy the oxalic acid, and recycled to the beginning of the extraction cycle. Figure 1.1-4 in the revised CAR shows the AP process.

1.1.1.3.2 MP Overview

The purified plutonium dioxide powder would be used in the MP where it would be blended with depleted uranium dioxide powder to make MOX fuel. The applicant proposes to use a micronized master blend (MIMAS) process, which has been used by COGEMA and Belgonucleaire to manufacture MOX fuel in Europe. The MOX fuel fabrication process consists of four major steps, (1) powder blending, (2) pellet production, (3) rod production, and (4) fuel assembly production.

In the first step, a master blend of plutonium dioxide and depleted uranium dioxide powder and recycled powder would be produced that consists of approximately 20 wt% plutonium dioxide The powder mixture would be ground in a ball mill and mixed with additional depleted uranium dioxide to produce a final mixture consisting of approximately 2–6 percent plutonium dioxide. The final blend would be homogenized to assure a uniform distribution of the plutonium dioxide. Lubricants and poreformers, to control density, would be added to the final mixture during homogenization.

In the second step, the final blend would be pressed to form pellets. These pellets are referred to as "green" pellets because they would not yet have been sintered in the furnace. The green pellets would then be sintered in a furnace, in which the atmosphere consists of a mixture of hydrogen and argon, to obtain the required ceramic properties. The sintering would also remove both organic products from the pellets, the lubricant and the poreformer. The sintered pellets would be ground to a specified diameter and sorted. Powder from the grinding operation and from discarded pellets would be recycled through a ball mill and reused in the powder processing.

In the third step of the MP, fuel rods would be loaded with the pellets to an adjusted pellet length column. The rods would be welded, pressurized with helium, and decontaminated in gloveboxes. The filled rods would then be removed from gloveboxes, placed on racks, and inspected.

In the fourth and final step, the filled rods would be pulled through the fuel assembly skeleton to form completed fuel assemblies. Each assembly would consist of a 17 x 17 square grid (fuel rods, control rod guide tubes, and instrument tubes). There are approximately 264 fuel rods (uranium or MOX fuel) per assembly. A variety of inspections are performed on the completed fuel assemblies before shipment. The completed assemblies would be stored for shipment to the mission reactors. Figure 1.1-5 in the revised CAR shows the MOX fuel fabrication process.

1.1.2 Evaluation Findings

The staff concludes that the facility and process overview descriptions provided by the applicant in Section 1.1 of the CAR are sufficient for the staff to obtain an introductory understanding of the facility and the processes. More detailed facility and process descriptions are provided in other sections of the CAR and are discussed in other chapters of this FSER.

1.1.3 References

(DCS, 2002) Ashe, K.L., Duke COGEMA Stone & Webster. Letter to U.S. Nuclear Regulatory Commission, RE: Construction Authorization Request and Environmental Report Change Pages. October 31, 2002 (including page changes through February 9, 2005).

1.2 Institutional Information

1.2.1 Conduct of Review

This chapter of the FSER contains the staff's review of institutional information described by the applicant in Chapter 1 of the revised CAR. The staff used Chapter 1 in NUREG-1718, "Standard Review Plan for the Review of an Application for a Mixed Oxide (MOX) Fuel Fabrication Facility," (NRC, 2000) as guidance in performing the review. The staff evaluated the institutional information provided by the applicant by reviewing Chapter 1 of the revised CAR, other sections of the revised CAR, and supplementary information provided by the applicant (DCS, 2002a; DCS 2002b).

1.2.1.1 Corporate Identity

The applicant for the proposed MFFF is DCS, a consortium of Duke Project Services Group, Inc. (DPSG), COGEMA, Inc., and Stone & Webster, Inc. The applicant is a limited liability company registered in the State of South Carolina and its principal offices are located in Charlotte, North Carolina.

The applicant is proposing to construct the facility at the DOE SRS near Aiken, South Carolina. This facility will produce MOX fuel from weapons-grade plutonium and depleted uranium using processes adapted from existing plants in France. Licensed activities will involve an AP process intended to remove impurities from the feed plutonium and fuel fabrication processes that include material mixing and blending, pelletizing, sintering, and fuel rod loading and inspection. The fuel is intended for use in commercial reactors in the United States.

The applicant provided the names of its principal corporate officers, all of whom are U.S. citizens. The applicant is using technical support from other companies in the design of the proposed facility. These companies include Belgonucleaire, SGN (a subsidiary of COGEMA), Framatome ANP, and Nuclear Fuel Services, Inc. The applicant also plans to use other contractors at the site, but these contractors have not yet been selected (DCS, 2002a; DCS 2002b).

1.2.1.2 Foreign Ownership, Control, or Influence

COGEMA, Inc., a U.S. company, owns 30 percent of DCS. COGEMA, Inc., is wholly owned by COGEMA, SA, a French company. The remainder of the corporation is owned by DPSG (40 percent) and Stone & Webster, Inc. (30 percent), which are both U.S. companies. The NRC has confirmed that DOE rendered a favorable foreign ownership, control, or influence determination (FOCI) for the applicant on March 12, 1999. In addition, DOE rendered a favorable FOCI determination for DPSG on June 10, 2002. The Department of Defense also rendered a favorable FOCI determination for Stone & Webster. On February 19, 2004, the NRC entered into a security cognizance agreement with DOE, whereby DOE will maintain security oversight of NRC interests at DPSG. The NRC accepts DOE FOCI determinations based on a memorandum of understanding between the NRC and DOE dated October 9, 1996.

1.2.1.3 Proposed License Information

The applicant provided information on its proposed operations and the type of license (including possession limits) it will later be requesting. The applicant is requesting authorization to construct an MFFF pursuant to 10 CFR 70.23(b). The applicant plans to request an NRC materials license pursuant to 10 CFR Part 30, "Rules of General Applicability to Domestic Licensing of Byproduct Material," 10 CFR Part 40, "Domestic Licensing of Source Material," and 10 CFR Part 70 for the following materials:

[Text removed under 10 CFR 2.390]

The term of the license to be requested to possess and use these materials is 20 years.

The applicant also identified special exemptions to be requested related to decommissioning funding and financial protection under the Price-Anderson Act. The applicant will request these exemptions as part of the license application.

1.2.2 Evaluation Findings

The staff evaluated the institutional information for approval to construct an MFFF at the SRS according to Section 1.2 of NUREG-1718 (NRC, 2000). This institutional information identifies the applicant's corporate structure, favorable FOCI determinations, proposed license possession limits, and special exemptions to be requested in the license application. This information is complete and accurate, is consistent with the recommendations in NUREG-1718 (NRC, 2000), and is, therefore, acceptable.

1.2.3 References

(DCS, 2002a) Hastings, P., Duke Cogema Stone & Webster. Letter to U.S. Nuclear Regulatory Commission, RE: Clarification of Responses to NRC Request for Additional Information. April 23, 2002.

(DCS, 2002b) Hastings, P., Duke Cogema Stone & Webster. Letter to U.S. Nuclear Regulatory Commission, RE: Requests for Additional Information, Clarifications, and Open Item Mapping into the Construction Authorization Request Revision. November 22, 2002.

(NRC, 2000) U.S. Nuclear Regulatory Commission. NUREG-1718, "Standard Review Plan for the Review of an Application for a Mixed Oxide (MOX) Fuel Fabrication Facility." Washington, DC, 2000.

1.3 Site Description

1.3.1 Conduct of Review

This chapter of the FSER contains the staff's review of the site description provided by the applicant in Chapter 1 of the revised CAR (DCS, 2002a). The objective of this review is to (1) ensure that site conditions, including site geography, demographics, meteorology, hydrology, and geology, are accurately described to properly define potential accident conditions, and (2) determine whether principal structures, systems, and components (PSSCs) and their design bases, identified by the applicant, provide reasonable assurance of protection against natural phenomena and the consequences of potential accidents. The staff evaluated the site description information provided by the applicant by reviewing Chapter 1 of the revised CAR, other sections of the revised CAR, supplementary information provided by the applicant, and relevant documents available at the applicant's offices but not submitted by the applicant. The review of the site description was closely coordinated with the natural phenomena accident sequences described in the safety assessment of the design bases (see Chapter 5 of this FSER) and the review of other plant systems.

The staff reviewed how the information in the revised CAR addresses the following regulations:

- 10 CFR 70.23(b) states, as a prerequisite to construction approval, that the design bases of the PSSCs and the quality assurance program be found to provide reasonable assurance of protection against natural phenomena and the consequences of potential accidents.

- 10 CFR 70.64, "Requirements for New Facilities or New Processes at Existing Facilities," requires that baseline design criteria (BDC) and defense-in-depth practices be incorporated into the design of new facilities. With respect to natural phenomena hazards,10 CFR 70.64(a)(2) requires that the design of new facilities must provide for adequate protection against such hazards, with consideration of the most severe documented historical events for the site.

Section 1.3 of the revised CAR discusses the geographical location of the MFFF and its environment, including demographic, meteorological, hydrological, geological, seismological, and geotechnical characteristics of the site and the surrounding area. It describes population distribution near the site, land and water uses, transportation routes, and nearby industrial facilities which potentially can affect the site. It also describes and evaluates site characteristics that affect the magnitude of natural phenomena (e.g., rain, snow, wind, and earthquakes) that may affect the site. This section also evaluates site characteristics with respect to safety and identifies assumptions and input that are needed to evaluate safety and the design bases in other evaluations in the revised CAR.

The staff evaluated site characteristics by reviewing Section 1.3 of the revised CAR, documents cited in the revised CAR, and other relevant literature. Where appropriate, findings of regulatory compliance are made for requirements that are fully addressed in Section 1.3 of the revised CAR. In some cases, regulatory compliance can only be determined by integrating the information in Section 1.3 of the revised CAR with information in other sections of the revised CAR. In these cases, evaluations of regulatory compliance are made in Chapter 5, "Safety Assessment of the Design Basis," of the revised CAR and findings of technical adequacy are made in Section 1.3 of this FSER.

1.3.1.1 Site Geography

In the application, the applicant provided information on the site location to include State, county, municipality, and topographic information; information on public and SRS roads, railroads, and waterways; nearby bodies of water; and significant geographical features.

The proposed site will be located in the F-Area of the SRS in southwest South Carolina near Aiken. The site is restricted and has few public roads. There are no unrestricted public roads in the vicinity of F-Area. A rail system is operated at the SRS by DOE. This rail system connects to commercial rail lines outside SRS boundaries. Nearby, the principal body of water is the Savannah River, which forms the SRS southwest boundary. The only river navigation that takes place is infrequent construction-related barge traffic. The significant physiographic features at the SRS are the Pleistocene Coastal Terraces and the Aiken Plateau. The applicant provided supplemental information about aircraft flights and airports in a letter dated March 8, 2002 (DCS, 2002a). The applicant stated that there are only two airports within 96.5

km (60 mi) of SRS that provide scheduled air passenger services. Six general aviation airports were identified by the applicant. Aircraft flight data were presented based on Federal Aviation Administration data. Aircraft hazards are discussed in Section 11.1 of this FSER.

This geographic information provided in the application is current and accurate, is appropriately referenced, and is consistent with information used in the safety assessments to support the design bases of PSSCs.

1.3.1.2 Demographics and Land Use

In the application, the applicant provided information on demographics and land use to include the 1990 Census data for the area and for minority and low-income populations; a description, distance, and direction to nearby population centers, public facilities, hospitals, and industrial facilities that could present potential hazards; residential, industrial, commercial, and agricultural land use data in the vicinity of the proposed site; and uses of nearby bodies of water.

There are 621,527 people living within 80 km (50 mi) of the proposed facility site based on the 1990 Census data. The population is expected to grow to slightly more than 1,000,000 in 2030. This population includes those living in the two metropolitan areas of Augusta, Georgia, and Aiken, South Carolina. Because the proposed site is on the SRS, there are no residents within 8 km (5 mi) of the proposed site. Within 8 km (5 mi) and 16 km (10 mi) of the site, 6528 people reside, the majority being in the towns of New Ellenton and Jackson, South Carolina.

Nearby industrial areas include other DOE SRS operations; several other Federal and State sponsored activities; Chem-Nuclear Systems, Inc., commercial low-level waste disposal and waste transportation activities in Barnwell County; Transnuclear, Inc., waste transportation activities in Aiken County; Carolina Metals, Inc., depleted uranium processing operations in Barnwell County; the Vogtle nuclear generating station across the Savannah River in Georgia; a fossil-fired electric generating plant 32 km (20 mi) north of the SRS; and the Fort Gordon Army post southwest of Augusta, Georgia.

Within the SRS, land use is controlled for the purposes of DOE operations and timber management. Forested areas within the SRS are managed by the U.S. Forest Service.

The Savannah River is used to supply domestic water, following treatment, for fish propagation, and for commercial and agricultural uses. Except for limited transportation of construction equipment, there is no commercial shipping performed on the river. Domestic uses of water from the Savannah River occur approximately 161 km (100 mi) downstream at treatment plants near Hardeeville, South Carolina, and Savannah, Georgia.

Ground water extracted near the SRS is used for domestic, industrial, and agricultural activities. Smaller communities, schools, and small commercial businesses also use local ground water. Nearly 133 million L/d (35 million gal/d) of ground water were pumped in 1985 by 56 communities and industries near the SRS.

This demographic and land use information provided in the application is accurate, is appropriately referenced, and is consistent with information in the safety assessments used to support the design bases of PSSCs.

1.3.1.3 Meteorology

In the revised CAR, the applicant provided meteorological information on temperatures; wind speeds and average and prevailing wind directions; amounts and form of precipitation; design basis values for maximum snow and ice loads and probable maximum precipitation; and types, magnitudes, and frequency of severe weather events, such as tornadoes, hurricanes, and lightning.

Temperature data for the SRS are presented in the revised CAR based on 30 years of measurements at the site. The annual average temperature is 18.2 °C (64.7 °F). Observed temperature extremes ranged from 41.7 to 19.4 °C (107 to 3 °F). Data for Augusta, Georgia, indicate that daytime high temperatures rarely fall below 0 °C (32 °F) during the winter. Temperatures are above 32.2 °C (90 °F) on more than half the days in the summer months.

Winds near the SRS are generally light to moderate, with the highest wind speeds occurring in the spring. The lightest winds occur in the summer and fall. The prevailing wind direction varies throughout the year, coming from the northwest in the winter, from the southeast in the late spring and early autumn, and from the southwest in the summer. The peak wind gust at Bush Field in Augusta, Georgia, was 96.5 km/h (60 mph) based on 10 years of data.

The average annual precipitation for the SRS from 1967 to 1996 is 126 cm (49.6 in.). The most rainfall during a 24-hr period was 19 cm (7.5 in.) in October 1990. During summer thunderstorms, rainfall rates of up to 5.1 cm/h (2 in./h) can occur. An average of 54 thunderstorm days per year have been observed. Hail storms occur infrequently, an average of once every 2 years.

Snowfalls of 2.5 cm (1 in.) or greater occur on the average once every 3 years. The greatest single snowfall recorded from 1951 to 1995 occurred in Augusta, Georgia, in 1973 when 35.6 cm (14.0 in.) of snow fell. The maximum ground snow load for a 100-yr recurrence period is 0.29 kPa (6 psf). Ice accumulates once every 2 years. The maximum accumulation for a 100-yr recurrence period is 1.7 cm (0.67 in.) or an ice load of 144 Pa (3 psf).

During a 30-yr period (1967–1997), 165 tornadoes occurred in the vicinity of the SRS. Five Fujita-scale 2 and four Fujita-scale 1 tornadoes have occurred onsite or in close proximity since site operations began. Damage was primarily to trees. One of these tornadoes produced wind speeds up to 241 km/h (150 mph). Design basis wind speeds for the DOE moderate hazard performance category (PC–3) facilities and high hazard performance category (PC–4) facilities are 290 km/h (180 mph) and 386 km/h (240 mph), respectively. The PSSCs are evaluated for a tornado recurrence interval of 2×10^{-6} per year and a design basis tornado with a 3-second tornado speed of 386 km/h (240 mph). For other extreme winds from hurricanes, tropical weather systems, thunderstorms, and winter storms, PSSCs will be evaluated based on a recurrence period of 1×10^{-4} per year for a 3-second wind speed of 209 km/h (130 mph). These extreme wind speeds are based on SRS meteorological data and data from National

Weather Service stations in Columbia, South Carolina, and Augusta, Macon, and Athens, Georgia.

During the period 1700–1992, 36 hurricanes caused damage in South Carolina. However, no hurricane-force winds of greater than 120 km/h (75 mph) have been measured at the SRS.

Extreme rainfalls generally occur during spring and summer thunderstorms and tropical storms. The design basis rainfall for PSSCs is evaluated for a recurrence interval of 1×10^{-5} for various rainfall durations (e.g., 9.9 cm (3.9 in.) for a 15-min rainfall and 58 cm (22.7 in.) for a 24-hr rainfall).

The number of lightning strikes is estimated at 10 strikes/km²·yr (26 strikes/mi²·yr). From 1989 to 1993, SRS data show an average of 4 strikes/km²·yr (10.4 strikes/km²·yr).

Meteorological information provided in the application was current and accurate, was appropriately referenced, and was consistent with information in the safety assessments used to support the design bases of PSSCs.

1.3.1.4 Hydrology

In the revised CAR, the applicant provided information on surface hydrology, including descriptions of nearby rivers, streams, and other water bodies; subsurface water hydrology, including water table depths, flow characteristics, potentiometric surfaces, and aquifer characteristics; and design basis floods.

The Savannah River forms the southwest boundary of the SRS and is the dominant body of surface water in the nearby area. The Savannah River Basin drains an area 27,394 km² (10,577 mi²) and extends 465 km (289 mi) from the Atlantic Ocean to the Blue Ridge Mountains. The principal streams that enter the Savannah River from the SRS are Upper Three Runs, Fourmile Branch, Pen Branch, Steel Creek, and Lower Three Runs. These streams discharge water from rainfall, subsurface waters, and various effluent streams from SRS operations. SRS surface water bodies include Par Pond and L Lake, created as cooling water reservoirs for production reactors, marshes, and natural basins, including Carolina bays.

The record historical Savannah River flood at Augusta, Georgia, in 1796 had a discharge of 10,000 m³/s (360,000 ft³/s). The peak Savannah River flow recorded by the U.S. Geological Survey was 9,900 m³/s (350,000 ft³/s) in 1929. There have been no major floods in the Augusta area since dams were constructed upstream of Augusta beginning in the 1950s. The estimated 50-yr maximum flow is now 2,100 m³/s (74,600 ft³/s). The probable maximum flood at the SRS is a water level of 68.4 m (224.5 ft) above mean sea level. The normal Savannah River flow elevation at the SRS boat dock is 25.9 m (85 ft). The design basis flood for the MFFF is 63.4 m (207.9 ft) above mean sea level with an annual recurrence interval of 1×10^{-5}. Because the facility is proposed to be located at an elevation of 82.9 m (272 ft), the probabilities of flooding the site were calculated to be less than 1×10^{-5} per year. A cascading failure of the Savannah River dams upstream of Augusta, Georgia, was estimated to produce a peak flow in the Savannah River of 28,000 m³/s (980,000 ft³/s) and a flood elevation of 43 m (141 ft) at the Vogtle station, which is directly across from the SRS on the Georgia side. Because the MFFF is at an elevation of 82.9 m (272 ft), this cascading failure and other events, such as ice flooding, wave surges, and seiches, will not affect the facility.

The ground water setting at the SRS is characterized by three aquifer systems that overlay the bedrock formations of the Southeastern Coastal Plain. The Southeastern Coastal Plain consists of sediments deposited from erosional processes of the Appalachian Mountains that lie to the west of the SRS. These sediments consist of water-bearing sandy materials and limestone and clayey confining units. In the F-Area, the confining units of the three aquifer systems become disjointed and have poor separation that allows flow between aquifer systems. In the uppermost Floridan Aquifer System, the Three Runs Aquifer overlays the deeper Gordon Aquifer. These aquifers are separated by a broken confining unit. Recharge of these aquifers is primarily through local precipitation, and discharge is primarily through local streams. Because Upper Three Runs Creek and the Savannah River incise the Floridan Aquifer System, there is a head reversal between it and the Crouch Branch Aquifer in the Dublin Aquifer System which lies just below the Floridan Aquifer System. This means that ground water from the lower system is under a greater head and flows up into the Floridan. This phenomenon tends to limit migration of contamination into the lower aquifer systems. The Midville Aquifer System is the deepest system and lies just above the bedrock formations.

At the proposed MOX site, the ground water table is nearly 15 m (50 ft) below the existing ground level. Potentiometric surface maps show that ground water in the uppermost Upper Three Runs Aquifer flows principally toward Upper Three Runs Creek and toward the unnamed creek located toward the northeast of the proposed site. The underlying Gordon Aquifer flows horizontally toward the Savannah River. The deeper Dublin and Midville Aquifer Systems flow to the southeast toward the Savannah River and the coast. The hydraulic conductivity of the Upper Three Runs Aquifer varies from less than 0.3 m/d (1.0 ft/d) to almost 10 m/d (33 ft/d), with an average of nearly 3 m/d (10 ft/d). At the MOX site, ground water is abundant, usually soft, slightly acidic, and low in dissolved solids. Ground water used in site operations from the F-Area is treated to raise the pH and remove iron.

The F-Area seepage basin, located to the west of the proposed MOX site, was remediated in 2000 according to a hazardous waste Part B, postclosure permit issued by the State of South Carolina. After remediating the site, boundary wells hydrologically downstream of the seepage basin were installed and samples were analyzed. The first set of analyses indicated that there is a contamination plume that exceeds the Environmental Protection Agency drinking water standards. The DOE staff is beginning an investigation to further evaluate this ground water contamination and to better understand the local hydrologic conditions that may exist at the proposed MOX site.

The applicant indicated that there is radioactive contamination in the Upper Three Runs aquifer from upgradient contamination sources in the F-Area, as well as the F-Area seepage basin. This ground water contamination consists of concentrations of gross alpha and beta activity, uranium, tritium, and trichloroethylene exceeding the maximum contamination limits for drinking water. The applicant also indicated that ground water contamination occurs at least 9.1 m (30 ft) below the deepest level of expected construction.

During site characterization activities, the applicant measured radioactivity levels of soils using Geiger-Mueller detector scans and gross alpha and beta measurements of soil samples. The applicant indicated that the sensitivity of the gross alpha and beta measurements was 200,000,000 pCi/kg and 100,000 pCi/kg, respectively. In a letter dated February 11, 2003, (DCS, 2003a), DCS stated that soil radioactivity measurement sensitivity (MDC) in the preconstruction environmental monitoring report (Fledderman, 2002) was much better than described in the CY2000 geotechnical investigations.

The NRC staff compared the results. The CY2000 geotechnical value was 200,000,000 pCi/kg gross alpha. The 2002 preconstruction environmental monitoring report measured values of actinides in soil which include a mean value of 12.5 pCi/kg ^{239}Pu, and a maximum of 4380 pCi/kg ^{239}Pu, for example. The SRS radiological soil guides for SRS worker protection is 248,000 pCi/kg (Jannik, 1995). Across the depth profile, the values are as follows:

depth	Pu-239 (pCi/kg) mean	max
0–3"	137	690
3–6"	87.1	1590
6–9"	154	4380
9–12"	121	4280

These values correspond to a potential maximum exposure of 0.003 mSv (0.3 mrem) to an exposed worker using the mean values and a maximum exposure of 0.033 mSv (3.3 mrem) using the maximum values (Fledderman, 2002). The 0.033-mSv (3.3-mrem) annual projected dose is acceptable because the NRC annual limit for members of the public in the controlled area is 1 mSv (100 mrem). The higher 50-mSv (5000-mrem) limit for workers does not apply, unless construction workers could receive an occupational dose.

The hydrologic information provided in the application is current and accurate, is appropriately referenced, and is consistent with information in the safety assessments used to support the design bases of PSSCs.

The planned construction of the facility will not penetrate into the upper ground water table that exists 15 m (50 ft) below grade level. The ground water contamination in the Upper Three Runs aquifer is, therefore, not expected to result in hazardous conditions that could affect construction.

1.3.1.5 Seismic Hazards

To assess the potential seismic hazard at the site, DCS established two sets of ground motion spectra in the revised CAR; one for the design of the surface facilities and one for soil stability analyses (liquefaction and dynamic settlements). Although the details of these spectra differ, analyses presented in the revised CAR show they are comparable. The design spectra (both vertical and horizontal) for the facility uses a spectrum included in Regulatory Guide (RG) 1.60, "Design Response Spectra for Seismic Design of Nuclear Power Plants" (NRC, 1973) anchored at 0.20g peak ground acceleration (PGA). The spectra are also used for the design of the nearby Vogtle Nuclear Power Plant (licensed under 10 CFR Part 50, "Domestic Licensing of Production and Utilization Facilities"). For soil stability, a spectrum was developed based on the existing DOE uniform hazard spectra developed for the SRS. Because the seismic design and analyses rely on these established spectra, much of the site-specific seismic hazard information presented in the revised CAR was developed to establish that these two proposed design and soil stability analysis spectra are adequate to meet the regulatory requirements of 10 CFR Part 70 and the performance guidelines in NUREG–1718.

The following areas concerning the seismic hazards applicable to the safety analysis and design of the proposed facility were reviewed:

- seismic source characterization
- ground motion attenuation
- seismic hazard calculations
- SRS-wide rock and surface response spectra
- site response and design ground motion
- surface faulting

1.3.1.5.1 Seismic Source Characterization

<u>Geological and Tectonic Setting</u>

The revised CAR provides a detailed description of the local and regional geological and tectonic settings. The revised CAR noted that the SRS is located on sediments of the Upper Atlantic Coastal Plain in South Carolina. These sediments consist of stratified, but generally unconsolidated, sands, silts, clays, and carbonaceous muds deposited in fluvial, deltaic, near-shore, and marine shelf environments. They range in age between Late Cretaceous (~100 Ma) and the present and reach a maximum thickness of approximately 1200 m (4000 ft). Similar to Coastal Plain sedimentary sequences along the entire Atlantic seaboard, the South Carolina Coast sediments rest unconformably on Precambrian to Paleozoic (~ 1.1 Ga to 245 Ma) metamorphic, metasedimentary, and igneous rocks of the Appalachian Orogen and on Triassic to Early Jurassic (~245–180 Ma) siliciclastic rocks associated with early rifting along the North American continental margin. Age and distribution of the rocks and strata provide an adequate geologic record to assess faulting and earthquake hazards.

Earthquakes that could impact safe operation of the proposed facility are associated with two seismic sources, a repeat of the Charleston 1886 earthquake within the Middle Place-Summerville Seismic Zone and small shallow earthquakes of the South Carolina Piedmont. Earthquake source characteristics associated with these seismic zones are consistent with information used in both the Electric Power Research Institute (EPRI) (EPRI, 1989) and Lawrence Livermore National Laboratory (LLNL) (NRC, 1994) seismic hazard studies for the eastern United States. As discussed in Section 1.3.1.5.2 of this report, the bedrock uniform hazard spectra for both the EPRI and LLNL form the basis for the site-wide DOE PC–3 and PC–4 hazard spectra (WSRC, 1997).

<u>Historical Seismicity</u>

The revised CAR provides a summary of the records of historical seismicity, including those from the cultural historical record (historical accounts date back to about 1698), as well as more recent instrumented earthquake records (the South Carolina Seismic network and the SRS network, both in operation since the mid-1970s). As noted in the revised CAR, the most significant earthquake source is a repeat of the 1886 Charleston, estimated to have a modified

Mercalli intensity (MMI) of X at Charleston, South Carolina, and an MMI at the SRS of VI–VII. Magnitude estimate of the 1886 Charleston earthquake is M 7.3 ± 0.3 (Johnston, 1996; Ambraseys, 1988). Other significant historical earthquakes felt at the SRS include the 1913 Union County earthquake (an MMI of VII at the epicenter and an MMI of II–III at Aiken, South Carolina); the 1811–12 New Madrid earthquakes (M > 8.0 at New Madrid, Missouri); and the 1897 Giles County, Virginia, earthquake (MMI of VII, M 5.6 at Pearisburg, Virginia).

Paleoliquefaction features indicate that the Charleston-type earthquake has reoccurred at least seven times in the last 6000 years (Talwani, 2001). These prehistoric earthquakes appear to be restricted to the Carolina Coastal Plain (Talwani, 2001). Two scenarios have been proposed to explain the distribution of the paleoliquefaction features. In the first scenario, the earthquakes occurred at Charleston, Georgetown, and Bluffton, South Carolina. In the second scenario, all the prehistoric earthquakes occurred at Charleston. Hu, et al. (Hu, 2002) conclude that the paleoliquefaction features were produced by earthquakes with magnitudes between 5.3 and 7.8.

No definitive geologic evidence has yet been discovered to tie the 1886 Charleston earthquake to a causative seismogenic fault. Tarr, et al. (Tarr, 1981) defined the Middleton Place-Summerville Seismic Zone to include the known distribution of seismicity and paleoseismicity associated with the Charleston-type earthquake. The Middleton Place-Summerville Seismic Zone is located 20 km (12 mi) northwest of the city of Charleston, South Carolina. Based on geological and geophysical data, Marple (Marple, 1994), Madabhushi and Talwani (Madabhushi, 1993), and Marple and Talwani (Marple, 2000) all inferred that complex and interactive strike slip and reverse faulting associated with the northwest trending Ashley River fault and the north-northeast trending Woodstock fault were the most likely causes of the Charleston earthquake. Recently, Weems and Lewis (Weems, 2002) concluded that the region around Charleston, South Carolina, is an active tectonic zone that accommodates differential movement between the Cape Fear arch and the Southeast Georgia embayment. All these models are consistent with the source characterization of the Charleston-type earthquake presented in the EPRI and LLNL probabilistic seismic hazard assessment (PSHA) seismic hazard studies.

Near the SRS, instrumented historical seismic records indicate that seismicity associated with the SRS and surrounding region is closely related to the earthquake activity within the South Carolina Piedmont (Bollinger, 1992). This activity is characterized by shallow, small magnitude, and infrequent earthquakes. Searches of the National Earthquake Information Center and Council of National Seismic System show that the vast majority of these earthquakes are M 3 or less. The largest magnitude earthquakes in the record are the 1974 M 4.9 and M 4.7 events. All instrumented earthquakes on the SRS itself were M 2.7 or less.

Earthquake Recurrence

The long repeat times (> 500 yr) and relatively brief historical record (< 350 yr), coupled with the absence of active surficial deformation, limit estimates of earthquake recurrence for a Charleston-type earthquake. The most complete record of the temporal and spatial distribution of large prehistoric earthquakes comes from identification of earthquake-induced liquefaction features called sand blows. Numerous sand blows have been identified throughout the South Carolina coastal area, but few if any outside this region (WSRC, 2000b). Recent reanalysis of the paleoliquefaction investigations in South Carolina and recalibrated ^{14}C ages suggest that there were as many as seven large-magnitude earthquakes in the Charleston region within the last 6000 years (Talwani, 2001). These results translate to a recurrence interval for the Charleston-type earthquake of 500 to 600 years. This estimated recurrence interval is conservative because it assumes the maximum number of possible paleoearthquakes using the age constraints derived from the ^{14}C age data. Talwani and Schaeffer (Talwani, 2001) used 1σ error ranges to develop their list of age-distinct paleoearthquakes. Overlap of the ^{14}C ages using 2σ error ranges, as advised by Tuttle (Tuttle, 2001) would result in a smaller number of age-distinct paleoearthquakes during this same 6000-yr interval and thereby increase the recurrence interval. Nevertheless, the 500–600-yr recurrence interval for the Charleston-type earthquake is consistent with the LLNL and EPRI PSHA studies.

Staff Review of Seismic Source Characterization

The staff reviewed the information about seismic sources presented in the revised CAR and found it sufficient because the applicant identified and assessed all of the potentially significant seismic sources related to the SRS (including, but not limited to, the Charleston seismic zone). The characterization of the tectonic setting and identification of capable seismic sources were based on extensive review of the published geological literature, regional and site geological and geophysical data, historical and instrumental seismicity data, regional stress field analysis, and geological investigations of prehistoric earthquakes. The information follows guidelines presented in RG 1.165, "Identification and Characterization of Seismic Sources and Determination of Safe Shutdown Earthquake Ground Motion," (NRC, 1997) and Section 2.5.2.2 of NUREG–0800, "Standard Review Plan for the Review of Safety Analysis Reports for Nuclear Power Plants" (NRC, 1987). Criteria used to assess capable fault and areal source zones include those outlined in Appendix A, "Seismic and Geologic Siting Criteria for Nuclear Power Plants" to 10 CFR Part 100, "Reactor Site Criteria," as well as those in DOE–STD–1022–94, "Natural Phenomena Hazards Characterization Criteria" (DOE, 1994).

Information provided by DCS to determine the tectonic setting of the facility was developed into a coherent, well-documented discussion that provides an adequate technical basis for evaluation of the seismic potential of the site. Specifically, documentation in the revised CAR is sufficient to determine the earthquake potential of geological structures and potential tectonic zones (i.e., regions of uniform earthquake potential). The information provided in the revised CAR is also sufficient to evaluate uncertainties associated with seismic source geometry (e.g., fault dip, width, segmentation, and depth of seismogenic crust) and recurrence models. Thus, the staff reviewed the information in the revised CAR and found it acceptable because the basic geologic and seismic characteristics of the site and vicinity are adequately described in detail to allow investigation of seismic characteristics at the facility.

1.3.1.5.2 Ground Motion Attenuation

Seismic hazards used to define bedrock uniform hazard spectra at the SRS are based on the LLNL and EPRI probabilistic seismic hazard studies. The LLNL and EPRI bedrock uniform hazard spectra were averaged and then broadened using the SRS-specific spectral shapes to develop bedrock response spectra.

Ground motion attenuation models contained in the LLNL and EPRI hazard studies incorporated a number of models developed by individuals and organizations for the southeastern United States. These models are considered to be representative of the state-of-the-art studies of ground motion attenuation characteristics in the southeastern United States and have captured diverse opinions in the scientific community.

The ground motion attenuation model used to develop site-specific spectral shapes was the Band Limited White Noise/Random Vibration Theory ground-motion model (Hanks, 1981; Boore, 1983). In applying this stochastic approach, the applicant used the layered crustal velocity model developed by Herrmann (Herrmann, 1986) with some modifications, the EPRI median site attenuation model (Q-model), and the range of the EPRI site-dependent parameter Kappa values (EPRI, 1989).

Staff Review of Ground Motion Attenuation

Ground motion attenuation models used in the LLNL and EPRI studies are representative of the current scientific understanding of ground motion attenuation in the southeastern United States. These attenuation models adequately capture uncertainty in ground motion estimates, including the potential for Moho bounce effects. For example, a recent ground motion attenuation model for the eastern United States (Campbell, 2003), which accounts for Moho bounce effects, yields ground motion estimates consistent with those derived using the LLNL and EPRI studies. Application of the LLNL and EPRI models to the SRS and, consequently, to the facility is considered acceptable. The NRC staff has previously accepted the LLNL and EPRI ground motion modeling (NRC, 1997) for sites in the central and eastern United States.

The use of the stochastic model or numerically simulated ground motions in the central and eastern United States, instead of recorded ground motions, is consistent with common practice and the state of knowledge because sufficient strong motion data are lacking in this tectonic regime as a result of low seismicity rates. The approach was accepted by the staff in its review of the PSHA for the Paducah Gaseous Diffusion Plant (CNWRA, 1999). In addition, the random vibration theory model has been shown to yield conservative results for eastern U.S. crustal conditions (Silva, 1989). Thus, the staff has determined that the applicant's ground motion attenuation modeling is acceptable because it provides reasonable assurance that ground motion attenuation modeling is accurate.

1.3.1.5.3 Seismic Hazard Calculations

The applicant used the seismic hazard results from the LLNL and EPRI probabilistic seismic hazard to define bedrock uniform hazard spectra at the Savannah River. No other probabilistic seismic hazard calculations were conducted specifically for the SRS or the facility. The LLNL and EPRI hazard studies include site-specific hazard calculations for the SRS.

Staff Review of Seismic Hazard Calculations

The LLNL and EPRI studies represent the state-of-the-art probabilistic hazard studies in the southeastern United States. Application of these results to the SRS and, consequently, to the facility, is considered acceptable. The NRC staff previously accepted the LLNL and EPRI data, seismic sources, seismic hazard methods, and results (NRC, 1997) for sites in the central and eastern United States. Thus, the staff determined using the LLNL and EPRI hazard results is technically sound.

1.3.1.5.4 SRS-Wide Rock and Surface Response Spectra

The SRS-wide rock response spectra were developed by Westinghouse Savannah River Company for the entire SRS (WSRC, 1997). These are site-specific uniform hazard spectra for bedrock from the LLNL and EPRI seismic probabilistic hazard studies, broadened by using site-specific spectral shapes. The rock response spectra were used as the bases for developing bedrock time histories as input into site response analyses for the facility and the SRS-wide surface response spectra.

The SRS-wide surface response spectra are not directly used in the design of structures or in soil stability analyses for the facility. However, they were used by the applicant to justify the sufficiency of the selected design spectra for the facility.

The SRS-specific rock uniform hazard spectra for bedrock were developed following the guidance and methodologies outlined in DOE–STD–1023-95, Change Notice #1, "Natural Phenomena Hazards Assessment Criteria," (DOE, 1995). Probabilistic hazards were developed according to DOE PC–3 and PC–4 spectra. The DOE PC–3 and PC–4 spectra were developed following seismic design and evaluation criteria in the DOE STD–1020–94, Change Notice No. 1, "Natural Phenomena Hazards Design and Evaluation Criteria for Department of Energy Facilities" (DOE, 1996) and DOE STD–1020–2002 (DOE, 2002). In DOE STD–1020–94, PC–3 and PC–4 categories have mean annual probabilities of exceedance for design ground motions at 5×10^{4} and 1×10^{4}, respectively. In terms of the annual return period ground motions, mean annual probabilities of exceedances of 5×10^{4} and 1×10^{4} correspond to mean 2,000-yr and 10,000-yr return period ground motions, respectively.

The development of the rock response spectra included the following procedures:

- The mean bedrock uniform hazard spectra were computed for two mean annual probabilities of exceedances, 5×10^{4} and 1×10^{4} (corresponding to performance categories of PC–3 and PC–4, respectively), by averaging the LLNL and EPRI mean uniform hazard spectra for the SRS.

- Site-specific spectral shapes were generated using EPRI mean magnitude and mean distance values based on the magnitude and distance deaggregation results at each probability of exceedance.

- The spectral shapes were then scaled to the corresponding mean bedrock uniform hazard spectrum at frequencies 1–2.5 and 5–10 Hz.

- The resulting three spectra (the averaged LLNL and EPRI uniform hazard spectrum and the 1–2.5-Hz- and 5–10-Hz-scaled site-specific spectra) were then enveloped and smoothed to obtain the broadened bedrock response spectra for the PC–3 and PC–4 hazards.

Site-Wide Surface Response Spectra

Site-wide surface response spectra were obtained by multiplying the broadened bedrock uniform hazard spectra by frequency-dependent site amplification factors to account for soil effects. In deriving site amplification factors, hypothetical bedrock spectra were vertically propagated through soil columns representative of the site soil conditions using the one-dimensional equivalent linear analysis procedure developed by Silva (Silva, 1989). The procedure was considered to be equivalent to SHAKE analyses summarized in Idriss and Sun (Idriss, 1992).

The hypothetical bedrock spectra were power spectral density functions and spectral accelerations for a suite of PGAs at the soil/bedrock interface (bedrock motions described previously) and were developed using the random vibration theory model (Boore, 1983). Three magnitude and distance dependent spectra were developed for each control motion acceleration representing the 5th, 50th, and 95th percentile contributions to the probability of exceedance. Again, the magnitude and distance pairs were obtained from the EPRI deaggregated hazard results.

The calculation of the site amplification factors considered SRS-wide variability in velocity profile, soil column thickness, bedrock velocity, and dynamic properties (WSRC, 1997). Soil conditions characterized in the most recent study of the site's geotechnical properties (DCS, 2003b) are consistent with subsurface conditions reported in the previous geotechnical reports of the site. The site-wide, uniform-hazard based response spectrum was taken as the envelope of all of the soil response spectra obtained by multiplying the broadened mean bedrock uniform hazard spectra by the site amplification factors for different soil/bedrock categories, scaling frequencies, and magnitude levels. As with the design ground motions, the site-specific soil spectra were shown to envelope the Charleston earthquake spectra.

Staff Review of SRS Rock and Surface Response Spectra

The LLNL and EPRI studies represent the state-of-the-art probabilistic hazard studies in the southeastern United States. Application of these probabilistic hazard results to the SRS and, consequently, to the facility is considered acceptable. The NRC staff previously accepted the LLNL and EPRI ground motion modeling (NRC, 1997) for sites in the central and eastern United States. In addition, broadening the LLNL and EPRI bedrock uniform hazard spectral shapes and the development of surface response spectra are consistent with the methodologies of DOE–STD–1023-95 (DOE, 1995). These methodologies and procedures are well established within the ongoing seismic program at the SRS. These site-specific adjustments have been extensively reviewed by the Westinghouse Savannah River Company and DOE. Thus, the staff determined that SRS-wide rock and surface response spectra are acceptable because they provide reasonable assurance that potential seismic hazards are sufficiently estimated.

1.3.1.5.5 Design Spectra and Site Response Analyses

Design Spectra

The applicant-proposed design basis ground motions for the surface facilities are an RG 1.60 spectra (NRC, 1973) anchored at 0.20g PGA, which is the same spectra used for the design of the nearby Vogtle Nuclear Power Plant (licensed under 10 CFR Part 50). More recently, regulations at 10 CFR 100.23, "Geologic and Seismic Siting Criteria," for nuclear power plants have been updated to include the application of probabilistic methods to the assessment of seismic hazards. RG 1.165 (NRC, 1997) provides general guidance for determining the safe-shutdown earthquake for new nuclear reactors based on a PSHA, consistent with the regulatory requirements of 10 CFR 100.23. RG 1.165 recommends a reference median annual probability of exceedance of 1×10^{5}. As shown by a similar analysis in Appendix C to DOE STD–1020–2002 (DOE, 2002), a median annual probability of exceedance of 1×10^{5} corresponds approximately to a mean annual probability of exceedance of 1×10^{4}.

Evaluations performed by the applicant in the safety analysis report (SAR) show that the 0.20g RG 1.60 spectra have mean annual exceedance probabilities that range between 1.6×10^{4} and 4.5×10^{5} (or equivalent return periods that range between 6,300 and 22,000 years; see Table 1 of Enclosure B of Ihde, 2001b). For frequencies between 2 and 10 Hz, the mean annual probabilities of exceedance are equal to or less than 1×10^{4} (or equivalent return periods greater than 10,000 years). For higher frequencies up to the PGAs, the mean annual probabilities of exceedance are equal to or sightly greater than 1×10^{4}. These mean annual exceedance probabilities are based on ground motions from the averaged EPRI (EPRI, 1989) and LLNL (NRC, 1994) seismic hazard results for the eastern United States.

RG 1.60 spectra anchored at 0.20g were selected by the applicant because they were deemed to be conservative. This seismic design spectrum is shown by the applicant to lie between the SRS-wide DOE PC–3 and PC–4 spectra. The current PC–3 and PC–4 site-wide spectra are based on ground motion spectra developed by Westinghouse Savannah River Company (WSRC, 1997) for the entire SRS. The PC–3 and PC–4 spectra were developed following seismic design and evaluation criteria in DOE STD–1020–94 (DOE, 1996) and DOE STD–1020–2002 (DOE, 2002), as discussed in more detail in Section 1.3.1.5.4 of this FSER.

To ensure safe operation of the structures, systems, and components (SSCs) beyond the design ground motions, DOE STD–1020–94 (DOE, 1996) and DOE STD–1020–2002 (DOE, 2002) developed performance goals associated with each performance category. The performance goals are defined in terms of the ability of the SSCs to perform essential safety functions during and after the natural hazard phenomena (in this case an earthquake). The acceptable behavior limit for normal use SSCs, such as buildings, is major damage, but limited in extent such that the occupants can safely exit the building. For more critical SSCs, such as nuclear containment structures, damage at the performance goal should be limited such that the containment is not compromised. In DOE STD–1020–94 (DOE, 1996), as well as in DOE STD–1020–2002 (DOE, 2002), the seismic ground motion performance goals for PC–3 and PC–4 SSCs were established with a mean annual probability of exceedance of 1×10^{4} and 1×10^{5}, respectively.

On page 27 of the MOX Fuel Fabrication Facility Site Geotechnical Report (DCS, 2001a), the applicant indicated that the desired performance goal probability is based on the approach recommended in DOE–STD–1020–94 (DOE, 1996) and DOE STD–1020–2002 (DOE, 2002). That assertion is supported by performance calculations (Enclosure B of DCS, 2002a) which show that many of the SSCs performed their safety functions to ground motion levels with a mean annual probability of exceedance of 1×10^{5} or less. These calculations support the conclusion that the design criteria—RG 1.60 (NRC, 1973) spectra anchored to the 0.20g PGA, which is significantly greater than the site-wide PC–3 spectra—is adequate for safe design of the facility.

The design spectra were also shown by the applicant to envelop the deterministic spectra for a repeat of the Charleston-type earthquake. This deterministic check analysis follows the requirements of DOE–STD–1023-95 (DOE, 1995) using the largest historic earthquakes within 121 km (75 mi) having a moment magnitude greater than 6. In this analysis, the deterministic median bedrock and soil spectra were generated for the 1886 Charleston earthquake using median source parameters, a source-to-site distance of 200 km (124 mi), and other parameters used in generating uniform-hazard-based response spectra.

In response to the staff's request for additional information, the applicant evaluated the vertical-to-horizontal seismic spectral ratios for the facility (DCS, 2002c). The results show that the vertical-to-horizontal spectral ratios could exceed the standard generally used at the SRS (normally the vertical is assumed to be two-thirds of the horizontal), particularly for frequencies greater than approximately 3 Hz. Thus, the applicant has agreed to use both the horizontal and vertical spectra in RG 1.60 (NRC, 1973) anchored at 0.20g PGA.

Site Response Analyses

The applicant indicated that the site-wide response spectra are intended for simple response analysis and are not appropriate for soil-structure interaction and soil stability analyses. It further indicated that the site-wide response spectra represent a surface response, not an embedded response. For soil stability and soil-structure interaction analyses, a one-dimensional, free-field site response analysis procedure was established by the applicant (DCS, 2001a). The control ground motions for site response analyses included the modified PC–3 motion and the 1886 Charleston motion. The modified PC–3 motion is the SRS-wide PC–3 rock response spectrum increased by a factor of 1.25 (PC–3+ rock spectrum) to yield a bedrock PGA of 0.14g, one that would achieve the design surface PGA of 0.20g at the facility through site response analyses. The 1886 Charleston motion is the 50th percentile attenuated rock motion at the actinide packaging and storage facility site. It was used by the applicant to evaluate the liquefaction potential associated with large, distant earthquakes. The spectrum-compatible acceleration time histories for both of these design motions were developed by Westinghouse Savannah River Company.

Site response analyses were conducted using PROSHAKE, a Windows version of SHAKE91 (Idriss, 1992). The design motion time histories were applied at the base of the soil column. Properties for the soil column were developed from geotechnical studies specific to the facility (WSRC, 2000a; DCS, 2001b; DCS, 2003b). Soil conditions characterized in the most recent study of the site geotechnical properties (DCS, 2003b) are consistent with subsurface conditions reported in the previous geotechnical reports of the site. The cyclic stress ratios computed from the site-response analyses were input into the dynamic soil-structure interaction analyses of the critical structures and into the liquefaction analyses.

Results from site response analyses show that the PC–3+ bedrock time history produces a surface PGA of 0.20g and a surface spectrum that correlates well to the RG 1.60 (NRC, 1973) surface spectrum anchored at 0.20g PGA. Thus, the applicant concluded that the PC–3+

bedrock spectrum satisfies the requirement for a bedrock time history that can be used for dynamic analysis at the facility.

Staff Review of Site Response and Design Ground Motion

Use of the RG 1.60 spectrum (NRC, 1973) anchored at 0.20 g PGA is deemed acceptable by the staff. Analyses performed by the applicant of several SSCs using the RG 1.60 design spectrum demonstrate that the performance objectives necessary for highly unlikely events with potentially high consequences, set forth in NUREG–1718 (NRC, 2000) are met. Similarly, the analysis procedures, input bedrock time histories, and soil column properties for soil stability analyses and soil-structure interaction analyses are also deemed acceptable by the staff. The resulting surface response spectrum exceeds the DOE PC–3 spectrum and is comparable to the design spectrum. The applicant further verified that the SRS-wide PC–3 spectrum is applicable to the design of the MFFF through examinations of the soil stratigraphy, soil column thickness, bedrock type, velocity profile, and geologic formations at the MFFF. Using response spectra that envelop the SRS-wide PC–3 spectrum for the analysis of soil and subsurface stability of the MFFF is therefore conservative.

1.3.1.5.6 Surface Faulting Hazard

The revised CAR summarizes the tectonic structures of interest in the SRS and surrounding region, including faults, folds, arches, basins, and paleoliquefaction features that resulted from past earthquakes. Many of these features are vestiges of the contractional tectonism that characterized the Appalachian Orogen from the Late Precambrian through the Late Paleozoic (~1.1 Ga to ~245 Ma) and rifting and extensional tectonism that characterized the break up of Pangea and the opening of the Atlantic Ocean in the Triassic and Early Jurassic periods (~245 to 180 Ma). Although reactivation of some of these features has been proposed to explain the origin of the Charleston-type earthquake (see discussion in Section 1.3.1.5.6 of the FSER), none of these features impact direct-faulting hazards at the SRS.

Faulting of the Atlantic Coastal Plain sediments is evident from geologic and geophysical data (e.g., Prowell, 1991). Most of the faults are moderately to steeply dipping reverse faults, although some small normal faults were noted in the Late Cretaceous and Early Tertiary strata (100 to 37 Ma). Maximum displacements are less than 80 m (250 ft), and displacements become progressively smaller in younger sediments suggesting that faulting was coeval with deposition.

At the SRS, the Pen Branch fault has been identified as the primary structural feature of interest to a potential faulting hazard. This fault appears to be an upward propagation of the boundary fault on the northern side of buried Dunbarton Basin, a Triassic to Early Jurassic rift feature. This boundary fault was originally a down-to-the-southeast normal fault, but was reactivated as an up-to-the-southeast reverse fault in the Late Cretaceous and Early Tertiary (100 to 37 Ma). Extensive geological and geophysical evidence summarized in the revised CAR documents that the Pen Branch fault was not active in the last 500,000 years and probably was not active in the Quaternary (last ~ 2 Ma). Thus, the Pen Branch fault is not deemed capable according to criteria established in Appendix A to 10 CFR Part 100.

Staff Review of Surface Faulting Hazard

The staff reviewed the information in the revised CAR and found it acceptable because the potential for surface faulting of the site and vicinity has been adequately assessed. There is sufficient evidence to conclude, with reasonable assurance, that surface faulting hazards do not exist at the SRS.

1.3.1.6 Stability of Subsurface Materials

The objective of the staff review in this section is to determine, with reasonable assurance, whether characterization of the stability of the subsurface materials for the facility is adequate for foundation design for the civil structural systems. The following areas concerning the subsurface material stability which were applicable to the safety analysis and design of the proposed facility were reviewed:

- soil liquefaction potential assessment
- soft zone characterization
- slope stability assessment

1.3.1.6.1 Soil Liquefaction Potential Assessment

The information regarding the paleoliquefaction at the SRS where the proposed facility will be located was provided in Section 1.3.5.3.4.3, "Post-Rift and Cenozoic Structures," of the revised CAR. The revised CAR indicates that no systematic reconnaissance surveys in search of paleoliquefaction evidence within the geomorphic and geologic environment of the SRS were performed in the past because of limited access, high water table conditions, dense vegetative cover, and few exposures.

For seismically induced liquefaction to occur and be identified, the following conditions have to be met (DCS, 2002a):

- presence of Quaternary-age unconsolidated deposits
- presence of a shallow ground water table
- proximity to potential seismogenic features
- quality and extent of exposure

According to these conditions, young fluvial terraces at or slightly above the level of the modern flood plain and Carolina bays may have the highest potential for generating and recording Holocene (last 10,000 years) and Quaternary (~ last 2 Ma) seismically induced liquefaction.

Limited investigation of the exposed young fluvial terraces along the Savannah River adjacent to the SRS suggests that most of the exposed deposits were clay and silt and thus have a low liquefaction potential. Although local clean sand deposits with a high liquefaction potential exist, evidence about the seismically induced liquefaction is not observed (DCS, 2002c). In general, these young fluvial deposits are historical in age. In historical times, no strong ground motions occurred in the SRS area. Consequently, evidence for seismically induced liquefaction in the young fluvial deposits may not exist.

According to the revised CAR, potential paleoliquefaction for the flood plain deposits at depth is likely. Evaluation of postdepositional features associated with the upland areas at the SRS, however, suggests that they are not related to seismically induced liquefaction (DCS, 2002a).

Liquefaction susceptibility at the facility was discussed in Section 1.3.7.1 of the revised CAR. The discussion was supported by detailed soil geotechnical testing data, as documented in three facility site geotechnical reports (DCS, 2001b; DCS 2001a; and DCS, 2003b). The site geotechnical reports present properties of soils, including soil classifications, particle-size distributions, water contents, plasticity indices, liquid limits, blow counts from standard penetration tests, tip shear resistances from cone penetration tests, and shear wave velocities.

The liquefaction potential of the facility site within the proximity of the MFFF and emergency generator buildings was evaluated using the cyclic stress approach described by the National Center for Earthquake Engineering Research (NCEER, 1997). This approach is acceptable to the staff for a liquefaction potential investigation because it represents the state-of-the-art procedure. This procedure is suited for evaluating liquefaction resistance of soils under level to gently sloping ground; the surface gradient at the proposed facility is gently sloping (as shown in Figure 1.3.1-2 of the revised CAR).

In the third geotechnical report (DCS, 2003b), liquefaction potential was evaluated for the 95 soil columns from cone penetration tests and 14 soil columns from standard penetration tests. Cyclic stress ratio and cyclic resistance ratio are two important parameters for assessing liquefaction. The cyclic stress ratios for the aforementioned penetration tests were estimated directly using the PROSHAKE computer program with a generalized soil profile. The liquefaction potential for 18 of the cone penetration tests was also estimated using the test-hole-specific soil profiles to compare with the results using the generalized soil profile. The revised CAR assumed full liquefaction was triggered if the factor of safety (cyclic stress ratio/cyclic resistance ratio) was equal to or smaller than 1.1. For factors of safety between 1.1 and 1.4, soil settlement may result because of the excessive water pressure buildup that reduces soil strength and stiffness.

The analysis results indicate that the liquefaction potential at the facility site is low. Only a few localized areas have been identified to be liquefiable or to have soil settlement potential because of excessive pore water pressure. The potentially liquefiable soils identified at the site are located in the lower Tertiary (~65 to 33 Ma) Tobacco Road, Dry Branch, and Santee formations. The revised CAR indicated that the analysis results were conservative because the effect of soil aging and the cohesiveness of the soils for the cone-penetration-based results are not considered in the analysis.

Staff Review of Soil Liquefaction Potential Assessment

The staff reviewed the information presented in the revised CAR and found reasonable assurance that paleoliquefaction at the SRS was sufficiently discussed to support the design of the PSSCs of the proposed facility. The staff review also concurred that the analysis of liquefaction potential at the proximity of the MFFF and emergency generator buildings demonstrated a conservative approach and is acceptable. Consideration of the effect of the seismically induced settlements caused by either liquefaction or excessive water pressure buildup in the development of design criteria for the PSSCs of the facility are evaluated in Section 11.1 of this FSER.

1.3.1.6.2 Soft Zone Characterization

Soft zones in the soils are unique features for the SRS. The origin of the soft zones is discussed in Section 1.3.5.1.5.5, "Carolina Bays," of the revised CAR. The discussion about the characterization of the soft zones at the facility is provided in Section 1.3.7.2, "Evaluation of Soft Zones," of the revised CAR and supported by site geotechnical data (DCS, 2001a; DCS 2003b).

The soft zones are often found in the Tinker/Santee Formation, particularly in the upper third of this section. These soft zones consist of weak material zones interspersed in stronger carbonate-rich matrix materials. The presence of soft zones may pose a concern for foundation design by developing undesirable soil settlement not accounted for in the design. In engineering terms, a soft zone is defined as a zone with a cone penetration test corrected tip resistance less than 1.44 MPa (15 tsf) or blow counts from a standard penetration test less than 5 over a continuous interval of at least 0.6 m (2 ft) (DCS, 2001a). In characterizing the soft zones, the applicant used these criteria to identify soft material zones not located in the Tinker/Santee Formation. The staff considered this approach prudent and acceptable.

The results of the site exploration program related to identifying soft material zones were documented in three site geotechnical reports (DCS, 2001b; DCS, 2001a; DCS, 2003b). The exploration hole spacing in the vicinity of an identified soft zone was generally 27 m (90 ft) or less. The lateral extent of soft zones was conveniently estimated to be half of the exploration spacing. The exploration program identified soft zones in the vicinity and beneath the MOX fuel fabrication building with limited lateral extent. The thickness of these soft zones ranges from 0.91 to 2.13 m (3 to 7 ft).

Staff Review of Soft Zone Characterization

Based on the review of the information concerning soft zones, the staff concluded the exploration program conducted by the applicant sufficiently characterized the soft zones at the facility to support design of the PSSCs. Consideration of the effect of the soft zones in the development of design criteria for the PSSCs of the facility is evaluated in Section 11.1 of this FSER.

1.3.1.6.3 Slope Stability

Slope stability was not specifically discussed in Section 1.3, "General Site Description," of the revised CAR. In evaluating the natural phenomena applicable to the site, however, debris avalanching and landslides were determined not to be applicable to the site because the site is relatively flat and no significant quantities of soil or rock are available in the surrounding area (see Table 5.5-5 in the revised CAR). An examination of topographic contours provided in Figure 1.3.4-6 confirms that the slopes at the facility site are relatively gentle in nature and therefore pose no threat for instability or landslide. The staff site visit further confirmed that slope stability is not a safety concern for the site.

1.3.2 Evaluation Findings

In Chapter 1 of its revised CAR, and in supplementary information, the applicant provided the geographic, demographic and land use, meteorologic, hydrologic, geologic, and seismic information relevant to the MFFF site. This information is generally current, appropriately referenced, and consistent with information in the safety assessments used to support the design bases of the PSSCs. The staff finds that the applicant has accurately described the site so as to allow potential accident conditions to be properly defined. Based on its review of the revised CAR and the relevant supplementary information provided by the applicant, the staff further finds that DCS has met the BDC set forth in 10 CFR 70.64(a)(2) for natural phenomena hazards. The staff concludes, pursuant to 10 CFR 70.23(b), that the design bases of the PSSCs evaluated in this FSER will provide reasonable assurance of protection against natural phenomena and the consequences of potential accidents.

1.3.3 References

(Ambraseys, 1988) Ambraseys, N.N. "Engineering Seismology," *Earthquake Engineering and Structural Dynamics* 17:1–105. 1988.

(Bollinger, 1992) Bollinger, G.A. "Specification of Source Zones, Recurrence Rates, Focal Depths, and Maximum Magnitudes for Earthquakes Affecting the Savannah River Site in South Carolina." *U.S. Geological Survey Bulletin*, 2017:1992.

(Boore, 1983) Boore, D.M. "Stochastic Simulation of High-Frequency Ground Motions Based on Seismological Models of the Radiated Spectra," *Bulletin of the Seismological Society of America,* 73:1865–1894. 1983.

(BNL, 1996) Brookhaven National Laboratory, "Description and Validation of the Stochastic Ground Motion Mode," Upton, New York, 1996.

(Campbell, 2003) Campbell, K. "Prediction of strong ground motion using the hybrid empirical method and its use in the development of ground-motion (attenuation) relations in eastern North America." *Bulletin of the Seismological Society of America*, 93:1012–1033. 2003.

(CNWRA, 1999) Center for Nuclear Waste Regulatory Analyses. "Review of the Probabilistic Seismic Hazard Analyses for the Paducah Gaseous Diffusion Plant—Final Report." San Antonio, Texas. 1999.

(DCS, 2001a) Duke Cogema Stone & Webster. Letter to U.S. Nuclear Regulatory Commission, RE: MOX Fuel Fabrication Facility CAR Supplemental Information—Site Geotechnical Report (DCS01–WRS–DS–NTE–G–00005– A). April 24, 2001.

(DCS, 2001b) Ihde, R., Duke Cogema Stone & Webster. Letter to U.S. Nuclear Regulatory Commission, RE: MOX Fuel Fabrication Facility Site Geotechnical Report (DCS01–WRS–DS–NTE–G–00005–C). August 10, 2001.

(DCS, 2002a) Duke Cogema Stone & Webster. Letter to U.S. Nuclear Regulatory Commission, RE: Clarification of Responses to NRC Request for Additional Information. March 8, 2002.

(DCS, 2002b) Duke Cogema Stone & Webster. Letter to U.S. Nuclear Regulatory Commission, RE: Clarification of Responses to NRC Request for Additional Information. April 23, 2002.

(DCS, 2002c) Duke Cogema Stone & Webster. Letter to U.S. Nuclear Regulatory Commission, RE: Submitting Revised Request for Authorization of Construction of the Mixed Oxide Fuel Fabrication Facility. October 31, 2002 (including page changes through February 9, 2005).

(DCS, 2003a) Hastings, P., Duke Cogema Stone & Webster. Letter to U.S. Nuclear Regulatory Commission, RE: Responses to Open Items/Additional NRC Questions on Construction Authorization Request Revision. February 11, 2003.

(DCS, 2003b) Duke Cogema Stone & Webster. MOX Fuel Fabrication Facility Site Geotechnical Report (DCS01–WRS–DS–NTE–G–00005–E). Report prepared for the U.S. Department of Energy, Chicago Operations Office. June 30, 2003.

(DOE, 1994) U.S. Department of Energy. DOE–STD–1022–94, "Natural Phenomena Hazards Characterization Criteria." Washington, DC, 1994.

(DOE, 1995) U.S. Department of Energy. DOE–STD–1023–95 Change Notice #1, "Natural Phenomena Hazards Assessment Criteria." Washington, DC, 1995.

(DOE, 1996) U.S. Department of Energy. DOE-STD–1020–94 Change Notice #1, "Natural Phenomena Hazards Design and Evaluation Criteria for Department of Energy Facilities." Washington, DC, 1996.

(DOE, 2002) U.S. Department of Energy. DOE–STD–1020–2002, "Natural Phenomena Hazards Design and Evaluation Criteria for Departments of Energy Facilities." Washington, DC, 2002.

(EPRI, 1989) Electric Power Research Institute. "Probabilistic Seismic Hazard Evaluations at Nuclear Power Plants in the Central and Eastern United States: Resolution of the Charleston Earthquake Issue," NP–6395–D. Palo Alto, California, 1989.

(Fledderman, 2002) Fledderman, P. D., Westinghouse Savannah River Company. "Plutonium Disposition Program (PDP) Preconstruction Environmental Monitoring Report," ESH–EMS–2002–1141. June 26, 2002.

(Hanks, 1981) Hanks, T.C. and R.K. McGuire. "The Character of High-Frequency Strong Ground Motion." *Bulletin of the Seismological Society of America.* 71:2071–2095, 1981.

(Herrmann, 1986) Herrmann, R.B. "Surface-Wave Studies of Some South Carolina Earthquakes." *Bulletin of the Seismological Society of America.* 76:111–121, 1986.

(Hu, 2002) Hu, K., S.L. Glassman, and P. Talwani. "Magnitudes of prehistoric earthquakes in the South Carolina Coastal Plain from geotechnical data." *Seismological Research Letters.* 73:979–991, 2002.

(Idriss, 1992) Idriss, I.M. and J.I. Sun. *User's Manual for SHAKE-91: A Computer Program for Conducting Equivalent Linear Seismic Response Analysis of Horizontally Layered Soil Deposits.* Center for Geotechnical Modeling, Department of Civil and Environmental Engineering, University of California, Davis, California, 1992.

(Jannik, 1995) Jannik, G.T., Westinghouse Savannah River Company. "Concentration Guidelines for the Initial Screening of Soil in Determining Onsite 'Soil Concentration Areas,'" WSRC–SRT–ETS–950155. 1995.

(Johnston, 1996) Johnston, A.C. "Seismic Moment Assessment of Earthquakes in Stable Continental Regions, III, New Madrid 1811–1812, Charleston 1886, and Lisbon 1755." *Geophysical Journal International.* 126:314–344, 1996.

(Madabhushi, 1993) Madabhushi, S. and P. Talwani. "Fault Plane Solutions and Relocations of Recent Earthquakes in Middleton Place Summerville Seismic Zone Near Charleston, South Carolina," *Bulletin of the Seismological Society of America.* 83:1442–1466, 1993.

(Marple, 1994) Marple, R.T. *Discovery of a Possible Seismogenic Fault System Beneath the Coastal Plain of South and North Carolina from an Integration of River Morphology and Geological and Geophysical Data.* Ph.D. Dissertation. University of South Carolina. Columbia, South Carolina, 1994.

(Marple, 2000) Marple, R.T. and P. Talwani. "Evidence for a Buried Fault System in the Coastal Plain of the Carolinas and Virginia—Implications for Neotectonics in the Southeastern United States." *Geological Society of America Bulletin.* 112:200–220, 2000.

(NCEER, 1997) National Center for Earthquake Engineering Research. Technical Report No. NCE ER–97–002. "Proceedings of the National Center for Earthquake Engineering Research Workshop on Evaluation of Liquefaction Resistance of Soils." State University of New York at Buffalo. Buffalo, New York, 1997.

(NRC, 1973) U.S. Nuclear Regulatory Commission., Regulatory Guide 1.60. "Design Response Spectra for Seismic Design of Nuclear Power Plants." Washington, DC, 1973.

(NRC, 1987) U.S. Nuclear Regulatory Commission. NUREG–0800. "Standard Review Plan for the Review of Safety Analysis Reports for Nuclear Power Plants." Washington, DC, 1987.

(NRC, 1994) U.S. Nuclear Regulatory Commission. NUREG–1488 "Revised Livermore Seismic Hazard Estimates for Sixty Nine Nuclear Power Plant Sites East of the Rocky Mountains." Washington, DC, 1994.

(NRC, 1997) U.S. Nuclear Regulatory Commission. Regulatory Guide 1.165. "Identification and Characterization of Seismic Sources and Determination of Safe Shutdown Earthquake Ground Motion." Washington, DC, 1997.

(NRC, 2000) U.S. Nuclear Regulatory Commission. NUREG–1718. "Standard Review Plan for the Review of an Application for a Mixed Oxide (MOX) Fuel Fabrication Facility." Washington, DC, 2000.

(Prowell, 1991) Prowell, D.C. and S. F. Obermeier. *Evidence of Cenozoic Tectonism. The Geology of the Carolinas.* University of Tennessee Press, Knoxville, Tennessee, 1991.

(Silva, 1989) Silva, W., et al. *Site Dependent Specification of Strong Ground Motion in Dynamic Soil Property and Site Characterization.* Workshop sponsored by the National Science Foundation and the Electric Power Research Institute. Palo Alto, California, 1989.

(Talwani, 2001) Talwani, P. and W.T. Schaeffer. "Recurrence Rates of Large Earthquakes in South Carolina Costal Plain Based on Paleoliquefaction Data." *Journal of Geophysical Research.* 106:6621–6642, 2001.

(Tarr, 1981) Tarr, A., et al. "Results of Recent South Carolina Seismological Studies." *Seismological Research Letters.* 71:1883–1902, 1981.

(Tuttle, 2001) Tuttle, M.P. "The use of liquefaction features in paleoseismology: lessons learned in the New Madrid seismic zone, central United States." *Journal of Seismology.* 5:361–380, 2001.

(Weems, 2002) Weems, R.E. and W.C. Lewis. "Structural and Tectonic Setting of the Charleston, South Carolina, Region: Evidence from the Tertiary Stratigraphic Record." *Geological Society of America Bulletin.* 114:24–42, 2002.

(WSRC, 1997) Westinghouse Savannah River Company. WSRC–TR–97–0085, Revision 0, "SRS Seismic Response Analysis and Design Basis Guidelines." Aiken, South Carolina, 1997.

(WSRC, 2000a) Westinghouse Savannah River Company. WSRC–TR–2000 00454, Revision 0, "Natural Phenomena Hazards (NPH) and Design Criteria and Other Characterization Information for the Mixed Oxide (MOX) Fuel Fabrication Facility at Savannah River Site (U)." Aiken, South Carolina, 2000.

(WSRC, 2000b) Westinghouse Savannah River Company. Report WSRC–TR–2000–00039, Revision 0. "Probability of Liquefaction for F-Area, Savannah River Site." Aiken, South Carolina, 2000.

(WSRC, 2001) Westinghouse Savannah River Company. WSRC–TR–2001–00342, Revision 0, "Development of MFFF-Specific Vertical-to-Horizontal Seismic Spectral Ratios." Aiken, South Carolina, 2001.

2. FINANCIAL QUALIFICATIONS

2.1 Conduct of Review

This chapter of the final safety evaluation report contains the staff's review of the financial qualifications presented by Duke Cogema Stone & Webster (DCS or the applicant) in Chapter 2 of the revised construction authorization request (CAR). Additionally, the staff evaluated the supplemental financial qualification information provided by the applicant (DCS, 2001, 2002, and 2003).

DCS is a limited liability company registered in the State of South Carolina and headquartered in Charlotte, North Carolina. DCS is a consortium of Duke Project Services Group, Inc. (DPSG), COGEMA, and Stone & Webster, Inc. The applicant is proposing to construct the Mixed Oxide (MOX) Fuel Fabrication Facility (MFFF or the facility) at the U.S. Department of Energy's (DOE) Savannah River Site near Aiken, South Carolina, under a cost-plus-fixed-fee contract with DOE. COGEMA, a French company, owns 30 percent of DCS. The remainder of the corporation is owned by DPSG (40 percent) and Stone & Webster, Inc. (30 percent), which are both U.S. companies.

2.1.1 Project Costs

In an August 31, 2001, proprietary response (DCS, 2001) to a staff request for additional information (RAI), the applicant provided financial statements for its fiscal year ending December 31, 2000, and an estimate of project costs. These costs include costs for facility design and construction. Design costs include licensing costs, contingencies, and escalation. The applicant, however, did not detail the project design cost estimate, which was provided as a single value. Although DCS is not a publicly held company—and is thus not required to submit Report 10-K to the U.S. Security and Exchange Commission—it provided, as part of its proprietary filing, an independent accountant's review report for the fiscal year ending December 31, 2000, and for the period of corporate inception (March 22, 1999) to December 31, 1999 (DCS, 2001). Balance sheets, cash flows, income statements, and equity statements were included in the accountant's report. The independent accountant concluded that it was not aware of any material modifications that should be made to the accompanying financial statements in order for them to be in conformity with accounting principles generally accepted in the United States.

In its March 8, 2002, clarification to the RAI responses (DCS, 2002), DCS stated that revised proprietary design cost information would be provided at a later date when available. In its initial draft safety evaluation report (NRC, 2002), the staff identified this as Open Item FQ-1. On February 13, 2003, NRC staff requested current financial statements and a commitment to provide annual updates of this information (NRC, 2003). In its subsequent February 18, 2003, filing (DCS, 2003), the applicant retracted its earlier commitment to provide revised proprietary design cost information.

2.1.2 Financial Qualifications

As indicated above, the applicant has a contract with DOE to design, construct, and operate the facility. The base scope of work of the contract covers design and engineering of the facility. An option, which has not yet been exercised, includes construction, functional testing, and preliminary startup. A second option, also not yet exercised, includes final startup and full operations. All funding for the facility will be from DOE. The applicant does not plan to self-finance the facility or seek external funding other than that under the DOE contract. In the event of cost overruns or funding shortfalls, the applicant would seek additional funding from DOE. If such funds were not provided, the applicant would stop any ongoing engineering, design, and construction activities.

As noted by DCS (DCS, 2003), the proposed MFFF is part of a fully funded U.S. Government project undertaken to carry out the national security mission to dispose of surplus plutonium. Other than temporary startup funding provided as part of DCS organizing as a company, DCS is not relying on its financial resources to provide money to cover engineering or construction costs. Instead, the funds necessary to cover such costs are provided directly through the DCS contract with the U.S. Government. Accordingly, DCS now takes the position that its financial qualifications are not relevant to the ability of the U.S. Government to carry out and fund the mission of surplus plutonium disposition (DCS, 2003).

2.1.3 Liability Insurance

Under Title 10, Section 140.13a, of the *Code of Federal Regulations* (10 CFR 140.13a), a holder of a license to possess and use plutonium at a plutonium processing and fuel fabrication plant is required to have and maintain public liability insurance in the amount of $200 million. This insurance is not required for MOX facility construction approval. The applicant stated that as a DOE contractor it is fully covered by the DOE nuclear liability protection under the Price-Anderson Act, as amended. Under Section 170d of the Atomic Energy Act, 42 U.S.C. 2210(d), the applicant and DOE have entered into an agreement that fully indemnifies the applicant and its contractors up to the statutory limit of liability. The applicant intends to request an exemption to the requirements of 10 CFR 140.13a in any possession and use license application it may later submit, and would use the DOE indemnification to meet its public liability insurance requirements.

2.2 Evaluation Findings

Based on its review of the requirements of 10 CFR Part 70, "Domestic Licensing of Special Nuclear Material," the staff has concluded that a finding regarding the adequacy of the DCS financial qualifications is not required in deciding whether the CAR should be approved. As a general matter, the issue of an applicant's financial qualifications may be relevant in considering applications for licenses to possess and use special nuclear material (see 10 CFR 70.23(a)(5)). However, this requirement is not applicable in deciding whether to approve the CAR because the CAR is not an application for a license to possess and use special nuclear material. Similarly, the financial assurance for decommissioning requirements, found in 10 CFR 70.22(a)(9), 10 CFR 70.25(b)(2), and 10 CFR 70.25(f)(4–5), are not applicable in deciding whether to approve the CAR. These requirements also apply only to applications for licenses to possess and use special nuclear material.

Pursuant to the above requirements, the staff may consider financial qualification issues as part of reviewing any application DCS may later submit for a license to possess and use licensed material. However, as noted by DCS (DCS, 2003), any funds necessary to cover MFFF-related costs would ultimately come from the U.S. Government. The staff need not decide these financial qualification issues now.

As discussed above in Section 2.1.3, as part of any possession and use license application it may later submit, DCS plans to request an exemption from the liability insurance requirements of 10 CFR 140.13a. The staff would consider any such exemption request in reviewing a DCS application for a license to possess and use licensed material at the proposed MFFF.

2.3 References

(DCS, 2001) Hastings, P., Duke Cogema Stone & Webster. Letter to U.S. Nuclear Regulatory Commission, RE: Proprietary DCS Financial Information. August 31, 2001.

(DCS, 2002) Hastings, P., Duke Cogema Stone & Webster. Letter to U.S. Nuclear Regulatory Commission, RE: Clarification of Responses to NRC Request for Additional Information. March 8, 2002.

(DCS, 2003) Hastings, P., Duke Cogema Stone & Webster. Letter to U.S. Nuclear Regulatory Commission, RE: Responses to Financial Qualification, Fire Safety, Chemical Safety, Aqueous Processing, Material Processing and Ventilation Open Items/Additional NRC Questions on Construction Authorization Request (CAR) Revision. February 18, 2003.

(NRC, 2002) U.S. Nuclear Regulatory Commission. "Draft Safety Evaluation Report on the Construction Authorization Request for the Mixed Oxide Fuel Fabrication Facility." Washington, DC, April 30, 2002.

(NRC, 2003) Persinko, A., U.S. Nuclear Regulatory Commission. Letter to Duke COGEMA Stone & Webster, RE: February 2003 Monthly Open Item Status Report. February 13, 2003.

This page intentionally left blank

3. PROTECTION OF CLASSIFIED MATTER

3.1 Conduct of Review

This chapter of the final safety evaluation report pertains to protection of classified matter and whether the applicant's treatment of classified matter will be sufficient to assure that it is adequately protected. The staff evaluated the information provided by the applicant in Chapter 3 of the revised construction authorization request (CAR), as well as supplementary information provided by the applicant.

The staff reviewed how the applicant has addressed the following regulations:

- Title 10, Part 25, "Access Authorization for Licensee Personnel," of the *Code of Federal Regulations* (10 CFR Part 25) specifies the requirements granting and terminating access authorizations to licensee personnel and its contractors who may require access to classified information.

- 10 CFR Part 95, "Facility Security Clearance and Safeguarding of National Security Information and Restricted Data," specifies the requirements for obtaining security facility approval and for safeguarding secret and confidential national security information and restricted data.

The staff used Chapter 3 in NUREG-1718, "Standard Review Plan for the Review of an Application for a Mixed Oxide Fuel Fabrication Facility," as guidance in performing the review.

Chapter 3 of the revised CAR states that the applicant is handling classified matter in accordance with applicable U.S. Department of Energy (DOE) requirements and that it will submit its standard practice procedures plan for the protection of classified matter along with its application for a license to possess and use licensed material.

The staff reviewed the applicant's commitment to provide its standard practice procedures plan for the protection of classified matter along with its license application and concludes that the revised CAR does not need to address procedures for the protection of classified matter because the design basis of the mixed oxide fuel fabrication facility, need not be evaluated for the protection of classified matter. However, the applicant will be required to submit a standard practice procedures plan for the protection of classified matter in accordance with the requirements of 10 CFR Part 25 and 10 CFR Part 95 as part of its license application.

The foreign ownership, control, or influence determination is discussed in Section 1.2.1.2 of the FSER.

3.2 Evaluation Findings

The staff concludes that it is acceptable for the applicant to exclude its standard practice procedures plan for the protection of classified matter as part of the revised CAR because it will be submitted along with the license application. The staff also concludes that it is acceptable for the applicant to handle classified matter based on a memorandum of understanding between the U.S. Nuclear Regulatory Commission and DOE dated October 9, 1996.

3.3 <u>References</u>

(NRC, 1999) U.S. Nuclear Regulatory Commission. "Standard Practice Procedures Plan Standard Format and Content for the Protection of Classified Matter for NRC Licensee, Certificate Holder and Others Regulated by the Commission." Washington, DC, October 1999, as revised.

4. ORGANIZATION AND ADMINISTRATION

4.1 Conduct of Review

This chapter of the final safety evaluation report (FSER) reviews the organization and administration information presented in Chapter 4 of the revised construction authorization request (CAR). The staff used Chapter 4 in NUREG-1718, "Standard Review Plan for the Review of an Application for a Mixed Oxide Fuel Fabrication Facility," (NRC, 2000), as guidance in performing the review. The objective of the review is to determine whether organizational and administrative functions have been identified which will enable the applicant to plan, implement, and control site activities in a manner that adequately ensures the safety of workers and individuals outside the controlled area, and which will also protect the environment. This review ensures that the qualifications for key management positions are adequate. Organizational information is also described in the applicant's quality assurance plan, which is discussed in FSER Chapter 15.

4.1.1 Organization

The applicant proposed a functional organization for engineering, design, and construction that includes lines of responsibility and control of engineering, design, construction, procurement, administrative services, environmental health and safety, licensing, and quality assurance. Proposed organization charts showing lines of responsibility and communications were provided.

The President of Duke Cogema Stone & Webster (DCS or the applicant) has overall responsibility for the project. Reporting to the president are managers responsible for licensing and safety analysis; environment, safety, and health; facility engineering; construction management; procurement; plant operations and startup; project services and administration; site integration; and quality assurance. The quality assurance manager has a direct line of communication to the president and is unencumbered with responsibilities for costs or schedules.

Management of process and facility design will be provided directly by the applicant. The applicant will be responsible for planning, managing, and controlling construction activities. Construction subcontractors will be managed by the construction management manager. At this time, no subcontractors have been selected for construction activities.

The organizational information provided by the applicant sets forth clear and unambiguous controls and communications between organizational groups responsible for designing and constructing the facility. Lines of communication, responsibility, and authority are clearly delineated in the organization chart. The president has overall responsibility for all design and construction activities.

4.1.2 Administration

As construction of the mixed oxide fuel fabrication facility proceeds, the applicant will transition from the design and construction phase of the proposed facility to the operating phase. During any construction of the proposed facility, the construction management manager would be assigned a greater work scope and level of resources. Concomitant with the increased scope of work for the construction management manager, a decreased work scope and level of resources would be assigned to the facilities design and process design managers. Toward the completion of any potential construction activities, the construction organization would shift to a manufacturing organization consisting of a corporate-level plant manager responsible for safety, plant operations, maintenance, and quality assurance. Reporting to the plant manager would be an operations manager, a regulatory management manager, a security manager, an administration manager, and a quality assurance manager. If plant systems are completed, acceptance tests would be performed followed by turnover to the operations organization. The operations organization would then be responsible for maintenance and configuration management of the systems.

Formal management controls described in Chapter 15, of this FSER would be applied to ensure that there is reasonable assurance that design bases are maintained during any transition between design and construction activities and operations.

4.1.3 Key Management Positions

The management positions described in Section 4.1.1, above, have responsibilities for activities involving the proposed facility. The applicant also provided the minimum qualifications for each of these positions.

The scope and number of each key management position are described appropriately for every management function involving the proposed facility. The qualification requirements for key management positions provide an adequate breadth and level of experience for their respective responsibilities and authorities. The staff filling key management positions will be available during the design and potential construction phases of the project.

4.2 Evaluation Findings

The staff reviewed the organization and administration for construction approval of the proposed facility in accordance with Chapter 4 of NUREG-1718 (NRC, 2000). The staff evaluated the proposed organization for design and construction; the administration of the project, including how the project would transition from design and construction to operations; and the responsibilities, qualifications, and authorities of key management positions. The proposed organization, administration, and key management position descriptions and qualifications are consistent with guidance in NUREG-1718 and are, therefore, acceptable.

The issue of whether the design bases of the principal structures, systems, and components of the proposed facility would provide reasonable assurance of protection against natural phenomena and the consequences of potential accidents are addressed in other chapters of this FSER.

4.3 References

(NRC, 2000) U.S. Nuclear Regulatory Commission. NUREG-1718, "Standard Review Plan for the Review of an Application for a Mixed Oxide Fuel Fabrication Facility." Washington, DC, 2000.

5. SAFETY ASSESSMENT OF THE DESIGN BASES

5.1 Conduct of Review

This chapter of the final safety evaluation report (FSER) contains the U.S. Nuclear Regulatory Commission (NRC) staff's review of the safety assessment of the design bases of the principal structures, systems, and components (PSSCs) performed by Duke Cogema Stone & Webster (DCS or the applicant) in Chapter 5 of the revised construction authorization request (CAR) (DCS, 2002b). The objective of this review is to determine whether the PSSCs and their design bases identified by the applicant provide reasonable assurance of protection against natural phenomena and the consequences of potential accidents. The staff evaluated the information provided by the applicant by reviewing Chapter 5 of the revised CAR, other sections of the revised CAR, supplementary information provided by the applicant, and relevant documents available at the applicant's offices, but not submitted by the applicant. The review of PSSCs and their design bases and strategies was closely coordinated with the review of evaluations performed in other chapters of the FSER.

The staff reviewed how the safety assessment information in the revised CAR addresses or relates to the following regulations:

- Title 10, Section 70.23(b), of the *Code of Federal Regulations* (10 CFR 70.23(b)) (CFR, 2004b) states, as a prerequisite to construction approval, that "the design bases of the PSSCs and the quality assurance program be found to provide reasonable assurance of protection against natural phenomena and the consequences of potential accidents."

- 10 CFR 70.61(b) and 10 CFR 70.61(c) set forth performance requirements addressing specified high-consequence events and intermediate-consequence events. For such events, controls must be identified and eventually implemented sufficient to either lessen the likelihood that such events will occur, and/or make the consequences of such events less severe.

- 10 CFR 70.62(c)(2) requires the applicant to have a team with expertise in engineering and process operations. The team must have at least one person who has experience and knowledge specific to each process being evaluated, as well as persons who have experience in nuclear criticality safety, radiation safety, fire safety, and chemical process safety. One member of the team must be knowledgeable in the specific analysis methodology being used. The guidance in Section 5.4.3.1 of NUREG-1718, "Standard Review Plan for the Review of an Application for the Construction of a Mixed Oxide Fuel Fabrication Facility," (NRC, 2000) recommends that a review of team qualifications be made during the safety assessment of the design bases (construction authorization review), as well as during review of the application for a license to possess and use licensed material.

- 10 CFR 70.64, "Requirements for New Facilities or New Processes at Existing Facilities," requires that baseline design criteria (BDC) and defense-in-depth practices be incorporated into the design of new facilities. It specifically addresses quality standards; natural phenomena hazards; fire protection; environmental conditions and dynamic effects; chemical protection; emergency capability; inspection, testing, and maintenance; criticality control; and instrumentation and controls.

The staff used Chapter 5 in NUREG-1718 as guidance in performing the review. NUREG-1718 states the following:

> The steps the applicant follows to develop the safety assessment for the design bases should be analogous to the steps that the applicant will use to develop the ISA; however the reviewer should expect the application of these steps to be adjusted according to the level of design when the applicant applies for construction approval.

NUREG-1718 also states that the description of PSSCs should include "the functional relationship of each principal SSC to the top-level safety function for a process...."

The review for this construction approval focused on the design basis of systems, their components, and other related information. For each PSSC, the staff reviewed information provided by the applicant for the safety function, system description, and safety analysis. The review also encompassed proposed design basis considerations, such as redundancy, independence, reliability, and quality. The staff reviewed descriptions of the systems to assure that the facility can be designed to meet the performance requirements of 10 CFR 70.61, Performance Requirements," during operation of the mixed oxide (MOX) fuel fabrication facility (MFFF or the facility). Much of the review was directed at the applicant's hazard assessment (including natural phenomena and external manmade events), the formulation of a strategy and identification of PSSCs to meet the performance requirements, and assuring that the design bases of these PSSCs are adequate with regard to the performance requirements of the regulation.

The safety assessment review was an integrated team approach. Team members with expertise in the various areas of technical review, such as engineering, nuclear science, and other disciplines, reviewed their respective revised CAR chapters, as well as the revised CAR Chapter 5. These revised CAR chapters or discipline reviews often identified issues that were referred to the safety assessment review at which point they were either resolved or carried as open items based on the hazard assessment or performance strategies.

5.1.1 Plant Site Description Relating to Safety Assessment of the Design Bases

The plant site description includes information to support the safety assessment of the design bases, including the following:

- Site description. The level of detail should be sufficient to allow an evaluation of natural phenomena and other external accidents. The site description is discussed in Section 1.3 of the FSER.

- Facility description. The level of detail should allow an understanding of the relationship between the design bases of the PSSCs and the facility. The facility description is discussed in Section 1.1 of the FSER.

- Process description. The process description should provide sufficient detail to allow the evaluation of the process design as it is established through the design bases. The process description is discussed in Section 1.1 of the FSER.

5.1.2 Safety Assessment Team Description

The safety assessment team is described in Section 5.2 of the revised CAR. The safety assessment team is described as a team of individuals experienced in hazard identification, hazard evaluation techniques, accident analysis, including dose consequence assessment, and probabilistic analysis. The team members possess operational experience at similar facilities, specific discipline knowledge (e.g., mechanical; electrical; and heating, ventilation, and air conditioning) and specific knowledge of the processes to be used in the facility. In addition, the team has safety analysis experience that is MOX process and aqueous polishing (AP) specific. The integrated safety analysis (ISA) manager is described as having overall responsibility for preparation of the safety assessment. The ISA manager reports to the facility's licensing and safety analysis manager. The ISA manager provides overall direction for the analysis, organizes and executes safety analysis activities, and facilitates team meetings. The technical analysis which supports the safety assessment is the responsibility of the ISA team leader who reports to the ISA manager. The ISA team leader is knowledgeable in the specific safety assessment methodologies chosen for the hazard and accident analyses and has an understanding of process operations and the hazards under evaluation. Based on the above description, the staff concludes that the applicant's safety assessment team satisfies the criteria of Section 5.4.3.1 of NUREG-1718 (NRC, 2000) at the construction authorization stage.

5.1.3 Chemical Standards and Consequences

DCS provided chemical concentration limits to evaluate the potential consequences to individuals outside the controlled area and workers for an accidental release of chemicals. The applicant based these limits on the acute exposure guideline level (AEGL) values and the emergency response planning guideline (ERPG) values. For chemicals which do not have AEGL or ERPG value, limits are based on temporary emergency exposure limits (TEELs) adopted by the U.S. Department of Energy (DOE), Subcommittee on Consequence Assessment and Protective Action. A discussion of the chemical consequences and the applicant's consequence analysis is provided in Section 8.1.2.3 of the FSER. A summary of the staff's review of chemical events is provided in Section 5.1.6.3.6 of the FSER.

5.1.4 10 CFR 70.61 Performance Requirements

As discussed in the introduction to the FSER, 10 CFR 70.61(b) sets forth the performance requirements for "credible high-consequence events" (i.e., potential accidents involving high levels of radiation or hazardous chemicals);, and its provisions pertain to the protection of both onsite workers and offsite individuals. As required by 10 CFR 70.61(b), the use of controls must be sufficient to either make the occurrence of such accidents "highly unlikely," or make the consequences of such accidents less severe than (1) an acute 1 Sv (100 rem) total effective dose equivalent (TEDE) to a "worker," as defined in 10 CFR 70.4, "Definitions," (2) an acute 0.25 Sv (25 rem) TEDE to a person outside the "controlled area," as defined in 10 CFR 20.1003, "Definitions," (3) an intake of 30 mg of soluble uranium to a person outside the controlled area, or (4) either an acute chemical exposure that could endanger the life of a worker or an acute chemical exposure that could lead to irreversible or other serious long-lasting health effects to a person outside the controlled area. (See 10 CFR 70.61(b)(1–4)).

As set forth in 10 CFR 70.61(c), the performance requirements for "credible intermediate-consequence events," (i.e., potential accidents involving lower levels of radiation or hazardous chemicals than referenced in 10 CFR 70.61(b)), and its provisions pertain to environmental protection, as well as to protecting the health of onsite workers and offsite individuals. This regulation also requires the use of controls sufficient to either make the occurrence of such accidents "unlikely," or to make the consequences of such accidents less severe than (1) an acute 0.25 Sv (25 rem) TEDE to a worker, (2) an acute 0.05 Sv (5 rem) TEDE to a person outside the controlled area, (3) a 24-hr average release of radioactive material outside the "restricted area," as defined in 10 CFR 20.1003, into the environment in concentrations exceeding 5000 times the values in Table 2 of Appendix B, "Radiation Protection Programs," to 10 CFR Part 20, "Standards for Protection Against Radiation," or (4) either a chemical exposure that could lead to irreversible or other serious long-lasting health effects to a worker or a chemical exposure that could cause mild transient health effects to a person outside the controlled area. (See 10 CFR 70.61(c)(1–4)).

Under both 10 CFR 70.61(b) and 10 CFR 70.61(c), properly identifying and implementing the required controls—which are later designated as items relied on for safety (IROFS) pursuant to 10 CFR 70.61(e) and 10 CFR 70.65(b)—ensures that an acceptable level of risk will be maintained during any operation of the proposed facility. This goal is met through a combination of limiting the chance that high-consequence or intermediate-consequence events would occur (prevention) and reducing the consequences of such events (mitigation).

The starting point for the applicant's demonstration of acceptable control over the risk of credible high-consequence and intermediate-consequence events, as well as the risk of nuclear criticality accidents, is its safety assessment of the facility design bases. In Section 5.1.6 of the FSER, the staff evaluates the hazards that have been addressed by DCS, and finds that these hazards are adequately controlled by the PSSCs designated by DCS.

5.1.5 Safety Assessment of Design Basis Methodology

The objective of the staff's review of the methodology was to determine if the safety assessment was complete by assuring that all appropriate natural phenomena and external manmade and internal process hazards were considered. The review of natural phenomena and external manmade hazards consisted of evaluating the DCS screening criteria to determine if they were appropriate for identifying all credible events. To evaluate whether the internal process hazards were sufficiently addressed, the staff reviewed the proposed plant processes, reviewed the operating experience and hazard analyses of other similar facilities, and considered feedback from the discipline-specific FSER reviews.

The DCS safety assessment of the design bases consisted of the identification and assessment of natural phenomena hazards, external manmade hazards, and internal process hazards. Pursuant to 10 CFR 70.61, high-consequence events must be highly unlikely and intermediate-consequence events must also be unlikely. In the revised CAR, the applicant has provided qualitative definitions of the terms "not unlikely," "unlikely," "highly unlikely," "credible," and "not credible". (Quantitative likelihood values are not required in 10 CFR Part70, "Domestic

Licensing of Special Nuclear Materials.") All initiating events were assumed to have a likelihood of "not unlikely," which the applicant has defined as events that may occur during the life of the facility. Because of the high probability associated with postulated events, PSSCs will be selected and designed such that the accident sequences with regard to the above threshold doses to the facility worker, site worker, and individuals outside the controlled area will be highly unlikely. The applicant has defined "highly unlikely" as "events originally classified as not unlikely or unlikely to which sufficient principal SSCs are applied to further reduce their likelihood to an acceptable level." The applicant has proposed deterministic design criteria to assure that the consequences from postulated events that will exceed the threshold limits of 10 CFR 70.61(c) are highly unlikely. These criteria include the following:

- application of the single-failure criterion or double-contingency principle

- application of NQA-1 in Appendix B, "Quality Assurance Criteria for Nuclear Power Plants and Fuel Reprocessing Plants," to 10 CFR Part 50, "Domestic Licensing of Production and Utilization Facilities"

- application of industry codes and standards

- management measures including IROFS failure detection (IROFS failure detection and repair or process shutdown capability)

In addition, the applicant committed to a supplemental likelihood assessment for event sequences that could exceed the 10 CFR 70.61(c) criteria for persons outside the controlled area boundary. Only the deterministic criteria would apply to facility workers and the environment. In the revised CAR, Section 5.4.3 (DCS, 2002b), the applicant stated that, "This supplemental likelihood assessment will be based on the guidance provided in the NUREG-1718 [NRC, 2000] and will demonstrate a target likelihood index comparable to a 'score' of -5 as defined in Appendix A of the SRP." The staff finds the applicant's definition of "highly unlikely" to be acceptable.

In regard to the environmental protection requirements of 10 CFR 70.61(c)(3), PSSCs will be selected and designed so as to ensure that the accident sequence is unlikely, as discussed in FSER Chapter 10. In the revised CAR, "unlikely" is defined as "events that are not expected to occur during the lifetime of the facility but may be considered credible." In that there were no accident sequences that resulted in over-the-threshold consequences for the environment only, this classification of events (i.e., unlikely) was not used in the safety assessment to demonstrate a compliance strategy with the performance requirements of 10 CFR 70.61.

Another definition provided by DCS in the revised CAR to support the safety assessment is "not credible." The applicant defined this term as "natural phenomena or external man-made events with extremely low initiating frequency and process events that are not possible." The application of this definition is explained in the revised CAR in the discussions of screening criteria for natural phenomena and external manmade external events. The staff found the applicant's definition of credible to be acceptable in its review of the natural and manmade event screening criteria in FSER Sections 5.1.5.1 and 5.1.5.2, respectively.

As indicated above, controlling facility risks entails the identification and assessment of potential facility hazards (i.e., accident scenarios). Based on this hazards assessment, PSSCs and the safety functions of the PSSCs can then be identified. The applicant's methodology for developing the PSSCs and their functions is presented graphically in the flowchart in the revised CAR Figure 5-4.1 The basic inputs to the selection process are the site description from which credible natural phenomena and external manmade hazards are determined from a screening process and preliminary design information from which credible internal hazards are identified. The results of the external event screening and internal hazard screening are inputs to a preliminary accident analysis.

An early step in the preliminary hazard analysis (PHA), the correlation of process units with facility workshops and process support units, is shown in the revised CAR Tables 5.5-1 and 5.5-2. The radioactive material inventory in each facility location is shown in the revised CAR Table 5.5-3a; the radioactive material inventory by fire area is shown in the revised CAR Table 5.5-3b. The summary hazard identification matrix of hazards versus workshops and process support groups is shown in the revised CAR Table 5.5-4. This segmentation and correlation with hazards allowed a comprehensive hazard identification for each individual area. A consequence analyses was then performed by the applicant to evaluate the bounding unmitigated consequences for each type of accident within a workgroup. If the unmitigated consequences exceeded the dose thresholds for 10 CFR 70.61(c)[2], then the group was further evaluated. For the event scenarios which exceed the 10 CFR 70.61(c) thresholds, a safety strategy for prevention or mitigation was established and PSSCs at the structure and system level were identified. The selection of safety strategies was facilitated by segregating events which had common features that would allow similar prevention or mitigation strategies into event groups. This simplified the analysis by allowing for the development of common safety strategies and PSSCs for multiple events, such that the PSSCs that cover bounding events also cover nonbounding events. In the context of the applicant's analysis, a bounding event is the event which results in the largest consequence in each group and the greatest risk because the likelihood of all of the events is considered to be the same.

After the PSSCs have been determined, their design bases are developed and, if the accident consequence is mitigated, the resulting bounding mitigated consequence is compared against the performance requirements of 10 CFR 70.61(c). If mitigation is successful at sufficiently reducing the consequences, or if the accident scenario is prevented, the developed PSSCs and support functions become input to the final design. If not, the evaluation is repeated with a different set of PSSCs (or a change in design-bases values). The applicant, in the revised CAR, Section 5.4.4.3, pursuant to 10 CFR 70.61(c)(3), also performed analyses of the potential radioactive release to the environment by calculating the 24-hr average effluent concentration of each radionuclide released in an accident sequence and comparing this to 5000 times the values specified in Table 2 of Appendix B, "Annual Limits on Intake (ALIs) and Derived Air Concentrations (DACs) of Radionuclides for Occupational Exposure; Effluent Concentrations; Concentrations for Release to Sewerage," to 10 CFR Part 20, "Standards for Protection Against Radiation" (CFR, 2004a). This is required to show compliance with 10 CFR 70.61(c)(3).

---

[2]Five rem total TEDE to any individual outside the controlled area, 25 rem TEDE to a facility worker, 24-hr concentrations exceeding 5000 times the values in Table 2 of Appendix B to 10 CFR Part 20, or the chemical safety criteria.

5.1.5.1 Natural Phenomena Hazards Methodology

Natural phenomena having a credible potential effect on facility operations were identified through a screening process in which natural phenomena hazards (NPHs) having a frequency of occurrence of less than 10^{-6} per year were designated as incredible and screened from further consideration. Deterministic methods were also used to screen out events that would not be physically appropriate for the site. For example, debris avalanching was ruled out because of the relatively level nature of the surrounding topography. The staff considers the NPHs screening methodology to be in accordance with the criteria of NUREG-1718, Section 5.4.5.2 (NRC, 2000) and is therefore acceptable for the purpose of meeting the performance requirements of 10 CFR 70.61. A comprehensive list of NPHs were initially evaluated and the rationale for further consideration or exclusion of each NPH is provided in the revised CAR Table 5.5-5.

5.1.5.2 External Manmade Events Methodology

The applicant considered external manmade hazards (EMMHs) to be those hazards that are caused by events originating from the operation of nearby public, private, government, industrial, chemical, nuclear, and military facilities and vehicles. The major categories of events that could result from EMMHs that were considered by DCS are as follows:

- a release of radioactive material resulting in exposures to facility personnel
- a release of hazardous chemicals resulting in exposures to facility personnel
- explosions or other events that directly impact facility PSSCs
- events that result in a loss of offsite power
- events that result in a fire (and/or resulting smoke) that spreads to the facility

Events in these categories were screened using applicable criteria from NUREG/CR-4839, "Methods for External Event Screening Quantification: Risk Methods Integration and Evaluation Program (RMIEP) Methods Development," Regulatory Guide (RG) 1.91, "Evaluations of Explosions Postulated to Occur on Transportation Routes Near Nuclear Power Plants," RG 1.78, "Assumptions for Evaluating the Habitability of a Nuclear Power Plant Control Room during a Postulated Hazardous Chemical Release," and NUREG-0800, "Standard Review Plan." A summary of the EMMH screening is provided in Table 5.5-8 of the revised CAR.

External manmade events that were evaluated and screened out as not applicable to the site or of having too low a probability for consideration include the following:

- roadway accidents
- rail accidents
- aircraft accidents
- barge/shipping traffic accidents
- industrial facility accidents, except for F-Area
- military facility accident

The staff performed an in-office review of the applicant's calculation of future flight activities over the life of the facility to confirm the aircraft accident analysis. The staff found these calculations and previously submitted calculations (DCS, 2002a; Enclosure A, DCS, 2002c) to be acceptable.

The staff considers the EMMHs screening methodology to be in accordance with NRC guidance and is acceptable for the purpose of meeting the performance requirements of 10 CFR 70.61.

5.1.5.3 Process Hazards Methodology

The applicant evaluated the potential for and consequences of process-related internal events. These events were divided into six major categories:

(1)   loss of confinement/dispersal of nuclear material events
(2)   fire events
(3)   load-handling events
(4)   explosion events
(5)   criticality
(6)   chemical

In the revised CAR, DCS only presented numerical radiological consequence values for the most severe event in each of the above major categories (except for the chemical category). Chemical consequences are discussed in Chapter 8 of the revised CAR and the staff's evaluation is provided in Chapter 8 of the FSER. The staff reviewed the accident scenarios developed by the applicant and has determined that these are a complete and bounding set based on the preliminary design and description of the processes projected for use at the proposed facility and described in Chapters 1, 7, 8, 9, 10, and 11 of the revised CAR (DCS, 2002b).

5.1.5.4 Baseline Design Criteria

As set forth in 10 CFR 70.64(a)(1–10), the following 10 BDC are applicable to the proposed facility:

(1)   quality standards and records, requiring that the facility design be developed and implemented in accordance with management measures and IROFS, and that records of the IROFS be maintained for the life of the facility

(2)   natural phenomena hazards, requiring that the facility design adequately protects against natural phenomena and takes into consideration the most severe historical events documented at the facility site

(3)   fire protection, requiring that the facility design adequately protects against fires and explosions

(4)   environmental and dynamic effects, requiring that the facility design adequately protects against internal environmental conditions and dynamic effects from normal operations, maintenance, testing, and accidents

(5)   chemical protection, requiring that the facility design adequately protects against specified chemical risks and internal conditions affecting the safety of licensed material

(6)   emergency capability, requiring that the facility be designed to provide emergency capability to control (i) licensed material and hazardous chemicals produced from licensed material, (ii) evacuation of onsite personnel, and (iii) specified onsite emergency facilities and services

(7)   utility services, requiring that the facility be designed so that essential utility services will continue to operate

(8)   inspection, testing, and maintenance, requiring that IROFS designs provide for adequate inspection, testing, and maintenance, to ensure that IROFS will be available and will reliably perform their functions when needed

(9)   criticality control, requiring that the facility be designed to provide for criticality control, including adherence to the double-contingency principle

(10)   instrumentation and controls, requiring that the facility be designed to provide for inclusion of instrumentation and control systems to monitor and control the behavior of IROFS

In the FSER sections referenced below, the staff states whether or not the facility's preliminary design satisfies the BDC outlined above, pursuant to 10 CFR 70.64(a). In the review of the revised CAR, meeting the 10 CFR 70.61 performance requirements was the primary criterion considered by the staff for satisfying the BDC.

(1)   Quality Standards and Records. The design must be developed and implemented in accordance with management measures to provide adequate assurance that IROFS will be available and reliable to perform their functions when needed. Appropriate records of these items would have to be maintained by or under the control of the licensee throughout the life of the proposed facility. This BDC is satisfied at the construction authorization stage, as discussed in FSER Section 15.1.

(2)   Natural Phenomena Hazards. The design must provide for adequate protection from natural phenomena with consideration of the most severe documented historical events for the site. This BDC is satisfied at the construction authorization stage, as discussed in FSER Sections 1.3, 5.1.6.1, 11.1, and 11.12.

(3)   Fire Protection. The design must provide for adequate protection against fires and explosions. This BDC is satisfied for fires at the construction authorization stage, as discussed in FSER Section 7.1.2. This BDC is also met for explosions, as discussed in Chapters 8 and 11.

(4)   Environmental and Dynamic Effects. The design must provide for adequate protection from environmental conditions and dynamic effects associated with normal operations, maintenance, testing, and postulated accidents that could lead to loss of safety functions. This BDC is satisfied at the construction authorization stage, as discussed in FSER Section 11.11.

(5)   Chemical Protection. The design must provide for adequate protection against chemical risks produced from licensed material, facility conditions which affect the safety of licensed material, and hazardous chemicals produced from licensed material. This BDC is met, as discussed in FSER Sections 8.2 and 11.2.2.

(6)   Emergency Capability. The design must provide for emergency capability to maintain control of the following:

   (i)   licensed material and hazardous chemicals produced from licensed material

   (ii)   evacuation of onsite personnel

   (iii)   onsite emergency facilities and services that facilitate the use of available offsite services

This BDC is satisfied at the construction authorization stage, as discussed in FSER Chapter 14.

(7)   Utility Services. The design must provide for continued operation of essential utility services. Based on the safety assessment of the proposed facility, the only utility with safety significance is electrical power. This BDC is satisfied at the construction authorization stage, as discussed in FSER Section 11.5.

(8)   Inspection, Testing, and Maintenance. The design of the IROFS must provide for adequate inspection, testing, and maintenance to ensure their availability and reliability to perform their function when needed. This BDC is satisfied at the construction authorization stage, as discussed in FSER Sections 15.1 and 15.3.

(9)   Criticality Control. The design must provide for criticality control including adherence to the double-contingency principle. This BDC is met at the CAR stage, as discussed in FSER Section 6.1.4.2.

(10)   Instrumentation and Controls. The design must provide for inclusion of instrumentation and control systems to monitor and control the behavior of the IROFS. This BDC is satisfied at the construction authorization stage, as discussed in FSER Section 11.6.

5.1.5.5 Defense-in-Depth

Under 10 CFR 70.64(b), the facility design and layout must be based on defense-in-depth practices. As used in 10 CFR 70.64, defense-in-depth practices at new facilities means a design philosophy, applied from the outset and through completion of the design, that is based on providing successive levels of protection, such that health and safety will not be wholly dependent upon any single element of the facility design. The net effect of incorporating defense-in-depth practices is a conservatively designed facility that will exhibit greater tolerance to failures and external challenges. As required by 10 CFR 70.64(b), to the extent practicable, the facility design must incorporate (1) a preference for engineered controls over administrative controls to increase overall system reliability, and (2) features that will enhance safety by reducing challenges to items which will be relied upon for safety.

In Section 5.5.5 of the revised CAR, the applicant described its general design philosophy and defense-in-depth practices. In this section, the applicant described a hierarchy of controls in its general philosophy of design that has been established as follows (the most favored control listed first):

- protection by a single passive safety device, functionally tested on a predetermined basis

- independent and redundant active engineered features, functionally tested on a predetermined basis

- single hardware system/engineered feature, functionally tested on a predetermined basis

- enhanced administrative controls

- simple administrative controls or normal process equipment

The staff has determined that the above hierarchy of controls demonstrates a preference for engineered controls, as required by 10 CFR 70.64(b)(1). In addition, the applicant's incorporation of additional protection features into the design of the proposed facility will enhance safety by reducing challenges to items which will be relied upon for safety, as required by 10 CFR 70.64(b)(2).

In addition, the applicant has described its defense-in-depth practices as consisting primarily of meeting double contingency (for protection against criticality events) and the single-failure criterion. The staff's evaluation of the applicant's implementation of double contingency is provided in Section 6.1.4.2 of the FSER. The applicant's implementation of the single-failure criterion, as described in Section 5.5.5.2 of the revised CAR, consists of (1) the use of redundant equipment or systems, (2) independence, (3) separation, and (4) the fail-safe principle.

The staff concludes that the applicant's strategy for defense-in-depth meets the requirements of 10 CFR 70.64.

5.1.6 Safety Assessment Results

The safety assessment methodology, as described above, resulted in the identification of accident scenarios and PSSCs and their functions. The PSSCs identified by the applicant through its safety assessment are summarized in Tables 5.1-1 and 5.1-2 of the FSER. Table 5.1-3 lists the design bases associated with the safety functions of the PSSCs and references the FSER sections that contain more detailed staff evaluations of the PSSCs.

5.1.6.1 Natural Phenomena Design Basis Events and Related PSSCs

As stated in FSER Section 5.1.5.1, the staff has determined the screening methodology for NPHs to be acceptable. The screening methodology identified design basis natural events and their related PSSCs. The likelihood of any such design basis event occurring should be sufficiently low to assure that any adverse consequences are highly unlikely because structural failures caused by natural phenomena were assumed to have the potential for high consequences. The adequacy of the PSSCs to prevent releases are evaluated using normally accepted industry practice as a criterion.

Natural phenomena that were not screened out included the following:

- extreme wind
- earthquake (including liquefaction)
- tornado (including tornado missiles)
- external fire
- rain, snow, and ice
- lightning
- temperature extremes

The design basis wind selected for the facility has an annual exceedance probability of 10^{-4} per year (DCS, 2002b, Section 5.5.2.6.5.1). The PSSCs identified to provide protection against the design basis wind are the MOX fuel fabrication building, emergency diesel generator building, associated missile barriers, and the waste transfer line. The safety function of the structures and missile barriers are to withstand design basis wind loads and wind-driven missiles and to provide protection for internal structures, systems, and components (SSCs). Although the exceedance probability of 10^{-4} does not, by itself, preclude a consequence which is highly unlikely, the design for wind loadings is controlled by the tornado at low frequencies of occurrence. Hence, the staff concludes that the design basis established by the applicant for extreme wind satisfies the performance requirements of 10 CFR 70.61.

The design basis earthquake selected for the facility also has an annual exceedance probability of 10^{-4} (DCS, 2002b, Section 5.5.2.6.5.2). The PSSCs identified to provide protection against the design basis earthquake are (1) the waste transfer line, (2) the MFFF building, (3) the emergency diesel generator building, (4) the fluid transport systems, as necessary, and (5) the seismic monitoring and associated isolation valves. The primary safety function of the first four PSSCs is to withstand the effects of the design basis earthquake and to assure that seismic effects on non-PSSCs will not result in the prevention of PSSCs from performing their safety function. The safety function of the seismic monitoring and isolation valves is to prevent fire and/or criticality as a result of an uncontrolled release of chemicals and water within the MFFF building. For NRC-licensed facilities, such as nuclear power plants, the conservatism between design and performance arises from factors such as prescribed analysis methods, specification of material strengths, and limits on inelastic behavior following nuclear design criteria and NRC Standard Review Plans (SRPs). Conservatism in the NRC seismic SRPs are not explicitly keyed to risk-reduction values. Nevertheless, the risk-reduction factors achieved by applying NRC guidelines to the evaluation of commercial reactor SSCs have been shown to be equal to or even higher than those prescribed by DOE STD-1020-94, Change Notice No. 1, "Natural Phenomena Hazards Design and Evaluation Criteria for Department of Energy Facilities" (DOE, 1994). For example, the average mean annual probability of exceedance (MAPE) for the design ground motions at existing nuclear power plants is approximately 1×10^{-4} (NRC, 1997), yet the mean annual seismic core damage frequency of nuclear power plants is estimated to range between 6×10^{-6} and 1×10^{-5} (Chen, 1991). Thus, an effective risk reduction for nuclear power plants is 10 times or greater. In additional information provided to NRC (DCS, 2002a; Enclosure B, DCS, 2002c), DCS provided the results of calculations which showed that, taking into account the building and component designs, the performance of SSCs in the facility will meet the availability (or failure) criteria necessary to make high consequences highly unlikely. The staff concludes that the applicant's selection of the design basis earthquake satisfies the performance requirements of 10 CFR 70.61.

The design basis tornado selected for the facility has an annual exceedance probability of 2×10^{-6} (DCS, 2002b, Section 5.5.2.6.5.3). The PSSCs identified to provide protection against the design basis tornado are the MFFF building, the emergency diesel generator building, the associated missile barriers, the waste transfer line, and the tornado dampers. The safety functions of these PSSCs are to withstand the design basis tornado wind loads, tornado-generated missiles, and differential pressure, as well as to provide protection for internal SSCs. The staff concludes that the applicant's selection of the design basis tornado satisfies the performance requirements of 10 CFR 70.61.

The design basis for external fire was assumed to be a forest fire near the facility (DCS, 2002b, Section 5.5.2.6.5.4). The plant exterior is designed to withstand a fire duration of at least 2 hours (further information may be found in Section 7.1.5.4 of the FSER). This is considered by the staff to be adequate, based on the availability of an onsite fire brigade and the fuel loading provided by natural growth around the building. The PSSCs identified to provide protection against the external fire are the MFFF building structure, the emergency generator building structure, the emergency control room air-conditioning system, and the waste transfer line. The safety functions of the PSSCs are to withstand the effects of the external fire, to provide protection for internal SSCs, and to ensure habitable conditions for operators as necessary. The staff concludes that the applicant's selection of the design basis external fire is an acceptable strategy for meeting the performance requirements of 10 CFR 70.61.

The design basis rainfall has an annual exceedance probability of 10^{-5} (DCS, 2002b, Section 5.5.2.6.5.5). This will meet the likelihood requirements for high- and intermediate-consequence events. The snow and ice loadings have an annual exceedance probability of 10^2 (DCS, 2002b, Section 5.5.2.6.5.5). The applicant has stated that effects of snow and ice loads with a lower annual exceedance probability are bounded by the design for other live loads. DCS determined that a 10,000-yr snow and ice load would be less than one-half of the design load for live loads. The PSSCs and design basis safety functions associated with rain, snow, and ice are the MOX fuel fabrication building structure and the emergency diesel generator building structure, which will be designed to withstand the effects of rain, snow, and ice without failing and will protect internal SSCs from the effects of rain, snow, and ice. The staff considers the applicant's strategy and selection of PSSCs to be acceptable for meeting the performance requirements of 10 CFR 70.61.

The design basis for lightning protection was in accordance with National Fire Protection Association (NFPA) 780-1997 (DCS, 2002b, Section 5.5.2.6.5.6). Design basis temperature extremes for the ventilation system were based on observed temperatures at the Savannah River Site (SRS) over a 35-yr period (1961 to 1996). Both of these design bases are appropriate because neither lightning nor severe temperature are expected to cause a significant consequence by themselves. No PSSCs are required for protection against lightning or extreme temperatures. The staff agrees with the applicant's rationale for not requiring additional PSSCs to protect against these events.

The staff considers the results of the evaluation of NPHs to be acceptable for meeting the performance requirements of 10 CFR 70.61. The PSSCs identified by DCS to control the risks of natural phenomena at the proposed facility will, if properly implemented, ensure that such risks will be acceptably low during any facility operation. In addition, the staff considers the applicant's evaluation to be adequate to satisfy the requirements of 10 CFR 70.64(a)(2) (BDC, NPHs) which state that the design must provide for adequate protection against natural phenomena with consideration of the most severe documented historical events for the site.

5.1.6.2 External Manmade Events and Related PSSCs

As stated in FSER Section 5.1.5.2, the staff has determined the screening methodology for external manmade hazards to be acceptable. The screening methodology identified manmade external design basis events and their related PSSCs. The likelihood of any such design basis event occurring should be sufficiently low to assure that any adverse consequences are highly unlikely because structural failures caused by such events were assumed to have the potential for high consequences. The adequacy of the PSSCs to prevent releases are evaluated using normally accepted industry practice as a criteria.

Manmade external events that were not screened out include:

- Potential hazardous chemical or radioactive releases from SRS facilities or vehicles. SRS documentation provides the radiological/chemical consequences of accidents at existing facilities. The applicant has reviewed these analyses and determined that there are no credible accidents that could potentially impact facility operations personnel. However, personnel in the emergency control room will be protected by the emergency control room air-conditioning system which is considered a PSSC for other evaluated events. In addition, based on existing DOE requirements, it is not expected that facilities to be designed and operated by DOE to support the MOX facility will present a significant risk for the facility. The staff will consider possible risks from these facilities as part of its review of any later application for a license to possess and use licensed material, when more information about the proposed DOE facilities is expected to be available.

- Potential explosions at a nearby facility or an explosion involving a vehicle, particularly one in the F-Area. The applicant stated that the main MOX building (BMF) and the emergency diesel generator buildings can withstand the impacts of explosions in the F-Area. The staff performed an in-office review of the applicant's calculations of vapor cloud explosions resulting from a tanker truck accident on an F-Area road nearest to the MFFF and an explosion from a large storage area more distant from the MFFF. The explosion calculations were based on conservative assumptions from RG 1.91. The staff made an independent check on the applicant's reflected overpressure calculations using Figure 2-5.14 on page 2-101 of the Society of Fire Protection Engineers (SFPE) *Handbook of Fire Protection Engineering*, First Edition, published September 1988. The staff determined that the applicant's calculations were close to the staff's values obtained from Figure 2-5.14 and were safely within the design overpressure limits of the structures.

- Loss of offsite power from EMMHs is considered similar in potential for consequences as loss of offsite power from NPHs. PSSCs requiring power are supplied with emergency power upon loss of offsite power and failure of the standby alternating current (ac) power system. The adequacy of the power supply in terms of the baseline design criteria and the performance requirements has been evaluated in FSER Section 11.5.1.3. The staff considers the applicant's strategy and selection of PSSCs to be acceptable for meeting the performance requirements of 10 CFR 70.61 with regard to loss of offsite power.

- External manmade fires are fires resulting from a vehicle crash, train crash/derailment, barge/shipping accident, or SRS facility fire that engulf neighboring grasslands or forests. This event has the same consequences as the design basis external fire listed as an NPH. The ability of the facility to withstand the effects of external fires is discussed in FSER Section 7.1.5.4. The staff also considers the applicant's strategy and selection of PSSCs to be acceptable for meeting the performance requirements of 10 CFR 70.61.

Based on its evaluation of the screening methodology (FSER Section 5.1.5.2) and the evaluated events as described above, the staff considers the results of the evaluation of manmade events to be acceptable for meeting the performance requirements of 10 CFR 70.61. The PSSCs identified by DCS to control the risks of such manmade events at the proposed facility will, if properly implemented, ensure that such risks will be acceptably low during any facility operation.

5.1.6.3 Internal Process Hazard Design Basis Events and Related PSSCs

As stated in FSER Section 5.1.5.3, the staff found that the screening methodology for internal process hazards was adequate. With respect to the internal process hazards which DCS evaluated, the staff's review of those hazards was primarily an evaluation of the strategy and PSSCs at a conceptual level with regard to their potential to guide the development of a design which will meet the 10 CFR 70.61 performance requirements. Criteria used in the staff evaluation consisted of a comparison against normally accepted industry practice, consideration of the applicant's design criteria, consideration of a probability index using the NUREG-1718, Table A-5 descriptions (primarily for protection of individuals outside the controlled area boundary (IOC)), and/or deterministic arguments primarily for protection of facility workers and/or the environment. Table A-5 of NUREG-1718 provides a table equating types of controls to approximate probabilities of failure on demand (PFOD). The following descriptions of controls were provided:

- exceptionally robust passive engineered control (PEC) or an inherently safe process (index -4 or -5; PFOD 10^{-4}–10^{-5})

- a single PEC or an active engineered control (AEC) with high dependability (index -3 or
- -4; PFOD 10^{-3}–10^{-4})

- a single AEC, an enhanced[1] administrative control, or an administrative control for routine planned operations (index -2 or -3; PFOD 10^{-2}–10^{-3})

- an administrative control that must be performed in response to a rare, unplanned demand (index -1 or -2; PFOD 10^{-1}–10^{-2})

---

[1]In its evaluation, the staff has considered administrative controls that consist of two or more independent actions, where the success of either will prevent the consequences, as enhanced administrative controls.

For the purposes of this review, the staff considered the revised CAR description of a high dependability AEC (such as the C4 confinement system) and assigned it an index of -4 or -5. In addition to the base dependability indices that the staff determined for the PSSCs based on their descriptions in the revised CAR, the staff also took into account the impact of surveillance intervals on the overall reliability. The assumption was made that as long as the PSSC was capable of being part of a surveillance program, surveillance intervals would be adjusted at the ISA preparation stage to achieve the desired dependability. In addition, enhanced administrative controls, such as combustible loading controls, are assigned a lower PFOD in the staff review because of the incorporation of features such as fire modeling, quantifiable margins, and surveillances in their implementation.

In addition to the dependability of the safety strategy, the staff also independently evaluated the applicant's consequence assessment for those event sequences in which the consequences for one or more of the potential receptors was determined by the applicant to be less than the 10 CFR 70.61(c) threshold value. These independent calculations are described in FSER Section 9.1.1.4.

Feedback from the technical reviews was also used to evaluate the practicality and appropriateness of the PSSCs or safety strategy to the event being evaluated. This feedback was used to assure that the proposed strategies did not significantly deviate from accepted nuclear industry practice, taking into account historical events as well as successful operation at other chemical or nuclear facilities.

An area of discussion with DCS was the protection of facility workers during accident events. It was the position of DCS that the index method and its implied numerical probability may not be applicable to protection of the facility worker. Reliance on worker actions for mitigation in many of the worker protection scenarios requires a deterministic, rather than a probabilistic evaluation. The accident scenarios which rely on prevention (such as most explosions, some fires, and some of the materials-handling accidents) do not require an evaluation for the facility worker separate from the one performed for protection of the environment and persons outside the controlled area boundary. If the specific prevention measures are considered sufficient to make the release of radioactive material from the accident sequence highly unlikely, dose to the worker does not need to be evaluated. Also, some of the fires are not prevented to a low probability of occurrence, but are considered sufficiently slow-growing events such that a worker may take a course of action such as leaving the area and/or donning a respirator that would make the worker's dose effectively zero or negligible. For some of the load-handling events, the staff questioned the ability of the worker to don a mask or vacate the area in sufficient time to keep the worker dose below the 10 CFR 70.61(c) threshold levels. In these cases, the staff requested the applicant to perform dose calculations which were reviewed on site (NRC, 2002). These were found to be acceptable, because the doses were low enough to allow the worker to take protective action within a reasonable time.

The following six subsections (5.1.6.3.1 through 5.1.6.3.6) discuss the accident sequences developed to evaluate internal process hazards. The staff finds that the evaluations of internal process hazards at the proposed facility meet the performance requirements of 10 CFR 70.61.

5.1.6.3.1 Confinement Events

Confinement of radioactive material at the facility is provided by static confinement boundaries in conjunction with ventilation systems and sealed confinement barriers (e.g., containers and fuel rods).

Thirty-one separate events with potentially significant consequences were analyzed by DCS to determine the bounding consequences from a potential loss of confinement event. These events were assigned to the following 12 groups with a unique prevention or mitigation strategy:

(1)   overtemperature
(2)   corrosion
(3)   glovebox breaches or backflows
(4)   leaks in the AP process vessels or pipes
(5)   backflow from a process vessel through utility lines
(6)   rod-handling operations
(7)   breaches in containers outside of gloveboxes as a result of handling
(8)   over- or underpressurization of glovebox
(9)   excess temperature caused by radioactive decay
(10)   glovebox dynamic exhaust failure
(11)   process fluid line leak in a C3 area outside a glovebox
(12)   sintering furnace confinement boundary failure

These 12 groups are discussed below:

Overtemperature (Confinement)

The bounding event for the overtemperature event group in the confinement events accident category was determined by the applicant to be excessive temperature of the AP electrolyzer resulting in high-temperature damage to, and breach of, the AP electrolyzer and damage to glovebox panels and dispersal of radioactive material. The material at risk was the maximum inventory of radioactive material in the electrolyzer glovebox. Such an event could be caused by control system failure, electrical isolation failure, or loss of cooling to process equipment. The applicant determined this to be an above the 10 CFR 70.61(c) threshold consequence event for facility workers, site workers, IOC, and the environment, and has opted to protect potential receptors through a strategy of prevention and mitigation. The PSSC identified for protection of the facility worker and the environment for this event is the process safety control subsystem, which will shut down process equipment before exceeding temperature safety limits. The PSSC for protection of site workers and IOC is the C3 confinement system, which will provide filtration to mitigate dispersions from the C3 area. In addition, the process safety control subsystem provides defense-in-depth for protection of the site worker and IOC. Based on the applicant's deterministic criteria and the NUREG-1718 Table A-5 descriptions[2] (high availability AEC), the

---

[2]NUREG-1718, Table A-5, is used to estimate a probability index number for the strategy based on the type of control(s) being described. The applicant's deterministic design criteria commitments were also considered. The staff considers these commitments sufficient to lower the PFOD and the probability of the accident sequence, often by at least an order of magnitude based on the type of control(s) proposed. If the probability index of the sequence can achieve a value of -5 or less, the staff considered the strategy

staff considers this to be an acceptable strategy for meeting the 10 CFR 70.61 performance requirements.

Corrosion (Confinement)

The corrosion event group is defined as catastrophic failure of a primary confinement boundary (i.e., a laboratory or an AP glovebox containing corrosive chemicals, AP fluid transport systems, a pneumatic transfer line, or ducting of the C4 confinement system) postulated to result from corrosion. Loss-of-confinement events caused by corrosion within process cells are included in the leaks of AP process vessels or pipes within process cells event group. Loss-of-confinement events caused by corrosion of pipes containing process fluids within C3 areas not enclosed within a glovebox are discussed in the process fluid leak in C3 area outside of glove box event group. The bounding event for the corrosion event group in the confinement events accident category was determined by the applicant to be corrosion of the pneumatic pipe automatic transfer system from corrosive chemicals resulting in a breach of confinement and dispersal of radioactive materials. The material at risk was the maximum inventory in the pneumatic pipe automatic transfer system. The applicant determined this to be an above the 10 CFR 70.61(c) threshold consequence event for the facility worker and the environment and a below the 10 CFR 70.61(c) threshold consequence event for the individuals outside the controlled area boundary. The staff independently evaluated this accident sequence and agrees to its categorization. The applicant has opted to protect the facility worker and the environment through a strategy of prevention and mitigation. The PSSCs identified for facility worker protection and protection of the environment are the material maintenance and surveillance programs, which will detect and limit the damage resulting from corrosion. No PSSCs are identified by the applicant as being necessary to protect the site worker and IOC. However, the C4 and C3 confinement systems, as well as the C2 confinement system passive boundary, provide defense-in-depth protection for site workers and IOC. Based on the nature of this event (mitigated by slow development or prevented by administrative controls), the staff considers this to be an acceptable strategy for meeting the 10 CFR 70.61 performance requirements.

Small Breaches in Glovebox Confinement Boundary (Confinement)

The bounding event for small breaches in a glovebox confinement boundary or backflow in the confinement events accident category was determined to be backflow through the interfacing gas line (e.g. nitrogen, helium) to the interfacing system followed by the opening of this interfacing system during a maintenance operation. The material at risk was the maximum inventory of radioactive material in a glovebox. Loss of gasflow through a supply line was listed as a possible cause. The applicant determined this to be an above the 10 CFR 70.61(c) threshold consequence event for the facility worker and the environment and a below the 10 CFR 70.61(c) threshold consequence event for the site worker and IOC. The staff independently evaluated this accident sequence and agrees to its categorization. The applicant has opted to protect the facility worker through a strategy of mitigation. The PSSC identified for protection of the facility worker and the environment is the C4 confinement system which maintains a negative glovebox pressure differential between the glovebox and the interfacing systems and will also maintain a minimum inward flow through small glovebox breaches. No

---

to be acceptable. This will be demonstrated by the applicant in a supplemental likelihood assessment to be conducted in conjunction with the ISA when specific IROFS are developed.

PSSCs are identified by the applicant as being necessary to adequately protect the site worker and IOC. In addition, the C3 confinement system provides defense-in-depth protection for the site worker and IOC. Based on the nature of this event (mitigated by a high-availability AEC), the staff considers this to be an acceptable strategy for meeting the 10 CFR 70.61 performance requirements.

<u>Leaks of AP Process Vessels or Pipes within Process Cells (Confinement)</u>

The bounding event for leaks in the AP process vessels or pipes within process cells in the confinement events accident category is a break or leakage of a tank/vessels inside the process cell containing the liquid waste reception unit. The material at risk was the maximum inventory of radioactive materials in the affected equipment in the AP process cell. Corrosion and mechanical failure were listed as potential causes. The applicant determined this to be an above the 10 CFR 70.61(c) threshold consequence event for the facility worker, environment, site worker, and IOC. The applicant has opted to protect the facility worker through a strategy of prevention. The PSSCs identified for facility worker protection are the process cells, which contain fluid leaks within the cells, and the process cell entry controls, which prevent the entry of personnel into process cells during normal operation. The environment, site worker, and IOC are protected through a strategy of mitigation. The PSSC identified for protection of the environment, site worker, and IOC is the process cell exhaust system. The function of the process cell exhaust system is to ensure that a negative pressure exists between the process cell areas and the C2 area and to ensure that the C2 exhaust is effectively filtered. The C2 confinement system passive boundary provides defense-in-depth protection for the environment, site worker, and IOC. Based on the applicant's deterministic criteria and the NUREG-1718 Table A-5 descriptions (AECs), the staff considers this to be an acceptable strategy for meeting the 10 CFR 70.61 performance requirements.

<u>Backflow from a Process Vessel through Utility Lines (Confinement)</u>

The bounding event for backflow from a process vessel through the utility lines confinement events accident category is backflow of radioactive material from a waste tank containing americium. This backflow was postulated to flow through an interfacing supply line that is subsequently breached or opened during a maintenance operation. The material at risk was the maximum radioactive material in the waste tank. Loss of gas flow through the supply lines and failure of pipes and/or valves are identified as potential causes. The applicant determined this to be an above the 10 CFR 70.61(c) threshold consequence event for the facility worker, environment, site worker, and IOC. The staff independently evaluated this accident sequence and agrees to its categorization. The applicant has opted to protect the facility worker, environment, site worker, and IOC through a strategy of prevention. The PSSC identified for protection of the facility worker and the environment, site worker, and IOC are backflow prevention features which will prevent process fluids from backflowing into interfacing systems. However, the C2 confinement system passive boundary provides defense-in-depth protection for the environment, site worker, and IOC. Based on the applicant's deterministic criteria and the NUREG-1718 Table A-5 descriptions (PEC), the staff considers this to be an acceptable strategy for meeting the 10 CFR 70.61 performance requirements.

Rod-Handling Operations (Confinement)

The bounding event for rod-handling operations in the confinement events accident category was determined to be the fracture of one or more fuel rods while utilizing fuel-rod-handling equipment resulting in a breach of confinement and dispersal of radiological materials. The material at risk was the maximum inventory of radioactive material in a tray of fuel rods. Human error or equipment failure were listed as potential causes. The applicant determined this to be an above the 10 CFR 70.61(c) threshold consequence event for the facility worker and a below the 10 CFR 70.61(c) threshold consequence event for the environment, site worker, and IOC. The staff independently evaluated this accident sequence and agrees to its categorization. The applicant has opted to protect the facility worker through a strategy of prevention and mitigation. The PSSCs identified for protection of the facility worker are facility worker actions which ensure that facility workers take proper actions to limit radiological exposure, materials-handling controls which ensure proper handling of primary confinements outside of gloveboxes, and materials-handling equipment to limit damage to fuel rods/assemblies during handling operations. No PSSCs were identified by the applicant as being necessary to adequately protect the environment, site worker, and IOC. The combination of rod cladding (primary confinement), materials-handling controls, and facility worker actions are intended to make the likelihood of above the 10 CFR 70.61(c) threshold consequences from the rod-handling accident sequence highly unlikely. However, because a release could occur without warning, the applicant provided dose calculations which were reviewed on site and found to be acceptable (NRC, 2002). The C2 confinement system passive boundary provides defense-in-depth protection for the environment, site worker, and IOC. Based on the nature of this event (mitigated by limited initial release and immediate worker responses), the staff considers this to be an acceptable strategy for meeting the 10 CFR 70.61 performance requirements.

Breaches in Containers outside of Gloveboxes (Confinement)

The bounding event for breaches in containers outside of gloveboxes caused by handling operations in the confinement events accident category was the failure of a 3013 canister, transfer container containing plutonium-bearing waste, or other primary confinement types within the C2 or C3 areas outside of a glovebox. The material at risk was the maximum inventory of radioactive material in the container. The applicant determined this to be an above the 10 CFR 70.61(c) threshold consequence event for the facility worker, environment, site worker, and IOC. The applicant has opted to protect the facility worker, environment, site worker, and IOC using a strategy of prevention and mitigation. PSSCs identified to protect the facility worker are materials-handling controls, a 3013 canister to withstand the effects of design basis drops without breaching, transfer container which will also withstand the effects of design basis drops without breaching, and facility worker controls for bag-out operations in C3 areas. The staff has requested the applicant to perform a dose calculation to determine the unmitigated dose to the worker from the drop of a container other than a 3013 canister. This calculation was provided and reviewed on site and found to be acceptable. PSSCs identified as required for protection of the environment, site worker, and IOC are materials-handling controls to ensure proper handling of primary confinement types outside of gloveboxes, as well as a transfer container, a 3013 canister, and the C3 confinement system, which will provide filtration to mitigate dispersions from the C3 areas. The C2 confinement system passive boundary and the preventive features utilized to reduce the risk to the facility worker and the environment provide defense-in-depth protection for the IOC and site worker. Based on the applicant's deterministic criteria and the

NUREG-1718 Table A-5 descriptions, (PEC, AEC, and administrative controls), the staff considers this to be an acceptable strategy for meeting the 10 CFR 70.61 performance requirements.

Over/Under Pressurization of Glovebox (Confinement)

The bounding event for over/under pressurization of a glovebox (i.e., C4 dynamic confinement) in the confinement events accident category was determined by the applicant to be a rapid overpressurization of the calcining furnace glovebox. The material at risk was the maximum inventory of radioactive material in the glovebox. Potential causes of this event were identified as the rupture of a high-flow or a high-pressure supply line or a clogged outlet high-efficiency particulate air (HEPA) filter. The applicant determined this to be an above the 10 CFR 70.61(c) threshold consequence event for the facility worker and the environment, site worker, and IOC. For a rapid over- or underpressurization event, a prevention strategy is used. The PSSC identified for protection of the facility worker is glovebox pressure controls, which will maintain glovebox pressure within design limits. For a slow pressurization event, a mitigation strategy is used. The PSSCs identified for facility worker protection are facility worker actions and the process safety control subsystem to warn operators of glovebox pressure discrepancies before exceeding differential pressure limits. The PSSCs identified for protection of the environment, site worker, and IOC are the C3 and C4 confinement systems. The C2 confinement system will provide defense-in-depth protection for the site worker and IOC. Based on the nature of this event (mitigated by a warning system and immediate worker responses or prevented by an AEC), the staff considers this to be an acceptable strategy for meeting the 10 CFR 70.61 performance requirements.

Excess Temperature before Decay Heat from Radioactive Materials (Confinement)

Thermal calculations have been performed to evaluate the effects of temperature on confinement structural materials. Thermal sources considered in the calculation include radioactive decay of nuclear materials, spontaneous heating of uranium dioxide caused by oxidation (burnback), operation of electrical/mechanical equipment, and process equipment (calcining furnace). However, only the 3013 storage area was found to require long-term cooling to mitigate the effects of decay heat. The material at risk was the maximum inventory of radioactive material in the powder storage area in the plutonium dioxide 3013 storage unit. The applicant determined this to be an above the 10 CFR 70.61(c) threshold consequence event for the facility worker, environment, site worker, and IOC. The applicant has opted to protect these receptors through a strategy of prevention. The PSSC identified for protection of the facility worker, environment, site worker, and IOC is the high depressurization exhaust system (part of the C3 confinement system) which will ensure that temperatures in the 3013 canister storage structure are maintained within design limits. Based on the applicant's deterministic criteria and the NUREG-1718 Table A-5 descriptions (high-availability AEC), the staff considers this to be an acceptable strategy for meeting the 10 CFR 70.61 performance requirements.

Glovebox Dynamic Exhaust Failure (Confinement)

The bounding event for the glovebox dynamic exhaust failure event group in the confinement event accident sequence was a loss of negative pressure or a flow perturbation involving the C4 dynamic confinement system resulting in a ventilation airflow reversal into a C3 area. The material at risk was the maximum inventory of airborne radioactive material in all connected gloveboxes. Potential causes of this event are loss of normal control system, loss of all power, or mechanical failure of the ventilation system. The applicant determined this to be an above the 10 CFR 70.61(c) threshold consequence event for the facility worker, environment, site worker, and IOC. The applicant has opted to protect all receptors through a strategy of prevention. The staff independently evaluated this accident sequence and agrees to its categorization. The PSSC identified for protection of the facility worker, environment, site worker, and IOC is the C4 confinement system. The C3 and C2 confinement system passive boundaries provide defense-in-depth protection of the environment and individuals outside the controlled area boundary. Based on the applicant's deterministic criteria and the NUREG-1718 Table A-5 descriptions (high-availability AEC), the staff considers this to be an acceptable strategy for meeting the 10 CFR 70.61 performance requirements.

Process Fluid Line Leak in a C3 Area outside of a Glovebox (Confinement)

This event was postulated to result from a leak from a line carrying a process fluid in a C3 area outside of a glovebox or process cell caused by corrosive chemicals or mechanical failure of AP piping. The material at risk was the maximum inventory of radiological material in a single AP vessel. The applicant determined this to be an above the 10 CFR 70.61(c) threshold consequence event for facility workers and the environment and a below the 10 CFR 70.61(c) threshold consequence event for the site workers and IOC. The staff independently evaluated this accident sequence and agrees to its categorization. The applicant has opted to use a strategy of prevention to protect the facility worker and the environment. The PSSC identified for this strategy is a double-walled pipe. No PSSCs were identified by the applicant as being necessary to adequately protect individuals outside the controlled area boundary. However, the C3 confinement system provides defense-in-depth protection for individuals outside the controlled area boundary. Based on the nature of this event (prevented by a PEC), the staff considers this to be an acceptable strategy for meeting the 10 CFR 70.61 performance requirements.

Sintering Furnace Confinement Boundary Failure

This event was postulated to result from a breach in the sintering furnace confinement boundary. The furnace is postulated to fail either through a slow leak through the seals or a rapid overpressurization event. These events could be caused by failure of the control system for the hydrogen/argon supply line, a failure in the sintering furnace exhaust system, or a sintering furnace seal failure. The material at risk was the maximum inventory of radiological material in both sintering furnaces. The applicant determined this to be an above the 10 CFR 70.61(c) threshold consequence event for the facility worker, environment, site workers, and IOC. For the rapid overpressurization event, the applicant has chosen a strategy of prevention to reduce the risk to the facility worker and the environment. The PSSCs for this event are the sintering furnace pressure controls and the sintering furnace which functions as a confinement boundary. For the seal failure event, the applicant has chosen a strategy of mitigation. The PSSC for this event is the sintering furnace which is designed to limit any postulated leakage. The PSSC identified by the applicant to adequately protect the site worker and IOC is the C3 confinement system. The C3 confinement system's passive boundary also provides defense-in-depth protection for the environment. The C2 confinement system's passive boundary provides defense-in-depth protection to the site worker and IOC. Based on the nature of this event (mitigated by a combination of PECs and AECs), the staff considers this to be an acceptable strategy for meeting the 10 CFR 70.61 performance requirements.

5.1.6.3.2 Fire Events

The potential consequences of fire events at the facility as listed by DCS include the following:

- destruction of confinement barriers
- destruction of civil structures
- destruction of equipment contributing to dynamic confinement
- failure or damage to utility equipment
- loss of criticality controls
- loss of other PSSCs

All of the above can lead to the release of nuclear and chemical material to the environment.

Potential causes for fire events within the facility identified by DCS include the following:

- short circuits or equivalent event involving electrical equipment
- ignition or combustion of fixed or transient combustibles
- equipment that operates at high temperatures
- ignition of a solvent or other flammable/reactive chemical

Thirty-five separate events with potentially significant consequences were analyzed by DCS to determine the bounding consequences from a potential fire event. These events were assigned to 13 groups as follows with a unique prevention or mitigation strategy:

The AP process cells
AP/MP C3 glovebox area
C1 and/or C2 areas
  — 3013 canister
  — 3013 transport cask
  — Fuel rod
  — MOX fuel transport cask
  — Waste container
  — Transfer container
  — Final C4 HEPA filter
Outside MOX fuel fabrication building
Facility-wide systems
Facility
Electrolyzer

These 13 groups are discussed below:

AP Process Cell (Fire)

The bounding event for the AP process cells events group in the fire events accident category was determined to be a fire in the cell containing the liquid waste reception unit tanks. The material at risk was taken to be the maximum inventory of radioactive material in the cell containing the tanks. A fire was postulated to occur in the process cell and consequences were evaluated. The applicant determined this to be an above the 10 CFR 70.61(c) threshold consequence event for the facility worker, environment, site worker, and IOC. The applicant has opted to protect these receptors through a strategy of prevention. The PSSC for protection of the facility worker was the process cell fire prevention features, the purpose of which is to ensure that fires in the process cells are highly unlikely. The process cell fire prevention features consist of the following:

- elimination of ignition sources, including electrical equipment and static electricity
- fire barriers to protect process cell areas
- elimination of all combustible materials from process cells containing aqueous solutions
- elimination of combustibles outside of process equipment in cells containing solvents
- maintenance of temperatures at levels to prevent creation of flammable vapors.

The process cell fire prevention features are also identified as the PSSC for protection of the environment, site worker, and IOC. In addition, the process cell ventilation system's passive boundary and the C2 confinement system's passive boundary provide defense-in-depth protection to mitigate the potential consequences to the environment, site worker, and IOC. Based on the applicant's deterministic criteria and the NUREG-1718 Table A-5 descriptions (enhanced administrative control with active and passive features), the staff considers this to be an acceptable strategy for meeting the 10 CFR 70.61 performance requirements.

AP/MP C3 Glovebox Area (Fire)

The bounding event for a fire in the AP/mixed oxide process (MP) C3 glovebox area in the fire events accident category was a fire within the plutonium dioxide buffer storage area. The material at risk was the maximum inventory of radioactive material within the fire area. The specific cause of a fire in this area was not addressed, but the bounding event in this event group was identified as a fire originating in a glovebox. The applicant determined this to be an above the 10 CFR 70.61(c) threshold consequence event for the facility worker, environment, site worker, and IOC. The applicant has opted to protect potentially affected workers, IOC, and the environment through a strategy of mitigation. The PSSCs identified for protection of the facility worker were facility worker actions and facility worker controls. The PSSCs identified for protection of the environment, site worker, and IOC were the C3 confinement system, the active portion of the C4 confinement system, and fire barriers. In addition, combustible loading controls is also identified as a PSSC for protection of the the environment, site worker, and IOC for fires involving storage gloveboxes. The primary protection of the worker will be early detection of the fire and the ability to evacuate the area before a release. Although not credited by the accident analysis, there will also be a fire suppression system in areas with dispersable radioactive material. This suppression system will be classified as a PSSC. Consideration of the warning time available from a fire before a breech in containment allows facility worker action to be an acceptable strategy for meeting the 10 CFR 70.61 performance requirements with regard to the facility worker. Based on the applicant's deterministic criteria and the NUREG-1718 Table A-5 descriptions (AEC, enhanced administrative control), the staff considers this to be an acceptable strategy for meeting the 10 CFR 70.61 performance requirements.

3013 Canister (Fire)

The bounding event for a fire affecting a 3013 canister event group in the C2 area is a fire in the 3013 storage area. The material at risk for this fire is the maximum inventory of radioactive material in the fire area. The cause of the fire was ignition of transient combustibles. The applicant determined this to be an above the 10 CFR 70.61(c) threshold consequence event for the facility worker, environment, site worker, and IOC. The applicant has opted to protect the facility worker, environment, site worker, and IOC through a strategy of prevention. The PSSC identified for protection of the facility worker, environment, site worker, and IOC is combustible loading controls, which are intended to limit the quantity of combustibles in a fire area containing 3013 canisters to ensure that the canisters are not adversely impacted by a fire. Based on the applicant's deterministic criteria and the NUREG-1718 Table A-5 descriptions (enhanced administrative control), the staff considers this to be an acceptable strategy for meeting the 10 CFR 70.61 performance requirements.

3013 Transport Cask (Fire)

The bounding event for a fire affecting a 3013 transport cask in the C1 or C2 area was determined to be a fire in the truck bay involving transport packages resulting in an energetic breach of the containers and the dispersal of radioactive materials. The type of fire postulated would be a fuel fire involving a truck. The material at risk was determined to be the maximum inventory in the transport packages. The applicant determined this to be an above the 10 CFR 70.61(c) threshold consequence event for the facility worker, environment, site worker, and IOC. The applicant has opted to protect the facility worker, environment, site worker, and IOC through a strategy of prevention. The PSSCs identified for protection of the facility worker, environment, site worker, and IOC are the 3013 transport cask, which will withstand the design basis fire without breaching, and combustible loading controls, which will limit the quantity of combustibles in a fire area containing 3013 transport casks to ensure that the cask design basis fire is not exceeded. In addition to the identified PSSCs, there will also be a fire suppression system which is considered an additional protective feature. Based on the applicant's deterministic criteria and the NUREG-1718 Table A-5 descriptions (PEC and enhanced administrative control), the staff considers this to be an acceptable strategy for meeting the 10 CFR 70.61 performance requirements.

Fuel Rods (Fire)

The bounding event for a fire affecting fuel rods in the fire event accident category is a fire in the fuel assembly storage area. The material at risk for this fire is the maximum inventory of radioactive materials in the assembly storage area. Combustible loading in this area is low, but the fire is still assumed to involve all of the radioactive materials in the storage area. The source of the fire is considered to be electrical equipment and transient combustibles. The applicant determined this to be an above the 10 CFR 70.61(c) threshold consequence event for the facility worker, the environment, site worker, and IOC and has opted to protect the potential receptors through a strategy of prevention. The PSSC identified for protection of these receptors is combustible loading controls which will limit the quantity of combustibles in a fire area containing fuel rods to ensure that the fuel rods are not adversely impacted by a fire. Based on the applicant's deterministic criteria and the NUREG-1718 Table A-5 descriptions (enhanced administrative control), the staff considers this to be an acceptable strategy for meeting the 10 CFR 70.61 performance requirements.

MOX Fuel Transport Cask (Fire)

The bounding event for a fire affecting the MOX fuel transport cask in the fire event accident scenario was determined to be a fire in the fuel assembly truck bay. The source of the fire was considered to be electrical equipment and transient combustibles. The material at risk was the radioactive material in the transport casks. The applicant determined this to be an above the 10 CFR 70.61(c) threshold consequence event for the facility worker, environment, site worker, and IOC. The staff independently evaluated this accident sequence and agrees to its categorization. The applicant has opted to protect potential receptors through a strategy of prevention. The PSSCs identified for protection of the facility worker, environment, site worker, and IOC are the MOX fuel transport cask, which will withstand the design basis fire without breaching, and combustible loading controls, which are intended to limit the quantity of combustibles in a fire area containing MOX fuel transport casks to ensure that the cask design basis fire is not exceeded. In addition to the identified PSSCs, there will also be a fire suppression system which is considered an additional protective feature. Based on the applicant's deterministic criteria and the NUREG-1718 Table A-5 descriptions (PEC and enhanced administrative control), the staff considers this to be an acceptable strategy for meeting the 10 CFR 70.61 performance requirements.

Waste Container (Fire)

The bounding event for a fire affecting a waste container in the C1, C2, or C3 area event group in the fire event accident category was determined to be a fire located in the assembly packaging area. The material at risk was the maximum inventory of radioactive material in the waste container. The source of the fire was considered to be electrical equipment and transient combustibles. The applicant determined this to be an above the 10 CFR 70.61(c) threshold consequence event for the facility worker, but a below the 10 CFR 70.61(c) threshold consequence event for the environment, site worker, and IOC. The staff independently evaluated this accident sequence and agrees to its categorization. The applicant has opted to limit the dose to the facility worker using a strategy of mitigation. The PSSC identified for protection of the facility worker is facility worker action to ensure that facility workers take proper actions to limit dose. No PSSCs were identified by the applicant as being necessary to adequately protect the environment, site worker, and IOC. Based on the nature of this event (mitigated by rapid detection of fire and immediate worker responses), the staff considers this to be an acceptable strategy for meeting the 10 CFR 70.61 performance requirements.

Transfer Container (Fire)

The bounding event for a fire affecting a transfer cask within the C1, C2, or C3 areas event group in the fire event accident category was determined to be a fire in either the air locks, corridors, stairways, safe areas, or liquid waste reception areas. The material at risk was the maximum inventory in a transfer container. The source of the fire was identified as electrical equipment, transient combustibles, or a HEPA filter. The applicant determined this to be an above the 10 CFR 70.61(c) threshold consequence event for the facility worker, environment, and site worker, but a below the 10 CFR 70.61(c) threshold consequence event for the IOC.

The staff independently evaluated this accident sequence and agrees to its categorization. The applicant has opted to limit dose to the facility worker, environment, and site worker using a strategy of prevention. The PSSC identified for protection of the facility worker, environment, and the IOC is combustible loading controls, which limit the quantity of combustibles in a fire area containing transfer containers to ensure that the containers are not adversely impacted by a fire. No PSSCs were identified by the applicant as being necessary to adequately protect the site worker. Based on the nature of this event (prevented by an enhanced administrative control), the staff considers this to be an acceptable strategy for meeting the 10 CFR 70.61 performance requirements.

<u>C4 HEPA Filter (Fire)</u>

The bounding event for a fire affecting the final C4 HEPA filter in the fire event accident category is a fire which breaches the HEPA filter housing and allows material from the HEPA filters to pass directly to the stack. The material at risk for this event is based on a conservative estimate of material present on the C4 HEPA filters. The applicant determined this to be an above the 10 CFR 70.61(c) threshold consequence event for the facility worker, environment, site worker, and IOC. The applicant has opted to limit dose to the facility worker, environment, site worker, and IOC using a strategy of prevention. The PSSC identified for protection of the facility workers, environment, site worker, and IOC are combustible loading controls. Based on the applicant's deterministic criteria and the NUREG-1718 Table A-5 descriptions (enhanced administrative control), the staff considers this to be an acceptable strategy for meeting the 10 CFR 70.61 performance requirements.

<u>Fire outside of MFFF Building (Fire)</u>

The bounding event for a fire originating outside of the MFFF building event group in the fire event accident category was determined to be a fire involving diesel fuel storage, gasoline storage, or the reagents processing building, such that the MFFF building structure is damaged and radioactive material is released. The material at risk was the maximum inventory of radioactive material in the MFFF that is susceptible to the effects of external fires. The applicant determined this to be an above the 10 CFR 70.61(c) threshold consequence event for the facility worker, environment, site worker, and IOC and has opted to meet the performance requirements using a strategy of prevention. The PSSCs identified to protect all receptors are the MFFF building structure, which is designed to maintain structural integrity and prevent damage to internal PSSCs from external fires, the emergency diesel generator building structure, which is designed to maintain structural integrity and prevent damage to internal PSSCs from fires external to the structure, the emergency control room air-conditioning system, which will ensure habitable conditions for operators, and the waste transfer line, which will prevent damage to the line from external fires. Based on the applicant's deterministic criteria and the NUREG-1718 Table A-5 descriptions (PECs and an AEC), the staff considers this to be an acceptable strategy for meeting the 10 CFR 70.61 performance requirements.

<u>Fire Affecting Facility-Wide Systems (Fire)</u>

The bounding event for a fire affecting facility-wide systems (fires involving systems that cross fire areas) in the fire event accident category was determined to be a fire involving the pneumatic pipe automatic transfer system which results in a breach of confinement and the dispersal of radioactive material. The material at risk was the maximum inventory of radioactive material in the pneumatic pipe automatic transfer system. The fire was postulated to be caused by electrical equipment and transient combustibles. The applicant determined this to be an above the 10 CFR 70.61(c) threshold consequence event for the facility worker and the environment, but a below the 10 CFR 70.61(c) threshold consequence event for the site worker and IOC. The staff independently evaluated this accident sequence and agrees to its categorization. The applicant has opted to limit the dose to the facility worker using a strategy of mitigation. The PSSCs identified for protection of facility workers are facility worker actions and combustible loading controls. The primary protection of the worker will be early detection of the fire and evacuation of the area before a release. The PSSC identified for protection of the environment is combustible loading controls. No PSSCs were identified by the applicant as being necessary to adequately protect the site worker and IOC. Based on the nature of this event (mitigated by enhanced administrative controls and immediate worker responses), the staff considers this to be an acceptable strategy for meeting the 10 CFR 70.61 performance requirements.

Facility (Fire)

The bounding event for a facility fire which involves more than one fire area in the fire event accident scenario was determined to be a fire in all process units and support units with radioactive materials present. The source term is the maximum inventory in the facility susceptible to a facility-wide fire. The applicant determined this to be an above the 10 CFR 70.61(c) threshold consequence event for the facility worker, environment, and IOC and has opted to meet the performance requirements using a strategy of mitigation and prevention. The PSSCs identified for protection of the facility worker are worker actions and fire barriers that will contain the fires within the fire area. The PSSC identified for protection of the the environment, site worker, and IOC is fire barriers. In addition to the identified PSSCs, there will also be a fire suppression system designated as a PSSC where dispersable radioactive material is present. Based on the applicant's deterministic criteria and the NUREG-1718 Table A-5 descriptions (PEC), the staff considers this to be an acceptable strategy for meeting the 10 CFR 70.61 performance requirements.

AP Electrolyzer (Fire)

The bounding event for a titanium fire in the AP electrolyzer was determined to be the energetic breach of the AP electrolyzer and the dispersal of radioactive materials. The material at risk was the maximum inventory in the AP dissolution units. The applicant determined this to be an above the 10 CFR 70.61(c) threshold consequence event for the facility worker, environment, and IOC and has opted to meet the performance requirements using a strategy of prevention. The PSSCs identified for the protection of the facility worker, environment, site worker, and IOC are maintenance activity controls, the process safety control subsystem, and specified electrolyzer components, such as the electrolyzer structure, sintered silicon nitride barrier, guide sleeves, and polytetrafluorethylene insulator (PTFE). The safety function of the maintenance activity controls is to isolate power from the electrolyzer when the electrolyzer is drained. The safety function of the process control subsystem is to monitor the electrolyzer for faults that could result in arcing or other imparting of electrical energy with the risk of titanium fire. The safety function of the specified electrolyzer components is to provide physical separation and electrical insulation between the electrolyzer components and structural integrity and stability to the electrolyzer system. The C3 confinement system, the C4 confinement system, and the fire suppression and detection system provide defense in depth protection to mitigate potential consequences to the environment, site worker, and IOC. Based on the applicant's deterministic criteria and the NUREG-1718 Table A-5 descriptions (AECs), the staff considers this to be an acceptable strategy for meeting the 10 CFR 70.61 performance requirements.

5.1.6.3.3 Load-Handling Events

Load-handling events may occur during the operation of load-handling or lifting equipment during normal operations or maintenance activities. Load-handling events may occur because of the failure of handling equipment to lift or support the load, failure to follow designated load paths, or toppling of loads. Consequences of load-handling events include possible damage to handled loads, resulting in dispersal of radioactive and/or chemical materials, possible damage to nearby equipment or structures, resulting in a loss of confinement and/or a loss of subcritical conditions, and possible damage to process equipment or structures relied on for safety.

Twenty-eight separate events with potentially significant consequences were analyzed by DCS to determine the bounding consequences from a potential load-handling event. These events were assigned to 12 groups as follows with a unique prevention or mitigation strategy:

The AP process cells
AP/MP C3 glovebox area
C1 and/or C2 areas
— 3013 Canister in the C2 confinement area
— 3013 Transport Cask
— Fuel rod in the C2 confinement area
— MOX fuel transport cask
— Transfer container
— Waste container
— Final C4 HEPA filter
C4 Confinement
Outside MFFF building
Facility-wide systems

These 12 groups are discussed below:

AP Process Cells (Load Handling)

The bounding event for load-handling events in the AP process cells event group in the load-handling event accident category was determined to be an event in the cell containing the liquid waste reception unit. The material at risk was the maximum inventory of radioactive material in the AP process cell containing the liquid waste reception unit. The load-handling event is postulated to result in a breach of the americium reception tank and subsequent release of americium in solution because of vessels in the process cell being impacted by a lifting device or a lifted load. The applicant determined this to be an above the 10 CFR 70.61(c) threshold consequence event for the facility worker, environment, site workers, and IOC. The applicant has opted to meet the performance requirements using a strategy of mitigation. The PSSCs for the protection of facility workers are the process cells which contain fluid leaks using drip trays and the process cell entry controls which will prevent entry during normal conditions and assure that worker dose limits are not exceeded during maintenance operations. The PSSC for protection of the environment, site worker, and IOC is the process cell exhaust system. The safety function of the process cell exhaust system is to ensure that a negative pressure exists between the process cell areas and the C2 areas, as well as to ensure that the process cell exhaust system is effectively filtered. The C2 confinement system's passive boundary provides defense-in-depth protection for the environment, site worker, and IOC. Based on the applicant's deterministic criteria and the NUREG-1718 Table A-5 descriptions (administrative control, AECs, and PECs), the staff considers this to be an acceptable strategy for meeting the 10 CFR 70.61 performance requirements.

AP/MP C3 Glovebox (Load Handling)

The bounding event for load-handling events in the AP/MP C3 glovebox area event group in the load-handling accident category was determined to be an event which occurs within the gloveboxes that contain jar storage and handling of the MOX powder workshop from a breach of the glovebox. The material at risk was the maximum inventory of radioactive material in the glovebox. The breach of the glovebox is from a lifting device or a lifted load. The applicant determined this to be an above the 10 CFR 70.61(c) threshold consequence event for the facility worker, environment, site worker, and IOC and has opted to meet the performance requirements using a strategy of prevention and mitigation. The PSSCs for protection of the facility worker and the environment are materials-handling controls, which are intended to prevent impacts to the glovebox during normal operations from loads outside or inside the glovebox that could exceed the glovebox design basis, materials-handling equipment, which is engineered to prevent impacts to the glovebox, the glovebox, which maintains confinement integrity for design basis impacts, and facility worker controls (facility worker during maintenance operations). An additional safety function of the materials-handling controls is to prevent potential overpressurization of the reusable plutonium dioxide cans caused by radiolysis or oxidation of plutonium (III) oxalate and its subsequent impact to the glovebox. The PSSC for protection of the site worker and IOC is the C3 confinement system. The C2 confinement system also provides defense-in-depth protection for the site worker and IOC. Based on the applicant's deterministic criteria and the NUREG-1718 Table A-5 descriptions (AEC, PEC, and administrative controls), the staff considers this to be an acceptable strategy for meeting the 10 CFR 70.61 performance requirements.

3013 Canister (Load Handling)

The bounding event for the 3013 canister event group (C2 area) in the load-handling accident category was the drop of one 3013 container onto another 3013 container, each containing unpolished plutonium dioxide in powder form. The material at risk was the amount of radioactive material in two 3013 canisters. The cause of the event would likely be human error or equipment failure during a hoisting operation. The applicant determined this to be an above the 10 CFR 70.61(c) threshold consequence event for the facility worker, environment, site worker, and IOC. The applicant has opted to meet the performance requirements using a strategy of prevention. The staff independently evaluated this accident sequence and agrees to its categorization. The PSSCs identified for protection of the facility worker, environment, site worker, and IOC are the 3013 canister and materials-handling controls. The C2 confinement system's passive boundary provides defense-in-depth for the environment, site worker, and IOC.

Based on the applicant's deterministic criteria and the NUREG-1718 Table A-5 descriptions (administrative control and PEC), the staff considers this to be an acceptable strategy for meeting the 10 CFR 70.61 performance requirements.

3013 Transport Cask (Load Handling)

The bounding event for the 3013 transport cask event group (C1 or C2 area) in the load-handling accident category was the drop of a 3013 transport cask containing unpolished plutonium dioxide in powdered form onto another 3013 transport cask. The material at risk was the maximum inventory of radioactive material in two 3013 transport canisters. The applicant determined this to be an above the 10 CFR 70.61(c) threshold consequence event for the facility worker, environment, site worker, and IOC. The applicant has opted to meet the performance requirements using a strategy of prevention. The staff independently evaluated this accident sequence and agrees to its categorization. The PSSCs identified for protection of the facility worker, environment, site worker, and IOC are the 3013 transport cask and materials-handling controls. The C2 confinement system's passive boundary provides defense-in-depth for the site worker, IOC, and the environment. Based on the applicant's deterministic criteria and the NUREG-1718 Table A-5 descriptions (PEC and administrative controls), the staff considers this to be an acceptable strategy for meeting the 10 CFR 70.61 performance requirements.

Fuel Rods in C2 Area (Load Handling)

The bounding event for the fuel rods in the C2 area event group in the load-handling accident category was the drop of a strongback containing three fuel assemblies containing MOX (6 percent). The material at risk was the maximum inventory of three fuel rod assemblies. The cause of this event would probably be human error or equipment failure. The applicant determined this to be an above the 10 CFR 70.61(c) threshold consequence event for the facility worker, but a below the 10 CFR 70.61(c) threshold consequence event for the environment, site worker, and IOC. The staff independently evaluated this accident sequence and agrees to its categorization. The applicant has opted to use a strategy of mitigation to protect the facility worker. The PSSC identified for protection of the facility worker is facility worker actions. No PSSCs were identified by the applicant as being necessary to adequately protect the environment, site worker, and IOC. However, the C2 confinement system passive boundary provides defense-in-depth for the environment, site worker, and IOC. Because a release could occur without warning, the applicant provided dose calculations which were reviewed on site and found to be acceptable (NRC, 2002). Based on the nature of this event (limited initial release and immediate worker response), the staff considers this to be an acceptable strategy for meeting the 10 CFR 70.61 performance requirements.

MOX Fuel Transport Cask (Load Handling)

The bounding event for the MOX fuel transport cask event group (C1 or C2 areas) in the load-handling accident category was determined to be the drop of one MOX fuel transport cask containing up to three MOX fuel assemblies. The cause of this event would probably be human error or equipment failure. The material at risk was determined to be the maximum inventory of one fuel assembly transport package. The applicant determined this to be an above the 10 CFR 70.61(c) threshold consequence event for the facility worker and the environment, but a below the 10 CFR 70.61(c) threshold consequence event for the site worker and IOC. The staff independently evaluated this accident sequence and agrees to its categorization. The applicant has decided to meet the performance requirements of 10 CFR 70.61 for the facility worker and the environment using a strategy of prevention. The PSSCs identified for protection of the facility worker and the environment are the MOX fuel transport cask and materials-handling controls. No PSSCs were identified by the applicant as being necessary to adequately protect the site worker and IOC. However, the MOX fuel transport cask provides defense-in-depth protection for the site worker and IOC. Based on the nature of this event (prevented by a PEC and administrative controls), the staff considers this to be an acceptable strategy for meeting the 10 CFR 70.61 performance requirements.

Waste Container (Load Handling)

The bounding event for the waste container event group (C1, C2, or C3 area) in the load-handling accident category is a damaged waste drum in the assembly packaging (truck bay) area caused by human error or equipment failure. The material at risk was determined to be the maximum inventory of radiological material in a waste container. The applicant determined this to be an above the 10 CFR 70.61(c) threshold consequence event for the facility worker, but a below the 10 CFR 70.61(c) threshold consequence event for the environment, site worker, and IOC. The staff independently evaluated this accident sequence and agrees to its categorization. The applicant has decided to meet the performance requirements of 10 CFR 70.61 for the facility worker using a strategy of mitigation. The PSSCs identified for protection of the facility worker are worker actions. No PSSCs were identified by the applicant as being necessary to adequately protect the environment, site worker, and IOC. For drops in the C2 area, the C2 confinement passive boundary provides defense-in-depth protection for the environment, site worker, and IOC. However, because a release could occur without warning, the applicant provided dose calculations which were reviewed on site and found to be acceptable (NRC, 2002). Based on the nature of this event (mitigated by limited initial release and immediate worker responses), the staff considers this to be an acceptable strategy for meeting the 10 CFR 70.61 performance requirements.

Transfer Container (Load Handling)

The bounding event for the transfer container event group (C2 area) in the load-handling accident category was the drop of a transfer container containing a HEPA filter with plutonium dioxide in powdered form. The material at risk was determined to be the maximum inventory in a HEPA filter. The applicant determined this to be an above the 10 CFR 70.61(c) threshold consequence event for the facility worker, environment, site worker, and IOC. The applicant has opted to meet the performance requirements using a strategy of prevention. The staff independently evaluated this accident sequence and agrees to its categorization. The PSSCs identified for protection of the facility worker, environment, site worker, and IOC are the transfer container and materials-handling controls. Based on the applicant's deterministic criteria and the NUREG-1718 Table A-5 descriptions (PEC and administrative control), the staff considers this to be an acceptable strategy for meeting the 10 CFR 70.61 performance requirements.

Final C4 HEPA Filter (Load Handling)

The bounding event in the final C4 HEPA filter event group in the load-handling accident category was determined to be the impacting of the final C4 filters by a load that breaches the

HEPA filter housing and allows material from the HEPA filters to pass directly to the stack. The cause of this event would probably be human error or equipment failure around the ventilation system. The material at risk was determined to be the radiological material contained in the HVAC system and filters. The applicant determined this to be an above the 10 CFR 70.61(c) threshold consequence event for the facility worker, environment, site worker, and IOC. The staff independently evaluated this accident sequence and agrees to its categorization. The applicant has decided to meet the performance requirements of 10 CFR 70.61 for the facility worker, environment, site worker, and IOC using a strategy of prevention. Materials-handling controls were identified as the PSSC for protection of these receptors. In addition, the applicant stated that in the current design and operations, there are no cranes or heavy equipment in the vicinity of the C4 final filters that could cause a load-handling event. Thus, there are no credible load-handling events during normal operations. During maintenance operations, maintenance will only be performed on out-of-service trains, which will prevent a release to the stack. The C2 confinement system's passive boundary provides defense-in-depth protection for the environment, site workers, and IOC for load-handling events that occur in the C2 areas where the final C4 filters are located. Based on the applicant's deterministic criteria and the NUREG-1718 Table A-5 descriptions (enhanced administrative control), the staff considers this to be an acceptable strategy for meeting the 10 CFR 70.61 performance requirements.

C4 Confinement (Load Handling)

The bounding event in the C4 confinement event group in the load-handling accident category was determined to be a spill of unpolished plutonium powder that occurs inside the glovebox, but does not result in a breach of the glovebox. The cause of this event would probably be human error or equipment failure during load-handling operations inside the glovebox. The material at risk would be the maximum inventory in the glovebox. The applicant determined this to be an above the 10 CFR 70.61(c) threshold consequence event for the facility worker, environment, site worker, and IOC. The applicant has decided to meet the performance requirements of 10 CFR 70.61 for the facility worker, environment, site worker, and IOC using a strategy of mitigation. The C4 confinement system is identified as the PSSC for protection of the facility worker, environment, site worker, and IOC. The safety functions of the C4 confinement system in this event are to ensure that the C4 exhaust is effectively filtered and to maintain a negative glovebox differential pressure between the glovebox and the interfacing systems. The C3 confinement system provides defense-in-depth protection for the environment, site worker, and IOC. Based on the applicant's deterministic criteria and the NUREG-1718 Table A-5 descriptions (high dependability AEC), the staff considers this to be an acceptable strategy for meeting the 10 CFR 70.61 performance requirements.

Outside MFFF Buildings (Load Handling)

The bounding event in the load-handling event category outside the MOX fuel fabrication building is an event involving the waste transfer line. The cause of this event would probably be human error or equipment failure. The material at risk was determined to be the maximum inventory in the waste tank. The applicant determined this to be an above the 10 CFR 70.61(c) threshold consequence event for the facility worker, environment, site worker, and IOC and has opted to meet the performance requirements using a strategy of prevention. The PSSC identified for protection of the facility worker, environment, site worker, and IOC is the waste transfer line which is double walled and buried. Based on the applicant's deterministic criteria and the NUREG-1718 Table A-5 descriptions (robust PECs), the staff considers this to be an acceptable strategy for meeting the 10 CFR 70.61 performance requirements.

Facility Wide (Load Handling)

The bounding event in the load-handling event category for the facility-wide event class is the breach of the facility structure from a heavy load resulting in a breach of primary confinement or in a breach of a container holding nuclear materials. The cause of this event would probably be human error or equipment failure. The material at risk was determined to be the maximum inventory in a container or primary confinement. The applicant determined this to be an above the 10 CFR 70.61(c) threshold consequence event for the facility worker, environment, site worker, and IOC and has opted to meet the performance requirements using a strategy of prevention. The PSSCs identified for protection of the facility worker, environment, site worker, and IOC are the MOX fuel fabrication building structure, which is designed to withstand the effects of load drops that could potentially impact radiological material, and materials-handling controls that would prevent load-handling events that could breach primary confinements. Based on the applicant's deterministic criteria and the NUREG-1718 Table A-5 descriptions (PEC and administrative control), the staff considers this to be an acceptable strategy for meeting the 10 CFR 70.61 performance requirements.

5.1.6.3.4 Explosion Events

Explosions are postulated to occur inside of the MOX fuel fabrication building from process operations, outside the MOX fuel fabrication building from nearby support facilities and the storage of chemicals on the facility site, and from laboratory operations. The following are considered to be the major consequences of explosions:

- release of radioactive materials or chemicals to the environment
- damage to a confinement boundary
- damage to equipment contributing to dynamic confinement
- loss of subcritical conditions
- damage to civil structures
- damage to other PSSCs

All of the above may result in the release of nuclear materials or chemicals to the environment.

Twenty-three separate events with potentially significant consequences were analyzed by DCS to determine the bounding consequences from a potential explosion event. These events were assigned to the following 15 groups with a unique prevention or mitigation strategy:

Hydrogen explosion
Steam over-pressure explosion
Radiolysis induced explosion
HAN explosion
Hydrogen peroxide explosion
Solvent explosion
Tributyl phosphate (TBP)-Nitrate (Red oils) explosion
AP vessel over-pressurization explosion

Pressure vessel over pressurization explosion
Hydrazoic acid explosion
Metal azide explosion
Pu (VI) Oxalate explosion
Electrolysis related explosion
Laboratory explosion
Outside explosion

These 15 groups are discussed below:

Hydrogen (Explosion)

The bounding event for the hydrogen explosion event group is the explosion of hydrogen and oxygen in a sintering furnace or sintering furnace room. The cause of this event would probably be excessive hydrogen in the furnace and air leakage into the furnace or hydrogen accumulation into the room. The applicant determined this to be an above the 10 CFR 70.61(c) threshold consequence event for the facility worker, environment, site worker, and IOC and has opted to meet the performance requirements using a strategy of prevention. The PSSC identified for prevention of this event is the process safety control subsystem with the safety function of preventing an explosive accumulation of hydrogen vapors. Specific control approaches may consist of hydrogen monitors outside the furnace, oxygen monitors inside the furnace, limiting the hydrogen content in the hydrogen-argon mixture, and crediting dilution flow from the high depressurization exhaust or very high depressurization exhaust systems to prevent an explosive mixture of hydrogen. The specific functions and systems selected will be identified as IROFS in the ISA. Based on the applicant's deterministic criteria and the NUREG-1718 Table A-5 descriptions (AEC), the staff considers this to be an acceptable strategy for meeting the 10 CFR 70.61 performance requirements.

Steam (Explosion)

A steam explosion is postulated to occur in the sintering furnace as a result of humidifier water in the inlet gas stream. The cause of the event is expected to be a failure of the water level controller in the humidifier. The material at risk was determined to be the maximum inventory of radiological materials in the sintering furnace. The applicant determined this to be an above the 10 CFR 70.61(c) threshold consequence event for the facility worker, environment, site worker, and IOC and has opted to meet the performance requirements using a strategy of prevention. The PSSC identified to protect the facility worker, environment, site worker, and IOC is the process safety control subsystem which will ensure the isolation of sintering furnace humidifier water flow on high-water level. Because of the design of the water cooling coils of the sintering furnace, water will not enter the sintering furnace through any other path. Based on the applicant's deterministic criteria and the NUREG-1718 Table A-5 descriptions (AEC), the staff considers this to be an acceptable strategy for meeting the 10 CFR 70.61 performance requirements.

Radiolysis (Explosion)

The bounding events in the radiolysis-induced explosion event group were explosions caused by radiolysis-induced hydrogen buildup in the vapor space of an AP vessel tank or piping, radiolysis-induced hydrogen buildup in the vapor space of a raffinates tank (in an AP process cell), and radiolysis-induced hydrogen accumulation in a waste container containing hydrocarbons. The applicant determined all three event sequences to be above the 10 CFR 70.61(c) threshold consequence events for the facility worker, environment, site worker, and IOC. The applicant has opted to meet the performance requirements using a strategy of prevention. The PSSCs identified for protection of the facility worker, environment, site worker, and IOC are the offgas treatment system, the instrument air system (scavenging system), and waste containers. The waste container contains a filter to allow the escape of radiolytic hydrogen before explosive mixtures occur as the PSSC to prevent an explosion inside the waste container. Based on the applicant's deterministic criteria and the NUREG-1718 Table A-5 descriptions (AECs and PEC), the staff considers this to be an acceptable strategy for meeting the 10 CFR 70.61 performance requirements.

HAN (Explosion)

The hydroxyl nitrate (HAN) explosions that could potentially occur within the facility may be characterized by one of the following two cases:

(1)   process vessels containing HAN and hydrazine nitrate without NO_x addition
(2)   process vessels containing HAN and hydrazine nitrate with NO_x addition

The applicant provided a hazard analysis and selected a PSSC for each case separately. The material at risk, however, was the same for both events and consisted of the maximum radiological inventory in AP vessels, tanks, and piping. The applicant has determined that the consequences from this event sequence for each of the cases are an above the 10 CFR 70.61(c) threshold consequence event for the facility worker, environment, site worker, and IOC and has opted to meet the performance requirements using a strategy of prevention. The applicant identified two PSSCs as required for the first case. These are the process safety control subsystem, which ensures that the temperature of the solution containing HAN is limited to temperatures that are within safety limits, and the chemical safety control, which ensures that the concentration of nitric acid, metal impurities, hydrazine nitrate, and HAN introduced in the process are within safety limits. In the second case, the chemical safety control is identified to limit the concentration of HAN, hydrazine nitrate, and hydrazoic acid and the offgas treatment system is identified to provide an exhaust path for offgases. Based on the applicant's deterministic criteria and the NUREG-1718 Table A-5 descriptions (enhanced administrative control and AEC), the staff considers this to be an acceptable strategy for meeting the 10 CFR 70.61 performance requirements.

Hydrogen Peroxide (Explosion)

The bounding event in the hydrogen peroxide explosion class was determined to be an event involving hydrogen peroxide in AP vessels, tanks, and piping which results in an energetic breach of the vessels, tanks, and piping and results in a loss of confinement and dispersal of nuclear materials. The applicant determined this to be an above the 10 CFR 70.61(c) threshold consequence event for the facility worker, environment, site worker, and IOC and has opted to meet the performance requirements using a strategy of prevention. The PSSC identified to protect the facility worker, environment, site worker, and IOC is the chemical safety control to ensure that explosive concentrations of hydrogen peroxide do not occur. This administrative control consists of a certified analysis by the manufacturer under an approved quality assurance plan, an analysis upon receipt at the facility, and an analysis after mixing and diluting. Based on the applicant's deterministic criteria and the NUREG-1718 Table A-5 descriptions (enhanced administrative control), the staff considers this to be an acceptable strategy for meeting the 10 CFR 70.61 performance requirements.

Solvent (Explosion)

The bounding event in the solvent explosion class was determined to be a process-related explosion involving solvents in AP vessels, tanks, and piping which results in an energetic breach of the vessels, tanks, and piping and a loss of confinement and dispersal of nuclear materials. The applicant determined this to be an above the 10 CFR 70.61(c) threshold consequence event for the facility worker, environment, site worker, and IOC. The applicant has opted to meet the performance requirements using a strategy of prevention. The staff independently evaluated this accident sequence and agrees to its categorization. The PSSCs identified to protect the facility worker, environment, site worker, and IOC are the process safety control subsystem, to ensure that the temperature of the solutions containing solvents do not exceed the temperature at which the gaseous phase becomes flammable, the process cell fire prevention features, to ensure that fires in process cells are highly unlikely; and the offgas treatment system, which will provide exhaust to ensure that an explosive buildup of explosive vapors does not occur. Based on the applicant's deterministic criteria and the NUREG-1718 Table A-5 descriptions (AECs and PECs), the staff considers this to be an acceptable strategy for meeting the 10 CFR 70.61 performance requirements.

TBP-Nitrate (Red Oils) (Explosion)

The bounding event in the TBP-nitrate (red oil) explosion class is a process-related chemical explosion involving red oil formation in the AP boiler, vessel, or tank and results in a loss of confinement and dispersal of nuclear materials. The applicant determined this to be an above the 10 CFR 70.61(c) threshold consequence event for the facility worker, environment, site worker, and IOC and has opted to meet the performance requirements using a strategy of prevention. The PSSCs identified for prevention of this event are the process safety control subsystem, chemical safety control, and offgas treatment system. The purpose of the process safety control subsystem is to ensure that the evaporator process temperature is maintained within safe limits and to control the residence time of organics in the presence of oxidizers, radiation fields, and high temperatures. The chemical safety control ensures that quantities of organics are limited from entering process vessels containing oxidizing agents and at potentially high temperatures and ensures that a diluent is used that is not very susceptible to either nitration or radiolysis. The offgas treatment system provides an exhaust path for the removal of gases in process vessels. Based on the applicant's deterministic criteria and the NUREG-1718 Table A-5 descriptions (PECs, AECs, and administrative controls), the staff considers this to be an acceptable strategy for meeting the 10 CFR 70.61 performance requirements.

AP Vessel Overpressurization (Explosion)

The bounding events in AP vessel overpressurization explosion class were determined to be the overpressurization of AP tanks, vessels, and piping postulated to result from increases in the temperature of exothermic chemical reactions of solutions into tanks or vessels within the facility. The applicant determined this to be an above the 10 CFR 70.61(c) threshold consequence event for the facility worker, environment, site worker, and IOC and has opted to meet the performance requirements using a strategy of prevention. The PSSCs identified for prevention of this event are the fluid transport systems, which will insure that vessels, tanks, and piping are designed to prevent process deviations from creating overpressurization events, the offgas treatment system, which will provide an exhaust path for the removal of gases in process vessels, and chemical safety controls to ensure control of the chemical makeup of the reagents and ensure segregation/separation of vessels/components from incompatible chemicals. Based on the applicant's deterministic criteria and the NUREG-1718 Table A-5 descriptions (PECs, AECs, and administrative controls), the staff considers this to be an acceptable strategy for meeting the 10 CFR 70.61 performance requirements.

Pressure Vessel Overpressurization (Explosion)

The bounding event in the pressure vessel overpressurization explosion class is an explosion related to the overpressurization of gas bottles, tanks, or receivers which could impact primary confinements and result in a release of radioactive material. The applicant determined this to be an above the 10 CFR 70.61(c) threshold consequence event for the facility worker, environment, site worker, and IOC and has opted to meet the performance requirements using a strategy of prevention. The PSSCs identified for prevention of this event are the pressure vessel controls, which ensure that primary confinement is protected from the impact of pressure vessel failures. Pressure vessels will be located away from PSSCs or otherwise protected so that a failure of any vessel would have no impact on the ability of the PSSC to perform its safety function. Based on the applicant's deterministic criteria and the NUREG-1718 Table A-5 descriptions (PECs and administrative controls), the staff considers this to be an acceptable strategy for meeting the 10 CFR 70.61 performance requirements.

Hydrazoic Acid (Explosion)

The bounding event in the hydrazoic acid explosion sequence was a process-related chemical explosion involving HAN/nitric acid in the AP vessels, tanks, and piping (in AP process cells or gloveboxes) which results in a breach of the AP vessels, tanks, and piping. The material at risk was the maximum inventory of radiological material in AP vessels, tanks, and piping. The applicant determined this to be an above the 10 CFR 70.61(c) threshold consequence event for the facility worker, environment, and individuals outside the controlled area boundary and has opted to meet the performance requirements using a strategy of prevention. The PSSCs identified for protection of the facility worker, environment, site worker, and IOC are the chemical safety control and the process safety control subsystem. The function of the chemical safety control is to (1) assure that the proper concentration of hydrazine nitrate is introduced to the system, limiting the quantity of hydrazoic acid produced, and (2) ensure that hydrazoic acid is not accumulated in the process or propagated into the acid recovery and oxalic mother liquors recovery units by either taking representative samples in upstream units or by crediting the neutralization process within the solvent recovery unit. The safety function of the process safety control subsystem is to limit the temperature of the solution, thereby limiting the evaporation rate and resulting vapor pressure of hydrazoic acid so that an explosive concentration of hydrazoic acid does not occur. Based on the applicant's deterministic criteria and the NUREG-1718 Table A-5 descriptions (enhanced administrative control and AEC), the staff considers this to be an acceptable strategy for meeting the 10 CFR 70.61 performance requirements.

Metal Azide Explosions

The bounding event in the metal azide explosion category was a process-related chemical explosion involving an azide (other than hydrazoic acid) in an AP boiler, vessel, or tank (in an AP cell or glovebox) that results in an energetic breach of the AP boiler, vessel, or tank. The material at risk was the maximum inventory of radiological material in AP vessels, tanks, and piping. The applicant determined this to be an above the 10 CFR 70.61(c) threshold consequence event for the facility worker, environment, site worker, and IOC and has opted to meet the performance requirements using a strategy of prevention. The PSSCs identified for protection of the receptors are the chemical safety control and the process safety control subsystem. The safety functions of the chemical safety control are to (1) ensure that metal azides are not added to high temperature process equipment, and (2) ensure that the sodium azide has been destroyed before transfer of the alkaline waste into the acidic high alpha waste of the waste recovery unit. The safety function of the process safety control subsystem is to ensure that metal azides are not exposed to temperatures that would supply sufficient energy to overcome the activation energy needed to initiate the energetic azide decomposition and limit and control conditions under which dryout can occur. Based on the applicant's deterministic criteria and the NUREG-1718 Table A-5 descriptions (AEC and enhanced administrative control), the staff considers this to be an acceptable strategy for meeting the 10 CFR 70.61 performance requirements.

Plutonium (VI) Oxalate Explosion

The bounding event for the plutonium (VI) oxalate explosion category was a process-related chemical explosion involving plutonium (VI) in the calcining furnace results in an energetic breach of the furnace and glovebox and the dispersal of radiological materials. The material at risk was the maximum inventory of radiological material in the AP vessels, tanks, and piping. The applicant determined this to be an above the 10 CFR 70.61(c) threshold consequence event for the facility worker, environment, site worker, and IOC and has opted to meet the performance requirements using a strategy of prevention. The PSSC identified for protection of the facility worker, site worker, environment, and IOC is the chemical safety control. The safety function of the chemical safety control is to perform a measurement of the valency of the plutonium before the addition of oxalic acid to the oxalic precipitation and oxidation unit to ensure that plutonium (IV) cannot be formed. In addition, the design basis for the calciner will assure that the rapid decomposition of any plutonium (VI) oxalate that may enter the calciner will not challenge the calciner vessel's integrity. Based on the applicant's deterministic criteria and the NUREG-1718 Table A-5 descriptions (enhanced administrative control), the staff considers this to be an acceptable strategy for meeting the 10 CFR 70.61 performance requirements.

Electrolysis-Related Explosion

The bounding event for the electrolysis-related explosion category was the explosion of hydrogen in the vapor space of the electrolyzer. The material at risk was the maximum inventory of radiological material in the AP vessels, tanks, and piping. The applicant determined this to be an above the 10 CFR 70.61(c) threshold consequence event for the facility worker, environment, site worker, and IOC and has opted to meet the performance requirements using a strategy of prevention. The PSSCs identified for protection of the facility worker, environment, site worker, and IOC are the process safety control subsystem and specified electrolyzer components, such as the electrolyzer structure, sintered silicon nitride barrier, and polytetrafluorethylene insulator (PTFE). The function of the process safety control subsystem is to ensure that the normality of the acid is sufficiently high to ensure that the offgas is not flammable and to limit excessive generation of hydrogen. The process safety control subsystem also has the function of monitoring the electrolyzer for electrical faults that could result in arcing or other imparting of electrical energy with the risk of a titanium fire and hydrogen explosion. The safety function of the specified electrolyzer components is to provide physical separation and electrical insulation between the electrolyzer components and structural integrity and stability to the electrolyzer system. Based on the applicant's deterministic criteria and the NUREG-1718 Table A-5 descriptions (AECs), the staff considers this to be an acceptable strategy for meeting the 10 CFR 70.61 performance requirements.

Laboratory (Explosion)

The bounding event for the laboratory explosion class is an explosion within the MFFF laboratory involving flammable, explosive, or reactive chemicals which results in a dispersal of radiological material. The radiological material assumed to be dispersed is the maximum inventory in the laboratory. The applicant determined this to be an above the 10 CFR 70.61(c) threshold consequence event for the facility worker, environment, site worker, and IOC and has opted to meet the performance requirements using a strategy of prevention and mitigation. The PSSCs for protection of the facility worker are the chemical safety control, laboratory material controls, and facility worker actions. The function of the chemical safety control is to ensure control of the chemical makeup of the reagents and ensure segregation/separation of vessels/components from incompatible chemicals. The safety function of the laboratory material controls is to minimize quantities of hazardous chemicals in the laboratory and to minimize quantities of radioactive materials in the laboratory. The function of facility worker actions is to ensure that facility workers take proper actions to limit radiological/chemical exposure. The PSSC identified for protection of the environment, site worker, and IOC is the C3 confinement system which provides filtration to mitigate dispersions from the C3 areas. Based on the applicant's deterministic criteria and the NUREG-1718 Table A-5 descriptions (enhanced administrative controls and AEC), the staff considers this to be an acceptable strategy for meeting the 10 CFR 70.61 performance requirements.

Outside (Explosion)

The bounding events for the outside explosion class were determined to be explosions in the reagent processing building, gas storage area, emergency diesel generator building, standby diesel generator building, and the access control building. The applicant determined this to be an above the 10 CFR 70.61(c) threshold consequence for the facility worker, environment, site worker, and IOC and has opted to meet the performance requirements using a strategy of prevention of a release. The PSSCs identified are the waste transfer line, which is designed to prevent damage to the line during an explosion, the MOX fuel fabrication building structure, which is designed to maintain structural integrity and prevent damage to internal PSSCs from explosions external to the structure, and the emergency diesel generator building structure, which is designed to maintain structural integrity and prevent damage to internal PSSCs from explosions external to the structure. Also identified as a PSSC is the hazardous material delivery controls which ensure that the quantity of delivered hazardous material and its proximity to the fuel fabrication building structure, emergency generator building structure, and the waste transfer line are controlled to within the bounds of the values used to demonstrate that the consequences of outside explosions are acceptable. Based on the applicant's deterministic criteria and the NUREG-1718 Table A-5 descriptions (robust PECs and administrative controls), the staff considers this to be an acceptable strategy for meeting the 10 CFR 70.61 performance requirements.

5.1.6.3.5 Criticality Events

A criticality event is characterized by a self-sustaining fission chain reaction and can potentially release a large amount of energy over a short period of time. When fissionable materials, such as ^{235}U or ^{239}Pu are present in sufficient quantities, a self-sustaining fission chain reaction may be attained depending on the size and shape of the fissionable materials, the nature of solvents or diluent, and the proximity of potential reflectors.

The most immediate potential consequences from a criticality event is direct radiation exposure to the facility worker. Distance from the event normally protects persons beyond the controlled area boundary. Shielding materials can also reduce the dose.

Criticality accidents may be caused by violation of safety limits such as the following:

- geometry control
- mass control
- density control
- isotopics control
- reflection control
- moderation control
- concentration control
- interaction control
- neutron absorber control
- volume control
- heterogeneity control
- process variable control

The applicant will use a strategy of prevention to protect the facility worker from a criticality accident. The applicant identified criticality control as the only PSSC for its prevention strategy. The staff's review and evaluation of the applicant's analysis of criticality events is discussed in Chapter 6 of the FSER.

5.1.6.3.6 Chemical Events

The applicant evaluated chemical event scenarios having the potential to affect radiological safety by considering a range of initial conditions and failure modes of storage containers and associated systems. The revised CAR provided the following release scenarios for chemical events:

- leaks and ruptures involving equipment vessels and piping
- evaporating pools formed by spills and tank failures
- flashing and evaporating liquified gases from pressurized storage

Fourteen chemical events with potentially significant radiological consequences were analyzed by DCS to determine the bounding consequences of such events. These events were assigned to two groups with a unique prevention or mitigation strategy. These two groups are discussed below.

Chemical Hazards

Events involving chemical hazards are chemical releases from vessels, tanks, pipes, or transport containers, either internal or external to the MOX fuel fabrication building. The primary potential for external releases would be from the reagent processing building (BRP). For this group of events, only chemical consequences that impact radiological safety or MFFF operations, and which may result in a radioactive material release, were considered. The applicant determined that the only chemical releases that could result in radiological releases would be chemical releases that could affect workers providing monitoring functions in the control room. The applicant has assumed that such radiological releases could be above threshold levels for receptors and has opted to meet the performance requirements through a strategy of prevention. The PSSC identified to prevent a radioactive release is the emergency control room air-conditioning system.

In addition to the potential for radiological releases through the incapacitation of workers performing critical functions (highly unlikely), there is also the potential for other facility workers to receive a chemical dose which could be injurious to their health and exceed the 10 CFR 70.61 performance requirements. These releases could occur from pipes and process vessels in the areas containing the dechlorination and dissolution unit electrolyzer and the dissolution unit chlorine offgas scrubbing column. The applicant has opted to meet the performance requirements through a strategy of mitigation. The PSSCs identified to implement this safety strategy are process cell entry controls, the C4 confinement system, and facility worker action. Based on the nature of this event (mitigated by ventilation system and administrative procedures), the staff considers this to be an acceptable strategy for meeting the 10 CFR 70.61 performance requirements.

Radiochemical Hazards

Events involving radiochemical hazards are postulated to occur inside the MFFF and involve the AP system. The applicant performed bounding consequence analysis to determine the unmitigated consequences to persons beyond the controlled area boundary. The applicant determined that under certain circumstances (e.g., a process failure), gases which would normally react with hydrazine, HAN, and hydrazoic acid could be released and result in an unacceptable dose to persons beyond the controlled area boundary. The applicant has opted to satisfy the performance requirements using a strategy of mitigation. The PSSC chosen to protect the site worker is the process safety control subsystem which will ensure that the flow of nitrogen dioxide/nitrogen tetroxide is limited to the oxidation column. With regard to the facility worker, the applicant stated that the same PSSCs that protect the worker from radioactive releases will also protect the facility worker from any radiochemical releases. Based on the applicant's deterministic criteria and the NUREG-1718 Table A-5 descriptions (AEC), the staff considers this to be an acceptable strategy for meeting the 10 CFR 70.61 performance requirements.

A fire inside the secured warehouse could also result in unacceptable consequences to the facility worker and persons beyond the controlled area boundary through the release of depleted uranium dioxide. The applicant has opted to meet the performance objectives using a strategy of mitigation. The PSSCs identified to protect the facility worker and persons beyond the controlled area boundary are facility worker action and combustible loading controls. The safety function of the combustible loading controls is to limit the quantity of combustibles in the secured warehouse to ensure that any fire that may occur will not encompass a large fraction of the stored depleted uranium dioxide. The function of the facility worker action is to ensure that facility workers take proper actions to limit chemical consequences as a result of a fire. Based on the applicant's deterministic criteria and the NUREG-1718 Table A-5 descriptions (administrative control and an enhanced administrative control), the staff considers this to be an acceptable strategy for meeting the 10 CFR 70.61 performance requirements.

5.1.6.4 Consequence Assessment

The applicant has performed an analysis of the bounding mitigated consequences of each event type. These analyses are derived from the hazard assessment performed to establish the PSSCs and represent the bounding accident from each event type. The event types considered are the same as those discussed earlier and consist of loss of confinement, internal fire, load-handling, explosion, and criticality. The calculated mitigated consequences from each event type were found to be below 10 CFR 70.61(c) threshold levels.

The bounding loss of confinement event is an event caused by a load-handling accident involving a glovebox in the jar storage and handling unit.

[Text removed under 10 CFR 2.390]

The bounding internal fire event is a fire in the fire area containing the final dosing unit. This unit contains polished plutonium powder for the purpose of down-blending the MOX powder to the desired blend for fuel rod fabrication.

[Text removed under 10 CFR 2.390]

The bounding explosion event is an event that involves the entire material at risk within a process cell. The cause of the explosion was not postulated.

[Text removed under 10 CFR 2.390]

The staff's independent evaluation of the applicant's bounding consequence determinations are provided in FSER Sections 9.1.1.4 and 10.1.3.

5.1.7 Description of PSSCs

The acceptability of the various strategies for preventing and/or mitigating identified hazards is also dependent on the design bases of the PSSCs. The specific designs or "design bases" of the PSSCs are determined through the DCS safety assessment and are discussed in the appropriate sections of this FSER. The FSER sections in which the design bases of each of the PSSCs are discussed is identified in Table 5.1-3.

5.2 Evaluation Findings

In Section 5 of the revised CAR, DCS provided a description of the safety analysis that it performed and the identified PSSCs for the proposed facility. Based on the staff's review of the revised CAR and supporting information provided by the applicant relevant to the safety analysis and the identified PSSCs, the staff concludes, pursuant to 10 CFR 70.23(b), that the design bases of the PSSCs identified by the applicant will provide reasonable assurance of protection against natural phenomena and the consequences of potential accidents.

A summary of the staff's evaluation findings in regards to the safety assessment portion of the review is as follows:

- The plant site description relating to safety assessment was found to be adequate.
- The safety assessment team description was found to be adequate.

The purpose of the safety assessment methodology review, as discussed in FSER Section 5.1.5, was to determine if the safety assessment was complete by assuring that all appropriate natural phenomena, external manmade hazards, and internal process hazards were considered. The methodology for the safety assessment of the design basis was found to be adequate.

The purpose of the safety assessment results review, as discussed in FSER Section 5.1.6, was to evaluate the appropriateness of natural phenomena and external manmade hazards selected for design, the adequacy of the PSSCs selected to protect against these events, and the adequacy of the strategy and identified PSSCs at a conceptual level for the internal process hazards. The results of the safety assessment of the design basis were found to be adequate and all open items have been closed.

5.3 <u>References</u>

(Chen, 1991) Chen, J.T, et al. "Procedural and Submittal Guidance for the Individual Plant Examination of External Events (IPEEE) for Severe Accident Vulnerabilities." Final Report 1-26. June 1991.

(CFR, 2004a) *Code of Federal Regulations.* Title 10, Energy, Part 20, "Standards for Protection Against Radiation."

(CFR, 2004b) *Code of Federal Regulations.* Title 10,Energy, Part 70, "Domestic Licensing of Special Nuclear Material."

(DCS, 2002a) Hastings, P., Duke Cogema Stone & Webster. Letter to U.S. Nuclear Regulatory Commission, RE: Clarification of Response to "Request for Additional Information." March 8, 2002.

(DCS, 2002b) Ihde, R., Duke Cogema Stone & Webster. Letter to Document Control Desk, U.S. Nuclear Regulatory Commission. Docket Number 070-03098 Duke Cogema Stone & Webster, Mixed Oxide (MOX) Fuel Fabrication Facility Construction Authorization Request, DCS-NRC-000114. October 31, 2002.

(DCS, 2002c) Ihde, R., Duke Cogema Stone & Webster. Letter to Document Control Desk, U.S. Nuclear Regulatory Commission, Docket NUMBER 070-03098, Duke Cogema Stone & Webster, Mixed Oxide (MOX) Fuel Fabrication Facility Construction Authorization Request. DCS-NRC-000120. November 22, 2002.

(DOE, 1994) U.S. Department of Energy. DOE-STD-1020-94, Change Notice No. 1, "Natural Phenomena Hazards Design and Evaluation Criteria for Department of Energy Facilities." Washington, DC, April 1994.

(NRC, 1997) U.S. Nuclear Regulatory Commission. Regulatory Guide 1.165, "Identification and Characterization of Seismic Sources and Determination of Safe Shutdown Earthquake Ground Motion." Washington, DC, 1997.

(NRC, 2000) U.S. Nuclear Regulatory Commission. NUREG-1718, "Standard Review Plan for the Review of an Application for a Mixed Oxide (MOX) Fuel Fabrication Facility." Washington, DC, 2000.

(NRC, 2002) Brown, D. U.S. Nuclear Regulatory Commission. Memorandum to Eric Leeds, U.S. Nuclear Regulatory Commission, "February 21–22, 2002, In-Office Review Summary: Review of Duke Cogema Stone and Webster Construction Authorization Request Supporting Documents for the Mixed Oxide Fuel Fabrication Facility." March 11, 2002.

Table 5.1-1 Safety Assessment Summary (Criticality, Natural Phenomena Hazards, External Manmade Hazards, and Chemical Hazards)

Event	Event Type	PSSC	Safety Function
Criticality	Criticality	Criticality Controls—Features required to ensure that design bases are fulfilled.	Maintain subcritical conditions in the process.
Natural Phenomena Hazards	Extreme Wind	MOX Fuel Fabrication Building	Withstand design basis wind loads and wind-driven missiles. Provide protection for internal SSCs.
		Emergency Generator Building	Withstand design basis wind loads and wind-driven missiles. Provide protection for internal SSCs.
		Missile Barriers	Withstand design basis wind loads and wind-driven missiles. Provide protection for internal SSCs.
		Waste Transfer Line	Withstand design basis wind loads and wind-driven missiles.
	Earthquake	MOX Fuel Fabrication Building	Withstand the effects of design basis earthquake.
		Emergency Generator Building	Withstand the effects of design basis earthquake.
		Waste Transfer Line	Withstand the effects of design basis earthquake.
		Fluid Transport Systems	Withstand the effects of design basis earthquake, as necessary.
		Seismic Monitoring and Associated Seismic Isolation Valves	Prevent fire and criticality as a result of an uncontrolled release of chemicals and water within the MFFF building.

Table 5.1-1 Safety Assessment Summary (Criticality, Natural Phenomena Hazards, External Manmade Hazards, and Chemical Hazards)

Event	Event Type	PSSC	Safety Function
Natural Phenomena Hazards (cont.)	Tornado	MOX Fuel Fabrication Building	Withstand design basis tornado wind loads and tornado-generated missiles.
			Provide protection to internal PSSCs.
		Emergency Generator Building	Withstand design basis tornado wind loads and tornado-generated missiles.
			Provide protection for internal PSSCs.
		Missile Barriers	Withstand design basis tornado wind loads and tornado-generated missiles.
			Provide protection for internal PSSCs.
		Waste Transfer Line	Withstand design basis tornado wind loads and tornado-driven missiles.
		Tornado Dampers	Protect MFFF ventilation systems from differential pressure effects of the tornado.
	External Fire	MOX Fuel Fabrication Building	Withstand the effects of the design basis external fire.
			Provide protection for internal SSCs from the effects of heat, fire, and smoke.
		Emergency Generator Building	Withstand the effects of the design basis external fire.
			Provide protection for internal SSCs from the effects of heat, fire, and smoke.
		Emergency Control Room Air-Conditioning System	Ensure habitable conditions for operators.
		Waste Transfer Line	Withstand the effects of external fires.

5-48

Table 5.1-1 Safety Assessment Summary (Criticality, Natural Phenomena Hazards, External Manmade Hazards, and Chemical Hazards)

Event	Event Type	PSSC	Safety Function
Natural Phenomena Hazards (cont.)	Rain, Snow, and Ice	MOX Fuel Fabrication Building	Withstand the effects of the design basis rain, snow, and ice loads.
			Provide protection for internal SSCs.
		Emergency Generator Building	Withstand the effects of the design basis rain, snow, and ice loads.
			Provide protection for internal SSCs.
		Waste Transfer Line	Withstand the effects of design basis rain, snow, and ice loads.
	Lightning	None	N/A
	Temperature Extremes	None	N/A
External Manmade	Release of Radioactive Material or Hazardous Chemicals	Emergency Control Room Air-Conditioning System	Ensure that the emergency control rooms are habitable.
	Direct Damage to PSSCs	MOX Fuel Fabrication Building (To be described in the ISA)	Withstand overpressure of explosions external to the MFFF area.
		Emergency Generator Building (To be described in the ISA)	Withstand overpressure of explosions external to the MFFF area.
	Loss of Offsite Power	Section 5.5.2.9, Section 5.4, Support Systems—Table 5.5-22 (Revised CAR)	Supply emergency power upon loss of offsite power.
	Fire	MOX Fuel Fabrication Building	Withstand the effects of the design basis external fire.
			Provide protection for internal SSCs from the effects of heat, fire, and smoke.

Table 5.1-1 Safety Assessment Summary (Criticality, Natural Phenomena Hazards, External Manmade Hazards, and Chemical Hazards)

Event	Event Type	PSSC	Safety Function
External Man-Made (cont.)	Fire (cont.)	Emergency Generator Building	Withstand the effects of the design basis external fire. Provide protection for internal SSCs from the effects of heat, fire, and smoke.
		Emergency Control Room Air-Conditioning System	Ensure habitable conditions for operators.
		Waste Transfer Line	Withstand the effects of external fires.
External Exposure	Operator is inadvertently exposed to excessive direct radiation.	Not Required	N/A
Chemical	A release of hazardous chemicals not produced from licensed materials.	Emergency Control Room Air-Conditioning System	Ensure habitable conditions for operators in the control room.
	A release of hazardous chemicals produced from licensed materials.	Process Cell Entry Controls	Prevent the entry of personnel into process cells during normal operations. Ensure that workers do not receive a chemical consequence in excess of limits while performing maintenance in the AP process cells.
		C4 Confinement Systems	Contain a chemical release within a glovebox and provide an exhaust path for removal of chemical vapors.

Table 5.1-1 Safety Assessment Summary (Criticality, Natural Phenomena Hazards, External Manmade Hazards, and Chemical Hazards)

Event	Event Type	PSSC	Safety Function
Chemical (cont.)	A release of hazardous chemicals produced from licensed materials (cont.).	Facility Worker Action	Ensure that facility workers take proper actions to limit chemical consequences for leaks occurring in C3 ventilated areas.
	A release of hazardous chemicals and radioactive materials.	Process Safety Control Subsystem (NO_2/N_2O_4)	Ensure that the flow rate of nitrogen dioxide/dinitrogen tetroxide is limited to the oxidation column of the purification cycle.
		Combustible Loading Controls (UO_2)	Limit the quantity of combustibles in the secured warehouse to ensure that any fire that may occur will not encompass a large fraction of the stored depleted uranium.
		Facility Worker Actions (UO_2)	Ensure that facility workers take proper actions to limit chemical consequences in the secured warehouse.

Table 5.1-2 Safety Assessment Summary (Process Hazards)

Event Type	Event Group	Bounding Event	PSSC (SW & IOC)	Safety Function	PSSC (facility worker)	Safety Function	PSSC (environment)	Safety Function
Loss of Confinement/ Dispersal of Nuclear Material	Over Temperature	Excessive temperature of the AP electrolyzer in a glovebox	C3 Confinement	Provide filtration to mitigate dispersions from C3 areas.	Process Safety Control Subsystem	Shutdown process equipment before exceeding a temperature safety limit.	Process Safety Control Subsystem	Shutdown process equipment before exceeding a temperature safety limit.
	Corrosion	Corrosion involving pneumatic transfer of corrosive chemicals	None	N/A	Material Maintenance and Surveillance Program	Detect and limit damage resulting from corrosion.	Material Maintenance and Surveillance Program	Detect and limit damage resulting from corrosion.
	Small Breaches in a Glovebox Confinement Boundary or Backflow from a Glovebox through Utility Lines	Backflow through the interfacing gas line to the interfacing system followed by the opening of this interfacing system during a maintenance operation.	None	N/A	C4 Confinement System	Maintain a negative glovebox differential pressure between glovebox and interfacing systems. Maintain a minimum inward flow through small glovebox breaches.	C4 Confinement System	Maintain a negative glovebox differential pressure between glovebox and interfacing systems. Maintain a minimum inward flow through small glovebox breaches

Table 5.1-2 Safety Assessment Summary (Process Hazards)

Event Type	Event Group	Bounding Event	PSSC (SW & IOC)	Safety Function	PSSC (facility worker)	Safety Function	PSSC (environment)	Safety Function
Loss of Confinement/ Dispersal of Nuclear Material (cont.)	Leaks of AP Process Vessels or Pipes Within Process Cells	Leak of tanks/vessels inside the process cell containing a portion of the purification cycle.	Process Cell Exhaust System	Ensure that a negative pressure exists between the process cells and the C2 area. Ensure that the process cell exhaust is effectively filtered.	Process Cells	Contain leaks within the process cells.	Process Cell Exhaust System	Ensure that a negative pressure exists between the process cells and the C2 area. Ensure that the process cell exhaust is effectively filtered.
					Process Cell Entry Controls	Prevent the entry of personnel into process cell during normal operation.		
	Backflow from a Process Vessel Through Utility Lines	Backflow of radioactive material from a waste tank containing americium through an interfacing supply line that is subsequently breached or opened during a maintenance operation.	Backflow Prevention Features	Prevent process fluids from backflowing into interfacing systems.	Backflow Prevention Features	Prevent process fluids from backflowing into interfacing systems.	Backflow Prevention Features	Prevent process fluids from backflowing into interfacing systems.

5-53

Table 5.1-2 Safety Assessment Summary (Process Hazards)

Event Type	Event Group	Bounding Event	PSSC (SW & IOC)	Safety Function	PSSC (facility worker)	Safety Function	PSSC (environment)	Safety Function
Loss of Confinement/ Dispersal of Nuclear Material (cont.)	Rod Handling Operations	Fracture of one or more fuel rods while utilizing fuel handling equipment resulting in a breach of confinement.	None	N/A	Facility Worker Action	Ensure that facility workers take proper actions to limit radiological exposure.	None	N/A
					Material Handling Controls	Ensure proper handling of primary confinements outside of gloveboxes.		
					Material Handling Equipment	Limit damage to fuel rods/assemblies during handling operations.		
	Breaches in Containers Outside the Gloveboxes from Handling Operations in C2 and C3 Areas	Container containing filters is breached while in C2 area.	Materials Handling and Control (C2 Areas)	Ensure proper handling of primary confinement types outside of gloveboxes. outside of gloveboxes.	Material Handling Controls	Ensure proper handling of primary confinement types outside the gloveboxes outside of gloveboxes.	Materials Handling and Control (C2 Areas)	Ensure proper handling of primary confinement types outside of gloveboxes outside of gloveboxes.
			3013 Canister (C2 Areas)	Withstand the effects of design basis drops without breaching.	3013 Canister	Withstand the effects of design basis drops without breaching.	3013 Canister (C2 Areas)	Withstand the effects of design basis drops without breaching.

Table 5.1-2 Safety Assessment Summary (Process Hazards)

Event Type	Event Group	Bounding Event	PSSC (SW & IOC)	Safety Function	PSSC (facility worker)	Safety Function	PSSC (environment)	Safety Function
Loss of Confinement/ Dispersal of Nuclear Material (cont.)	Breaches in Containers Outside the Gloveboxes from Handling Operations in C2 and C3 Areas (cont.)	Container containing filters is breached while in C2 area (cont.)	Transfer Container (C2 Areas)	Withstand the effect of design basis drops without breaching.	Transfer Container	Withstand the effects of design basis drops without breaching.	Transfer Container (C2 Areas)	Withstand the effect of design basis drops withou breaching.
			C3 Confinement System (C3 Areas)	Provide filtration to mitigate dispersion from the C3 areas.	Facility Worker Controls (C3 Areas)	Ensure that facility workers take proper actions before bag out operations to limit radiological exposure.	C3 Confinement System (C3 Areas)	Provide filtration to mitigate dispersion from the C3 areas.
	Over or Underpressurization of Glovebox	Rapid overpressurization of the calcining furnace	C3/C4 Confinement System	Provide filtration to mitigate dispersion from C3/C4 areas.	Facility Worker Action (slow pressurization)	Ensure that facility workers take proper actions to limit radiological exposure.	C3/C4 Confinement System	Provide filtration to mitigate dispersion from C3/C4 areas.
					Process Safety Control Subsystem (slow pressurization)	Warn operators of glovebox pressure discrepancies before exceeding differential pressure limits.		
					Glovebox Pressure Controls (rapid or slow pressurization)	Maintain glovebox pressure within design limits.		

Table 5.1-2 Safety Assessment Summary (Process Hazards)

Event Type	Event Group	Bounding Event	PSSC (SW & IOC)	Safety Function	PSSC (facility worker)	Safety Function	PSSC (environment)	Safety Function
Loss of Confinement/ Dispersal of Nuclear Material (cont.)	Excessive Temperature from Decay Heat from Radioactive Materials	Excessive temperature (from decay heat) of C2 storage area (PuO₂ powder 3013 storage unit)	High Depressurization Exhaust System (C3 Confinement System)	Provide exhaust to ensure that temperatures in the 3013 canister storage structure are maintained within design limits.	High Depressurization on Exhaust System (C3 Confinement System)	Provide exhaust to ensure that temperatures in the 3013 canister storage structure are maintained within design limits.	High Depressurization on Exhaust System (C3 Confinement System)	Provide exhaust to ensure that temperatures in the 3013 canister storage structure are maintained within design limits
	Glovebox Dynamic Exhaust Failure	Loss of negative pressure or a flow perturbation involving the C4 dynamic confinement system resulting in a ventilation airflow reversal into a C3 area	C4 Confinement System	Operate to ensure that a negative pressure differential exists between the C4 glovebox and the C3 areas. Ensure that the C4 exhaust is effectively filtered.	C4 Confinement System	Operate to ensure that a negative pressure differential exists between the C4 glovebox. Ensure that the C4 exhaust is effectively filtered.	C4 Confinement System	Operate to ensure that a negative pressure differential exists between the C4 glovebox. Ensure that the C4 exhaust is effectively filtered.
	Process Fluid Line Leak in a C3 Area Outside the Glovebox	A leak from a line carrying a process fluid in a C3 area outside of a glovebox or process cell caused by corrosive chemicals or mechanical failure of AP piping	None	N/A	Double Walled Pipe	Prevent leaks from pipes containing process fluids from leaking into C3 areas.	Double Walled Pipe	Prevent leaks from pipe containing process fluids from leaking into C3 areas.

Table 5.1-2 Safety Assessment Summary (Process Hazards)

Event Type	Event Group	Bounding Event	PSSC (SW & IOC)	Safety Function	PSSC (facility worker)	Safety Function	PSSC (environment)	Safety Function
Loss of Confinement/Dispersal of Nuclear Material (cont.)	Sintering Furnace Confinement Boundary Failure	Rapid overpressurizati on of the sintering furnace and a slow leakage through seals	C3 Confinement System	Provide filtration to mitigate dispersion from the C3 areas	Sintering Furnace	Provide primary confinement boundary against leaks into C3 areas. Minimize consequences of leak from seal failure.	Sintering Furnace	Provide primary confinement boundary against leaks into C3 areas
					Sintering Furnace Pressure Controls	Maintain sintering furnace pressure within	Sintering Furnace Pressure Controls	Maintain sintering furnace pressure within design limits design limits.
Fire	AP Process Cells	Fire in the AP process cells containing the dissolution tanks	Process Cell Fire Prevention Features	Ensure that fires in the process cells are highly unlikely.	Process Cell Fire Prevention Features	Ensure that fires in the process cells are highly unlikely.	Process Cell Fire Prevention Features	Ensure that fires in the process cells are unlikely.

Table 5.1-2 Safety Assessment Summary (Process Hazards)

Event Type	Event Group	Bounding Event	PSSC (SW & IOC)	Safety Function	PSSC (facility worker)	Safety Function	PSSC (environment)	Safety Function
Fire (cont.)	AP/MP C3 Glovebox Areas	Fire inside or outside the glovebox (fire areas containing process gloveboxes)	C3/C4 Confinement Systems	Remain operable during design basis fire and effectively filter any release.	Facility Worker Action	Ensure that facility workers take proper actions to limit radiological exposure.	C3/C4 Confinement Systems	Remain operable during design basis fire and effectively filter any release.
		Fire within the PuO$_2$ buffer storage area (Gloveboxes that store radiological materials)	Combustible Loading (For storage gloveboxes only)	Limit the quantity of combustibles in fire areas containing a storage glovebox such that any fire that may occur will not encompass a large fraction of the stored radiological material.	Facility Worker Controls	Ensure that facility workers take proper actions before maintenance activities to limit radiological exposure.	Combustible Loading (For storage gloveboxes only)	Limit the quantity of combustibles in fire areas containing a storage glovebox such that any fire that may occur will not encompass a large fraction of the stored radiological material.
	C1 and/or C2 3013 Canister	Fire affecting 3013 canister	Combustible Loading Controls	Limit the quantity of combustibles in a fire area containing 3013 canisters to ensure that the canisters are not adversely impacted by a fire.	Combustible Loading Controls	Limit the quantity of combustibles in a fire area containing 3013 canisters to ensure that the canisters are not adversely impacted by a fire.	Combustible Loading Controls	Limit the quantity of combustibles in a fire area containing 3013 canisters to ensure that the canisters are not adversely impacted by a fire.

Table 5.1-2 Safety Assessment Summary (Process Hazards)

Event Type	Event Group	Bounding Event	PSSC (SW & IOC)	Safety Function	PSSC (facility worker)	Safety Function	PSSC (environment)	Safety Function
Fire (cont.)	C1 and/or C2 3013 Transport Cask	Fire in the truck bay involving transport packages	3013 Transport Cask	Withstand the design basis fire without breaching.	3013 Transport Cask	Withstand the design basis fire without breaching.	3013 Transport Cask	Withstand the design basis fire without breaching.
			Combustible Loading Controls	Limit the quantity of combustibles in a fire area containing 3013 transport casks to ensure that the cask design basis fire is not exceeded.	Combustible Loading Controls	Limit the quantity of combustibles in a fire area containing 3013 transport casks to ensure that the cask design basis fire is not exceeded.	Combustible Loading Controls	Limit the quantity o combustibles in a fire area containing 3013 transport casks to ensure that the cask design basis fire is not exceeded
	C1 and/or C2 Areas Fuel Rod	Fire in the fuel assembly storage area	Combustible Loading Controls	Limit the quantity of combustibles in a fire area containing fuel rods to ensure that the fuel rods are not adversely impacted by a fire.	Combustible Loading Controls	Limit the quantity of combustibles in a fire area containing fuel rods to ensure that the fuel rods are not adversely impacted by a fire.	Combustible Loading Controls	Limit the quantity o combustibles in a fire area containing fuel rods to ensure that the fuel rods are not adversely impacted by a fire.
	C1 and/or C2 Areas MOX Fuel Transport Cask	Fire affecting the MOX fuel transport cask	MOX Fuel Transport Cask	Withstand the design basis fire without breaching.	MOX Fuel Transport Cask	Withstand the design basis fire without breaching.	MOX Fuel Transport Cask	Withstand the design basis fire without breaching.

Table 5.1-2 Safety Assessment Summary (Process Hazards)

Event Type	Event Group	Bounding Event	PSSC (SW & IOC)	Safety Function	PSSC (facility worker)	Safety Function	PSSC (environment)	Safety Function
Fire (cont.)	C1 and/or C2 Areas MOX Fuel Transport Cask (cont.)	Fire Affecting the MOX fuel transport cask (cont.)	Combustible Loading Controls	Limit the quantity of combustibles in a fire area containing MOX fuel transport casks to ensure that the cask design basis fire is not exceeded.	Combustible Loading Controls	Limit the quantity of combustibles in a fire area containing MOX fuel transport casks to ensure that the cask design basis fire is not exceeded.	Combustible Loading Controls	Limit the quantity of combustibles in a fire area containing MOX fuel transport casks to ensure that the cask design basis fire is not exceeded.
	C1 and/or C2 Areas Waste Container	Fire in the assembly packaging area	None	N/A	Facility Worker Action	Ensure that facility workers take proper actions to limit radiological exposure.	None	N/A
	C1 and/or C2 Areas Transfer Container	Fire in e the airlocks, corridors, stairways, safe areas, or liquid waste reception areas	Combustible Loading Controls (For site worker only no PSSCs required for IOC)	Limit the quantity of combustibles in a fire area containing transfer containers to ensure that the containers are not adversely impacted by a fire.	Combustible Loading Controls	Limit the quantity of combustibles in a fire area containing transfer containers to ensure that the containers are not adversely impacted by a fire.	Combustible Loading Controls	Limit the quantity of combustibles in a fire area containing transfer containers to ensure that the containers are not adversely impacted by a fire.

Table 5.1-2 Safety Assessment Summary (Process Hazards)

Event Type	Event Group	Bounding Event	PSSC (SW & IOC)	Safety Function	PSSC (facility worker)	Safety Function	PSSC (environment)	Safety Function
Fire (cont.)	C1 and/or C2 Areas Final C4 HEPA Filter	Fire impacting final C4 HEPA filters breaches the HEPA filter house allowing the material to pass directly to the stack.	Combustible Loading Controls	Limit the quantity of combustibles in the filter area to ensure that the final C4 HEPA filters are not impacted by a fire in the filter room.	Combustible Loading Controls	Limit the quantity of combustibles in the filter area to ensure that the final C4 HEPA filters are not adversely impacted by a fire in the filter room.	Combustible Loading Controls	Limit the quantity of combustibles in the filter area to ensure that the final C4 HEPA filters are not impacted by a fire in the filter room.
	Outside the MOX Fabrication Building	Fire involving diesel fuel storage, gasoline storage, or the reagents processing building such that the MFFF building structure is damaged and radioactive material is released.	MOX Fuel Fabrication Building Structure	Maintain structural integrity and prevent damage to internal PSSCs from external fires.	MOX Fuel Fabrication Building Structure	Maintain structural integrity and prevent damage to internal PSSCs from external fires.	MOX Fuel Fabrication Building Structure	Maintain structural integrity and prevent damage to internal PSSCs from external fires.
			Emergency Generator Building Structure	Maintain structural integrity and prevent damage to internal PSSCs from fires external to the structure.	Emergency Generator Building Structure	Maintain structural integrity and prevent damage to internal PSSCs from fires external to the structure.	Emergency Generator Building Structure	Maintain structural integrity and prevent damage to internal PSSCs from fires external to the structure.

5-61

Table 5.1-2 Safety Assessment Summary (Process Hazards)

Event Type	Event Group	Bounding Event	PSSC (SW & IOC)	Safety Function	PSSC (facility worker)	Safety Function	PSSC (environment)	Safety Function
Fire (cont.)	Outside the MOX Fabrication Building (cont.)	Fire involving diesel fuel storage, gasoline storage, or the reagents processing building such that the MFFF building structure is damaged and radioactive material is released (cont.)	Emergency Control Room Air Conditioning System	Ensure habitable conditions for operators.	Emergency Control Room Air Conditioning System	Ensure habitable conditions for operators.	Emergency Control Room Air Conditioning System	Ensure habitable conditions for operators.
			Waste Transfer Line	Prevent damage to line from external fires.	Waste Transfer Line	Prevent damage to line from external fires.	Waste Transfer Line	Prevent damage to line from external fires.
	Facility Wide Systems	Fire Involving the pneumatic pipe automatic transfer system	None	N/A	Facility Worker Action	Ensure that facility workers take proper actions to limit radiological exposure.	Combustible Loading Controls	Limit the quantity of combustibles in areas containing the pneumatic transfer system to ensure this system is not adversely impacted.
					Combustible Loading Controls	Limit the quality of combustibles in a fire area containing a pneumatic system to ensure that this system is not adversely impacted by a fire.		

5-62

Table 5.1-2 Safety Assessment Summary (Process Hazards)

Event Type	Event Group	Bounding Event	PSSC (SW & IOC)	Safety Function	PSSC (facility worker)	Safety Function	PSSC (environment)	Safety Function
Fire (cont.)	Facility	Fire in all process cell units and support units with radioactive material present	Fire Barriers	Contain fires within a single fire area.	Fire Barriers	Contain fires within a single fire area.	Fire Barriers	Contain fires within a single fire area.
					Facility Worker Action	Ensure that facility workers take proper actions to limit radiological exposure.		
	AP Electrolyzer	Energetic breach of the AP electrolyzer and the dispersal of radioactive materials	Maintenance Activity Controls	Isolation of power to the electrolyzer when the electrolyzer is drained.	Maintenance Activity Controls	Isolation of power to the electrolyzer when the electrolyzer is drained.	Maintenance Activity Controls	Isolation of power to the electrolyzer when the electrolyzer is drained.
			Electrolyzer structure, sintered silicon nitride barrier, guide sleeves, and polytetrafluorethylene insulator (PTFE)	To provide physical separation and electrical insulation between electrolyzer components and structural integrity and stability to the electrolyzer system.	Electrolyzer structure, sintered silicon nitride barrier, guide sleeves, and PTFE	To provide physical separation and electrical insulation between electrolyzer components and structural integrity and stability to the electrolyzer system.	Electrolyzer structure, sintered silicon nitride barrier, guide sleeves, and (PTFE)	To provide physical separation and electrical insulation between electrolyzer components and structural integrity and stability to the electrolyzer system.
			Process Safety Control Subsystem	Monitor the electrolyzer for electrical faults that could result in arcing or other imparting of electrical energy with the risk of titanium fire.	Process Safety Control Subsystem	Monitor the electrolyzer for electrical faults that could result in arcing or other imparting of electrical energy with the risk of titanium fire.	Process Safety Control Subsystem	Monitor the electrolyzer for electrical faults that could result in arcing or other imparting of electrical energy with the risk of titanium fire.

Table 5.1-2 Safety Assessment Summary (Process Hazards)

Event Type	Event Group	Bounding Event	PSSC (SW & IOC)	Safety Function	PSSC (facility worker)	Safety Function	PSSC (environment)	Safety Function
Load Handling	AP Process Cells	AP process cell containing the dissolution tanks (Event results in a breach of the AP dissolution tanks and subsequent release of unpolished PuO_2 solution.)	None	N/A	Process Cells	Contain fluid leaks within the process cells.	Process Cell Ventilation System Passive Boundary	Provide filtration to limit the dispersion of radioactive material.
					Process Cell Entry Controls	Prevent the entry of personnel into process cells during normal operations. Ensure that facility workers do not receive a radiological dose in excess of limits while performing maintenance.		

Table 5.1-2 Safety Assessment Summary (Process Hazards)

Event Type	Event Group	Bounding Event	PSSC (SW & IOC)	Safety Function	PSSC (facility worker)	Safety Function	PSSC (environment)	Safety Function
Load Handling (cont.)	AP/MP C3 Glovebox Areas	Gloveboxes that contain jar storage and handling of the MOX powder workshop (Event results in a breach of a glovebox and subsequent release of radiological material.)	C3 Confinement System	Provide filtration to mitigate dispersion from the C3 Areas.	Material Handling Controls	Prevent impacts to the glovebox during normal operations from loads handled either outside or inside the glovebox that could exceed the glovebox design basis.	Material Handling Controls	Prevent impacts to the glovebox during normal operations from loads handled either outside or inside the glovebox that could exceed the glovebox design basis.
					Glovebox	Maintain confinement integrity for design basis impacts.	Glovebox	Maintain confinement integrity for design basis impacts.
					Material Handling Equipment	Prevent impacts to the glovebox through the use of engineered equipment.	Material Handling Equipment	Prevent impacts to the glovebox through the use of engineered equipment.
					Facility Worker Controls	Ensure that facility workers take proper actions before maintenance activities to limit radiological exposure.		

Table 5.1-2 Safety Assessment Summary (Process Hazards)

Event Type	Event Group	Bounding Event	PSSC (SW & IOC)	Safety Function	PSSC (facility worker)	Safety Function	PSSC (environment)	Safety Function
Load Handling (cont.)	C1 and/or C2 Areas 3013 Canister	Drop of one 3013 canister onto another 3013 canister each containing unpolished PuO_2 in powder form	3013 Canister	Withstand the effects of the design basis drop without breaching.	3013 Canister	Withstand the effects of the design basis drop without breaching.	3013 Canister	Withstand the effects of the design basis drop without breaching.
			Material Handling Controls	Ensure that the design basis lift height of the 3013 canister is not exceeded.	Material Handling Controls	Ensure that the design basis lift height of the 3013 canister is not exceeded.	Material Handling Controls	Ensure that design basis lift height of the 3013 canister is not exceeded.
	C1 and/or C2 Areas 3013 Transport Cask	Drop of a 3013 transport cask containing unpolished PuO_2 in powder form onto another transport cask	3013 Transport Cask	Withstand the effects of design basis drops without release of radioactive materials.	3013 Transport Cask	Withstand the effects of design basis drops without release of radioactive materials.	3013 Transport Cask	Withstand the effects of design basis drops without release of radioactive materials.
			Material Handling Controls	Ensure that the design basis lift height of the 3013 transport cask is not exceeded.	Materials Handling Controls	Ensure that the design basis lift height of the 3013 transport cask is not exceeded.	Material Handling Controls	Ensure that the design basis lift height of the 3013 transport cask is not exceeded.
	C1 and/or C2 Areas Fuel Rod	Drop of a fuel assembly onto another fuel assembly each containing MOX (6%)	None	N/A	Facility Worker Action	Ensure that facility workers take proper actions to limit radiological exposure.	None	N/A

5-66

Table 5.1-2 Safety Assessment Summary (Process Hazards)

Event Type	Event Group	Bounding Event	PSSC (SW & IOC)	Safety Function	PSSC (facility worker)	Safety Function	PSSC (environment)	Safety Function
Load Handling (cont.)	C1 and/or C2 Areas MOX Fuel Transport Cask	Drop of one MOX fuel transport cask containing up to three MOX fuel assemblies	None	N/A	MOX Fuel Transport Cask	Withstand the effects of design basis drops without release of radioactive material.	MOX Fuel Transport Cask	Withstand the effects of design basis drops withou release of radioactive material.
					Material Handling Design Controls	Ensure that the design basis lift height of MOX fuel transport cask is not exceeded.	Material Handling Design Controls	Ensure that design basis lift height of MOX fuel transport cask is not exceeded.
	C1, C2, and/or C3 Areas Waste Container	A damaged waste drum in the assembly packaging (truck bay) Area	None	N/A	Facility Worker Action	Ensure that facility workers take proper actions to limit radiological exposure.	None	N/A
	C1 and/or C2 Areas Transfer Container	Drop of a transfer container containing a HEPA filter with PuO$_2$ in powder form	Transfer Container	Withstand the effects of design basis drops without breaching.	Transfer Container	Withstand the effects of design basis drops without breaching.	Transfer Container	Withstand the effects of design basis drops without breaching.
			Material Handling Controls	Ensure that the design basis lift height of the transfer container is not exceeded.	Material Handling Controls	Ensure that the design basis lift height of the transfer container is not exceeded.	Material Handling Controls	Ensure that the design basis lift height of the transfer container is not exceeded.

Table 5.1-2 Safety Assessment Summary (Process Hazards)

Event Type	Event Group	Bounding Event	PSSC (SW & IOC)	Safety Function	PSSC (facility worker)	Safety Function	PSSC (environment)	Safety Function
Load Handling (cont.)	C1 and/or C2 Areas Final C4 HEPA Filter	Impact to final C4 HEPA filters breaching the HEPA filter housing and allowing the material from the HEPA filters to pass directly to the stack	Material Handling Controls	Ensure that load handling activities that could potentially lead to a breach in the final C4 HEPA filters do not occur.	Material Handling Controls	Ensure that load handling activities that could potentially lead to a breach in the final C4 HEPA filters do not occur.	Material Handling Controls	Ensure that load handling activities that could potentially lead to a breach in the final C4 HEPA filters do not occur.
	C4 Confinement	Spill of unpolished plutonium powder inside a glovebox but does not result in a breach of the glovebox	C4 Confinement System	Maintain a negative glovebox pressure differential between the glovebox and the interfacing systems.	C4 Confinement System	Maintain a negative glove box pressure differential between the glovebox and the interfacing systems.	C4 Confinement System	Maintain a negative glove box pressure differential between the glovebox and the interfacing systems.
	Outside the MOX Fuel Fabrication Building	Load handling event involving the waste transfer line	Waste Transfer Line	Ensure that the waste transfer line is protected from activities taking place outside the MOX fuel fabrication building.	Waste Transfer Line	Ensure that the waste transfer line is protected from activities taking place outside the MOX fuel fabrication building.	Waste Transfer Line	Ensure that the waste transfer line is protected from activities taking place outside the MOX fuel fabrication building
	Facility wide	Breach of the MFFF structure from a heavy load resulting in a breach of confinement or in a breach of container holding nuclear materials	MOX Fuel Fabrication Building Structure	Withstand the effects of load drops that could potentially impact radiological material.	MOX Fuel Fabrication Building Structure	Withstand the effects of load drops that could potentially impact radiological material.	MOX Fuel Fabrication Building Structure	Withstand the effects of load drops that could potentially impact radiological material.

Table 5.1-2 Safety Assessment Summary (Process Hazards)

Event Type	Event Group	Bounding Event	PSSC (SW & IOC)	Safety Function	PSSC (facility worker)	Safety Function	PSSC (environment)	Safety Function
Load Handling (cont.)	Facility wide (cont.)	Breach of the MFFF structure from a heavy load resulting in a breach of confinement or in a breach of container holding nuclear materials (cont.)	Material Handling Controls	Prevent load handling events that could breach primary confinements.	Material Handling Controls	Prevent load handling events that could breach primary confinements.	Material Handling Controls	Prevent load handling events that could breach primary confinements.
Explosion	Hydrogen Explosion	Explosion of hydrogen and oxygen in a sintering furnace or sintering furnace room	Process Safety Control Subsystem	Prevent formation of an explosive mixture of hydrogen within the MFFF associated with the use of the hydrogen argon gas.	Process Safety Control Subsystem	Prevent formation of an explosive mixture of hydrogen within the MFFF associated with the use of the hydrogen argon gas.	Process Safety Control Subsystem	Prevent formation of an explosive mixture of hydrogen within the MFFF associated with the use of the hydrogen argon gas.
	Steam Overpressurization Explosion	Water entry into the sintering furnace from failure of the water level controller in the humidifier, from the sintering furnace, results in a steam explosion	Process Safety Control Subsystem	Ensure isolation of the sintering humidifier water flow on high water level.	Process Safety Control Subsystem	Ensure isolation of the sintering humidifier water flow on high water level.	Process Safety Control Subsystem	Ensure isolation of the sintering humidifier water flow on high water level.

Table 5.1-2 Safety Assessment Summary (Process Hazards)

Event Type	Event Group	Bounding Event	PSSC (SW & IOC)	Safety Function	PSSC (facility worker)	Safety Function	PSSC (environment)	Safety Function
Explosion (cont.)	Radiolysis Induced Explosion	Buildup in the vapor space of an AP vessel tank or piping	Offgas Treatment System	Provide an exhaust path for the removal of the diluted hydrogen gas in process vessels.	Offgas Treatment System	Provide an exhaust path for the removal of the diluted hydrogen gas in process vessels.	Offgas Treatment System	Provide an exhaust path for the removal of the diluted hydrogen gas in process vessels.
		Hydrogen buildup in the vapor space of a raffinates tank	Instrument Air System (Emergency Scavenging Air)	Provide sufficient scavenging air to dilute the hydrogen generated during radiolysis.	Instrument Air System (Emergency Scavenging Air)	Provide sufficient scavenging air to dilute the hydrogen generated during radiolysis.	Instrument Air System (Emergency Scavenging Air)	Provide sufficient scavenging air to dilute the hydrogen generated during radiolysis.
		Hydrogen accumulation in waste container containing hydrocarbons	Waste Containers	Ensure that hydrogen buildup in excess of explosive limits does not occur while providing appropriate confinement of radioactive material.	Waste Containers	Ensure that hydrogen buildup in excess of explosive limits does not occur while providing appropriate confinement of radioactive material.	Waste Containers	Ensure that hydrogen buildup in excess of explosive limits does not occur while providing appropriate confinement of radioactive material.
	Hydroxylamine Nitrate (HAN) Explosion Process Vessels containing HAN and Hydrazine without NO$_x$ Addition	Explosion in AP process vessels containing HAN	Process Safety Control Subsystem	Ensure that the temperature of the solution containing HAN is limited to temperatures that are within safety limits.	Process Safety Control Subsystem	Ensure that the temperature of the solution containing HAN is limited to temperatures that are within safety limits.	Process Safety Control Subsystem	Ensure that the temperature of the solution containing HAN is limited to temperatures that are within safety limits.

Table 5.1-2 Safety Assessment Summary (Process Hazards)

Event Type	Event Group	Bounding Event	PSSC (SW & IOC)	Safety Function	PSSC (facility worker)	Safety Function	PSSC (environment)	Safety Function
Explosion (cont.)	Hydroxylamine Nitrate (HAN) Explosion Proc ess Vessels containing HAN and Hydrazine Nitrate without NO$_x$ Addition (cont.)	Explosion in AP process vessels containing HAN (cont.)	Chemical Safety Control	Ensure that concentration of nitric acid, metal impurities, and HAN introduced in the process are within safety limits.	Chemical Safety Control	Ensure that concentration of nitric acid, metal impurities, and HAN introduced in the process are within safety limits.	Chemical Safety Control	Ensure that concentration of nitric acid, metal impurities, and HAN introduced in the process are within safety limits.
	Hydroxylamine Nitrate Explosion Proc ess Vessels Containing HAN and Hydrazine Nitrate with NO$_x$ Addition	Explosion in AP process vessels containing HAN	Chemical Safety Control Offgas Treatment System	Ensure that concentrations of HAN, hydrazine nitrate, and hydrazoic acid are within the safety limits. Provide an exhaust path for the removal of offgases generated during the decomposition of HAN, hydrazine nitrate, and hydrazoic acid. Provide heat transfer/ pressure relief for affected process vessels.	Chemical Safety Control Offgas Treatment System	Ensure that concentrations of HAN, hydrazine nitrate, and hydrazoic acid are within the safety limits. Provide an exhaust path for the removal of offgases generated during the decomposition of HAN, hydrazine nitrate, and hydrazoic acid. Provide heat transfer/ pressure relief for affected process vessels.	Chemical Safety Control Offgas Treatment System	Ensure that concentrations of HAN, hydrazine nitrate, and hydrazoic acid are within the safety limits. Provide an exhaust path for the removal of offgases generated during the decomposition of HAN, hydrazine nitrate, and hydrazoic acid. Provide heat transfer/ pressure relief for affected process vessels.

Table 5.1-2 Safety Assessment Summary (Process Hazards)

Event Type	Event Group	Bounding Event	PSSC (SW & IOC)	Safety Function	PSSC (facility worker)	Safety Function	PSSC (environment)	Safety Function
Explosion (cont.)	Hydrogen Peroxide Explosion	Hydrogen peroxide explosion in AP vessels, tanks, and piping which results in an energetic breach of the vessels, tanks, and piping and in a loss of confinement and dispersal of nuclear materials	Chemical Safety Control	Ensure that explosive concentrations of hydrogen peroxide do not occur.	Chemical Safety Control	Ensure that explosive concentrations of hydrogen peroxide do not occur.	Chemical Safety Control	Ensure that explosive concentrations of hydrogen peroxide do not occur.
	Solvent Explosion	Solvents in AP vessels, tanks, and piping which results in an energetic breach of the vessels, tanks, and piping and in a loss of confinement and dispersal of nuclear materials	Process Safety Control Subsystem	Ensure the temperature of the solution containing solvents is limited within the safety limits.	Process Safety Control Subsystem	Ensure the temperature of the solution containing solvents is limited within the safety limits.	Process Safety Control Subsystem	Ensure the temperature of the solution containing solvents is limited within the safety limits.
			Process Cell Fire Prevention Features	Ensure that fires in process cells are highly unlikely.	Process Cell Fire Prevention Features	Ensure that fires in process cells are highly unlikely.	Process Cell Fire Prevention Features	Ensure that fires in process cells are highly unlikely.
			Offgas Treatment System	Provide an exhaust path for removal of gases in process vessels.	Offgas Treatment System	Provide an exhaust path for removal of gases in process vessels.	Offgas Treatment System	Provide an exhaust path for removal of gases in process vessels.

Table 5.1-2 Safety Assessment Summary (Process Hazards)

Event Type	Event Group	Bounding Event	PSSC (SW & IOC)	Safety Function	PSSC (facility worker)	Safety Function	PSSC (environment)	Safety Function
Explosion (cont.)	TBP Nitrate Explosion (Red Oil)	Process related chemical explosion involving red oil formation in the AP boiler, vessel, or tank results in loss of confinement and dispersal of nuclear materials	Offgas Treatment System	Provide an exhaust path for aqueous phase evaporative cooling in process vessels, thereby providing a mechanism for heat removal. Provide venting of vessels/equipment that potentially contain TBP and its associated by products to prevent over pressurization in the case of excessive oxidation of TBP and/or its degradation products.	Offgas Treatment System	Provide an exhaust path for aqueous phase evaporative cooling in process vessels, thereby providing a mechanism for heat removal. Provide venting of vessels/equipment that potentially contain TBP and its associated by products to prevent over pressurization in the case of excessive oxidation of TBP and/or its degradation products.	Offgas Treatment System	Provide an exhaust path for aqueous phase evaporative cooling in process vessels, thereby providing a mechanism for heat removal. Provide venting of vessels/ equipment that potentially contain TBP and it associated by products to prevent over pressurization in the case of excessive oxidation of TBP and/or its degradation products.

Table 5.1-2 Safety Assessment Summary (Process Hazards)

Event Type	Event Group	Bounding Event	PSSC (SW & IOC)	Safety Function	PSSC (facility worker)	Safety Function	PSSC (environment)	Safety Function
Explosion (cont.)	TBP Nitrate Explosion (Red Oil) (cont.)	Process related chemical explosion involving red oil formation in the AP boiler, vessel, or tank results in loss of confinement and dispersal of nuclear materials (cont.)	Process Safety Control Subsystem	Ensure that the temperature of solutions containing organics is restricted to temperatures within safety limits in order to limit the rate of energy generation. Ensure that the design basis heatup rate is not exceeded. Limit the residence time of organics in process vessels containing oxidizing agents and potentially exposed to high temperatures and in radiation fields.	Process Safety Control Subsystem	Ensure that the temperature of solutions containing organics is restricted to temperatures within safety limits in order to limit the rate of energy generation. Ensure that the design basis heatup rate is not exceeded. Limit the residence time of organics in process vessels containing oxidizing agents and potentially exposed to high temperatures and in radiation fields.	Process Safety Control Subsystem	Ensure the temperature of solutions containing organics is restricted to temperatures within safety limits in order to limit the rate of energy generation. Ensure that the design basis heatup rate is not exceeded. Limit the residence time of organics in process vessels containing oxidizing agents and potentially exposed to high temperatures and in radiation fields.
			Chemical Safety Control	Ensure a diluent is used that does not contain cyclic chain hydrocarbons.	Chemical Safety Control	Ensure a diluent is used that does not contain cyclic chain hydrocarbons.	Chemical Safety Control	Ensure a diluent is used that does not contain cyclic chain hydrocarbons.

Table 5.1-2 Safety Assessment Summary (Process Hazards)

Event Type	Event Group	Bounding Event	PSSC (SW & IOC)	Safety Function	PSSC (facility worker)	Safety Function	PSSC (environment)	Safety Function
Explosion (cont.)	AP Vessel Overpressurizati on Explosion	Overpressurizati on of AP tanks, vessels, and piping as a result of an increase of exothermic chemical reactions of solutions	Fluid Transport Systems	Ensure that process vessels, tanks, and piping are designed to prevent process deviations from creating overpressurizatio n events.	Fluid Transport Systems	Ensure that process vessels, tanks, and piping are designed to prevent process deviations from creating overpressurizatio n events.	Fluid Transport Systems	Ensure that process vessels, tanks, and piping are designed to prevent process deviations from creating overpressurization events.
			Offgas Treatment Systems	Provide an exhaust path for removal of gases in process vessels.	Offgas Treatment Systems	Provide an exhaust path for removal of gases in process vessels.	Offgas Treatment Systems	Provide an exhaust path for removal of gases in process vessels.
			Chemical Safety Controls	Ensure control of the chemical makeup of the reagents. Ensure segregation/ separation of vessels/ components from incompatible chemicals.	Chemical Safety Controls	Ensure control of the chemical makeup of the reagents. Ensure segregation/ separation of vessels/ components from incompatible chemicals.	Chemical Safety Controls	Ensure control of the chemical makeup of the reagents. Ensure segregation/ separation of vessels/ components from incompatible chemicals.

Table 5.1-2 Safety Assessment Summary (Process Hazards)

Event Type	Event Group	Bounding Event	PSSC (SW & IOC)	Safety Function	PSSC (facility worker)	Safety Function	PSSC (environment)	Safety Function
Explosion (cont.)	Pressure Vessel Over pressurization Explosion	Overpressurization of gas bottles, tanks, or receivers which could impact primary confinement and result in a release of radioactive material	Pressure Vessels Controls	Ensure that primary confinements are protected from the impact of pressure vessels failures.	Pressure Vessels Controls	Ensure that primary confinements are protected from the impact of pressure vessels failures.	Pressure Vessels Controls	Ensure that primary confinements are protected from the impact of pressure vessels failures.
	Hydrazoic Acid (HN₃) Explosion	Process related chemical explosion involving HAN/nitric acid vessels, tanks, and piping which results in the breach of AP vessels, tanks, and piping	Chemical Safety Control	Assure the proper concentration of hydrazine nitrate is introduced into the system. Ensure that hydrazoic acid is not accumulated in the process or propagated into the acid recovery and oxalic mother liquors recovery units.	Chemical Safety Control	Assure the proper concentration of hydrazine nitrate is introduced into the system. Ensure that hydrazoic acid is not accumulated in the process or propagated into the acid recovery and oxalic mother liquors recovery units.	Chemical Safety Control	Assure the proper concentration of hydrazine nitrate is introduced into the system. Ensure that hydrazoic acid is not accumulated in the process or propagated into the acid recovery and oxalic mother liquors recovery units.
			Process Safety Control Subsystem	Ensure that the temperature solutions potentially containing hydrazoic acid is limited to prevent an explosive concentration of hydrazoic acid.	Process Safety Control Subsystem	Ensure that the temperature solutions potentially containing hydrazoic acid is limited to prevent an explosive concentration of hydrazoic acid.	Process Safety Control Subsystem	Ensure that the temperature solutions potentially containing hydrazoic acid is limited to prevent an explosive concentration of hydrazoic acid.

Table 5.1-2 Safety Assessment Summary (Process Hazards)

Event Type	Event Group	Bounding Event	PSSC (SW & IOC)	Safety Function	PSSC (facility worker)	Safety Function	PSSC (environment)	Safety Function
Explosion (cont.)	Metal Azide Explosions	Chemical explosion involving an azide (other than hydrazoic acid in an AP boiler, vessel, or tank that results in an energetic breach of the AP boiler, vessel, or tank	Chemical Safety Control	Ensure that metal azides are not added to high temperature process equipment. Ensure that the sodium azide has been destroyed before transfer of the alkaline waste into the high alpha waste of the waste recovery unit.	Chemical Safety Control	Ensure that metal azides are not added to high temperature process equipment. Ensure that the sodium azide has been destroyed before transfer of the alkaline waste into the high alpha waste of the waste recovery unit.	Chemical Safety Control	Ensure that metal azides are not added to high temperature process equipment. Ensure that the sodium azide has been destroyed before transfer of the alkaline waste into the high alpha waste of the waste recovery unit.
			Process Safety Control Subsystem	Ensure that metal azides are not exposed to temperatures that can allow the energetic decomposition of azides. Limit and control conditions in which dryout can occur.	Process Safety Control Subsystem	Ensure that metal azides are not exposed to temperatures that can allow the energetic decomposition of azides. Limit and control conditions in which dryout can occur.	Process Safety Control Subsystem	Ensure that metal azides are not exposed to temperatures that can allow the energetic decomposition of azides. Limit and control conditions in which dryout can occur.

Table 5.1-2 Safety Assessment Summary (Process Hazards)

Event Type	Event Group	Bounding Event	PSSC (SW & IOC)	Safety Function	PSSC (facility worker)	Safety Function	PSSC (environment)	Safety Function
Explosion (cont.)	Pu (IV) Oxalate Explosion	Chemical explosion involving Pu (IV) oxalate in the calcining furnace results in an energetic breach of the furnace and glovebox and the dispersal of radioactive material	Chemical Safety Control	Measure the valence of plutonium before adding oxalic acid to the oxalic precipitation and oxidation unit to ensure that plutonium (IV) cannot be formed.	Chemical Safety Control	Measure the valence of plutonium before adding oxalic acid to the oxalic precipitation and oxidation unit to ensure that plutonium (IV) cannot be formed.	Chemical Safety Control	Measure the valence of plutonium before adding oxalic acid to the oxalic precipitation and oxidation unit to ensure that plutonium (IV) cannot be formed.
	Electrolysis Related Explosion	Explosion of hydrogen in the vapor space on the electrolyzer	Process Safety Control Subsystem	Limit the generation of hydrogen to ensure that the normality of the acid is sufficiently high to ensure that the offgas is not flammable. Monitor the electrolyzer for electrical faults that could result in arcing or other imparting of electrical energy with the risk of titanium fire.	Process Safety Control Subsystem	Limit the generation of hydrogen to. ensure that the normality of the acid is sufficiently high to ensure that the offgas is not flammable. Monitor the electrolyzer for electrical faults that could result in arcing or other imparting of electrical energy with the risk of titanium fire.	Process Safety Control Subsystem	Limit the generation of hydrogen to ensure that the normality of the acid is sufficiently high to ensure that the off gas is not flammable. Monitor the electrolyzer for electrical faults that could result in arcing or other imparting of electrical energy with the risk of titanium fire.

Table 5.1-2 Safety Assessment Summary (Process Hazards)

Event Type	Event Group	Bounding Event	PSSC (SW & IOC)	Safety Function	PSSC (facility worker)	Safety Function	PSSC (environment)	Safety Function
Explosion (cont.)	Electrolysis Related Explosion (cont.)	Explosion of hydrogen in the vapor space on the electrolyzer (cont.)	Electrolyzer Structure, Sintered Silicon Nitride Barrier, and PTFE insulator	To provide physical separation and electrical insulation between electrolyzer components and structural integrity and stability to the electrolyzer system. Ensures that the normality of the acid is sufficiently high to ensure that the offgas is not flammable.	Electrolyzer Structure, Sintered Silicon Nitride barrier, and PTFE insulator	To provide physical separation and electrical insulation between electrolyzer components and structural integrity and stability to the electrolyzer system. Ensures that the normality of the acid is sufficiently high to ensure that the offgas is not flammable.	Electrolyzer Structure, Sintered Silicon Nitride Barrier, and PTFE insulator	To provide physical separation and electrical insulation between electrolyzer components and structural integrity and stability to the electrolyzer system. Ensures that the normality of the acid is sufficiently high to ensure that the offgas is not flammable.
	Laboratory Explosion	Explosion within the BMF laboratory involving flammable, explosive, or reactive chemicals which results in a dispersal of radiological material	C3 Confinement	Provide filtration to mitigate dispersions from the C3 areas.	Chemical Safety Control	Ensures control of the chemical makeup of the laboratory reagents. Ensures segregation/separation of vessels/components from incompatible chemicals.	C3 Confinement	Provides filtration to mitigate dispersions from the C3 areas.

Table 5.1-2 Safety Assessment Summary (Process Hazards)

Event Type	Event Group	Bounding Event	PSSC (SW & IOC)	Safety Function	PSSC (facility worker)	Safety Function	PSSC (environment)	Safety Function
Explosion (cont.)	Laboratory Explosion (cont.)	Explosion within the BMF laboratory involving flammable, explosive, or reactive chemicals which results in a dispersal of radiological material (cont.)			Laboratory Materials Controls	Minimize the quantities of hazardous chemical/ radiological materials in the laboratory		
					Facility Worker Action	Ensures that facility workers take proper actions to limit radiological/ chemical exposures.		
	Outside Explosion	Explosions in the reagent processing building, gas storage area, emergency diesel generator building, and the access building	MOX Fuel Fabrication Building Structure	Maintain structural integrity and prevent damage to internal SSCs.	MOX Fuel Fabrication Building Structure	Maintain structural integrity and prevent damage to internal SSCs.	MOX Fuel Fabrication Building Structure	Maintain structural integrity and prevent damage to internal SSCs.
			Emergency Generator Building Structure	Maintain structural integrity and prevent damage to internal SSCs from external explosions.	Emergency Generator Building Structure	Maintain structural integrity and prevent damage to internal SSCs from external explosions.	Emergency Generator Building Structure	Maintain structural integrity and prevent damage to internal SSCs from external explosions.
			Waste Transfer Line	Prevent damage to the line from outside explosions.	Waste Transfer Line	Prevent damage to the line from outside explosions.	Waste Transfer Line	Prevent damage to the line from outside explosions

5-80

Table 5.1-2 Safety Assessment Summary (Process Hazards)

Event Type	Event Group	Bounding Event	PSSC (SW & IOC)	Safety Function	PSSC (facility worker)	Safety Function	PSSC (environment)	Safety Function
Explosion (cont.)	Outside Explosion (cont.)	Explosions in the reagent processing building, gas storage area, emergency diesel generator building, and the access building (cont.)	Hazardous Materials Delivery Controls	Ensures that the quantity of delivered hazardous material and its proximity to the outside explosion's PSSCs are controlled within the bounds of the values used to demonstrate that the consequences of outside explosions are acceptable.	Hazardous Materials Delivery Controls	Ensures that the quantity of delivered hazardous material and its proximity to the outside explosion's PSSCs are controlled within the bounds of the values used to demonstrate that the consequences of outside explosions are acceptable.	Hazardous Materials Delivery Controls	Ensures that the quantity of delivered hazardous materia and its proximity to the outside explosion's PSSCs are controlled within the bounds of the values used to demonstrate tha the consequences of outside explosions are acceptable.

Table 5.1-3 PSSCs and Design Basis Functions and Values Developed from the Safety Assessment

PSSC	Design Basis Safety Function	Design Basis Values	FSER Section
3013 Canister	Withstand the effects of design basis drops without breaching	DOE STD 3013 2000 Outer canister designed to withstand 9.14 m (30 ft) drop while remaining leak tight. Inner container designed to withstand 1.22 m (4 ft) drop while remaining leak tight. Outer container designed to withstand 4927 kPa (699 psig), inner container withstands 790 kPa (100 psig). Qualified 50 yr life.	11.7.1.2
3013 Transport Cask	Withstand design basis fire	Thermal design per 10 CFR 71.73, 800 °C (1472 °F) for 30 minutes.	7.1.5.2
	Withstand design basis drop	Designed for free drop, crushing, and puncture per 10 CFR 71.73	11.7.1.2
Backflow Prevention features	Prevent process fluids from backflowing into interfacing systems	ASME B31.3	11.8
C2 Confinement System Passive Barrier	Limit the dispersion of radioactive material	Two HEPA filter banks before discharge HEPA filter design temperature of 232 °C (450 °F) Automatic and manual fire rated dampers between designated fire areas In place HEPA filter testing for final discharge filtration assemblies System design in accordance with RG 3.12, except heat removal is by airflow dilution HEPA filter design, HEPA filter housing design, construction and testing, and HEPA filter housing isolation dampers in accordance with ASME N509; HEPA filter design and testing, HEPA filter housing design and testing; ductwork and pipe flexible connections, and fan design, construction, and testing in accordance with ASME AG 1 Sheet metal ductwork design, construction, and testing; "bubble tight" isolation damper construction; and testing; HEPA filter housing testing; and HEPA filter testing in accordance with ERDA 76 21; Filter testing in accordance with ASME N510 with each HEPA stage having a leakage efficiency of 99.95% Tornado dampers Final filters and downstream ductwork remain structurally intact during and after tornadoes and design basis earthquakes	11.4.1.3

5-82

Table 5.1-3 PSSCs and Design Basis Functions and Values Developed from the Safety Assessment

PSSC	Design Basis Safety Function	Design Basis Values	FSER Section
C3 Confinement System	Provide filtration to mitigate dispersions from the C3 areas	C3 zone pressure maintained at negative pressure with respect to atmosphere during normal operation and transients Two 100% capacity fans in C3 confinement system System design in accordance with RG 3.12, except heat removal is by airflow dilution HEPA filter design; HEPA filter housing design, construction and testing; and HEPA filter housing isolation dampers in accordance with ASME N509 HEPA filter design and testing; HEPA filter housing design and testing; ductwork and pipe flexible connections; and fan design, construction, and testing in accordance with ASME AG 1 Sheet metal ductwork design, construction, and testing; "bubble tight" isolation damper construction and testing; HEPA filter housing testing; and HEPA filter testing in accordance with ERDA 76 21 Filter testing in accordance with ASME N510 with each HEPA stage having a leakage efficiency of 99.95% Tornado dampers Fan power from normal (non PSSC), standby (non PSSC), and emergency (PSSC) supplies Remains operational after facility fires, tornadoes, and design basis earthquakes	11.4.1.3
	Remain operable during design basis fire and effectively filter any release	Spark arrestors (roughing filters) in each filtration assembly upstream of HEPA filters Fire rated dampers between designated fire areas In place HEPA filter testing for final discharge filtration assemblies HEPA filter design temperature of 232 °C (450 °F)	7.1.5.5 11.4.1.3
	Limit the dispersion of radioactive material	Designed to maintain exhaust safety function assuming single active component failure HEPA filter assembly release fraction of 1×10^4 Two, 100% capacity redundant assemblies of two HEPA filter banks before discharge	11.4.1.3
	Provide exhaust to ensure that temperature in the 3013 canister storage structure is maintained within design limits.	Maintain ambient temperatures with sufficient airflow in the canister storage structure Reliability design bases are as described above	11.4.1.3
	Provide cooling air exhaust from designated electrical rooms	Maintain ambient temperatures with sufficient airflow in the designated electrical rooms Reliability design bases are as described above	11.4.1.3
C4 Confinement System	Remain operable during design basis fire and effectively filter any release	Fire isolation valves where process room exhausts exit fire areas In place HEPA filter testing for final discharge filtration assemblies HEPA filter design temperature of 232 °C (450 °F)	7.1.5.5 11.4.1.3

Table 5.1-3 PSSCs and Design Basis Functions and Values Developed from the Safety Assessment

PSSC	Design Basis Safety Function	Design Basis Values	FSER Section
C4 Confinement System (cont.)	Maintain negative glovebox pressure between glovebox and interfacing systems	Same as above, as appropriate C4 zone pressure maintained at negative pressure with respect to C3 zone during normal operation and transients Redundant pressure sensors to maintain C4 pressures	11.4.1.3
	Maintain minimum inward flow through small glovebox releases	Same as above, as appropriate High capacity flow system [38.1 m/min (125 ft/min) in the event of glovebox breach to maintain negative pressure	11.4.1.3
	Ensure that C4 exhaust is effectively filtered	Same as above, as appropriate In place HEPA filter testing for final discharge filtration assemblies System design in accordance with RG 3.12, except heat removal is by airflow dilution HEPA filter design; HEPA filter housing design, construction and testing; and HEPA filter housing isolation dampers in accordance with ASME N509 HEPA filter design and testing; HEPA filter housing design and testing, ductwork and pipe flexible connections; and fan design, construction, and testing in accordance with ASME AG 1 Filter testing in accordance with ASME N510 with each HEPA stage having a leakage efficiency of 99.95%	11.4.1.3
	Operate to ensure that a negative pressure differential exists between the C4 glovebox and the C3 area.	Fan power from normal (non PSSC), standby (non PSSC), emergency (PSSC), and uninterruptible (PSSC) supplies Remains operational during facility fires and tornadoes and design basis earthquakes Four, 100% capacity fans in C4 discharge system Piping, valves, and fittings associated with gloveboxes in accordance with ASME B31.3	11.4.1.3
	Contain a chemical release within a glovebox and provide an exhaust path for removal of the chemical vapors	Contain chemicals within C4 and exhaust so that moderate chemical consequence limits are not exceeded outside	8.1.2.4.1
Chemical Safety Controls	Ensure that explosive concentrations of hydrogen peroxide do not occur.	Limit the received H_2O_2 solution concentrations to 35% or less	8.1.2.5.4
	Ensure a diluent is used that is not very susceptible to either nitration or radiolysis	Diluent does not contain cyclic hydrocarbons	8.1.2.5.5

5-84

Table 5.1-3 PSSCs and Design Basis Functions and Values Developed from the Safety Assessment

PSSC	Design Basis Safety Function	Design Basis Values	FSER Section
Chemical Safety controls (cont.)	Ensure that hydrazoic acid is not accumulated in the process or propagated to units that might lead to explosive conditions.	Maximum hydrazine concentration of 0.14M. Hydrazine yield of 39.3% or less. No hydrazoic acid accumulation into acid recovery and OML recovery units.	8.1.2.5.3
	Ensure metal azides are not introduced into high temperature process equipment	No addition to high temperature equipment. Tanks potentially containing azides not allowed to dry out.	8.1.2.5.3.3
	Ensure the sodium azide has been destroyed before the transfer of the alkaline waste to the waste recovery unit.	Azides completely destroyed (OM) before acidification	8.1.2.5.3.3
	Ensure the valance of the plutonium before oxalic acid addition is not VI.	Pu(VI) concentration will be low actual value to be derived at ISA stage	8.1.2.5.6
	Ensure that nitric acid, metal impurities, and HAN concentrations are controlled and maintained to within safety limits.	$[HNO_3] < 6$ M $[N_2H_4] \geq 0.1$ M $[HAN] < 2.5$ M	8.1.2.5.3.1
	Ensure concentrations of HAN, hydrazine nitrate, and hydrazoic acid are controlled to within safety limits.		8.1.2.5.3
	Ensure the proper concentration of hydrazine nitrate is introduced into the system.	Hydrazine is not added at concentrations exceeding 35% (as N_2H_4)	8.1.2.5.3.1

Table 5.1-3 PSSCs and Design Basis Functions and Values Developed from the Safety Assessment

PSSC	Design Basis Safety Function	Design Basis Values	FSER Section
Chemical Safety controls (cont.)	Ensure control of the chemical makeup of the reagents and ensure segregation/separation of vessels/components from incompatible chemicals.	Separation and prevention of mixing of incompatible reagents	8.1.2.5
Combustible Loading Controls	Limit combustibles in C2 filter area to ensure that the C4 final HEPA filters are not adversely impacted by a filter room fire.	Based on defense in depth principles and multiple layers of protection. Includes control of fixed combustibles by design and control of transient combustibles by design and during operations (through worker training, regular surveillance, and postings). Utilizes NFPA 801.	7.1.5.1
	Limit the quantity of combustibles in fire areas containing a storage glovebox and the secured warehouse such that any fire that may occur will not encompass a large fraction of the stored radiological material.	Same as above	7.1.5.1
	Limit combustible in areas containing 3013 canisters	Same as above	7.1.5.1
	Limit combustibles in a fire area containing 3013 transport casks	Same as above	7.1.5.1
	Limit combustibles in a fire area containing fuel rods.	Same as above	7.1.5.1
	Limit combustibles in areas containing MOX fuel transport casks.	Same as above	7.1.5.1
	Limit the quantity of combustibles in areas containing transfer containers.	Same as above	7.1.5.1

5-86

Table 5.1-3 PSSCs and Design Basis Functions and Values Developed from the Safety Assessment

PSSC	Design Basis Safety Function	Design Basis Values	FSER Section
Combustible Loading Controls (cont.)	Limit the quantity of combustibles in areas containing the pneumatic transfer system to ensure this system is not adversely impacted.	Same as above	7.1.5.1
Criticality controls	Prevent criticality events	1. Design of facility operations shall comply with the double contingency principle, as stated in ANSI/ANS 8.1. Nuclear criticality shall be made "highly unlikely" and the failure of each leg of double contingency shall be "unlikely." 2. Computer calculations shall not exceed a maximum k_{eff}, taking all uncertainties and biases into account. Description of calculational methods and their validation, or means of establishing subcritical margins if parameter limits are not based on computer calculations. 3. Facility operations shall be designed to be subcritical under both normal and credible abnormal conditions. Normal conditions will be considered to be those when all controlled parameters are at their controlled values and uncontrolled parameters at their worst credible values. Abnormal conditions shall consider the worst case upset for each loss of a control or controlled parameter. 4. Dominant nuclear criticality safety controlled parameters shall be specified for each major process and in their order of preference. 5. Design approach shall prefer engineered over administrative controls and passive over active engineered controls. 6. The facility shall have a criticality accident alarm system that complies with the requirements of 10 CFR 70.24. The detection system and its operating characteristics shall be described. 7. The management measures and how they are applied to each controlled parameter shall be described, along with the safety grades for criticality IROFS and the criteria used to assign these IROFS to individual safety grades. 8. A description of the organization and administration for NCS and the key elements of the NCS Program (including those in SRP Section 6.4.3.2). 9. A description of the technical practices used to determine limits and controls on each controlled parameter, in criticality safety evaluations, including what ANSI/ANS standards are being committed to in whole or in part. 10. Where moderation control is required for subcriticality, a description of the approach to designing the facility to meet both fire safety and criticality safety requirements (including presence and type of fire suppression).	6.1.3.4.1 6.1.4.3
Double Walled Pipe	Prevent leaks from pipes containing process fluids from leaking into C3 areas.	ANSI/ASME B31.3	11.8

Table 5.1-3 PSSCs and Design Basis Functions and Values Developed from the Safety Assessment

PSSC	Design Basis Safety Function	Design Basis Values	FSER Section
Electrolyzer Structure	Provides support for the electrolysis pot.	Seismically designed, designed to withstand turbulent flow, maintains geometry	11.2.1.3.4
Emergency AC Power System	Provide ac power to emergency dc system battery charger	Overall design per IEEE Std 308 1991 and RG 1.32 (Rev.2). Environmental and seismic qualification per ANSI/AISC N690 1994, ASCE 4 98, IEEE Std 323 1983, IEEE Std 344 1987, RG 1.61, and RG 1.100 (Rev.2). Designed for single failure per IEEE Std 379 1994. Electrical independent and separation per IEEE 384 1992 and RG 1.75 (Rev. 2). Periodic testing per IEEE Std 338 1987 and RG 1.118 (Rev. 3). Electrical cables in open trays qualified per IEEE Std 383 1974. Equipment protection based on IEEE Std 741 1997. Battery design and installation per IEEE Std 484 1996. Emergency diesel generators with overall design per IEEE Std 387 1995 and RG 1.9 (Rev. 3) and fuel oil per ANSI/ASTM D975 94. Overall design of uninterruptible power supplies per IEEE Std 944 1986.	11.5.1.3.1
	Provide ac power to emergency diesel generator fuel oil system	Same as above	11.5.1.3.1
	Provide ac power to high depressurization exhaust system	Same as above	11.5.1.3.1
	Provide ac power to emergency control room air conditioning system	Same as above	11.5.1.3.1
	Provide ac power to C4 confinement system	Same as above	11.5.1.3.1
	Provide ac power to emergency diesel ventilation system	Same as above	11.5.1.3.1
	Provide ac power to emergency control system	Same as above	11.5.1.3.1
	Provide ac power to seismic monitoring and trip system and seismic isolation valves.	Same as above	11.5.1.3.1
	Provide ac Power to process cell exhaust system	Same as above	11.5.1.3.1

5-88

Table 5.1-3 PSSCs and Design Basis Functions and Values Developed from the Safety Assessment

PSSC	Design Basis Safety Function	Design Basis Values	FSER Section
Emergency Control Room Air Conditioning System	Ensure habitable conditions for operators	One, 100% capacity filtration stage (using prefilter stage, two HEPA filter stages, and chemical filters) for each control room air supply	11.4.1.3
		One, 100% capacity air conditioning unit for each engineering control room and for each emergency electrical battery room	8.1.2.6
		Two, 100% capacity exhaust fans for each emergency battery room	7.1.5.4
		Designed to maintain protection assuming single component failure	
		HEPA filter design temperature of 232 °C (450 °F)	
		Tornado dampers prevent pressurization in supply air system	
		In place HEPA filter testing for final discharge filtration assemblies	
		System design in accordance with RG 3.12	
		HEPA filter design; HEPA filter housing design, construction and testing; and HEPA filter housing isolation dampers in accordance with ASME N509	
		HEPA filter design and testing; HEPA filter housing design and testing; ductwork and pipe flexible connections; and fan design, construction, and testing in accordance with ASME AG 1	
		Sheet metal ductwork design, construction, and testing; "bubble tight" isolation damper construction and testing; HEPA filter housing testing; and HEPA filter testing in accordance with ERDA 76 2;	
		Filter testing in accordance with ASME N510 with each HEPA stage having a leakage efficiency of 99.95%	
		Fan power from normal (non PSSC), standby (non PSSC), and emergency (PSSC) supplies	
		Remains operational during and after facility fires and after tornadoes and design basis earthquakes	
		Fresh air inlets are located so that the presence of contaminants are minimized (NFPA 801)	
Emergency Control System	Provide controls for high depressurization exhaust system	Two redundant, separate, and independent trains. Fundamental design as per IEEE 603 1998. Electrical independence and separation as per IEEE 384 1992 and RG 1.75 (Rev. 2). Single failure criteria as per IEEE 379 1994. Instrument setpoints as per ANSI/ISA 67.04.01 2000, and RG 1.105 (Rev. 3). Designed to function during design basis event as per ANSI/AISC N690 1994, ASCE 4 98, IEEE Std 323 1983, IEEE Std 344 1987, and RG 1.100 (Rev.2). Software programmable electronic systems per EPRI Topical Report TR 106439 (with NRC safety evaluation), IEC 61131 3 (1993 03), IEEE Std 7 4.3.2 1993, IEEE Std 730 1998, IEEE Std 828 1998, IEEE Std 830 1998, IEEE Std 1012 1998, IEEE Std 1028 1997, IEEE Guide 1042 1987, IEEE Std 1074 1997, IEEE Std 1228 1994, NUREG/CR 6090, NUREG/CR 6463, RG 1.168, RG 1.169, RG 1.172, and RG 1.173. Human system interface per IEEE Std 338 1987, NUREG 1023 1988 and NUREG 0700 (Branch Technical Position HICB 17), and RG 1.118 (Rev. 3). Reduction of electromagnetic and radiofrequency interference per IEEE Std 518 1982, IEEE St Std 1050 1996, and RG 1.180 with the design of data communications networks per ANSI/IEEE 802.3.	11.6.1.3.1

Table 5.1-3 PSSCs and Design Basis Functions and Values Developed from the Safety Assessment

PSSC	Design Basis Safety Function	Design Basis Values	FSER Section
Emergency Control System (continued)	Provide controls for C4 confinement system	Same as above	11.6.1.3.1
	Provide controls for emergency control room air conditioning system	Same as above	11.6.1.3.1
	Provide controls for emergency ac system	Same as above	11.6.1.3.1
	Provide controls for emergency dc system	Same as above	11.6.1.3.1
	Provide controls for emergency generator ventilation system	Same as above	11.6.1.3.1
	Provide controls for emergency diesel generator fuel oil system	Same as above	11.6.1.3.1
	Shutdown process on loss of power	Same as above	11.6.1.3.1
	Shut down and isolate process and systems (as necessary) in response to an earthquake	Same as above for seismic monitoring and trip system	11.6.1.3.1
	Provide controls for process cell exhaust system	Same as above	11.6.1.3.1
Emergency DC System	Provide dc power for high depressurization exhaust system	Overall design per IEEE Std 308 1991, IEEE Std 946 1992, and RG 1.32 (Rev.2). Environmental and seismic qualification per ANSI/AISC N690 1994, ASCE 4 98, IEEE Std 323 1983, IEEE Std 344 1987, and RG 1.100 (Rev.2). Designed for single failure per IEEE Std 379 1994. Electrical independent and separation per IEEE 384 1992 and RG 1.75 (Rev. 2). Periodic testing per IEEE Std 338 1987, IEEE Std 450 1995, and RG 1.118 (Rev. 3). Battery design and installation per IEEE Std 484 1996, IEEE Std 485 1997, and NFPA 111.	11.5.1.3.2
	Provide dc power for C4 confinement system	Same as above	11.5.1.3.2

Table 5.1-3 PSSCs and Design Basis Functions and Values Developed from the Safety Assessment

PSSC	Design Basis Safety Function	Design Basis Values	FSER Section
Emergency DC System (continued)	Provide dc power for emergency ac power system controls	Same as above	11.5.1.3.2
	Provide dc power for emergency control room air conditioning system	Same as above	11.5.1.3.2
	Provide dc power for emergency control system	Same as above	11.5.1.3.2
	Provide dc power for emergency generator ventilation system	Same as above	11.5.1.3.2
	Provide dc power to process cell exhaust system	Same as above	11.5.1.3.2
Emergency Diesel Generator Structure	Maintain structural integrity and prevent damage to internal SSCs from external fires, external explosions, earthquakes, extreme winds, tornadoes, missiles, rain, and snow and ice loadings	[Text removed under 10 CFR 2.390]	7.1.5.4 11.1.1.3
Emergency Diesel Generator Ventilation System	Provide emergency diesel generator ventilation	One, 100% capacity air conditioning unit for each switchgear room One , 100% capacity roof ventilator for engine room cooling during standby (engine fan cools room duing engine operation) Fan power from normal (non PSSC), standby (non PSSC), and emergency (PSSC) supplies Remains operational after facility fires, tornadoes, and design basis earthquakes;	11.4.1.3
Emergency Diesel Generator Fuel Oil System	Provide emergency diesel generator fuel oil for the emergency diesels	7 days plus margin fuel storage tank, day tanks 2498 L (660 gal), dual 100% transfer pumps, strainers, dual cartridge filters, isolation and maintenance valves. IEEE 344 1987, RG 1.100 (Rev. 2), IEEE 308 1991, ANS 59 51 1997, ASTM D75 94, NFPA 37, NFPA 110	11.9.1.1, 11.9.1.3
Facility Worker Action	Ensure that facility workers take proper action to limit chemical and/or radiological exposure	Facility worker response to exit the affected area.	9.1.2.4

Table 5.1-3 PSSCs and Design Basis Functions and Values Developed from the Safety Assessment

PSSC	Design Basis Safety Function	Design Basis Values	FSER Section
Facility Worker Controls	Ensure that facility workers take proper actions before bagout operations to limit radiological exposure	Facility worker prejob preparation to prevent and/or limit dose during tasks involving transient primary confinements or maintenance	9.1.2.4
	Ensure that facility workers take proper actions before maintenance activities to limit radiological exposure	Same as above	9.1.2.4
Fire Barriers	Contain fires within fire area	Minimum rating of 2 hours. Constructed in accordance with NFPA 221 1997. Fire doors are designed in accordance with NFPA 80 1999. Fire damper per UL 555 1995. Barrier selection and penetration seal program per NFPA 801 1998.	7.1.5.6
Fire Detection and Suppression	Support fire barriers as necessary	Detection and alarm per NFPA 72 1996 Suppression per NFPA 2001 1996 (clean agent) where dispersable fissile material is present	7.1.5.7
Fluid Transport System	Prevent overpressurization	ASME Section Boiler & Pressure Vessel Code VIII ASME B31.3 Effectiveness for reactive chemicals (HAN, red oil) not specified	11.8.1.3
	Withstand as necessary the effects of the DBE such that confinement of radionuclides is maintained	Seismic Category I design as per seismic qualification program	11.8
Glovebox	Maintain confinement integrity for design basis impacts	Leak integrity 2.5×10^3 vol/hr @ 500 Pa. Impact resistant windows, glovebox floor designed to withstand load drops. Internal guides and barriers to prevent fall of containers. Have pressure relief devices. Welding per AWS D9.1 1998.	11.7.1.2
Glovebox Pressure controls	Maintain glovebox pressure within design limits	Redundant pressure sensors to monitor differential pressures and provide alarm Remains operational after facility fires in nonaffected areas, tornadoes, and design basis earthquakes	11.4.1.3
Guide Sleeves	Provides insulation/separation between anode and titanium shell of electrolyzer	Capable of withstanding environmental conditions in electrolyzer	11.2.1.3.4

Table 5.1-3 PSSCs and Design Basis Functions and Values Developed from the Safety Assessment

PSSC	Design Basis Safety Function	Design Basis Values	FSER Section
Hazardous Material Delivery Controls	Ensure that the quantity of delivered hazardous material and its proximity to the MFFF building structure, EDG building structure, and the waste transfer line are controlled to within the bounds of the values used to demonstrate that the consequences of outside explosions are acceptable	Limit quantities and distances of deliveries so that they are within the bounds found acceptable in the safety analyses performed for potential explosives in F area	5.1.6.2
Instrument Air System (Emergency Scavenging Air)	Provide sufficient scavenging airflow to dilute the hydrogen produced by radiolysis such that an explosive condition does not occur	Limiting hydrogen concentration to 1% or less. Two, 100% capacity banks of compressed air available. Will be constructed to ASME B&PV and B31.3 standards. Also RG 1.100 or IEEE 344.	11.9.1.1, 11.9.1.2, 11.2.1.11, 8.1.2.5.1.2
Laboratory Material Controls	Minimize quantities of hazardous chemicals in the laboratory	Procedures will be established to limit sample size, number, and reagent quantity, in accordance with safe laboratory operating practices	8.1.2.1.3
	Minimize quantities of radioactive materials in the laboratory	Procedures will be established to limit sample size, number, and reagent quantity, in accordance with safe laboratory operating practices	8.1.2.1.3
Maintenance Activity Controls	Isolation of power to the electrolyzer when electrolyzer is drained	Administrative controls	11.2.1.3.4
Material Handling Controls	Ensure proper handling of primary confinement types outside of gloveboxes	Management measures include training and qualification of personnel, approved procedures, including precautions and limitations, use of proper equipment, testing, and surveillances	11.7.1.2
	Ensure that design bases lift heights are not exceeded	MOX fresh fuel casks 9.14 m (30 ft) drop MOX waste containers 1 m (3.28 ft) drop 3013 outer can 9.14 m (30 ft) drop	11.7.1.2

Table 5.1-3 PSSCs and Design Basis Functions and Values Developed from the Safety Assessment

PSSC	Design Basis Safety Function	Design Basis Values	FSER Section
Material Handling Controls (cont.)	Prevent load handling activities that could potentially lead to a breach in the final C4 HEPA filters	PSSC for structural protection is C4 confinement system	11.4.1.3
	Prevent impacts to the inside or outside of glovebox during normal operations	Engineered equipment used to reduce likelihood of failures causing glovebox breaches	11.7.1.2
	Prevent potential overpressurization of the reusable plutonium dioxide cans caused by radiolysis or oxidation of Pu (III) oxalate and its subsequent impact to the glovebox	The reusable can is designed to withstand the maximum pressure attainable from radiolysis and plutonium (III) oxalate reactions, plus 10%	11.3.1.2.3
	Prevent load handling events that could breach primary confinements	During normal operations; material handling equipment, material handling controls, and the glovebox will prevent breaches. During maintenance operations the above plus training and procedures will be used.	11.7.1.2
Material Handling Equipment	Limit damage to fuel rods/assemblies during handling	Designed using hardware stops, limit switches, speed controllers, bumpers to limit travel of equipment. Will fail to safe condition upon loss of power.	11.7.1.1, 11.7.1.2
	Prevent impacts to the glovebox through the use of engineered equipment	Same as above	same
Material Maintenance and Surveillance Programs	Detect and limit the damage resulting from corrosion	Corrosion allowance in accordance with rational code practices	11.8.1.3
BMF Tornado Dampers	Protect BMF ventilation systems from differential pressure effects of the tornado	Designed per ASME AG 1	11.4.1.3

Table 5.1-3 PSSCs and Design Basis Functions and Values Developed from the Safety Assessment

PSSC	Design Basis Safety Function	Design Basis Values	FSER Section
Missile Barriers	Protect BMF, BEG, and UEF building internal SSCs from damage caused by tornado or wind driven missiles	Designed for impacts from: [Text removed under 10 CFR 2.390]	11.1.1.3.2.1
MOX Fuel Fabrication Building Structure (including vent stack)	Maintain structural integrity and prevent damage to internal SSCs from external events	[Text removed under 10 CFR 2.390]	7.1.5.4 11.1.1.3.2.2
	Withstand the effects of load drops that could potentially impact radiological material	NUREG 0612	11.1.1.3.2.1
MOX Fuel Transport Cask	Withstand the design basis fire	Thermal design per 10 CFR 71.73, 800 °C (1,472 °F) for 30 minutes	7.1.5.2
	Withstand the effects of design basis drops without release of radioactive material	Mechanical design per 10 CFR 71.73, certified to withstand 9.14 m (30 ft) drop	11.7.1.2
Offgas Treatment System	Provide an exhaust path for the removal of gases in process vessels	Process vessels do not pressurize Final filtration units in accordance with Table 11.4 1 of FSER	11.4.1
	Provide an exhaust path for aqueous phase evaporative cooling in process vessels, thereby providing a mechanism for heat removal	Relieve 1.2 times the combination of energy generation and energy input to the system	8.1.2.5.5
	Provide venting of vessels/equipment that potentially contain TBP and its associated byproducts to prevent overpressurization in the case of excessive oxidation of TBP and/or its degradation products	Vent sizes > 8 × 10^3 mm^2/g (5.6 × 10^{-3} in^2/lb) of organic material present in the system	8.1.2.5.5

5-95

Table 5.1-3 PSSCs and Design Basis Functions and Values Developed from the Safety Assessment

PSSC	Design Basis Safety Function	Design Basis Values	FSER Section
Polytetrafluoro ethylene Insulator (PTFE)	Provides insulation/separation between cathode and anode structures, and anode and ground of electrolyzer	Capable of withstanding environmental conditions in electrolyzer	11.2.1.3.4
Pressure Vessel Controls	Ensure that primary confinements are protected from the impact of pressure vessel failures (bulk gas, breathing air, service air, and instrument air systems)	Limited by ASME Section VIII and ASME B31.3 code design	11.8.1.3 11.9.1.2
Process Cells	Contain fluid leaks within process cells	Fully welded, designed to handle maximum inventory of largest vessel in cell	11.7
Process Cell Entry Controls	Prevent the entry of personnel into process cells during normal operations	Engineered and administrative controls, including radiation work permits, signs and postings, and barricades to restrict access.	9.1.2.3
	Ensure that workers do not receive a dose in excess of limits while performing maintenance	Same as above and procedures which implement 10 CFR 20.1602 controls for very high radiation areas	9.1.2.3
Process Cell Exhaust System	Effectively filter process cell exhaust	Fire protection and filtration design bases as per Table 11.4 of this FSER	11.4.1.3
	Operation to ensure that a negative pressure exists between the process cell areas and the C2 areas	Process cell pressure maintained subatmospheric during normal and transient operation. Designed to maintain a safety function assuming single component failure.	11.4.1.3
Process Cell Fire Prevention Features	Ensures that fires in process cells are highly unlikely	Combustible loading fire controls per NFPA 801 Ignition source controls	7.1.5.3

5-96

Table 5.1-3 PSSCs and Design Basis Functions and Values Developed from the Safety Assessment

PSSC	Design Basis Safety Function	Design Basis Values	FSER Section
Process Safety Control Subsystem	Prevent the formation of an explosive mixture of hydrogen within the MFFF associated with the use of the hydrogen argon gas.	I & C aspects same as emergency control system 25% of the LFL	11.6.1.3.1 8.1.2.5.1.1
	Ensure isolation of sintering furnace humidifier waterflow on high water level.	I & C aspects same as emergency control system No overflow or liquid water into furnace	11.6.1.3.1 11.3.1.2.4
	Ensure the temperature of solutions containing HAN is limited to temperatures within safety limits	I & C aspects same as emergency control system Control flow of HAN/hydrogen solution such that chemical consequence limits are not exceeded in the oxidation column Maintain temperature of HAN within safety limits	11.6.1.3.1 11.3.1.2.4 8.1.2.5.3.1
	Ensure the temperature of solutions containing organics is limited to temperatures within safety limits	I & C aspects same as emergency control system Temperature limited to less than 135 °C (275 °F)	11.6.1.3.1 8.1.2.5.5
	Limit the residence time of organics in process vessels containing oxidizing agents and potentially exposed to high temperatures and in radiation fields	I & C aspects same as emergency control system Residence time of concern is on the order of years. Actual limit will be developed in the ISA.	11.6.1.3.1 8.1.2.5.5
	Ensure the temperature of solutions potentially containing hydrazoic acid is limited to prevent an explosive concentration of hydrazoic acid from developing	I & C aspects same as emergency control system Temperature not exceeding 60 °C (140 °F)	11.6.1.3.1 8.1.2.5.3
	Limit and control conditions under which dryout can occur	I & C aspects same as emergency control system Tanks potentially containing azides not left dry	11.6.1.3.1 8.1.2.5.3

5-97

Table 5.1-3 PSSCs and Design Basis Functions and Values Developed from the Safety Assessment

PSSC	Design Basis Safety Function	Design Basis Values	FSER Section
Process Safety Control Subsystem (cont.)	Ensure the temperature of solutions potentially containing metal azides is insufficient to overcome the activation energy needed to initiate the energetic decomposition of the azide	I & C aspects same as emergency control system Temperature is not to exceed 140 °C (284 °F).	11.6.1.3.1 8.1.2.5.3.3
	Ensure the normality of the nitric acid is sufficiently high to ensure that the offgas is not flammable and to limit excessive hydrogen production	I & C aspects same as emergency control system Limit hydrogen concentration to under 25% of LFL; nitric acid normality not specified Electrical design basis not identified	11.2.1.5
	Warn operators of glovebox pressure discrepancies before exceeding differential pressure limits	I & C aspects same as emergency control system Redundant pressure sensors monitor differential pressure with respect to the process room and alert the operators to upset conditions. The instruments remain operational following facility fires in unaffected areas, tornadoes, and design basis earthquakes.	11.6.13.1 11.4.1.3
	Shut down process equipment before exceeding temperature safety limits	I & C aspects same as emergency control system Temperature safety limits specified in CAR	11.6.1.3.1 8.1.2.5.2.3.1
	Ensure the temperature of solutions containing solvents is limited to temperatures within safety limits	I & C aspects same as emergency control system Temperature limits will control maximum vapor concentrations	11.6.1.3.1 8.1.2.5.2
	Monitor the electrolyzer for electrical faults that could result in arcing or other imparting of electrical energy with the risk of titanium fire	Detection of 10 ma	11.2.1.3.4
	Ensure the flow rate of nitrogen dioxide/dinitrogen tetroxide is limited to the oxidation column of the purification cycle	I & C aspects same as emergency control system Concentration of NO_x at site worker location < TEEL 2	11.6.1.3.1 8.1.2.4.1

5-98

Table 5.1-3 PSSCs and Design Basis Functions and Values Developed from the Safety Assessment

PSSC	Design Basis Safety Function	Design Basis Values	FSER Section
Seismic Monitoring System and Associated Seismic Isolation valves	Prevent fire and criticality as a result of an uncontrolled release of hazardous material and water within the BMF building in the event of an earthquake	Seismic monitoring and trip system are same as emergency control system	11.6.1.3
Sintered Silicon Nitride Barrier	Physically separate the cathode from the anode in the nitric acid solution by serving as a dielectric barrier in the electrolyzer	Nonconductive, high dielectric strength, sufficient porosity to allow ionic current transfer but insufficient porosity to allow mixing of the anode and cathode solutions	11.2.1.3.4
Sintering Furnace	Provide a primary confinement boundary against leaks into the C3 areas	Seals designed for peak temperature of 316 °C (601 °F). Furnace shell and airlocks designed to withstand an overpressure of 2,552 g/cm^2 (36.3 psi). Leak tightness is 5×10^{5} leaked vol/hr at 155 g/cm^2 (2.2 psi). Controls prevent overpressure. Furnace is designed to maintain confinement function during design basis earthquake.	11.4.1.3
Sintering Furnace Pressure Controls	Maintain sintering furnace pressure within design limits	Same as process safety control subsystem	11.6.1.3
Supply Air System	Provide unconditioned emergency cooling air to the storage vault and designated electrical rooms	Provide supply air for emergency cooling HEPA filter stages for static confinement HEPA filter design temperature of 232 °C (450 °F) System design in accordance with RG 3.12 Sheet metal ductwork design, construction, and testing; "bubble tight" isolation damper construction and testing; HEPA filter housing testing; and HEPA filter testing in accordance with ERDA 76 21 Filter testing in accordance with ASME N150 with each HEPA stage having a leakage efficiency of 99.95%	11.4.1.3
Transfer Container	Withstand the effects of design basis drops without breaching	Designed to withstand 9.14 m (30 ft) drop; DOE STD 3013 2000	11.7.1.27

Table 5.1-3 PSSCs and Design Basis Functions and Values Developed from the Safety Assessment

PSSC	Design Basis Safety Function	Design Basis Values	FSER Section
Waste Container	Ensure that hydrogen buildup in excess of limits does not occur while providing appropriate confinement of radioactive particles	Meet 49 CFR 178.350 requirements for certification; withstand 1 m (3.3 ft) drop	11.7.1.3
Waste Transfer Line	Ensure that the waste transfer line is protected from activities taking place outside of the BMF	Double walled SS piping with leak detection designed to DBE (0.2g hor, 0.13g vertical) RG 3.10, ANSI N13.10 1974, ANSI N317 1980, ASME B31.3	11.8.1.3
	Prevent damage to the line from external events	Same as above	11.8.1.3

5-100

6. NUCLEAR CRITICALITY SAFETY

6.1 Conduct of Review

This chapter of the final safety evaluation report (FSER) contains the staff's review of the nuclear criticality safety (NCS) analysis performed by the applicant, Duke Cogema Stone & Webster (DCS), in Chapter 6 of the revised construction authorization request (CAR) (DCS, 2002d). The objective of this review is to (1) ensure that special nuclear material (SNM) storage and processing remains subcritical under normal and credible abnormal conditions during all operations, transfers, and storage at the mixed oxide (MOX) fuel fabrication facility (MFFF or the facility) and (2) determine whether the principal structures, systems, and components (PSSCs) and their design bases identified by the applicant provide reasonable protection against natural phenomena and the consequences of potential accidents. The staff evaluated the information provided by the applicant for NCS by reviewing Chapter 6 of the revised CAR, other sections of the revised CAR, and supplementary information provided by the applicant. The review of NCS design bases and strategies was closely coordinated with the review of chemical process safety, the safety assessment of the design bases (see Chapter 5 of this FSER), and the review of other plant systems.

The staff reviewed how the NCS information in the revised CAR addresses or relates to the following regulations, which are the top-level criticality safety requirements to be factored into the technical practices used in designing the facility:

- Title 10, Section 70.23(b), of the *Code of Federal Regulations* (10 CFR 70.23(b)) states, as a prerequisite to construction approval, that the design bases of the PSSCs and the quality assurance program be found to provide reasonable assurance of protection against natural phenomena and the consequences of potential accidents.

- 10 CFR 70.24, "Criticality Accident Requirements," requires licensees authorized to possess specified quantities of SNM to have a criticality accident alarm system (CAAS).

- 10 CFR 70.61(d) requires that the risk of nuclear criticality accidents be limited by assuring that, under normal and credible abnormal conditions, all nuclear processes are subcritical, including use of an approved margin of subcriticality for safety. Preventive controls and measures must be the primary means of protection against nuclear criticality accidents.

- 10 CFR 70.64, "Requirements for New Facilities or New Processes at Existing Facilities," requires that baseline design criteria (BDC) and defense-in-depth practices be incorporated into the design of new facilities. With respect to NCS, the design of new facilities must provide for criticality control, including adherence to the double contingency principle (DCP) pursuant to 10 CFR 70.64(a)(9).

The review for this construction approval focused on the design bases of the aqueous polishing (AP) and MOX process (MP) systems, their components, and other related information. The review also encompassed proposed design basis considerations, such as redundancy and independence. The staff used Chapter 6.0 in NUREG-1718, "Standard Review Plan for the Review of an Application for Mixed Oxide (MOX) Fuel," issued August 2000 (NRC, 2000), as guidance in performing the review. Sections 6.3.1 to 6.3.3 of NUREG-1718 discuss NCS organization and administration, management measures, and technical practices.

6.1.1 Organization and Administration

Sections 6.3.1 and 6.5.1 of NUREG-1718 state that the staff should review the applicant's commitment to establish organization and administration methods to determine whether the applicant has identified the responsibilities and authorities for organizations and individuals implementing the NCS program. For construction authorization, the design bases include a description of the roles and responsibilities of the NCS function related to the design of the facility, as well as NCS staff education and experience levels. Certain aspects of the NCS program would only be relied on during operation of the facility, such as audits and assessments, event investigations, and emergency response, while other aspects are relied on during design. However, because specific items relied on for safety (IROFS) need not be identified at the construction authorization stage, significant reliance is placed on the NCS program to provide reasonable assurance that adequate controls will be established to protect against an accidental criticality, in the event that the facility is licensed to operate.

Revised CAR Section 6.1.1 (DCS, 2002d) contains information relating to the organization and administration of the NCS program during the design phase of the proposed facility, including a description of its relevant roles and responsibilities. The roles and responsibilities during design basically consist of establishing design criteria, supporting the integrated safety analysis (ISA) process, validating calculational methodologies, performing analyses, and designing criticality control systems. Section 6.4.3.1 of NUREG-1718 states that the NCS function should be independent of operations to the extent practical. However, during the construction phase, the staff considers it appropriate that the NCS function be part of the design team to ensure that criticality safety features are designed into the proposed facility.

Revised CAR Section 6.1.2 only discusses the NCS program during the operations phase, not the construction phase. The staff considered only commitments relating to the NCS program during the construction phase to be part of the design bases of the facility and therefore did not review the material in CAR Section 6.1.2.

Commitments to specific American National Standards Institute/American Nuclear Society (ANSI/ANS) NCS standards (ANSI/ANS-8.1-1983 (R1988), "Nuclear Criticality Safety in Operations with Fissionable Materials Outside Reactors," revised 1988 (ANSI, 1988), and ANSI/ANS-8.19-1996, "Administrative Practices for Nuclear Criticality Safety," issued 1996 (ANSI, 1996)) are discussed in FSER Section 6.1.4.

The applicant provided qualifications for key NCS positions during design (DCS, 2002d, 2003c). The staff evaluated the proposed qualifications (education and experience levels) for the NCS function manager, senior NCS engineer, and NCS engineer. The staff concluded that the education levels for these positions are sufficient to provide the requisite skills and knowledge to perform their technical duties and equivalent to the education level of many of the practicing specialists in the NCS community.

Much of criticality safety is practiced by "skill-of-the-craft" and requires an intuitive understanding of the neutron physics and margins of safety for the materials and systems being evaluated.

6-2

Moreover, comparable industry experience is likely to consist of experience at low- and high-enriched uranium (HEU) fuel fabrication and enrichment plants. The staff also recommended that plutonium- or MOX-specific experience be sought for at least senior NCS positions. The applicant provided a summary of its criticality safety experience base involved in the design work (DCS, 2003b). This included significant participation by COGEMA, Inc., and SGN (a subsidiary of COGEMA) staff with over 20 years of experience with plutonium and MOX processing at MELOX and La Hague, France, as well as staff with experience in the domestic nuclear industry. In addition, the applicant has committed to training the criticality staff in the appropriate characteristics of plutonium and MOX materials. This commitment, combined with the significant in-house experience with criticality safety associated with plutonium and MOX processing, should ensure a sufficient knowledge base suitable for the design of the facility. Therefore, the staff finds the qualifications and training requirements for new NCS staff, and the experience of current NCS staff, adequate for the design of the facility.

6.1.2 Management Measures

Sections 6.3.2 and 6.5.1 of NUREG-1718 (NRC, 2000) state that the staff should review the applicant's commitment to establish management measures in support of the applicant's ability to implement and maintain the NCS program and to ensure the continued availability and reliability of IROFS. The applicant is not required to identify specific IROFS in the revised CAR (DCS, 2002d); therefore, the specific management measures to be applied to them cannot yet be specified. However, the staff reviewed the quality level definitions and the associated statements pertaining to quality levels of individual IROFS in Section 2.2, "Graded Quality Assurance," of the applicant's MOX Process Quality Assurance Plan (MPQAP). As discussed in FSER Chapter 15, the staff approved Revision 2 of the MPQAP in its safety evaluation report issued on October 1, 2001, and approved Revision 3 of the MPQAP in its letter to DCS dated January 10, 2003. The staff considered a description of the management measures and quality levels to be part of the design basis of the facility. These quality-level definitions were approved by the U.S. Nuclear Regulatory Commission (NRC) (NRC, 2001b). Questions remained, however, concerning the examples of how these quality levels would be applied in practice (i.e., application to different classes of IROFS), as discussed below.

The controlled parameters have been identified as part of the design basis of the proposed facility for NCS, but the specific controls to be relied on have not. They will be specified later as IROFS by the applicant and identified in the ISA summary as part of the license application (to be submitted if the revised CAR is approved). Therefore, the means of providing for criticality control (as required by 10 CFR 70.64(a)(9)) are described at the parameter level for the construction review. The only specific structure, system, or component (SSC) addressed in the revised CAR is the criticality alarm system required by 10 CFR 70.24 (discussed in FSER Section 6.1.3.2). Because there are no specific controls identified, the application of specific management measures (with quality levels as approved in Section 2.2 of the MPQAP) to IROFS will be reviewed later as part of the license application.

Specific administrative management measures applicable to NCS (e.g., training, procedures, and audits and assessments) are not applicable to the design and construction of the facility. Commitments to specific NCS ANSI/ANS-8 standards (ANSI/ANS-8.1-1983 (R1988) (ANSI, 1988) and ANSI/ANS-8.19-1996 (ANSI, 1986)) are discussed in Section 6.1.4 of this FSER.

6.1.3 Technical Practices

Sections 6.3.3 and 6.5.1 of NUREG-1718 (NRC, 2000) state that the staff should review the applicant's commitment to designing and operating the proposed facility in accordance with NCS technical practices to ensure that, if the facility is later authorized to operate, NCS requirements will be met. For construction authorization, only those technical practices related to the design of criticality safety PSSCs are applicable. Design basis information includes (1) commitments describing the design philosophy for meeting the performance requirements of 10 CFR 70.61(d) and BDC of 10 CFR 70.64, (2) the technical practices related to the determination of criticality safety limits, including calculational methods and criticality code validation methods, and (3) the technical practices related to the determination of controls, including the preferred hierarchy of controls and measures to ensure control reliability and availability. Many of these technical practices are contained in ANSI/ANS-8 standards; therefore, the revised CAR section on technical practices (DCS, 2002d) contains a description of the codes and standards to which the applicant is committing. Commitments to specific ANSI standards (ANSI/ANS-8.1-1983 (R1988) and ANSI/ANS-8.19-1996) are discussed in Section 6.1.4 of this FSER.

6.1.3.1 Commitment to Baseline Design Criteria

The revised CAR states that the design of the proposed facility will adhere to the DCP as required by 10 CFR 70.64(a)(9). The DCP will be used to meet the performance requirements in an ISA summary supporting a license application. This revised CAR section also describes the process for flowing down NCS evaluation (NCSE) requirements into the ISA as IROFS. These commitments are part of the design basis of the facility.

Compliance with the DCP will be demonstrated in an ISA summary supporting a license application by identifying two or more process conditions which are relied on to ensure a subcritical configuration. The facility NCSEs will evaluate both normal and credible abnormal conditions. Common-mode failures and system interactions will also be considered. The NCSEs will use geometry controls (with fixed neutron absorbers, as necessary) as the preferred controlled parameter for plant systems. To enhance the reliability of criticality controls, the following preferred hierarchy of controls will be used:

- Passive engineered features will be preferred over active engineered features.
- Engineered features will be preferred over administrative controls.
- Enhanced administrative controls will be preferred over simple administrative controls.

Where controlled parameters rely on physical measurements, representative sampling and analysis will be used. The sampling and analysis requirements will be identified as IROFS along with appropriate management measures. The staff concludes that these programmatic commitments provide reasonable assurance that an adequate safety basis will be established and documented.

6.1.3.2 MFFF CAAS

The staff reviewed the facility CAAS described in revised CAR Section 6.3.2 based on the requirements in 10 CFR 70.24 and guidance in Section 6.3.3 of NUREG-1718 (NRC, 2000). Although the applicant has not listed the CAAS as a PSSC, it stated that the CAAS will be designed in accordance with 10 CFR 70.24, and the staff thus finds the CAAS to be part of the design basis of the facility. Specific technical commitments regarding the CAAS are found in the discussion of ANSI/ANS-8.3-1997, "Criticality Accident Alarm System," issued 1997 (ANS, 1997), in FSER Section 6.1.4.3.

The revised CAR describes the CAAS as a monitoring system composed of groups of detectors called monitoring units (detector network, data processing, and alarm actuation units) that will activate audible and visual alarms (network of audible and visual alarms, data processing capability) in case of a criticality accident. The CAAS will be designed to detect both gamma and neutron radiation and to actuate within 0.5 second of detector recognition of a criticality accident. The range and design features of the alarm will also follow the guidance provided in ANSI/ANS-8.3-1997. Each area in which alarms are required will be covered by two alarm units. The applicant committed that any areas requiring exemption from CAAS coverage would be identified and justified in the license application or separate exemption request (DCS, 2003c). These commitments are consistent with the requirements of 10 CFR 70.24 and ANSI/ANS-8.3-1997 (as endorsed by Regulatory Guide (RG) 3.71, "Nuclear Criticality Safety Standards for Fuels and Material Facilities," issued 1972 (NRC, 1972)) and are, therefore, acceptable.

6.1.3.3 Criticality Safety Control Design Criteria

Section 6.3.3 of NUREG-1718 (NRC, 2000) includes the expectation that "the technical practices to ensure that sufficient NCS controls, developed in the criticality safety evaluations (CSEs) and flowed into the ISA, are identified for each process." The staff considers the description of the technical practices for each controlled parameter and the design criteria, including the preferred hierarchy of controls, to be part of the design basis of the proposed facility.

Revised CAR Section 6.3.3.1 defines passive and active engineered controls and enhanced and simple administrative controls and describes the preferred hierarchy of controls. Tolerances will be conservatively taken into account and neutron interaction fully evaluated. Revised CAR Section 6.3.3.2, "Available Methods of Control," discusses the criticality control modes and available methods of control to be used in the proposed facility.

Revised CAR Section 6.3.3.2 also describes the requirements for geometry, mass, density, isotopics, reflection, moderation, concentration, interaction, neutron absorber, volume, heterogeneity, and process variable controls. Technical practices associated with these controlled parameters were found to be acceptable, based on standard industry practice and a comparison with the acceptance criteria of NUREG-1718. The following identifies additional clarifications regarding specific controlled parameters:

- Revised CAR Section 6.3.3.2.1, "Geometry Control," states that "tolerances on nominal design dimensions are treated conservatively." The applicant expanded on this, stating the following:

    The design approach with respect to criticality...for each controlled parameter, assumes the credible optimal condition (i.e., most reactive condition physically possible) for the parameter, or calculates the allowed range for the parameter. Criticality calculations and nuclear criticality safety evaluations are performed assuming the most reactive physical condition to ensure that the process remains subcritical under all normal and abnormal conditions, in accordance with 10 CFR 70.61(d). (DCS, 2001a, 2002e)

The NRC staff considers the applicant's commitment acceptable to ensure that the most reactive combination of tolerances will be used.

- Revised CAR Section 6.3.3.2.4 states that "isotopics control includes both the $^{235}U/U$ concentration (enrichment) and the concentration of fissile and nonfissile plutonium isotopes...as well as the relative abundance of plutonium to uranium." The plutonium isotopics will be represented as 96 wt% ^{239}Pu and 4 wt% ^{240}Pu in criticality models. The revised CAR also states the following:

  > The presence of ^{240}Pu (5% to 9%) and ^{242}Pu (<0.02%) offsets any contribution from ^{241}Pu (<1%) such that it can be neglected for ^{239}Pu ranges from 90% to 95%....This will be demonstrated in the criticality calculation to be referenced in the Nuclear Criticality Safety Evaluations (NCSEs). Justification will be provided in NCSEs and ISA summary.

The applicant also indicated, in a footnote to revised CAR Table 6-1, that the bounding isotopics would be sufficient to bound the slight amount ($^{235}U/Pu$ < 2 wt%) of HEU in the incoming plutonium stream. To confirm that the bounding assumption (96 wt% ^{239}Pu, 4 wt% ^{240}Pu) is more reactive than any combination of incoming feed material, the staff performed a series of independent calculations. These calculations were done using a near-critical sphere of plutonium nitrate solution with full water reflection. The isotopic abundance of ^{239}Pu, ^{240}Pu, and ^{241}Pu was varied within the specification range above (the small presence of ^{242}Pu was neglected). The calculations show that the most reactive combination occurs when ^{240}Pu and ^{241}Pu are at their maximum values (i.e., 94 wt% ^{239}Pu, 5 wt% ^{240}Pu, and 1 wt% ^{241}Pu). This is still less reactive than the bounding assumption. Further, the effect of increasing both ^{240}Pu and ^{241}Pu by the same amount (1 wt%) is a net decrease in k_{eff} ($|\partial k/\partial\, ^{240}Pu| > |\partial k/\partial\, ^{241}Pu|$), again indicating that the assumed plutonium isotopics are bounding. Subsequent to completion of its independent evaluation, the staff reviewed an additional analysis prepared by the applicant in the supplement to the validation report, DCS 01 ZJJ DS CALC H 35075 A, "MFFF Validation Report Supplemental Information," dated January 31, 2005 (DCS, 2005; NRC, 2005). The NRC considers this to be part of the licensing basis in support of the validation effort for the MFFF. In this analysis, DCS modeled an infinite cylinder containing a plutonium dioxide-water mixture at five different H/X ratios, assuming four fissile media with different plutonium isotopic vectors. The isotopic vectors include both the bounding reference isotopics used in criticality calculations (96 wt% ^{239}Pu, 4 wt% ^{240}Pu) and three different combinations of on-spec material, including the combination determined to be the most reactive by the NRC staff (94 wt% ^{239}Pu, 5 wt% ^{240}Pu, 1 wt% ^{241}Pu). This combination contains the minimum ^{240}Pu and maximum ^{241}Pu content and is expected to be conservative based on the relative values of the cross sections. The applicant's calculations concluded that the bounding reference case was as reactive or more reactive than the three combinations of on-spec material across the entire spectrum of H/X values. Therefore, the staff concludes that DCS has demonstrated that the bounding reference case is conservative.

- Revised CAR Section 6.3.3.2.5 states that a minimum 1-in. tight-fitting water reflector would be assumed, with the amount of reflection justified in the NCSE. In addition, the applicant stated the following:

    It is MFFF criticality calculation practice that the ranges of water reflection up to and including 12 inches (30 cm) be employed. The calculations as referenced by the NCSEs shall provide a demonstration of the full range of reflection. Any exceptions to this principle, such as in the moderation controlled, normally dry areas, shall be fully justified in the NCSEs. (DCS, 2001a, 2002e)

    This approach is conservative and corresponds to common industry practice for modeling both incidental and full reflection.

- Revised CAR Section 6.3.3.2.6, "Moderation Control," states that, in moderation-controlled areas, hydrogenous firefighting materials will not be allowed. The applicant's approach to balancing the combined risk of criticality and fire during the facility design is part of the design basis of the facility. This requires an integrated approach to safety, and, while this is not limited conceptually to the interaction between fire protection and criticality safety, this has historically been one of the most significant areas of overlap. The commitment to avoid moderating fire suppressants is acceptable to the staff because it removes a credible source of water from plant areas relying on moderation control for criticality safety.

- Revised CAR Section 6.3.3.2.9 contains commitments regarding the use of fixed neutron absorbers. The applicant did not consider other types of neutron absorbers as inherently reliable as fixed neutron absorbers and did not envision them to be employed in any facility process unit or area (see also FSER Section 6.1.4 on ANSI standards). Moreover, the applicant stated that, wherever neutron absorber control will be used, it will be part of the geometry and fixed by design. Thus, this excludes the use of removable (raschig rings) and soluble absorbers. Because fixed neutron absorbers are inherently more reliable than removable or soluble absorbers, the NRC staff finds this acceptable.

The staff considers the information in this section of the revised CAR to be in broad agreement with the regulatory acceptance criteria in the standard review plan and finds, in accordance with 10 CFR 70.23(b), that it provides an acceptable basis for design.

6.1.3.4 Criticality Safety Process Description

Revised CAR Section 6.3.4 (DCS, 2002d) contains a description of the applicable safety principles to be applied to the design, as well as a preliminary description of how the controlled parameters will be applied to the proposed facility. The staff reviewed this material, as well as the process descriptions in revised CAR Sections 11.2, "MOX Process Description," and 11.3, "Aqueous Polishing Process Description," along with the tables of associated criticality control units (CCUs), revised CAR Tables 6-1 and 6-2. Revised CAR Chapter 11 provides an overview of both the MP and AP process. Revised CAR Tables 6-1 and 6-2 list each of the 55 AP and 25 MP CCUs and their controlled parameters.

In reviewing the controlled parameters for the MP and AP process, the staff reviewed the applicant's use of the preferred design approach. Revised CAR Section 6.3.3.1, "Criticality Control Modes," commits to the preferred use of engineered over administrative controls and the preferred use of geometry control, where practicable. Revised CAR Section 6.3.4.2, "Applicable Safety Principles," commits to the preferred use of passive engineered over active engineered controls, engineered over administrative controls, and enhanced over simple administrative controls, as well as dual- over single-parameter controls.[3] Reliance on diverse control modes is important to minimize the potential for common-mode failure. The applicant's commitment to the DCP ensures that, regardless of the parameter or type of control, at least two independent and unlikely process changes must occur before criticality is possible. However, accident sequences in which there are no credible means to achieve criticality (as may occur in certain cases involving geometry control) do not require dual independent controls to comply with the DCP.

Although specific controls (i.e., IROFS) have not been identified in the revised CAR, the design bases include the dominant controlled parameters. Specific values for these parameters were not considered necessary in most cases, although some assumed values were provided (e.g., powder density and isotopics) when they represented bounding constraints on the design. This information is presented in revised CAR Tables 6-1 and 6-2. The staff's review in the following sections is based on revised CAR Tables 6-1 and 6-2 (DCS, 2004d).

6.1.3.4.1 NCS—AP Process

The AP process, as described in revised CAR Table 6-1 and revised CAR Section 11.3.2 (DCS, 2004d), consists of 16 major areas that have been subdivided into 55 CCUs. The plutonium dioxide powder is received from off site in cans and then batched to the dissolution unit and/or dissolution and dechlorination unit, where it is electrolytically dissolved into plutonium nitrate (primarily $Pu(NO_3)_4$) solution. Following this, it is purified in pulsed solvent extraction, scrubbing, and stripping columns. During these stages, the plutonium nitrate chemical form and process equipment favorable geometry are the dominant controlled parameters. Following purification, the plutonium nitrate is precipitated and calcined to form a purified plutonium dioxide powder used as feed for the MP, all still in favorable geometry process equipment. Unprecipitated plutonium dioxide is filtered and then recovered in the oxalic mother liquor recovery unit. Additional process units recover solvent, nitric acid, offgases, and liquid waste and recycle raffinate produced in the solvent extraction. These latter units primarily rely on dual independent concentration controls because they are designed not to contain plutonium under normal process conditions.

For each CCU in revised CAR Table 6-1, the dominant controlled parameters are listed along with some parameter ranges. The chemical form through most of this process is primarily aqueous plutonium oxide, plutonium nitrate solution,[4] or plutonium oxalate. Plutonium oxide is

---

[3] While this commitment does not preclude the use of administrative controls or control modes other than geometry, these are to be preferred over other types of controls; the process descriptions in revised CAR Sections 11.2 and 11.3, and the revised CAR Tables 6-1 and 6-2, were reviewed to confirm that the facility will be designed in accordance with this design philosophy to the greatest extent practical (e.g., passive geometry control used on a majority of processes).

[4] $Pu(NO_3)_4$ is the most common nitrate, but the applicant stated that it will assume $Pu(NO_3)_3$ in the analysis because it is more reactive.

considered more reactive than plutonium nitrate. (The applicant also determined that plutonium oxalate is bounded by plutonium oxyfluoride.) The solution is converted to plutonium dioxide during calcination. For those areas in which plutonium nitrate is the chemical form, the applicant identified physicochemical form as a criticality control mode. For units relying on chemical form, the applicant has committed to take into account both nominal conditions and possible process upsets in accordance with the DCP. The bounding nature of these compounds will be demonstrated in NCSEs.

According to revised CAR Table 6-1, most of the process is conducted in favorable geometry equipment, with the general geometrical shape but not the dimensions specified. Typically, units in the AP process will consist of cylindrical columns or slab/annular tanks. Dimensions will be based on validated calculational methods, standards (such as ANSI/ANS-8.1-1983 (R1988) (ANSI, 1988)), or handbook data. This will ensure that favorable geometry units will be subcritical under normal and all credible abnormal conditions. When using methods that determine favorable dimensions for a specific material composition (e.g., chemical form, isotopics), the commitment to the DCP provides assurance that process upsets capable of changing the form of the material will be considered.

Revised CAR Table 6-1 also assumes specific powder densities and isotopic compositions. The AP process downstream of the plutonium dioxide filter assumes a chemical form of $Pu(NO_3)_3$, which is less reactive than plutonium dioxide. Dual independent filters preclude the presence of plutonium dioxide beyond this point. For powder density, different bounding values are presented. In some CCUs within the decanning and milling units, powder density is controlled by upstream measurement by a variety of means (revised CAR Sections 11.3.2.1.2 and 11.3.2.2.2). For these units, density must be confirmed to be less than 7 g/cm^3. This is considered very high for plutonium dioxide because of its highly porous nature.

Downstream of the calcination furnace in the oxalic precipitation and oxidation unit, density is limited to less than 3.5 g/cm^3. These values were derived based on La Hague experience and will be confirmed during startup testing. The applicant stated that, wherever the physical form of the material changes, density will be confirmed by direct measurement or otherwise justified (DCS, 2003b). Based on domestic nuclear industry experience, the staff accepts that a density of 7 g/cm^3 is very high for plutonium dioxide or MOX powder and that the described measurement techniques are reasonable to confirm this. For lower density values, appropriate controls on density, and on any process variables that can affect powder density nonconservatively, will be specified and justified in the NCSEs. The staff will review the justification in the NCSEs to be submitted as part of any later DCS application for a license to possess and use licensed material.

In several CCUs, the applicant identified neutron absorption as a means of control. The staff reviewed the application of neutron absorption and concludes that it is appropriate. For several cases, neutron absorption is combined with geometry control (i.e., the presence of fixed absorbers is assumed in determining subcritical geometric limits). The use of fixed absorbers is recognized in industry practice as a highly reliable means of control, especially in conjunction with geometry control, and thus complies with the preferred hierarchy of controls and is acceptable.

The applicant proposed concentration control for those units not relying on favorable geometry, where fissile material is not expected under normal conditions. Revised CAR Table 6-1 indicates only that concentration is controlled upstream of these processes. However, a previous communication from the applicant had stated that this would be through inline monitors and/or sampling measurements (DCS, 2001a, RAI 84; DCS, 2002e). Revised CAR Section 6.3.3.2.7 contains the requirements for both sampling and instrumentation, including the use of dual independent sampling methods. This method for preventing plutonium intrusion into these units is in accordance with established industry practice and the DCP and, therefore, is acceptable to the staff.

Revised CAR Table 6-1 also describes the design bases for auxiliary systems connected to the main AP process. The offgas treatment unit removes plutonium from the gaseous effluents in the AP process. The liquid waste reception unit treats liquid effluents, including high-alpha waste. High-enriched uranium is present in the incoming plutonium stream (assumed to be 100 wt% ^{235}U). This is isotopically diluted with depleted uranyl nitrate in two steps, first to less than or equal to 30 wt% ^{235}U (evaluated at 35 wt% ^{235}U), and then to less than or equal to 1 wt% ^{235}U, before being separated from the plutonium in the purification unit. Control of uranium assay is required because the high-alpha waste assay must be less than 2 wt% ^{235}U for criticality safety. The applicant identified uranium isotopics as a control mode for the downstream processes in revised CAR Table 6-1; the staff, therefore, finds this acceptable.

For other auxiliary systems for which plutonium intrusion must be precluded (such as chemical and water addition), the applicant stated that a similar approach will be used (DCS, 2003b). The control methods will include passive design features and dual, independent sampling. These were not included in the revised CAR tables as they are not process units but will be addressed in an auxiliary system NCSE. This conforms to common industry practice and is, therefore, acceptable to the staff. (This will apply to auxiliary systems supporting both the MP and AP process.)

The staff reviewed the application of the preferred design approach (i.e., the preferred use of passive over active and engineered over administrative controls, and the preferred reliance on favorable geometry) in the AP process and concluded that it was appropriate because of the observed predominance of passive engineered controls. In addition, two or more parameters were identified for almost every CCU. This diversity in the control modes is desirable because it reduces the potential for common-mode failure. Given the higher inherent risk of the AP process relative to the MP, because of the forms and types of materials being processed, the favorable geometry design of the facility is appropriate. Most of the process relies on favorable geometry, with the exception of the concentration-controlled units. Moreover, fixed neutron absorbers are used in the process as part of the design. The identified parameters provide reasonable assurance that the design of the facility will be based primarily on passive design features, in accordance with the preferred control hierarchy. The only notable exceptions to this are the density and isotopic composition, which have bounding values defined that are not passively controlled (the basis for bounding densities will be provided with any license application). Coupled with the commitment to follow the preferred design approach during the design of actual IROFS, revised CAR Table 6-1 provides reasonable assurance that the design of the AP process will be in agreement with the regulatory acceptance criteria.

6.1.3.4.2 NCS—MP

The MP, as described in revised CAR Table 6-2 and revised CAR Section 11.2.2 (DCS, 2004d), consists of five major areas that have been subdivided into 25 CCUs. In the receiving area, depleted uranium dioxide powder, as well as purified plutonium dioxide powder (in 3013 containers) is received and stored before being used as feed in the powder area. Depleted uranium poses no criticality concerns. The plutonium dioxide powder is stored in closed 3013 containers.[5] Geometry and spacing in the plutonium dioxide 3013 storage pit and buffer storage unit are controlled, along with the moderation inside containers. In the powder area, the plutonium dioxide containers are emptied inside a glovebox onto a conveyor supplying the dosing unit, where the depleted uranium dioxide and the plutonium dioxide powder is blended, homogenized, and milled to form the master blend. The relative proportion of uranium dioxide and plutonium dioxide is carefully controlled for product specification and criticality purposes. The master blend has a composition of less than or equal to 22 wt% plutonium; the final blend consists of less than or equal to 6.3 wt% plutonium. These values are the conservative modeled values for maximum plutonium content that are used in the criticality safety analyses. During this process, mass, isotopic composition, and moderation are the dominant controlled parameters. The powder process ends with the addition of a controlled amount of poreformer and pelletizing of the powder into green pellets. The remaining process areas (pellet process area, fuel rod process area, and assembly area) consist of sintering and handling the green pellets, loading the pellets into rods, and loading the rods into assemblies, in a fashion similar to other fuel manufacturing processes. During these latter processes, the chemical and geometric form, isotopic composition, and moderation are the dominant controlled parameters.

For each CCU in revised CAR Table 6-2, the dominant controlled parameters are listed along with parameter ranges for various parameters. The chemical form throughout this process is either purified plutonium dioxide or MOX. Powder in the receiving area is handled in fixed geometry 3013 containers and stored in fixed storage arrays. Once the powder is removed from the containers, criticality safety is based on mass and moderation control in lieu of geometry control. Favorable geometry is also a dominant control in the processing and handling of pellets, rods, and fuel assemblies once the fuel has been processed into a fixed configuration.

Following primary dosing, the isotopic composition of the MOX powder is controlled for criticality safety. The primary blend consists of MOX with less than or equal to 22 wt% plutonium and is stored in J60 jars; the final blend consists of MOX with less than or equal to 6.3 wt% plutonium and is stored in J80 jars. These are controlled by geometry; in addition, the MOX operations are batch processes that rely largely on mass control for criticality safety.

Moderation is relied on for criticality safety in much of the MP, including primarily the powder handling and storage operations. The moderator is limited to less than or equal to 1 wt% water for incoming plutonium dioxide powder, is limited by the controlled addition of organic additives following the primary dosing CCU, and is limited further downstream by controlling the amount of poreformer added.

---

[5]There are some inconsistencies between revised CAR Table 6-2 and revised CAR Section 11.2.2 (e.g., primary dosing and plutonium dioxide container handling are in the receiving area according to revised CAR Table 6-2, but in the powder area according to revised CAR Section 11.2.2). This is not significant because all plant areas are covered in both sections. The areas defined in revised CAR Section 11.2.2 are used in this discussion.

Heterogeneity is controlled by ensuring a homogeneous mixture of uranium dioxide and plutonium dioxide powder to prevent the formation of high-plutonium hot spots in the fuel.

Revised CAR Table 6-2 also assumes specific powder densities and isotopic compositions. For powder densities, different bounding values are presented. Powder density is controlled by upstream measurement.

There are several other units, however, where the density of powder is limited to less than 3.5 g/cm^3 or 5.5 g/cm^3, or the density of scrap to less than 4.6 g/cm^3. These values were derived based on MELOX experience and will be confirmed during startup testing. Following pelletizing, full theoretical density (11.46 g/cm^3) is assumed for NCS purposes. The applicant stated that, wherever the physical form of the material changes, density will be confirmed by direct measurement or otherwise justified (DCS, 2003b). Appropriate controls on density, and on any process variables that can affect powder density nonconservatively, will be specified and justified in the NCSEs. The staff will review the justification in the NCSEs to be submitted as part of any later DCS application for a license to possess and use licensed material.

The staff reviewed the application of the preferred design approach (i.e., the preferred use of passive over active and engineered over administrative controls, and the preferred reliance on favorable geometry) in the MP and concluded that it was appropriate because of the observed predominance of passive engineered controls. In addition, two or more parameters were identified for almost every CCU. This diversity in the control modes is desirable because it reduces the potential for common-mode failure. The majority of CCUs in the MP utilize favorable geometry for criticality control. Where geometry control is not used, mass and moderator control is typically used. As stated above, these operations consist of batch processes inside gloveboxes designed to exclude moisture. The relative risk of the MP is low compared with that of the AP process (the risk is especially low following downblending of the plutonium dioxide into MOX powder and subsequent pelletizing). The identified parameters provide reasonable assurance that the design of the facility will be based primarily on passive design features, in accordance with the preferred control hierarchy. Coupled with the commitment to follow the preferred design approach during the design of actual IROFS, the information presented in revised CAR Table 6-2 provides reasonable assurance that the design of the MP will be in agreement with the regulatory acceptance criteria.

Although diverse control modes were described for both the MP and AP process in revised CAR Tables 6-1 and 6-2, the extent to which diverse control modes were used in meeting the DCP could not be determined. This was simply because the tables contain no information on the structure of individual accident sequences. While it appeared that dual parameter control had been implemented, during the January 16, 2003, public meeting on criticality safety, the applicant made the following statement:

> For the MFFF, double contingency, in most cases, is based on 2 controls or barriers to prevent a change in one controlled parameter. A loss of one of these controls or barriers does not cause a change in the controlled parameter and therefore does not change the k_{eff} value.

The applicant further clarified that the use of two-parameter control is not typically practicable for MFFF operations because of the assumed ^{239}Pu content (DCS, 2003e). As clarified in the July

31, 2003, public meeting, for systems with high ^{239}Pu content (96 wt% ^{239}Pu/Pu and greater than or equal to 6.3 wt% $PuO_2/(UO_2 + PuO_2)$), mass limits are sufficiently low that limiting operations to the minimum critical mass would be infeasible. At the masses used in the process, only a loss of moderation control is needed for criticality. In addition, upon loss of mass control, criticality can be possible without additional moderation beyond what is required by the process. Therefore, criticality may occur if control over either mass or moderation is lost, and two independent controls are required on each of these parameters in order to meet the DCP. The staff acknowledges that this may be the case in large portions of the plant but that a determination of whether dual-parameter control is practical must be made on a case-by-case basis. Dual-parameter control is recognized as generally more reliable than single-parameter control because of the increased susceptibility to common-mode failures when controlling only one parameter. However, in CAR Section 6.3.4.2, the applicant committed to the use of dual-parameter control where practical, and it further committed that the determination of the practicality of following the design principles will be made in the development of the NCSEs and subjected to numerous technical and managerial reviews (DCS, 2001a, RAI 67). Based on these commitments, the staff understands that the applicant will provide justification in the NCSEs whenever single-parameter control is necessary. While this is less desirable than dual-parameter control for meeting the DCP, the applicant has also stated (DCS, 2003e) that, in both cases, controls must be demonstrated to be sufficiently robust and independent to prevent criticality. The staff therefore finds these commitments to be acceptable for the purpose of construction authorization. The staff will review the justification for the use of single-parameter control and the adequacy of specific control measures to be submitted as part of any later DCS application for a license to possess and use licensed material.

6.1.3.4.3 Single-Parameter Limits

In revised CAR Tables 6-3 and 6-4, the applicant provided a list of the single-parameter limits for various parameters, such as subcritical masses, dimensions, and volumes, for use in the facility design. Revised CAR Section 6.3.4.5 and the footnote to these tables indicate that they are not to be used in support for criticality calculations or NCSEs. These are only intended as typical, order-of-magnitude estimates. Therefore, the staff did not evaluate the adequacy or derivation of the values in these revised CAR tables as part of the revised CAR review.

6.1.3.5 Nuclear Criticality Analysis and Safety Evaluation Methods

The staff considers a description of the calculational methodology, including the computer code validation methodology, to be part of the design basis of the facility. The staff also considers the design basis to include the maximum k_{eff} or upper subcritical limit (USL). The applicant has committed to ANSI/ANS-8.1-1983 (R1988) (ANSI, 1988) as part of the design basis of the facility. The standard requires that processes be shown to be subcritical under both normal and credible abnormal conditions and also contains requirements for the validation of calculational methodologies. The staff evaluated the applicant's submitted validation report (DCS, 2003f, 2003h), which sets forth the applicant's proposed methodology for determining the margin of subcriticality for safety. To approve the revised CAR, the staff must, pursuant to 10 CFR 70.61(d), approve the margin of subcriticality for safety. The applicant has committed to the methodology described below to ensure that there will be an adequate margin of subcriticality for safety during any operation of the proposed facility.

The ANSI/ANS-8.1-1983 (R1988) standard provides single and multiparameter limits to be referenced in criticality safety calculations, as well as guidance used in the performance of criticality analysis method validation. In revised CAR Section 6.3.5, the applicant described the criticality analysis methodology to be used in facility design activities and facility safety programs, as also discussed in ANSI/ANS-8.1-1983 (R1988). The applicant stated that the chosen computational methods are those using the evaluated nuclear data file (ENDF) cross-section libraries (i.e., the Standardized Computer Analyses Licensing Evaluation-(SCALE)-4.4 and 4.4a code system with the 238-group cross-section library derived from the ENDF/B-V file of evaluated nuclear physics data and the Monte Carlo Neutron Photon code, with ENDF60 cross-section libraries). Use of these codes and cross section libraries is accepted industry practice and has been recognized by the NRC as generally acceptable (NUREG/CR-6361, "Criticality Benchmark Guide for Light-Water-Reactor Fuel in Transportation and Storage Packages").

In the revised CAR, the applicant discussed the method of validation and calculated k_{eff} design limits to be used in the facility. Consistent with ANSI/ANS-8.1-1983 (R1988), the applicant's code validation methodology uses critical experiment benchmarks to determine bias and bias uncertainties. Critical experiments are selected to represent the neutronic and physical characteristics of the systems analyzed in specific design applications. The derived biases are then applied to calculations on systems with neutronic and physical characteristics that are bounded by those of the selected set of experimental benchmarks. When the neutronic and physical characteristics of the system analyzed are not bounded by those of the experimental benchmarks, additional adjustments for bias and uncertainties will be employed as appropriate, by making use of trends in the bias. The applicant stated that criticality safety calculations would be performed for each of the stages of process operations at the facility (ranging from receipt of plutonium dioxide and uranium dioxide powders through fabricated MOX fuel assembly storage and shipment).

The validation approach to be used by the applicant is similar to those already approved by the NRC in previous license application approvals for a broad range of applications. This validation approach is described in NUREG/CR-6361, NUREG/CR-6689, "Guide for Validation of Nuclear Criticality Safety Calculational Methodology," and NUREG/CR-6655, "Sensitivity and Uncertainty Analyses Applied to Criticality Safety Validation." The *International Handbook of Evaluated Criticality Safety Benchmark Experiments* of the International Criticality Safety Benchmark Evaluation Project provides a large compilation of benchmark criticality experiment descriptions.

The staff reviewed the information contained in revised CAR Section 6.3.5, "Nuclear Criticality Analysis and Safety Evaluation Methods," and the code validation report (DCS, 2003f) submitted separately to determine that an acceptable subcritical margin will be provided for the design of the facility. The following sections describe the validation methodology and the application of that methodology to determine design basis k_{eff} limits for each of five areas of applicability (AOAs).

6.1.3.5.1 NCS Validation Report

Pursuant to 10 CFR 70.64(a)(9), the design of the proposed MFFF must provide for criticality control, including adherence to the DCP. The ANSI/ANS-8.1 standard (ANSI, 1988) requires both adherence to the DCP and that processes be subcritical under both normal and credible abnormal conditions. The applicant has chosen to commit to this standard to meet, in part, 10 CFR 70.64 and has identified it as a part of the design basis of the MFFF. Determining the bias, uncertainty in the bias, and an administrative margin are all necessary tasks in setting a USL that ensures an adequate margin of subcriticality. For the purpose of showing compliance with the USL, $k_s + \Delta k_s \leq$ USL, where k_s is the calculated k_{eff} for the specific application. As the applicant stated (DCS, 2003g), for criticality calculations using the SCALE code, $\Delta k_s = 2\sigma_k$, where σ_k is the Monte Carlo standard deviation.

The applicant used the statistical methods in NUREG/CR-6361, which describes the two methods for determining the USL, including (1) confidence band with administrative margin and (2) lower tolerance band, also known as the single-sided uniform width closed interval approach. These approaches are summarized in the validation report.

Method 1 requires the choice of an administrative margin, which the applicant set to 0.05. Method 2 requires the choice of the confidence α that some fraction P of all future calculations below the USL are subcritical. The applicant used values of $\alpha = 0.95$ and $P = 0.999$ throughout the validation. These values are conservative with respect to the guidance in NUREG/CR-6698, "Guide for Validation of Nuclear Criticality Safety Calculation Methodology," and therefore acceptable to the staff. The choice of administrative margin is discussed in FSER Section 6.1.3.5.2. The linear regression and statistical analysis required by these methods is performed by the USLSTATS code distributed with the SCALE code package.

An important caution is that the aforementioned techniques can only be used on data that are normally distributed. To verify normality, the applicant's validation methodology uses the χ^2 (chi-squared) test for normality included with the USLSTATS code. The guidance in NUREG/CR-6361 advises that normality can usually only be assured when there are more than 25 data points. In those cases when data are not normally distributed, the applicant committed to using nonparametric methods, as discussed in NUREG/CR-6698. This requires basing the USL on the lowest observed value, with additional margin based on the number of benchmark experiments.

The aforementioned methodology is acceptable to the staff, as it entails techniques that have been endorsed in various NUREG publications and have been used to validate criticality codes at fuel facilities and across the domestic nuclear industry.

The applicant submitted its validation report in three separate parts by letter dated January 8, 2003 (DCS, 2003a). The latest revision to Parts I and III was submitted by letter dated July 2, 2003 (DCS, 2003f). The latest revision to Part II was submitted by letter dated October 10, 2003 (DCS, 2003h). Each part of the validation report addresses one or more separate AOAs. Part I includes AOA(1), covering plutonium nitrate solutions, and AOA(2), covering MOX fuel pellets, rods, and assemblies. Part II includes AOA(3), covering plutonium dioxide powder systems, and AOA(4), covering MOX powder systems. Part III includes AOA(5), covering solutions of plutonium compounds, such as plutonium oxalate and plutonium oxyfluoride. For each AOA, the applicant used the validation methodology described in this section of the FSER.

In a previous revision to Part II of the validation report, the applicant made use of the sensitivity/uncertainty (S/U) methodology developed recently at Oak Ridge National Laboratory (ORNL). This method provides a more quantitative basis for choosing benchmark experiments than the traditional qualitative approach, by comparing dependencies on the underlying cross section data. The sensitivity portion of the code calculates the sensitivity of k_{eff} to changes in the underlying cross section data for a particular system. The uncertainty portion of the code determines the degree of correlation between candidate benchmarks and hypothetical design applications by combining the sensitivity information with known uncertainties in the cross section data. Following the staff's questions on the use of the S/U method, the applicant provided Revision 3 to Part II of the validation report by letter dated October 10, 2003 (DCS, 2003h). In Revision 3 to Part II, the applicant used a more traditional approach, relying on parametric screening criteria rather than on the S/U approach. However, the staff did use the TSUNAMI modules (TSUNAMI-1D, TSUNAMI-3D, and TSUNAMI-IP) of the SCALE 5 code package to test the physics assumptions used in each part of the validation report and to obtain additional information on the applicability of chosen benchmarks. As a result of these independent TSUNAMI calculations, the NRC staff was able to conclude that the applicant's selected benchmarks and statistical analyses were appropriate, with the exception of certain benchmarks in AOA(4). At the time the NRC staff performed these analyses, ORNL had not yet released the SCALE 5 code for general use. While SCALE 5 was made available to the NRC, it was not made available to the applicant; therefore, the applicant was unable to perform the type of S/U analysis relied on by the NRC staff to reach these conclusions.

Because of the developing nature of the TSUNAMI methodology and the staff's questions on its use, the staff does not consider it appropriate to base acceptability solely on the application of the S/U method.

Except for remaining questions on the AOA(4) benchmarks used in Revision 3 to Part II, the NRC staff determined that the code validation methodology described generally in CAR Section 6.3.5 was appropriate. The validation of specific AOAs intended to cover anticipated AP and MP operations is discussed in the following sections. For the purpose of construction authorization review, only the results of the validation (i.e., determination of the benchmark experiments and parametric range for each AOA, and computation of the bias and uncertainty) were reviewed. The staff expects that the applicant will demonstrate the applicability of the validation results to specific design applications in NCSEs and will ensure that the design calculations use the code in a manner consistent with that used in the validation (e.g., similar statistics and use of code options, such as albedo and biasing). This will be reviewed as part of the staff's review of any later DCS application for a license to possess and use licensed material.

6.1.3.5.1.1 AOA(1)—Plutonium Nitrate Solutions. To validate AOA(1), the applicant modeled 191 critical experiments using the KENO-VI code with the 238-group cross section library and the INFHOMMMEDIUM cross section treatment option. The bias was determined for both the SCALE-4.4 code package on the SGN Sun hardware platform and the SCALE-4.4a code package on the SGN PC hardware platform. The applicant's chosen experiments included plutonium nitrate systems modeled as spheres, cylinders, and slabs with a variety of reflector conditions (no reflector, water, concrete, and cadmium). The staff examined the chosen benchmarks and determined that they were appropriate for a validation of systems covered by AOA(1), as they consist of plutonium nitrate benchmarks with parameters close to the parametric range of AOA(1).

The applicant defined the validated range of parameters for AOA(1) in Table 5-2 of Part I of the validation report. The validated AOA is described in Table 6.1-1.

Table 6.1-1 Definition of AOA(1)

Geometry	Cylinder, slab, annular cylinders, arrays of cylinders
Reflectors	Full water, cadmium*/water, borated concrete* (1) Cd limited to 0.05-cm- (0.02-in.-) thick sheet surrounding 4.5–9.5-cm- (1.8–3.7-in.-) thick slab tanks. (2) Borated concrete limited to 15 cm (5.9 in.) inside and outside 7–7.5-cm- (2.76–2.95-in.-) thick annulus and separated from tank by 1.8–2-cm (0.71–0.79-in.-) gap, with the composition below: $^{10}B = 1.59 \times 10^{-3}$ atoms/b cm (2.61×10^{22} atoms/in.^{3}) $^{11}B = 7.04 \times 10^{-3}$ atoms/b cm (1.15×10^{23} atoms/in.^{3}) $Ca = 4.65 \times 10^{-3}$ atoms/b cm (7.62×10^{22} atoms/in.^{3}) $Fe = 5.01 \times 10^{-4}$ atoms/b cm (8.21×10^{21} atoms/in.^{3}) $Si = 1.66 \times 10^{-4}$ atoms/b cm (2.72×10^{21} atoms/in.^{3}) $H = 2.17 \times 10^{-2}$ atoms/b cm (3.56×10^{23} atoms/in.^{3}) $Al = 1.96 \times 10^{-3}$ atoms/b cm (3.21×10^{22} atoms/in.^{3}) $O = 3.25 \times 10^{-2}$ atoms/b cm (5.33×10^{23} atoms/in.^{3})
Chemical form	Plutonium nitrate solution
Pu/(U + Pu)	100 wt%
^{240}Pu	4 wt%
H/Pu	100–200
gPu/l	125–237
EALF	0.14–0.25 eV
*Subject to the limitations of this section	

The applicant ran SCALE/KENO-VI for the chosen experiments, calculated k_{eff}, and plotted the values of USL-1 and USL-2 (determined by USLSTATS) versus the trending parameters of energy of average lethargy causing fission (EALF), the moderator to fuel ratio (H/Pu), and the ^{240}Pu content. Over the range of applicability, the minimum USL-1 and USL-2 were calculated as a function of these parameters and the lowest such value chosen as the USL for the entire AOA. Because the data were found not to be normally distributed, nonparametric estimates of the USL were also made. Based on the number of benchmarks included, no additional nonparametric margin (NPM) was needed. The lowest value determined using USLSTATS was more conservative than that using the NPM, so this was used as the final USL for the AOA.

The staff compared the range of parameters covered by the benchmark data to the applicant's AOA and determined that there was a sufficient number of experiments and that they provided adequate coverage of the defined AOA across the entire range, except for cases involving some absorber materials.

For certain absorber materials, sufficient benchmarks were not available. Therefore, in Attachments 5 and 6 to Part I of the validation report, the applicant included an analysis that attempted to show that the presence of borated concrete and cadmium has a negligible effect on the bias. Because the applicant did not identify any plutonium nitrate systems containing boron, it analyzed several uranyl nitrate benchmarks with borated concrete, plaster, or water to determine if the presence of the boron had an effect on the bias for these systems. Although the applicant did not observe any effect on the bias, the staff considered the number of benchmarks containing boron insufficient to make a definitive conclusion. The applicant also stated that the conclusions regarding the effect of these absorbers on the bias for uranyl nitrate systems could be applied to plutonium nitrate systems. This determination was based on similarities in the neutron absorption spectra between the uranium benchmarks and the same cases with plutonium substituted for uranium. The applicant also analyzed several additional plutonium nitrate cases containing cadmium to demonstrate that cadmium has a negligible effect on the bias. These cases were not included in determining the bias for AOA(1) because the ^{240}Pu content was outside the target range. They also did not show any statistically significant effect of the absorber on the bias. Therefore, the applicant concluded that the existing benchmarks were sufficient to validate calculations for plutonium nitrate systems containing borated concrete and cadmium.

The staff had concerns regarding the number of experiments containing boron and cadmium, as well as using uranyl nitrate benchmarks as part of the basis for this conclusion. The staff, therefore, performed an independent analysis to evaluate the applicant's conclusions regarding the use of absorbing materials in AOA(1). Perturbation of the number densities of ^{10}B, ^{1}H, and ^{239}Pu show that the most highly poisoned benchmarks are relatively insensitive to the ^{10}B cross sections compared to the sensitivity to ^{1}H and ^{239}Pu (i.e., a small change in the ^{10}B content produces a negligible change in the overall system k_{eff}). Therefore, any bias resulting from errors in the boron cross section is expected to be correspondingly small. Subsequent to completion of its independent evaluation, the staff reviewed an additional analysis prepared by the applicant in the supplement to the validation report, DCS01 ZJJ DS CALC H 35075 A, "MFFF Validation Report Supplemental Information," dated January 31, 2005 (DCS, 2005; NRC, 2005). The NRC considers this to be part of the licensing basis in support of the validation effort for the MFFF. In this analysis, DCS modeled three different configurations containing colemanite concrete as a neutron absorber. The three cases considered (two fissile solution tanks and a drip tray) correspond to actual process calculations. The applicant varied the boron content in the concrete to determine the k_{eff} sensitivity to boron number density and concluded that a 50-percent reduction in the boron density would produce only an approximately 1 percent Δk_{eff} for these three cases. These cases correspond to thicknesses and arrangements of the colemanite concrete reflector described in Table 6.1-1 above. Therefore, the staff concludes that even a very large error in the ^{10}B cross section will not be sufficient to significantly affect the code bias. The staff concludes that DCS has demonstrated that cases involving borated concrete reflectors are within the scope of AOA(1). In addition because of the high sensitivity to the ^{239}Pu cross sections, the staff expects that there is little relevance of the uranyl nitrate benchmarks to the behavior of plutonium nitrate systems.

Because the experiments considered appear to show little sensitivity to the absorbers, the staff concluded that the lack of an observed trend in the bias could result from either the poor statistics or low reactivity worth of the absorbers in the cases considered. Because the absorber worth can depend strongly on the geometrical configuration and size of the absorbers, system leakage, and other physical characteristics of the system, these results may not be indicative of the behavior of systems with much higher absorber worth. Therefore, the NRC requested the applicant to describe and justify "the range of boron and cadmium absorber loading for which AOA(1) is considered valid" (NRC, 2003a). The footnotes to Table 5-2 in Part I of the validation report provide this detailed information on the dimensions and compositions of borated concrete and cadmium to be used in anticipated design applications. The applicant confirmed that the footnotes to Table 5-2 represent the maximum loading of these materials (DCS, 2003g). The behavior of these design applications with regard to their sensitivity to perturbations in the absorber number density is comparable to that in the uranyl nitrate study discussed above. In addition, the applicant stated that, for AOA(1), no more than 5 percent of the borated concrete is required to ensure subcriticality (maintain k_{eff} below the USL) (DCS, 2003g). This information is useful because it indicates the amount of error that would be required to cause the k_{eff} limits to be exceeded. While comparable information was not initially provided for cadmium, the staff's independent calculations indicate that it would require at least a 50-percent decrease in the cadmium number density before a significant change in the system k_{eff} occurred, assuming a fully reflected slab tank with a cadmium sheet as described in the footnote to Table 5-2 of Part I. This appears to occur because cadmium is a nearly black absorber of thermal neutrons at the thicknesses described. The staff considers the likelihood of the ^{10}B and cadmium cross sections being in error by such a large proportion (95 percent for ^{10}B and approximately 50 percent for cadmium) to be very low, as both boron and cadmium are commonly used absorbing materials, and errors of this magnitude would have been noticed during the long history of using these materials in criticality calculations. Subsequent to completion of its independent evaluation, the staff reviewed an additional analysis prepared by the applicant in the supplement to the validation report, DCS01 ZJJ DS CALC H 35075 A, "MFFF Validation Report Supplemental Information," dated January 31, 2005 (DCS, 2005; NRC, 2005). The NRC considers this to be part of the licensing basis in support of the validation effort for the MFFF. In this analysis, DCS modeled two different configurations containing cadmium sheets as neutron absorbers. The cases considered (slab tanks) correspond to actual process calculations. The applicant performed two series of calculations in which it (1) maintained the cadmium sheet thickness but varied the cadmium density and (2) maintained the cadmium density but varied the cadmium sheet thickness. For both applications, the two methods yield nearly identical results and show that a 60-percent decrease in the cadmium density would be required to produce an approximately 1 percent Δk_{eff} for these cases. These cases correspond to thicknesses and arrangements of the cadmium reflector described in Table 6.1-1 of the FSER. Therefore, the staff concludes that even a very large error in the cadmium cross section will not be sufficient to significantly affect the code bias.

Based on the foregoing considerations, the staff has reasonable assurance that systems meeting the criteria in Table 6.1-1 are within the scope of AOA(1) and will be safely subcritical with a USL as defined in Part I of the validation report (USL = 0.9370).

6.1.3.5.1.2 AOA(2)—MOX Pellets, Fuel Rods, and Fuel Assemblies. To validate AOA(2), the applicant modeled 36 critical experiments using the KENO-VI code with the 238-group cross section library and the LATTICECELL cross section treatment option. The bias was determined for both the SCALE-4.4 code package on the SGN Sun hardware platform and the SCALE-4.4a code package on the SGN PC hardware platform. The applicant's chosen experiments include lattices of MOX fuel rods in water, with a variety of different moderator-to-fuel ratios (v^m/v^f) and plutonium contents. The staff examined the chosen benchmarks and determined that they are appropriate for the validation of systems covered by AOA(2), as they consist of water-moderated MOX fuel rod lattices with parameters close to the parametric range of AOA(2).

The applicant defined the validated range of parameters for AOA(2) in Table 5-4 of Part I of the validation report. The validated AOA is described in Table 6.1-2.

Table 6.1-2 Definition of AOA(2)

Geometry	Heterogeneous, rectangular lattices
Reflectors	Water
Chemical form	MOX fuel
Pu/(U + Pu)	6.3 wt%
^{240}Pu	4 wt%*
v^f/v^m	1.9–10
EALF	0.1–0.66 eV
*See discussion below	

The applicant ran SCALE/KENO-VI for the chosen experiments, calculated k_{eff}, and plotted the values of USL-1 and USL-2 (determined by USLSTATS) versus the trending parameters of EALF, the moderator to fuel ratio (v^m/v^f), and the plutonium content ($PuO_2/(PuO_2 + UO_2)$). Over the range of applicability, the minimum USL-1 and USL-2 were calculated as a function of these parameters, and the lowest such value was chosen as the USL for the entire AOA. The benchmark data tested normal, although USLSTATS indicated that the amount of data may not be sufficient for this normality test. However, no anomalies in the k_{eff} histogram indicating a deviation from normality were apparent.

The staff compared the range of parameters covered by the benchmark data to the applicant's AOA and determined that there was a sufficient number of experiments to provide adequate coverage of the defined AOA across the entire range, except for the ^{240}Pu content.

The staff noted that the isotopic composition considered validated in Table 5-4 (4 wt% ^{240}Pu) was outside the range of the benchmark experiments (8–22 wt% ^{240}Pu), and no trending of the bias as a function of ^{240}Pu content was performed for AOA(2). The applicant justified this on the basis that LA-12683, "Forecast of Criticality Experiments and Experimental Programs Needed to Support Nuclear Operations in the United States of America: 1994–1999," supports a ±4 wt% extension to the defined AOA with regard to ^{240}Pu content. However, while this extrapolation may be permissible, the applicant has committed to extend the bias beyond the range of experimental data using the guidance in NUREG/CR-6698. This guidance states that extensions to the AOA should take into account trends in the bias and any increased uncertainty in the bias. This was not done.

Therefore, staff used the USLSTATS program to determine the effect of extrapolating the bias and the uncertainty in the bias to systems with 4 wt% ^{240}Pu content. This was done to evaluate whether additional margin should have been included to account for the ^{240}Pu content being outside the screening range. This analysis shows that there is a very weak correlation between the bias and the ^{240}Pu content and no significant change in the bias as a result of extending the AOA down to 4 wt% ^{240}Pu. Subsequent to completion of its independent evaluation, the staff reviewed an additional analysis prepared by the applicant in the supplement to the validation report, DCS01 ZJJ DS CALC H 35075 A, "MFFF Validation Report Supplemental Information," dated January 31, 2005 (DCS, 2005; NRC, 2005). The NRC considers this to be part of the licensing basis in support of the validation effort for the MFFF. In this analysis, DCS performed a statistical analysis of the benchmark k_{eff} values as a function of ^{240}Pu content, using the ORNL code USLSTATS. Using the results from the confidence band technique generated by USLSTATS, the applicant determined the functional form of both the calculated k_{eff} and its uncertainty, k(x) and w(x), as a function of ^{240}Pu content. The applicant then used this information to extend the trend, including allowance for uncertainty, down to 4 wt% ^{240}Pu content, and showed that the originally determined USL is still valid. Therefore, the staff concludes that the definition of AOA(2) may be extended down to 4 wt% ^{240}Pu content without the need for additional margin.

Therefore, based on the foregoing considerations, the staff has reasonable assurance that systems meeting the criteria in the table above are within the scope of AOA(2) and will be safely subcritical with a USL, as defined in Part I of the validation report (USL = 0.9321).

6.1.3.5.1.3 <u>AOA(3)—Plutonium Oxide Powder</u>. To validate AOA(3), the applicant modeled 70 critical experiments using the KENO-VI code with the 238-group cross section library. The applicant modeled some of the benchmarks using the INFHOMMEDIUM cross section treatment option and some using the LATTICECELL treatment option. The validation report does not clearly state which code options (including cross section treatment) were considered validated. The bias was determined for the SCALE-4.4a code package on the SGN PC hardware platform. The applicant's chosen experiments include reflected and bare plutonium-metal systems and arrays of plutonium dioxide- and MOX-polystyrene compacts. The applicant selected this set of experiments from a much larger suite of 308 candidate benchmarks by comparing the most important physical and neutronic parameters of the candidate benchmark experiments to those of anticipated design applications. To do this, the applicant chose a set of screening criteria consisting of allowable ranges of important physical parameters, as described in NUREG/CR-6698. The staff examined the chosen benchmarks to determine whether they were appropriate for a validation of systems covered by AOA(3). The evaluation of the chosen benchmarks is discussed below.

The applicant defined the validated range of parameters for AOA(3) in Table 5-2 of Part II of the validation report. The validated AOA is described in Table 6.1-3.

Table 6.1-3 Definition of AOA(3)

Geometry	Parallelepipeds, arrays of cylinders, spheres
Reflectors	Water
Chemical form	PuO_2 powder
Pu/(U + Pu)	100 wt%
^{240}Pu	4 wt%
H/Pu	0–15
EALF	5 eV–266 keV

The applicant ran SCALE/KENO-VI for the chosen experiments, calculated k_{eff}, and plotted the value of k_{eff} versus the trending parameters of EALF, the moderator to fuel ratio (H/Pu), and the [240]Pu content. Because the data were not normally distributed, a non-parametric estimate of the USL was made, assuming an administrative margin of 0.05. Based on the number of benchmarks included, no additional NPM was needed. The non-parametric estimate was used as the USL for AOA(3).

The staff compared the range of parameters covered by the benchmark data to the applicant's AOA and determined that the benchmarks provide adequate coverage of the defined AOA across the entire range.

To select the benchmarks, the applicant followed the steps in Section 2.5 of NUREG/CR-6698, which consist of (1) identifying key parameters associated with normal and upset conditions of systems to be evaluated, (2) based on these key parameters, establishing a screening AOA for critical experiments, (3) identifying critical experiments that are within this screening AOA, and (4) from the scope of selected critical experiments, determining the detailed AOA covered by the experiments. A final step, consisting of showing that design applications are within the detailed AOA, will be conducted as part of the overall final design process (as indicated in Figure 3-1 of Part II of the validation report) and is outside the scope of the construction validation review.

With regard to the implementation of this methodology, the staff noted that the screening criteria used were significantly broader in some cases than those in Table 2.3 of NUREG/CR-6698. The values in this table are the result of consensus among criticality experts and acknowledged to be conservative. The staff questioned the ranges of parameters in the screening criteria and the justification for extending them beyond the ranges in Table 2.3 of NUREG/CR-6698. In addition, the parameters used in the primary screening criteria ([239]Pu content, [240]Pu content, H/Pu, and EALF) did not include the total plutonium content (wt% of plutonium in MOX), which is known as one of the parameters to which k_{eff} is most sensitive. Since it could not resolve all the technical concerns with the applicant's implementation of the NUREG/CR-6698 methodology, the staff examined all the chosen benchmarks on a case-by-case basis to make an independent determination regarding their inclusion for a validation of AOA(3). There were two broad categories of benchmarks, those involving plutonium metal and those involving plutonium oxide powder, which needed additional justification. The staff's review of the applicability of these benchmarks is described below.

Plutonium-Metal Benchmarks

The staff included several sets of plutonium-metal benchmarks (PU-MET-FAST-001, -003, -016, -017, -033, and -037) in the benchmarks for AOA(3). At the November 13, 2003, meeting, the applicant stated that these were included to validate plutonium dioxide powder storage array applications at the front end of the AP process (NRC, 2003b). The applicant stated that, for some array cases, dry plutonium dioxide powder at full theoretical density (i.e., with H/Pu = 0) very closely resembles plutonium metal and that it is the most reactive case. This is relevant because the most reactive normal condition case primarily determines the subcritical limits for the proposed process. During the in-office review (NRC, 2004a), the staff examined proposed calculations for the 3013 storage and the plutonium buffer storage units and determined that these plutonium dioxide powder storage arrays are physically very similar to the configuration of several of the plutonium-metal benchmarks (particularly, PU-MET-FAST-003, -016, -017, and -037). For both these plutonium dioxide storage applications and the plutonium-metal benchmarks, the configuration consists of bare or reflected arrays of high-density plutonium encased in steel cans, with similar isotopics, absorber materials, geometric arrangement, and neutron fission spectra. Thus, the most important parameters characterizing these systems are very similar to those of the chosen benchmarks.

To confirm that the dry case is the most reactive, the applicant performed a sensitivity study showing that plutonium dioxide powder at full theoretical density and an H/Pu = 0 results in a higher k_{eff} than plutonium dioxide powder with lower densities and a moisture content up to 1 wt% water (H/Pu approximately 0.3) (NRC, 2004a). The staff performed its own confirmatory calculations that show that the dry case is the most reactive, provided that the moisture content does not exceed approximately 1.25 wt% (at which point k_{eff} will approach that of the dry case). Because normal-condition moisture content is limited to 1 wt% water content (as stated in CAR Section 6.3.4.3.2.6), the dry case is expected to be bounding for some array calculations under normal conditions, and thus it is appropriate to include H/Pu = 0 cases in the definition of AOA(3).

There are two differences between the material composition of plutonium metal and plutonium dioxide powder that can theoretically affect the bias—(1) differing material density and (2) differing chemical composition. The staff investigated the importance of these two effects relative to other differences between the systems. The staff performed independent calculations with different-sized spheres containing various plutonium dioxide-water mixtures, using the XSDRNPM discrete ordinates code. To gain further insights, the staff also performed S/U calculations using the TSUNAMI sequences of the SCALE 5 code package. These results all show that k_{eff} is much less sensitive to density than to other system parameters; what sensitivity exists appears because of changes in the energy spectrum and the amount of leakage that occurs with changing density. To confirm this, the staff performed an additional sensitivity study of the variation of k_{inf} as a function of water content and EALF for several different values of density (3–10 g/cm^3(0.11–0.36 $lb/in.^3$)). The model consists of an infinite fissile medium modeled with the KENO-Va Monte Carlo code. When k_{inf} is plotted as a function of H/Pu or EALF at several different densities, the curves at different densities coincide, indicating that k_{eff} is relatively insensitive to powder density when leakage can be neglected. (The leakage is related to the neutron energy and system geometry, which the TSUNAMI codes show could be considered parameters of secondary importance for criticality code validation.) The S/U calculations discussed above corroborate this, as there was a very high correlation coefficient between the plutonium-metal benchmarks and dry plutonium dioxide powder cases with several different densities. Subsequent to completion of its independent evaluation, the staff reviewed an additional analysis prepared by the applicant in the supplement to the validation report, DCS01 ZJJ DS CALC H 35075 A, "MFFF Validation Report Supplemental Information," dated January 31, 2005 (DCS, 2005; NRC, 2005). The NRC considers this to be part of the licensing basis in support of the validation effort for the MFFF. In this analysis, DCS modeled 83 different plutonium dioxide-water mixtures, ranging from 0–0.999 weight fraction of water, for 20 different values of powder density, ranging from theoretical density down to 5 percent of theoretical density. For each density value, the powder and water density were both scaled by the same density factor to maintain a constant H/X ratio. The K_{inf} values were chosen for the comparison because there is a strong correlation between density, neutron leakage, and energy for a finite system. The K_{inf} values were used to isolate the direct effect of density on k_{eff}, in the absence of leakage. When k_{inf} is plotted on a three-dimensional graph as a function of both water content and density, k_{inf} appears to be totally independent of density. As further demonstration, the residuals $(k_{eff}–k_{REF})/k_{REF}$ were plotted on a quantile-quantile plot and observed to be normally distributed, such that the variation can be inferred to be caused only by Monte Carlo statistical fluctuation. The neutron spectra causing fission for cases with the same water fraction are also identical. Therefore, the staff concludes that variations in powder density have an insignificant effect on the bias and may be ignored for the purpose of choosing validation benchmarks.

In addition, differences in the chemical form between plutonium dioxide powder and plutonium metal are not expected to be significant because ^{16}O is essentially transparent to neutrons across most of the energy range. In addition to evaluating these differences in composition, the staff also evaluated the effects of heterogeneous geometry. The applicant performed calculations showing that the effect of resonance self-shielding in the plutonium-metal array calculations is small (NRC, 2004a). When the LATTICECELL cross section treatment option was used instead of INFHOMMEDIUM, the result was a very small difference in k_{eff} (less than or equal to 0.3 percent) (NRC, 2004a). This confirmed the staff's expectation that the cross-section treatment option would have a small effect. This expectation resulted from the presence of a very fast spectrum and therefore very little absorption in the resonance region.

Based on these foregoing considerations, the staff has reasonable assurance that the assumed design applications of AOA(3) include storage applications involving dry, full-density plutonium dioxide powder and that the plutonium-metal benchmarks are applicable to the validation of these cases in AOA(3).

Plutonium Dioxide Powder Benchmarks

PU-COMP-MIXED-001: This benchmark set consists of bare arrays of plutonium dioxide-polystyrene compacts. These benchmarks are moderated internally by polystyrene, which has similar moderating effects to water. The staff found the geometrical arrangement (a single homogeneous fuel region with vacuum boundary conditions) comparable to that of anticipated design applications. The staff compared the values of benchmark parameters to the screening criteria and concluded that they were sufficiently close to those of anticipated design applications. The staff noted that, in some cases, the ^{240}Pu content is somewhat outside the range endorsed in NUREG/CR-6698. This is not considered significant because the staff's independent calculations and bias trending show that the bias is relatively insensitive to the ^{240}Pu proportion. Based on this review, the staff considers the three experiments chosen from the PU-COMP-MIXED-001 benchmark set to be applicable for validation of AOA(3).

PU-COMP-MIXED-002: This benchmark set consists of water-reflected arrays of plutonium dioxide-polystyrene compacts. These benchmarks are moderated internally by polystyrene, which has similar moderating effects to water. The staff found the geometrical arrangement (a single homogeneous fuel region with external hydrogenous reflector) comparable to that of anticipated design applications. The staff compared the values of benchmark parameters to the screening criteria and concluded that they are sufficiently close to those of anticipated design applications. The staff noted that, in some cases, the ^{240}Pu content is somewhat outside both the applicant's screening criteria and the range endorsed in NUREG/CR-6698. This is not considered significant because the staff's independent calculations and bias trending show that the bias is relatively insensitive to the ^{240}Pu proportion. Subsequent to completion of its independent evaluation, the staff reviewed an additional analysis prepared by the applicant in the supplement to the validation report, DCS01 ZJJ DS CALC H 35075 A, "MFFF Validation Report Supplemental Information," dated January 31, 2005 (DCS, 2005; NRC, 2005). The NRC considers this to be part of the licensing basis in support of the validation effort for the MFFF. In this analysis, DCS performed a statistical evaluation of the benchmark k_{eff} values as a function of ^{240}Pu content, using the ORNL code USLSTATS. This was done for the PU-COMP-MIXED-001 and -002 benchmark experiments in AOA(3), which cover the range in ^{240}Pu content from approximately 2–18 wt%. Therefore, no extrapolation was needed, and this analysis shows that the originally determined USL is still valid. The staff concludes that the definition of AOA(3) adequately covers cases down to 4 wt% ^{240}Pu content. The staff considers the 22 experiments chosen from the PU-COMP-MIXED-002 benchmark set to be applicable for validation of AOA(3).

Low Plutonium Assay Powder Benchmarks: Combining the metal and powder benchmarks discussed above gives a minimum of 49 applicable critical experiments. Based on statistical techniques employed in the validation report with 49 benchmarks, no additional NPM is required. The applicant included additional low-plutonium benchmarks; however, the staff did not make any determination as to the applicability of these benchmarks (experiment sets BNWL2129T4, NSE55T5, and PU-29) for validation of AOA(3). The non-parametric technique bases calculation of the USL on the lowest observed k_{eff}; thus, including more experiments can only decrease the USL. Therefore, the acceptance of these experiments is irrelevant to the determination of whether there is sufficient margin of subcriticality.

Combination of Metal and Powder Benchmarks in AOA(3)

The staff questioned, however, whether the plutonium-metal benchmarks were sufficiently similar to the plutonium dioxide-polystyrene benchmarks to be combined in the same AOA. Because the non-parametric method applies a statistical margin that is a function only of the total number of experiments, combining disparate benchmarks together could cause the margin to be artificially decreased. There is also almost no guidance on when it is appropriate to combine different types of benchmarks in a single AOA. The staff, therefore, had to develop a criterion for determining when it is appropriate to combine different types of benchmarks in a single AOA. Many of the powder benchmarks have been shown to apply to slightly moderated plutonium dioxide applications, and the metal benchmarks have been shown to apply to certain dry plutonium dioxide storage applications. The finding that each of these sets of experiments is acceptable implies that the range of parameters covered by each group is not too broad for the experiments to be evaluated together in validation. Therefore, the staff determined that it would consider combining the two sets of benchmarks in a single AOA to be acceptable if the difference between the two sets is less than or comparable to the difference within each set. Conversely, if the difference between the two sets is much greater than the difference within each set, they should be considered as constituting two separate AOAs. To make this determination, the staff used several different types of comparisons between the plutonium-metal benchmarks and plutonium dioxide powder benchmarks.

The staff first compared the ranges of the most important physical parameters covered by each set of benchmarks (i.e., H/(U + Pu), plutonium content, and EALF), including trends of k_{eff} as a function of these parameters. The staff noted that there is significant overlap between many of the parametric ranges covered by the two sets of benchmarks. Where the parametric ranges do not overlap, they are contiguous (i.e., there are no discernable gaps between them). The staff did not note any significant differences in trends, with the exception that there appears to be a different trend in EALF for the plutonium-metal and plutonium dioxide powder benchmarks. The somewhat different behavior of the two sets may be responsible for their not following a normal distribution. This uncertainty in the distribution is taken into account in the nonparametric treatment.

Because of the apparent difference in bias trends, the staff compared the spectra for the two sets of benchmarks in more detail. The staff collapsed the 238 energy groups into 30 groups and divided the entire energy range into three regions, based on the three categories in Table 2.3 of NUREG/CR-6698 (i.e., thermal = 0–1 eV, intermediate = 1 eV–100 keV, and fast = 100 keV–20 MeV). The rebinning was done to improve statistical accuracy. The staff compared histograms of the fast, intermediate, and thermal fission fractions for the two benchmark sets. The distributions of fast, intermediate, and thermal fission fractions showed a significant degree of overlap or contiguity. The staff then compared the fission spectra of the two benchmark sets across the entire energy range to determine how much difference there was in the fission spectra within each benchmark set and between the benchmark sets. For each benchmark set, the maximum, minimum, and average fission contribution in each energy group was then tabulated and graphed. The results show that the spread within the plutonium-metal and within the plutonium dioxide powder benchmark sets is comparable to or greater than (for each energy group) the spread between the averages of the two sets. That is, there is as much or more variation in the fission spectra when comparing different plutonium-metal benchmarks (or different plutonium dioxide powder benchmarks) than when comparing the typical plutonium-metal and typical plutonium dioxide powder benchmark. The bands formed by the minimum and maximum fission contributions in each energy group show a significant degree of overlap or contiguity. Subsequent to completion of its independent evaluation, the staff reviewed an additional analysis prepared by the applicant in the supplement to the validation report, DCS01 ZJJ DS CALC H 35075A,"MFFFValidation Report Supplemental Information," dated January 31, 2005 (DCS, 2005; NRC, 2005). The NRC considers this to be part of the licensing basis in support of the validation effort for the MFFF. In this analysis, DCS compared the neutron spectra causing fission for the two groups of benchmarks in AOA(3) (the 24 plutonium-metal benchmarks and the 25 plutonium dioxide-polystyrene benchmarks). The spectra corresponding to plutonium-metal benchmarks were graphed in red and the spectra corresponding to plutonium-powder benchmarks in blue. When these were overlaid, there was significant overlap across the entire neutron energy spectrum (i.e., fission fraction as a function of the EALF). In addition, the applicant plotted the maximum, minimum, and average fission fraction within each energy group for each of the two classes of benchmarks. In addition to considering the spectra using the 238-group structure, the applicant also collapsed the data into 30 energy groups to obtain better statistics within each group. These analyses all show that, on an energy-group basis, the spread in the fission fraction within either the metal or powder benchmark set is at least as great as the difference between the average values for each benchmark set. Therefore, the staff concludes that the difference between these two sets is smaller than or comparable to the variation within each set and that there is sufficient overlap in the spectra that they may be combined for validation purposes.

In addition, use of the TSUNAMI sequences in the SCALE 5 code package shows that, for fast systems, the reactivity is dominated by ^{239}Pu fission. This observation is particularly important for the acceptance of plutonium-metal benchmarks for this AOA. Subsequent to completion of its independent evaluation, the staff reviewed an additional analysis prepared by the applicant in the supplement to the validation report, DCS01 ZJJ DS CALC H 35075 A, "MFFF Validation Report Supplemental Information," dated January 31, 2005 (DCS, 2005; NRC, 2005). The NRC considers this to be part of the licensing basis in support of the validation effort for the MFFF. In this analysis, DCS compared the macroscopic cross sections for dominant reactions that contribute to the reactivity of a plutonium dioxide-water system in the high-energy range. The nuclides considered were ^{239}Pu, ^{240}Pu, ^{16}O, and ^{1}H. The reactions evaluated were fission (n,f), radiative capture (n,γ), elastic scattering (n,n), and neutron knockout (n,2n). Three different fissile media were considered, corresponding to different plutonium dioxide densities and moderation conditions. It is not necessary to consider higher values of moderation because this demonstration is limited to the high-energy (defined as 100 keV–10 MeV) range. Based on a comparison of the kinematics for the neutron energy loss in elastic scattering, it was determined that elastic scattering is only an important contribution from ^{16}O and ^{1}H. The microscopic cross sections for these nuclide-reaction pairs were multiplied by appropriate number densities to determine the macroscopic cross sections. This shows that, in the highest energy range, ^{239}Pu(n,f) dominates all other reactions by at least an order of magnitude. In addition, the relative ranking of nuclide-reaction pairs is consistent with the ranking of sensitivity coefficients determined by the NRC staff using TSUNAMI. Therefore, the staff concludes that ^{239}Pu fission is the dominant reaction for AOA(3) systems in the high-energy limit.

As systems become increasingly moderated, however, absorption becomes increasingly important and the sensitivity to other materials (e.g., ^{240}Pu, hydrogen, and absorbers) increases. The applicant provided justification as to why the fission spectrum is more sensitive to differences between physical systems than the absorption or leakage spectrum (NRC, 2004a). The basis for concluding that the leakage spectrum is relatively insensitive to differences between systems is that (1) for fully reflected systems, there is essentially no leakage and (2) for bare or nominally reflected systems, the leakage spectrum is preferentially skewed towards fast neutron energies (and thus fission dominates). The staff considers this reasonable based on the known behavior of systems with these boundary conditions. In addition, for fully reflected systems, the absorption spectrum is greatly dominated by parasitic absorption in the external reflector and thus relatively insensitive to changes in the interior physical system.

With regard to the absorption spectrum, the applicant provided graphs of the fission, absorption, and leakage spectra as a function of benchmark number (NRC, 2004a). These graphs show that there is considerably more variation in the fission spectrum than in either the absorption or leakage spectrum. This leads to the conclusion (at least for these experiments) that comparisons of the fission spectra are more indicative of similarity between systems than comparisons of the absorption or leakage spectra. As an additional check, the staff performed the same analysis described above for the absorption spectrum (i.e., comparison of the fast, intermediate, and thermal absorption fractions, and comparison of the spectra on a group-by-group basis). The staff analysis shows very similar results to the analysis done on the fission spectra because the difference in absorption characteristics between the two sets of benchmarks is comparable to or less than the difference within each set.

The staff noted that there was a weak correlation between the metal and powder benchmarks when the TSUNAMI code sequences were used to compare all the benchmarks in pair-wise fashion. In light of the results of the spectral comparison described above, the staff surmises this is likely because of parametric differences (most importantly H/Pu ratio and plutonium content). As previously noted, however, the ranges in these parameters covered by the two benchmark sets have considerable overlap or are reasonably contiguous.

Therefore, based on the relative insensitivity of k_{eff} to density and chemical form (presence of ^{16}O), physical considerations about the relative importance of fission, absorption, and leakage, and a comparison of parametric ranges and energy spectra for the two sets of benchmarks, the staff concludes that it is appropriate to validate AOA(3) using both the plutonium-metal and plutonium dioxide powder benchmarks. The combination of these two benchmark sets results in 49 benchmarks found acceptable by the staff for validating AOA(3). In accordance with the non-parametric method described in Part II of the validation report and NUREG/CR-6698, no additional NPM is required.

Therefore, based on the foregoing considerations, the staff has reasonable assurance that the systems meeting the criteria in the table above are within the scope of AOA(3) and will be safely subcritical with a USL, as defined in Part II of the validation report (USL = 0.9345).

6.1.3.5.1.4 <u>AOA(4)—Mixed Oxide Powder</u>. To validate AOA(4), the applicant modeled 59 critical experiments using the KENO-VI code with the 238-group cross-section library. The applicant modeled some of the benchmarks using the INFHOMMEDIUM cross section treatment option, and some using the LATTICECELL treatment option. The validation report does not clearly state which code options (including cross section treatment) were considered validated. The bias was determined for the SCALE-4.4a code package on the SGN PC hardware platform. The applicant's chosen experiments included reflected arrays of plutonium dioxide- and MOX-polystyrene compacts. The applicant selected this set of experiments from a much larger suite of 308 candidate benchmarks by comparing the most important physical and neutronic parameters of candidate benchmark experiments to those of anticipated design applications. To do this, the applicant chose a set of screening criteria, consisting of allowable ranges of important physical parameters, as described in NUREG/CR-6698. The staff examined the chosen benchmarks to determine whether they were appropriate for a validation of systems covered by AOA(4). The evaluation of the chosen benchmarks is discussed below.

The applicant defined the validated range of parameters for AOA(4) in Table 5-4 of Part II of the validation report. The validated AOA is described in Table 6.1-4.

The applicant ran SCALE/KENO-VI for the chosen experiments, calculated k_{eff}, and plotted the value of k_{eff} versus the trending parameters of EALF, the moderator to fuel ratios (H/(U + Pu) and H/Pu), and the ^{240}Pu and total plutonium content. Because the data were not normally distributed, a non-parametric estimate of the USL was made, assuming an administrative margin of 0.05. Based on the number of benchmarks included, no additional NPM was needed. The non-parametric estimate was used as the USL for AOA(4).

The staff compared the range of parameters covered by the benchmark data to the applicant's AOA and determined that the benchmarks provided adequate coverage of the defined AOA across the entire range (except as discussed below).

Table 6.1-4 Definition of AOA(4)

Geometry	Parallelepipeds, spheres
Reflectors	Water, depleted uranium*
Chemical form	MOX powder
Pu/(U + Pu)	8–22 wt%
^{240}Pu	4 wt%
H/(U + Pu)	2.8–15*
EALF	0.63–92.6 eV*
*Except as described below	

To select the benchmarks, the applicant followed the steps in Section 2.5 of NUREG/CR-6698, which consist of (1) identifying key parameters associated with normal and upset conditions of systems to be evaluated, (2) based on these key parameters, establishing a screening AOA for critical experiments, (3) identifying critical experiments that are within this screening AOA, and (4) from the scope of selected critical experiments, determining the detailed AOA covered by the experiments. The final step, showing that design applications are within the detailed AOA, will be conducted as part of the overall final design process (as indicated in Figure 3-1 of Part II of the validation report) and is outside the scope of the construction validation review.

With regard to the implementation of this methodology, the staff noted that the screening criteria are significantly broader in some cases than those in Table 2.3 of NUREG/CR-6698. The values in Table 2.3 of NUREG/CR-6698 are the result of consensus among criticality experts and may be considered to be conservative (i.e., they are narrowly defined to provide a very high degree of confidence of the benchmark applicability). The staff questioned the ranges of parameters in the screening criteria and the justification for extending them beyond the ranges in Table 2.3 of NUREG/CR-6698. In addition, the parameters used in the primary screening criteria (^{239}Pu content, ^{240}Pu content, H/(U + Pu), and EALF) do not include the total plutonium content (wt% of plutonium in MOX), which is known to be one of the parameters to which k_{eff} is most sensitive. Since the staff could not resolve all the technical concerns with the applicant's implementation of the NUREG/CR-6698 methodology, the staff examined all the chosen benchmarks on a case-by-case basis to evaluate whether they should be included for a validation of AOA(4). The staff's review of the applicability of these benchmarks is described below.

NSE55T5: This benchmark set consists of plexiglass-reflected arrays of MOX-polystyrene compacts. These benchmarks are moderated internally by polystyrene and reflected by plexiglass, which have similar moderating and reflecting effects to water. The staff found the geometrical arrangement (a single homogeneous fuel region with external hydrogenous reflector) comparable to that of anticipated design applications. The staff compared the values of benchmark parameters to the screening criteria and concluded that, in most cases, they are sufficiently close to those of anticipated design applications. The staff noted that the H/(U + Pu) and ^{240}Pu content are somewhat outside the range endorsed in NUREG/CR-6698. The H/(U + Pu) was considered slightly high. Despite this, the EALF of these benchmarks, which is strongly correlated to the H/(U + Pu) ratio, remain within appropriate bounds. Bias trending showed that there is an unexpectedly strong correlation between k_{eff} and the ^{240}Pu content. This was unexpected because calculations using the TSUNAMI code sequences have shown that the system sensitivity to the ^{240}Pu content is typically low. The trend is such that including high ^{240}Pu benchmarks results in a lower k_{eff} and therefore reduces the overall USL. Subsequent to completion of its independent evaluation, the staff reviewed an additional analysis prepared by the applicant in the supplement to the validation report, DCS01 ZJJ DS CALC H 35075 A, "MFFF Validation Report Supplemental Information," dated January 31, 2005 (DCS, 2005; NRC, 2005). The NRC considers this to be part of the licensing basis in support of the validation effort for the MFFF. In this analysis, DCS performed a statistical analysis of the benchmark k_{eff} values as a function of ^{240}Pu content, using the ORNL code USLSTATS. Using the results from the confidence band technique generated by USLSTATS, the applicant determined the functional form of both the calculated k_{eff} and its uncertainty, k(x) and w(x), as a function of ^{240}Pu content. The applicant then used this information to extend the trend, including allowance for uncertainty, down to 4 wt% ^{240}Pu content and showed that the originally determined USL is still valid. Therefore, the staff concludes that the definition of AOA(4) may be extended down to 4 wt% ^{240}Pu content without the need for additional margin. Because of outstanding questions concerning the H/(U + Pu) ratio and ^{240}Pu content, the staff considers the 10 experiments chosen from the NSE55T5 benchmark set to be reasonably applicable for validation of AOA(4).

PU-29: This benchmark set consists of plexiglass-reflected arrays of MOX-polystyrene compacts. These benchmarks are moderated internally by polystyrene and reflected by plexiglass, which have similar moderating and reflecting effects to water. The staff found the geometrical arrangement (a single homogeneous fuel region with external hydrogenous reflector) comparable to that of anticipated design applications. The staff compared the values of benchmark parameters to the screening criteria and concluded that, in most cases, they are sufficiently close to those of anticipated design applications. The staff noted that the H/(U + Pu) and ^{240}Pu content are somewhat outside the range endorsed in NUREG/CR-6698, as discussed above for NSE55T5. Because of outstanding questions concerning the H/(U + Pu) ratio and ^{240}Pu content, the staff considers the nine experiments chosen from the PU-29 benchmark set to be reasonably applicable for validation of AOA(4).

BNWL2129T4: This benchmark set consists of plexiglass-reflected arrays of MOX-polystyrene compacts. These benchmarks are moderated internally by polystyrene and reflected by plexiglass, which have similar moderating and reflecting effects to water. The staff found the geometrical arrangement (a single homogeneous fuel region with external hydrogenous reflector) comparable to that of anticipated design applications. The staff compared the values of benchmark parameters to the screening criteria and concluded that, in most cases, they are sufficiently close to those of anticipated design applications. The staff noted that the H/(U + Pu) and ^{240}Pu content are somewhat outside the range endorsed in NUREG/CR-6698, as discussed above for NSE55T5. Because of outstanding questions concerning the H/(U + Pu) ratio and ^{240}Pu content, the staff considers the 19 experiments chosen from the BNWL2129T4 benchmark set to be reasonably applicable for validation of AOA(4).

PU-8: This benchmark set consists of plexiglass-reflected arrays of MOX-polystyrene compacts. These benchmarks are moderated internally by polystyrene and reflected by plexiglass, which have similar moderating and reflecting effects to water. The staff found the geometrical arrangement (a single homogeneous fuel region with external hydrogenous reflector) comparable to that of anticipated design applications. The staff compared the values of benchmark parameters to the screening criteria and concluded that, in most cases, they are sufficiently close to those of anticipated design applications. The staff noted that the H/(U + Pu) and ^{240}Pu content are outside the ranges endorsed by NUREG/CR-6698. The H/(U + Pu) was considered well outside the screening range in NUREG/CR-6698, although the EALF of these benchmarks, which is strongly correlated to the H/(U + Pu) ratio, remains within appropriate bounds. The situation with regard to the ^{240}Pu content is as discussed above for NSE55T5. Because of the high H/(U + Pu) values, the staff does not consider the four experiments chosen from the PU-8 benchmark set to be demonstrated applicable for validation of AOA(4).

PU-COMP-MIXED-002: This benchmark set consists of water-reflected arrays of plutonium dioxide-polystyrene compacts. These benchmarks are moderated internally by polystyrene, which has similar moderating effects to water. The staff found the geometrical arrangement (a single homogeneous fuel region with external hydrogenous reflector) comparable to that of anticipated design applications. The staff compared the values of benchmark parameters to the screening criteria and concluded that they are significantly different from those of anticipated design applications. The staff found that, in several cases, the H/(U + Pu) ratio is significantly outside the range endorsed in NUREG/CR-6698. The total plutonium content (100 wt% plutonium) is also significantly outside the range endorsed in NUREG/CR-6698. The applicant's screening criteria specified a range of allowable plutonium content of 1.2 to 52 wt% Pu/(U + Pu). The staff performed independent calculations using the TSUNAMI sequences of the SCALE 5 code package to obtain more information about the bias sensitivity to plutonium content. This was done by comparing a reference case with a fixed plutonium content to a series of test cases with varying plutonium content. The calculations consist of series of MOX-water spheres at the same H/X but with different plutonium content. These calculations show that the correlation between MOX reference cases with the same plutonium content as AOA(4) (6.3–22 wt% plutonium) and test cases with arbitrary plutonium content drops off substantially for test cases in the high plutonium range (greater than or equal to 60 wt% plutonium). Similarly, the correlation between pure plutonium dioxide reference cases and MOX test cases drops off substantially for test cases in the low plutonium range (less than or equal to 30 wt% plutonium).

To justify the inclusion of 100 wt% plutonium benchmarks in the data set for AOA(4), the applicant provided two sets of calculations (NRC, 2004a). The first consists of a comparison of the fission spectra for 30 and 100 wt% plutonium cases, and the second consists of a sensitivity study of the effect on k_{eff} of ^{238}U removal. With regard to fission spectra, the results show that there is a high degree of similarity between the spectra for the 30 and 100 wt% cases. The staff remarked during the in-office review that the TSUNAMI results suggest that similarity would decrease significantly below approximately 30 wt%. In response to this, the applicant provided some additional calculations comparing 22 and 100 wt% plutonium cases that showed similar behavior. The staff examined the TSUNAMI results described above in greater detail to gain additional insights into the similarities and differences between low-plutonium MOX and 100 wt% plutonium systems. These calculations showed that, for very fast (low-moderation) plutonium dioxide systems, the sensitivity is dominated by ^{239}Pu fission, but that as systems become increasingly moderated and the plutonium content decreases, other nuclides and reactions become increasingly important. For example, for systems with plutonium content in the range covered by AOA(4), the dominant nuclide-reaction pairs include ^{238}U fission and absorption.

Comparing only the ^{239}Pu fission spectra does not take into account the effect of these competing nuclide-reaction pairs.

In the ^{238}U-removal study, the applicant performed calculations to determine the change in k_{eff} when all of the ^{238}U was removed from 30 wt% plutonium cases. This shows that the ^{238}U produces a Δk_{eff} up to approximately 8 percent. Based on this, the applicant concluded that the presence of ^{238}U at these levels is not significant to validation. However, the applicant did not justify that Δk_{eff} corresponds to an acceptable degree of similarity. Because both ^{238}U fission and absorption were shown to be significant in the low plutonium range and they produce opposite effects on k_{eff}, simply studying their combined effect on k_{eff} may mask the significance of either fission or absorption; an error in either the absorption or fission cross section could produce a bias in k_{eff}. Because of the outstanding questions concerning the sufficiency of the comparison of the ^{239}Pu fission spectra and the ^{238}U-removal study, the staff does not consider the applicability of 100 wt% plutonium content benchmarks to be demonstrated for applications in AOA(4).

In addition, for some of the cases, the ^{240}Pu content is somewhat outside the range endorsed in NUREG/CR-6698, as discussed above for NSE55T5. Because of the high H/(U + Pu) values and high plutonium content, the staff does not consider the 17 experiments chosen from the PU-COMP-MIXED-002 benchmark set to be demonstrated applicable for validation of AOA(4).

Based on the preceding review, the staff determined that, out of the 59 benchmarks selected for AOA(4), only 38 are marginally acceptable without further justification (i.e., those in the NSE55T5, PU-29, and BNWL2129T4 benchmark sets). The non-parametric technique from NUREG/CR-6698 followed by the applicant specifies that, with 38 applicable benchmarks that are not normally distributed, an NPM of 1 percent in k_{eff} is appropriate. The staff therefore considers an NPM of 1 percent in k_{eff} necessary to provide reasonable assurance that MOX powder systems will be subcritical under normal and credible abnormal conditions.

In accordance with the methodology of NUREG/CR-6698, the staff examined the parametric range covered by the 38 benchmark experiments and determined that they do not adequately cover the entire validated AOA defined in Table 5-2 of Part II of the validation report. In particular, the 38 benchmarks do not cover the highest and lowest EALF ranges or the high H/(U + Pu) range. Based on the span of the data, the staff considers the range of parameters in Table 6.1-5 validated with margin as described above.

In addition to restrictions on H/(U + Pu) and EALF, the staff noted that the 38 benchmarks did not contain depleted uranium reflectors. However, the staff's independent calculations using the TSUNAMI sequences of the SCALE 5 code package show a reasonably high correlation between bare systems and those reflected with 60 cm of depleted uranium (with the same H/(U + Pu) ratio). The staff also notes that all design applications covered by AOA(4) contain a large proportion of depleted uranium as uranium dioxide in the fissile region (78 and 93.5 wt% uranium dioxide). This is suggestive that cases containing depleted uranium reflectors may be within the scope of AOA(4). Subsequent to completion of its independent evaluation, the staff reviewed an additional analysis prepared by the applicant in the supplement to the validation report, DCS01 ZJJ DS CALC H 35075 A, "MFFF Validation Report Supplemental Information," dated January 31, 2005 (DCS, 2005; NRC, 2005). The NRC considers this to be part of the

Table 6.1-5 Definition of AOA(4)

Geometry	Parallelepipeds, spheres
Reflectors	Water, depleted uranium Depleted uranium up to a reflector of 60 cm thickness
Chemical form	MOX powder
Pu/(U + Pu)	8–22 wt%
^{240}Pu	4 wt%
H/(U + Pu)	2.77–2.79
EALF	1.49–43.5 eV

licensing basis in support of the validation effort for the MFFF. In this analysis, DCS performed a series of sensitivity calculations to determine the effect of depleted uranium reflectors on the calculated k_{eff}. The first series of runs consisted of spheres containing MOX powder with 0, 1, 3, and 5 wt% water content, surrounded by spherical shells of depleted uranium of varying thicknesses. The results of these calculations show that the presence of depleted uranium results in a significant increase in the system reactivity up to approximately 20 cm thickness, but beyond that, the system k_{eff} is nearly constant. A reduction of 67 percent in the depleted uranium cross sections would thus be required to have a significant effect on the system k_{eff}.

DCS then performed a series of sensitivity calculations applicable to the ball-milling operation in the MP. This series consisted of a heterogeneous fissile region containing a spherical core of plutonium dioxide and water surrounded by a spherical shell of depleted uranium dioxide and water. This arrangement was reflected by a shell containing 500 kg of depleted uranium, the same mass as in the ball milling operation. These results show that, unlike the homogeneous case, there is only a Δk_{eff} of approximately 5 percent in going from 2 to 6 cm of depleted uranium. The staff reviewed the calculation for the ball milling operation, DCS01 NBX CG CAL H 05629D, "Criticality Safety of Ball Milling Units NBX and NBY," which demonstrates that the heterogeneous two-zone configuration is more reactive than the homogeneous case and thus was used to determine criticality safety limits for this operation. The applicant also performed a calculation in which a depleted uranium packing fraction of approximately 74 percent (corresponding to a hexagonal close-package array) was used instead of 100 percent. This is because the depleted uranium will not realistically form a contiguous shell around the fissile material. This calculation shows that there is an approximately 2-percent conservatism from assuming that the depleted uranium formed a uniform shell.

DCS also provided a comparison of the fission spectra for the bare, water-reflected, and depleted uranium-reflected calculations and showed that the spectra were nearly identical, except with regard to the small contribution under the thermal peak. The fact that systems covered under AOA(4) will have a significant quantity of depleted uranium mixed in with the fissile medium (78 percent depleted uranium for the master blend) and that the spectra are nearly identical shows that the depleted uranium cross sections are well-represented by MOX benchmarks applicable to AOA(4). The staff believes that the presence of ^{238}U and the similarity in the spectra are responsible for the high degree of correlation evidenced in its TSUNAMI calculations, even though k_{eff} exhibits significant sensitivity to the presence of depleted uranium reflectors of up to 20-cm thickness.

The staff concluded from the analysis provided by the applicant that (1) it would take a 2/3 error in the depleted uranium cross sections to produce a significant effect on k_{eff}, (2) the heterogeneous case is bounding and is used as the basis for criticality safety limits, (3) the heterogeneous case is much less sensitive to the presence of depleted uranium than the homogeneous case, (4) the highest-plutonium MOX applications still have 78 percent depleted uranium in the fissile medium, and (5) there is considerable overlap between the spectra of bare, water-reflected, and depleted uranium-reflected cases covered by AOA(4). Therefore, based on the staff's review of the additional analysis prepared by the applicant and the TSUNAMI results described above, the staff has reasonable assurance that design applications containing depleted uranium with no greater worth in k_{eff} than a 60-cm external reflector are within the scope of AOA(4).

Based on the foregoing review, the staff has reasonable assurance that systems meeting the criteria in the preceding table are within the scope of AOA(4) and will be safely subcritical with a USL of 0.9249. This is based on a value of 0.9349 as determined in Part II of the validation report, plus an additional NPM of 0.01 in k_{eff}.

6.1.3.5.1.5 <u>AOA(5)—Solutions of Plutonium Compounds</u>. To validate AOA(5), the applicant modeled 119 critical experiments using the KENO-VI code with the 238-group cross-section library and the INFHOMMEDIUM cross-section treatment option. The bias was determined for the SCALE-4.4a code package on the SGN PC hardware platform. The applicant's chosen experiments include arrays of bare and reflected plutonium dioxide-polystyrene compacts and plutonium nitrate solutions. The applicant separated these benchmark experiments into two groups and determined the bias for each group. Group 1 consists of intermediate energy plutonium dioxide-polystyrene compacts, while Group 2 consists of highly thermal plutonium nitrate solution systems. Thus, Group 1 (with 32 experiments) covers the higher energy, low-moderated portion of AOA(5), while Group 2 (with 87 experiments) covers the low-energy, highly moderated portion of AOA(5). The staff examined the chosen benchmarks to determine whether they are appropriate for the validation of systems covered by AOA(5). The evaluation of the chosen benchmarks is discussed below.

The applicant defined the validated range of parameters for AOA(5) in Table 5-2 of Part III of the validation report. The validated AOA is described in Table 6.1-6.

The applicant ran SCALE/KENO-VI for the chosen experiments, calculated k_{eff}, and plotted the values of USL-1 and USL-2 (determined by USLSTATS) versus the trending parameters of EALF and the moderator to fuel ratio (H/Pu) for each group. Over the range of applicability, the minimum USL-1 and USL-2 were calculated as a function of these parameters and the lowest such value chosen as the USL for the entire AOA. The lowest USL (which occurred for Group 1) was used as the USL for the entire AOA. The benchmark data tested normal, though USLSTATS indicated that the amount of data may not be sufficient for this normality test. The staff noted that the histogram in Figure 6-1 shows a double-humped distribution of k_{eff} values. FSER Reference 6.3.18 calculated the USL using the non-parametric technique, with an additional margin of 0.01 based on the number of experiments in Group 1. This results in essentially the same limit as that derived in the validation report, and, therefore, the staff considers the USL to be adequate.

The staff compared the range of parameters covered by the benchmark data to the applicant's AOA and determined that there are a sufficient number of experiments and that they provide adequate coverage of the defined AOA across the entire range, except for the cases discussed below. Additional justification of portions of this table was required for the chemical form of the fissile material, the ^{240}Pu content, and the reflecting materials.

Table 6.1-6 Definition of AOA(5)

Geometry	Parallelepipeds, arrays of cylinders, spheres
Reflectors	Water, cadmium*, borated concrete* (1) Cd limited to 0.05 cm (0.02 in.) thick sheet surrounding 4.5–9.5-cm- (1.8–3.7-in.-) thick slab tanks. (2) Borated concrete limited to 15 cm (5.9 in.) inside and outside 7–7.5-cm- (2.76–2.95-in.-) thick annulus and separated from tank by 1.8–2-cm (0.71–0.79-in.) gap, with the composition below: ^{10}B = 1.59×10^{-3} atoms/b cm (2.61×10^{22} atoms/in.^{3}) ^{11}B = 7.04×10^{-3} atoms/b cm (1.15×10^{23} atoms/in.^{3}) Ca = 4.65×10^{-3} atoms/b cm (7.62×10^{22} atoms/in.^{3}) Fe = 5.01×10^{-4} atoms/b cm (8.21×10^{21} atoms/in.^{3}) Si = 1.66×10^{-4} atoms/b cm (2.72×10^{21} atoms/in.^{3}) H = 2.17×10^{-2} atoms/b cm (3.56×10^{23} atoms/in.^{3}) Al = 1.96×10^{-3} atoms/b cm (3.21×10^{22} atoms/in.^{3}) O = 3.25×10^{-2} atoms/b cm (5.33×10^{23} atoms/in.^{3})
Chemical form	PuO_2F_2 solution
Pu/(U + Pu)	100 wt%
^{240}Pu	4 wt%
H/Pu	30–50, 85–210
EALF	0.685–4900 eV, 0.135–0.551 eV
*Except as described below	

The applicant ran SCALE/KENO-VI for the chosen experiments, calculated k_{eff}, and plotted the values of USL-1 and USL-2 (determined by USLSTATS) versus the trending parameters of EALF and the moderator to fuel ratio (H/Pu) for each group. Over the range of applicability, the minimum USL-1 and USL-2 were calculated as a function of these parameters and the lowest such value chosen as the USL for the entire AOA. The lowest USL (which occurred for Group 1) was used as the USL for the entire AOA. The benchmark data tested normal, though USLSTATS indicated that the amount of data may not be sufficient for this normality test. The staff noted that the histogram in Figure 6-1 shows a double-humped distribution of k_{eff} values. FSER Reference 6.3.18 calculated the USL using the non-parametric technique, with an additional margin of 0.01 based on the number of experiments in Group 1. This results in essentially the same limit as that derived in the validation report, and, therefore, the staff considers the USL to be adequate.

The staff compared the range of parameters covered by the benchmark data to the applicant's AOA and determined that there are a sufficient number of experiments and that they provide adequate coverage of the defined AOA across the entire range, except for the cases discussed below. Additional justification of portions of this table was required for the chemical form of the fissile material, the ^{240}Pu content, and the reflecting materials.

The applicant stated that a chemical form of PuO_2F_2 solution would be used in any criticality calculations. To demonstrate that this compound is sufficient to bound the various compounds present (plutonium oxalates and nitrates), the applicant performed sensitivity studies to determine whether PuO_2F_2 is more reactive than these compounds over the entire H/Pu range. The study shows that PuO_2F_2 is more reactive than any compound considered except plutonium dioxide in water. Therefore, the staff considers modeling the various miscellaneous compounds covered by AOA(5) as PuO_2F_2 solution to be conservative within the limits on H/Pu, provided that the presence of plutonium dioxide is precluded.

In addition, the validated range of 4 wt% ^{240}Pu is somewhat outside the range of the Group 2 (plutonium nitrate) benchmarks (which cover 4.23–4.67 wt% ^{240}Pu) but within the range of the Group 1 (plutonium dioxide-polystyrene) benchmarks. It was not apparent that the validated AOA therefore covered highly thermal systems (with an H/Pu = 85–210) with less than 4.23 wt% ^{240}Pu content. The staff used the USLSTATS program to determine the effect of extrapolating the bias and the uncertainty in the bias to systems with 4 wt% ^{240}Pu content. This showed that extending the AOA down to 4 wt% ^{240}Pu does not result in exceeding the USL of 0.9328 and that the USL is sufficient to account for any necessary extrapolation. Subsequent to completion of its independent evaluation, the staff reviewed an additional analysis prepared by the applicant in the supplement to the validation report, DCS01 ZJJ DS CALC H 35075 A, "MFFF Validation Report Supplemental Information," dated January 31, 2005 (DCS, 2005; NRC, 2005). The NRC considers this to be part of the licensing basis in support of the validation effort for the MFFF. In this analysis, DCS performed a statistical analysis of the benchmark k_{eff} values as a function of ^{240}Pu content, using the ORNL code USLSTATS. Using the results from the confidence band technique generated by USLSTATS, the applicant determined the functional form of both the calculated k_{eff} and its uncertainty, k(x) and w(x), as a function of ^{240}Pu content, performing this analysis for the Group II benchmarks in AOA(5). The benchmarks in Group I cover the range in ^{240}Pu content from approximately 2–18 wt%, while those in Group II require a slight extrapolation down to 4 wt% ^{240}Pu content. The applicant then used this information to extend the trend, including allowance for uncertainty, down to 4 wt% ^{240}Pu content and showed that the originally determined USL is still valid. Therefore, the staff concludes that the definition of AOA(5) may be extended down to 4 wt% ^{240}Pu content without the need for additional margin.

Finally, the applicant referenced the analysis in Attachments 5 and 6 to Part 1 of the validation report in justifying the inclusion of cadmium and borated concrete absorbers in AOA(5). However, no justification for applying the results of that analysis to cover PuO_2F_2 systems, especially in the low H/Pu range, was provided. Therefore, further justification for including these absorbers in the materials covered by AOA(5) was reviewed, as described below.

To demonstrate the applicability of the selected benchmarks in these two groups to postulated design applications, the applicant conducted several sensitivity studies to determine the effect of different chemical forms, geometrical shapes, and reflector conditions, on the average neutron energy (EALF). The chemical compositions considered include plutonium dioxide, hydrides of

Pu (III) nitrate ($Pu(NO_3)_3 \cdot 5H_2O$) and plutonium oxalate ($Pu(C_2O_4)_2 \cdot 6H_2O$), and plutonium oxyfluoride (PuO_2F_2), which the applicant also referred to as standard salt. The geometrical shapes considered include infinite critical slabs and cylinders. The reflectors considered include water, plexiglass, colemanite (borated) concrete, regular concrete, and cadmium/steel. For each of these sensitivity studies, the applicant plotted EALF versus H/Pu for the different configurations. The applicant concluded from these sensitivity studies (i.e., comparison of EALF values) that the plutonium dioxide-polystyrene benchmarks are sufficient to validate PuO_2F_2 design applications at low H/Pu and that the plutonium nitrate benchmarks are sufficient at high H/Pu. The applicant further concluded that AOA(5) adequately encompasses both cylindrical and slab geometries, as well as systems with a variety of external reflectors.

While the difference in EALF between the models was considered diminished at high H/Pu (i.e., for highly thermal, dilute systems), an increasing divergence in EALF was observed at low H/Pu. The range over which this effect could be seen extended significantly higher than the minimum H/Pu of 30 in some graphs. The applicant has not addressed the impact of this on the validation in the low H/Pu limit. The applicant also has not justified using this technique as a test for benchmark applicability or provided a criterion as to what constitutes an acceptable difference in EALF. The use of an integral parameter such as EALF may not account for all important characteristics of the systems modeled; systems with the same EALF can have significantly different material specification and energy spectra.

To test the appropriateness of using the EALF comparison as a sufficiency test for benchmark applicability, the staff performed an independent analysis to evaluate the effect on the bias to changes in the chemical form of the fissile material, geometry, and reflectors.

Chemical Form

The staff constructed several models consisting of fully water-reflected spheres containing (PuO_2 or PuO_2F_2)-H_2O mixtures at various H/Pu ratios and calculated k_{eff} using the XSDRNPM discrete ordinates code. The largest difference in k_{eff} for the two systems is expected at the lowest H/Pu because the concentration of nuclides in the fissile compound is at a maximum in the low H/Pu limit. This is consistent with the observed behavior in the sensitivity studies performed by the applicant. The difference between the plutonium dioxide and PuO_2F_2 cases at the same H/Pu is caused by the presence of ^{19}F in the design application, as well as differences in the number densities of plutonium, hydrogen, and oxygen. The substitution of ^{19}F for ^{16}O is expected to have a small effect on the system k_{eff}. These nuclides have very small absorption and scattering cross sections over the entire energy range of interest (up to about 10^4 eV), with elastic scattering the dominant mechanism. Because these nuclides have similar scattering cross sections and similar atomic masses, they are expected to result in similar neutron moderation. Because the mass is large and the scattering cross section small compared to that for hydrogen, moderation by hydrogen (which is common to both cases) will dominate. Differences in k_{eff} and neutron energy primarily result from differences in density, rather than the presence of different chemical elements. Therefore, little significant difference in the underlying physics between the plutonium dioxide benchmarks and the PuO_2F_2 design applications is expected.

Perturbation of the ^{19}F and ^{16}O number densities confirms that the presence of these nuclides has a small effect on the bias. Removing all of the ^{19}F from the model resulted in a Δk_{eff} of just over 2 percent; removing all the ^{16}O resulted in a Δk_{eff} of about 1.5 percent. Subsequent to completion of its independent evaluation, the staff reviewed an additional analysis prepared by the applicant in the supplement to the validation report, DCS01 ZJJ DS CALC H 35075 A, "MFFF Validation Report Supplemental Information," dated January 31, 2005 (DCS, 2005; NRC, 2005). The NRC considers this to be part of the licensing basis in support of the validation effort for the MFFF. In this analysis, DCS modeled plutonium dioxide-water and PuO_2F_2-water mixtures at six different H/X ratios to cover the full range in moderation covered by AOA(5) (H/X = 30–210). For each modeled configuration, DCS performed sensitivity studies in which the dioxide and diflouride in the fissile medium was removed and the change in k_{eff} determined. The applicant used the CIGALES V2.0 code to derive the number densities of the solutions. The staff examined the report entitled Sec/T/03.023, "Density of Actinides Nitric Solutions and their Mixtures: Uranium (VI), Uranium (IV), Plutonium (IV), Plutonium (III), Americium (III) and Thorium (IV)," and determined that the report demonstrates good agreement between the code's predictions and experimental measurements.

The results show that the largest change in k_{eff} occurs for PuO_2F_2 at low H/X, consistent with the results of staff's calculations. The largest change is a Δk_{eff} of approximately 1.6 percent, demonstrating that the presence or absence of oxygen and fluorine in the fissile medium has a small effect on the calculated k_{eff}. Removal of all the oxygen and fluorine represents a condition which greatly exceeds the effect of the worst credible error in the material cross sections. Therefore, the staff concludes that plutonium dioxide benchmarks may be used to validate PuO_2F_2 design applications in the parametric range covered by AOA(5).

Geometry

Historically, geometry has been recognized as less important to code validation than material specification or neutron energy (NUREG/CR-6698, Table 2.3). The system geometry is only important to the extent that it affects the problem boundary conditions and neutron energy spectrum. Table 2.3 of NUREG/CR-6698 contains the guidance that the neutron energy spectrum should cover the same energy range as the design application (i.e., thermal, intermediate, or fast spectrum). This table also contains broad tolerances on the dimensions of geometric bodies. The staff expects that systems with similar geometric shapes and sizes, but with otherwise identical material and neutronic characteristics, will be mutually applicable. In addition, system geometry is not one of the parameters ordinarily used to trend the bias. The staff's calculations show that system shape and dimensions have a direct correlation to the energy spectrum. Because the bias is trended as a function of EALF, the potential effect of system geometry on the bias has been taken into account. Therefore, the staff has reasonable assurance that the benchmark experiments selected, which consist of arrays of plutonium dioxide-polystyrene compacts, can be used to validate design applications consisting of cuboidal, cylindrical, or spherical arrangements of the same materials.

Reflector Materials

The results of the staff's evaluation of different absorbers in Part I of the validation report for AOA(1) indicate that the effect of external reflectors on benchmark applicability is expected to depend strongly on the type, geometry, and thickness of the reflector (see FSER Section 6.1.3.5.1.1). Unreflected benchmark experiments are generally considered to be applicable to applications containing low-worth, relatively thin absorbers. Based on this, the staff expected similar qualitative behavior for externally reflected applications covered by AOA(5).

A comparison of the EALF values in these modeled cases shows that the reflector materials can be divided into two categories, including (1) hydrogenous reflectors, consisting of water, plexiglass, and regular concrete, and (2) strongly absorbing reflectors, including cadmium, boron, and carbon steel. The energy of applications with the first type of reflector is significantly lower than those with the second type because of hydrogenous moderation. Hydrogenous reflectors return a large proportion of the incident neutrons back into the fissile region with significant softening of the neutron spectrum. Strong absorbers result in a much smaller proportion of neutrons being returned to the fissile region and with less softening of the spectrum. Since the benchmark set contains both bare and reflected cases, in general, the bare benchmarks are expected to be most applicable to cases with strong absorbers and the water-reflected benchmarks to cases with hydrogenous absorbers. Borated concrete is expected to exhibit characteristics of both groups, as it is a mixture of a strong absorber and hydrogenous reflector.

To determine the potential effect of the presence of different absorber materials on the bias, the staff performed several independent calculations using the XSDRNPM discrete ordinates code. The models consist of slab or cylindrical regions with reflectors of the same thickness and composition as those in Table 5-1 of Part I of the validation report. The applicant stated that the borated concrete and cadmium reflectors used are similar to those for AOA(1) (DCS, 2003g). The fissile material consists of a plutonium dioxide-water mixture at an H/Pu of 30. As with the sensitivity studies for chemical form and geometry, the greatest differences are expected at the low H/Pu range of the AOA. Perturbation of the number densities of the absorber (^{10}B or cadmium), ^{1}H, and ^{239}Pu shows that, for cases with these absorber loadings, the models are relatively insensitive to the ^{10}B or cadmium cross sections compared to the sensitivity to ^{1}H and ^{239}Pu (i.e., a small change in the ^{10}B or cadmium cross sections produces a small change in the overall system k_{eff}). While the sensitivity to the absorber is greater than in AOA(1), direct perturbation shows that the absorber is responsible for only a few percent of the total system sensitivity. The staff notes that AOA(5) has a much lower minimum H/Pu ratio (H/Pu = 30) than AOA(1) (H/Pu = 100), which is likely to be responsible for the difference in sensitivity. Therefore, any bias resulting from errors in the boron or cadmium cross sections is expected to be correspondingly small.

Because the absorber worth can depend strongly on the geometrical configuration and size of the absorbers, system leakage, and other physical characteristics of the system, these results may not be indicative of the behavior of systems with much higher absorber worth. Therefore, the specification of the applicable absorber loading is necessary in the definition of the AOA.

The staff also performed additional sensitivity calculations in which the number density of the ^{10}B in the borated concrete and cadmium were reduced. These calculations indicate that, in both cases, it would require a significant decrease in the absorber number density before there would a significant change in the system k_{eff}, assuming reflector conditions as described above. Especially in the case of cadmium, it would require at least a 50-percent decrease in the number density for a 1-percent change in k_{eff}. The applicant stated that, for AOA(5), no more than 30 percent of the borated concrete is required to ensure subcriticality (maintain k_{eff} below the USL) (DCS, 2003g). While comparable information was not provided for cadmium, calculations show that the change in k_{eff} is less than that observed for boron for a proportional change in number density. The staff considers the likelihood of the ^{10}B and cadmium cross sections being in error by such a large proportion to be very low because both boron and cadmium are commonly used absorbing materials, and errors of this magnitude would have been noticed in other applications. Subsequent to completion of its independent evaluation, the staff reviewed an additional analysis prepared by the applicant in the supplement to the validation report, DCS01 ZJJ DS CALC H 35075 A, "MFFF Validation Report Supplemental Information," dated January 31, 2005 (DCS, 2005; NRC, 2005). The NRC considers this to be part of the licensing basis in support of the validation effort for the MFFF. In this analysis, DCS modeled three different configurations containing borated concrete and cadmium as neutron absorbers. The three cases considered (an annular tank with borated concrete inside and outside, and a slab tank and precipitator with cadmium sheets) corresponds to actual process calculations. The applicant varied the boron and cadmium content in increments of 5 percent from 95 percent to 0 percent of their nominal values. These nominal values correspond to the thicknesses and arrangements of the absorbers in Table 6.1-6 above. The applicant concluded that a 40-percent reduction in the boron content would be required to produce an approximately 1 percent Δk_{eff}. In addition, a 70-percent and 55-percent reduction in the cadmium density for the precipitator and slab tank, respectively, would be required for a noticeable change in k_{eff}. Therefore, the staff concludes that even a very large error in the ^{10}B and cadmium cross sections will not be sufficient to significantly affect the code bias.

The staff determined that the chosen benchmarks are appropriate for the validation of systems covered by AOA(5) based on the considerations presented above. Therefore, based on the foregoing considerations, the staff has reasonable assurance that systems meeting the criteria in Table 6.1-1 are within the scope of AOA(5) and will be safely subcritical with a USL as defined in Part III of the validation report (USL = 0.9328).

6.1.3.5.2 Determination of Normal and Abnormal Subcritical Margins

The applicant selected an administrative margin of 0.05 for both normal and credible abnormal conditions for each AOA covered in the validation report. The administrative margin is used to determine, along with the bias and uncertainty (as documented in FSER Section 6.1.3.5.1), the maximum k_{eff} or USL, which is part of the design basis of the MFFF. This section discusses the selection of an appropriate administrative margin, which is part of the margin of subcriticality for safety as required in 10 CFR 70.61(d).

This margin represents a minimum margin of subcriticality to account for any uncertainties not explicitly treated in the validation process and is applied in Method 1 (confidence band with administrative margin) and in the nonparametric method. This margin is not included as a separate term in Method 2 (lower tolerance band). However, applying the criterion that USL-1 is less than USL-2 practically means that the USL is determined on the basis of USL-1 (or on the basis of the nonparametrically determined USL). Justification for the chosen administrative margin is provided in each part of the validation report and is based on three arguments, including (1) historical nuclear industry practice, (2) NRC guidance, and (3) a comparison of USL-1 (including a 0.05 administrative margin) with USL-2. The staff reviewed these justifications and determined that they are insufficient by themselves to support using an administrative margin of 0.05 for all potential operations involving plutonium or MOX fuel, for the reasons stated below.

Historical fuel facility subcritical margin, as put forth by the applicant, does not support a minimum subcritical margin of 0.05 for the MFFF because this precedent is not applicable to systems at the MFFF. Systems involving high-^{239}Pu SNM (e.g., 96 wt% ^{239}Pu oxide and solutions) exhibit a higher sensitivity of k_{eff} to changes in the underlying physical and neutronic parameters than systems encountered at many other fuel facilities. The effect of any unknowns can reasonably be expected to be magnified in systems displaying this higher degree of sensitivity. Therefore, in general, a larger margin of subcriticality for safety is appropriate for operations with high-^{239}Pu SNM than those with lower fissile content (e.g., 6 wt% plutonium-content MOX, low-enriched uranium (LEU)).

Guidance from the NRC (NUREG-1718) does state that an administrative margin of 0.05 is "generally considered to be acceptable without additional justification when both the bias and its uncertainty are determined to be negligible." However, the applicant has not demonstrated that these conditions have been met. When the bias and its uncertainty are not negligible, it is difficult to justify that the underlying physics is sufficiently well-understood to justify such a low administrative margin.

Finally, the comparison of USL-1 with USL-2 does not demonstrate the adequacy of the administrative margin. This condition that USL-1 is less than USL-2 is necessary, but not sufficient, to show that adequate margin as used in Method 1 has been provided. Method 1 and Method 2 are two different statistical treatments of the data, and a comparison between them can only demonstrate whether the administrative margin is sufficient to bound statistical uncertainties included in Method 2 but not included in Method 1. There may also be other statistical or non-statistical errors in the calculation of k_{eff} that are not handled in the statistical treatments.

For the reasons stated above, the staff did not agree with the applicant's stated justification for the use of a 0.05 administrative margin.

Therefore, staff considered other factors, which are described below, before approving the margin. In general, an administrative margin of 0.05 has been historically accepted for most fuel cycle operations. However, additional margin has historically been considered necessary for operations involving HEU (and, by extension, high-^{239}Pu SNM) than that for those involving LEU. In addition, more margin has historically been considered necessary for normal conditions than for abnormal conditions[6] (NRC, 2001c). Therefore, the staff considers an additional administrative margin of subcriticality for safety to be necessary for proposed operations at the MFFF.

This margin may be provided in terms of either margin in k_{eff} or margin in system parameters. The staff has concluded that the applicant's proposed k_{eff} limits (including an administrative margin of 0.05), together with margin inherent in the parameters associated with proposed MFFF operations, provide reasonable assurance that there will be a sufficient margin of subcriticality for safety for proposed operations at the MFFF.

---

[6]Historically, some applicants have been licensed with separate limits for normal and abnormal conditions, while others have been licensed with a single limit. The use of a single limit is acceptable if judged sufficient to cover both normal and abnormal conditions. Either approach would in principle have been acceptable for the MFFF.

This determination regarding margin in proposed MFFF operations is based on specific commitments in the CAR and associated documentation, as described below.

Normal Conditions

As stated above, the staff considers an additional margin of subcriticality for safety necessary for proposed operations at the MFFF. This margin is provided by the following commitments:

- The applicant stated the following:

  > All criticality applications in the design of the MFFF show that the abnormal conditions cases are bounding—by an appreciable equivalent margin in k_{eff}—over the normal condition cases....Operations are rarely expected to be conducted at the subcritical value.[7] (DCS, 2003g).

  The staff reviewed the description of technical practices contained in CAR Section 6.3.4 to verify this statement.

The staff determined, based on this review, that the technical practices provide considerable margin in many of the criticality parameters (especially with regard to isotopics, reflection, density, concentration, and physicochemical form). With regard to these specific criticality parameters—

- CAR Section 6.3.4.3.2.4 states the following:

  > The assumptions used in the criticality calculations are, typically, as follows:
  >
  > ^{239}Pu/Pu$_{total}$ = 96%, ^{240}Pu/Pu$_{total}$ = 4% and ^{235}U/Pu$_{total}$ = 0%

  This is also stated throughout CAR Tables 6-1 and 6-2 ("^{240}Pu $\geq$ 4%"). CAR Section 6.3.4.3.2.4 also states the following:

  > Incoming plutonium will respect the following conditions (for the main plutonium isotopes):
  >
  > 90% $\leq$ ^{239}Pu/Pu$_{total}$ $\leq$ 95%, 5% $\leq$ ^{240}Pu/Pu$_{total}$ $\leq$ 9%

  This CAR section also states that there may be up to 1 percent ^{241}Pu and up to 2 percent U/Pu total present in the incoming material and that "calculations have demonstrated that increasing the ^{239}Pu content by 1.0 wt% while decreasing the ^{240}Pu content by a corresponding amount is sufficient to offset any reactivity effect from ^{241}Pu and ^{235}U." The staff's confirmatory calculations show that the assumed composition is sufficient to

---

[7]During the public meeting held on July 31, 2003, the applicant clarified that "normal condition cases" in this context refers to the actual physical state of plant operations ("normal operating" conditions), as opposed to conditions defined by the controlled parameter limits.

bound the worst-case combination of possible isotopics for incoming plutonium, with a Δ k_{eff} between 0.3 and 0.8 percent.

- CAR Section 6.3.3.2.5 states the following:

> At a minimum, reflection conditions equivalent to 2.54 cm (1 in)...tight-fitting water jacket are assumed to account for personnel and other transient incidental reflectors....In cases when reflection control is not indicated, full water reflection of process stations or fissile units is represented by a minimum of 30.48 cm (12 in)...of tight-fitting water jacket.

The applicant also stated that "there is no active, controllable reflection...control used in the MFFF" (DCS, 2001a, RAI 101). Based on these two quotations, the use of full water reflection is expected in the majority of the MFFF criticality calculations (except when some other level of interstitial moderation is more reactive). This is supported by a review of CAR Tables 6-1 and 6-2. Tight-fitting full water reflection is known to be an extremely conservative way to model hydrogenous reflection around process equipment. The staff's confirmatory calculations show that there is a significant Δk_{eff} resulting from even a small gap between the fissile material and reflector, as well as from small changes in reflector thickness. The staff expects, based on these calculations, that the use of tight-fitting full water reflection will result, in most cases, in conservatism of at least several percent in Δk_{eff}.

- CAR Section 6.3.3.2.3 states, "Conservative assumptions are always made about the density of the fissile material." CAR Section 6.3.4.3.2.3 states that plutonium dioxide incoming to the dissolution unit will be limited to less than or equal to 7 g/cm^3 and polished plutonium dioxide will be limited to less than or equal to 3.5 g/cm^3. In addition, this section states the following:

> The assumed density of PuO_2 powder being dissolved (of $\leq$ 7 gm/cm^3) is quite high and, based upon experience, would not actually be expected...the storage of material initially received will be shown to be sub-critical for maximum theoretical density material—11.46 g/cm^3.

Moseley (1965) states that powder density depends primarily on the chemical reactions and temperature of decomposition used in its formation. The density is typically lowest for powder prepared from plutonium oxalate, as in the MFFF. Based on the density measurements in this reference, the staff concludes that the applicant's assumed densities for most processes appear conservative. The staff's confirmatory calculations show that there is a significant Δk_{eff} in going from the densities reported in Moseley (1965) (1–2.3 g/cm^3) up to the applicant's assumed densities (ranging from 3.5–11.46 g/cm^3). The difference in density produced conservatism in Δk_{eff} from approximately 5 percent up to approximately 75 percent.

- CAR Section 6.3.3.2.7 states the following:

Concentration control...involves the use of concentration-based single-parameter limits established based upon worst-case geometry (i.e., spherical) and SNM fissile composition unless these parameters are controlled by IROFS....Single-parameter limits for concentration are established in a manner that ensures an adequate margin of subcriticality (including margins to protect against uncertainties in process variables and against limits being accidentally exceeded)....

CAR Sections 6.3.3.2.7 and 6.3.4.3.2.7 state that concentration control is generally used only in systems with "very low fissile material concentrations (secondary streams)." Systems with very low concentration will also be deeply subcritical. For those systems, the commitments above are expected to provide a large margin of subcriticality.

For the large majority of wet processes not relying on concentration control, CAR Section 6.3.5.5 applies. This section states the following:

Credible optimum conditions (i.e., most reactive conditions physically possible) for each controlled parameter are assumed in criticality safety calculations and NCSEs unless specified controls are implemented to limit the controlled parameter to a specified value or range.

Based on these statements, the staff expects that optimal concentration will be assumed in most of the wet portions of the AP process. To gauge the degree of conservatism, the staff compared nominal concentration values (based on the plutonium inventories and vessel volumes in CAR Section 11.3) with the optimum values for $Pu(NO_3)_4$ solutions contained in ARH-600, "Criticality Handbook." The ARH-600 curves for the subcritical cylindrical diameter (i.e., the primary controlled parameter in the process) as a function of the plutonium concentration show that the assumption of optimal rather than nominal concentration results in conservatism of at least several percent in Δk_{eff}.

In addition to commitments with regard to specific criticality parameters, there is conservatism in the physicochemical form:

- CAR Section 6.3.4.3.1.1 states, "In the MP process, no chemical transformations take place. As a consequence, the oxide form of the fissile medium (PuO_2 or UO_2 as applicable) is always assumed." It is well known from calculations and criticality handbooks that the uranium dioxide and plutonium dioxide oxide forms are typically the most reactive physicochemical form besides metal.

  CAR Section 6.3.4.3.1.1 also states the following:

  For the AP process, a conservative assumption concerning the chemical form of the fissile matter is made in each step of the process, taking into account not only the nominal conditions but also the possible process upsets....The different chemical forms used in the criticality analyses are as follows...PuO_2...$Pu(NO_3)_4$...$Pu(NO_3)_3$...plutonium oxalate.

For certain wet processes in the AP process, Footnote 3 to CAR Table 6-1 states, "Actual chemical form of Pu Nitrate is $Pu(NO_3)_4$, for most process steps, which is less reactive than $Pu(NO_3)_3$." The staff calculated the magnitude of this effect by changing the number densities of atoms in the fissile medium and showed that the effect of this on k_{eff} is negligible. However, the process description in CAR Chapter 11 shows that much of the AP process (such as dissolution and the plutonium purification cycle) involves the presence of excess nitric acid, which acts as a soluble neutron poison. CAR Table 6-1 shows that excess nitric acid is not considered in the criticality calculations. The previously referenced graphs in ARH-600 show that there is a varying degree of margin in the critical cylinder diameter resulting from neglecting the nitric acid. The actual amount of margin depends on the amount of nitric acid present and was not quantified by the staff.

In addition, following oxalic precipitation, PuO_2F_2 is assumed as the reference fissile medium. For these units, Footnote 4 to CAR Table 6-1 states, "Actual chemical form is a mixture of Pu Oxalate and Pu Nitrate. Either chemical form is less reactive than PuO_2F_2." As documented in FSER Section 6.1.3.5.1.5, the staff's analysis has determined that PuO_2F_2 bounds both plutonium oxalate ($Pu(C_2O_4)_2$) and plutonium nitrate (assumed $Pu(NO_3)_4$). As shown by sensitivity calculations in support of the validation of AOA(5), over the range of interest in H/Pu, PuO_2F_2 bounds pure plutonium oxalate by approximately 0.5–4 percent in Δk_{eff} and bounds pure plutonium nitrate by approximately 1–6 percent in Δk_{eff}.

In addition to these specific commitments, the CAR contains preliminary process information (primarily in CAR Tables 6-1 and 6-2, and CAR Chapter 11) that shows how this margin is applied in each of the proposed process steps, as described below.

The following relates to the AP process.

- The decanning, milling, and recanning units rely primarily on mass and moderation control modes. Because the geometry is not controlled, the staff expects that spherical geometry will be assumed in the derivation of mass limits. CAR Section 6.3.3.2.2 states, "Mass control involves the use of mass-based single-parameter limits established based upon worst-case geometry (i.e., spherical)...unless these parameters are controlled by principal SSCs/IROFS." Typically, this provides a large degree of conservatism, as there is no known mechanism for loose powder to spontaneously take on a spherical shape. Conservative assumptions regarding plutonium isotopics, reflection, and density are also made. In particular, CAR Table 6-1 states that before measurement of density, full theoretical density is assumed.

- For dissolution/dechlorination and dissolution units, conservative assumptions are made with regard to plutonium isotopics, reflection, density, and physicochemical form (i.e., neglecting of excess nitric acid). In addition, the actual nominal plutonium concentration will be less than optimal.

- For the purification cycle, conservative assumptions are made with regard to plutonium isotopics, reflection, and physicochemical form (i.e., neglecting excess nitric acid). In addition, the actual nominal plutonium concentration will be significantly less than optimal. In addition, for the uranium stripping and diluent washing mixer settlers, conservative uranium isotopics are assumed (i.e., 35 wt% ^{235}U used in calculations, whereas processes will be limited to less than or equal to 30 wt% ^{235}U). The staff's calculation indicate that this will result in a Δk_{eff} of approximately 2 percent.

- For the oxalic precipitation and oxidation unit, conservative assumptions are made with regard to plutonium isotopics, reflection, and physicochemical form (i.e., $Pu(NO_3)_3$ without excess nitric acid before precipitation; plutonium dioxide diflouride following precipitation). The calcination furnace is treated below.

- For the homogenization unit, canning unit, and calcination furnace, conservative assumptions are made with regard to plutonium isotopics and reflection. These units rely primarily on geometry in combination with neutron absorber control, mass, and density. The density still appears to be conservative relative to the values in Moseley (1965), although it is lower (3.5 g/cm^3) than for upstream units. Moderation is not controlled, but powder should be dry under normal conditions following calcining.

- For the oxalic mother liquor recovery unit, conservative assumptions are made with regard to plutonium isotopics, reflection, and physicochemical form (i.e., PuO_2F_2).

- The remaining units in the AP process contain only low levels of plutonium under normal conditions, relying mainly on concentration control. Under normal conditions, they are expected to be deeply subcritical.

The following relates to the MP:

- For the receiving area, conservative assumptions are made with regard to plutonium isotopics and reflection. In addition, some processes in this area rely on mass control without controlling geometry and therefore must assume spherical geometry. There is also a variety of densities assumed; all of these densities are conservative relative to the values in Moseley (1965).

- For the remaining Units in the MP, the plutonium content of powder and fuel is limited to 22 and 6.3 wt% plutonium dioxide. The neutronic behavior of these systems (especially at 6.3 wt%) is more indicative of LEU fuel than that of other processes at the MFFF. Especially for systems with 6.3 wt% plutonium dioxide content (comprising AOA(2) and part of AOA(4)), an administrative margin comparable to that historically allowed at licensed LEU facilities is appropriate.

There is also some conservatism in density and reflection. While the density is much greater following pelletizing and sintering, pellets are all produced at the lower assay of 6.3 wt% plutonium dioxide. There is some conservatism in the plutonium isotopics for the powder and pellet areas. CAR Section 6.3.4.3.2.4 states the following regarding the MP:

> At each step of the process (pure PuO_2, master blend, final blend, and pellets), a conservative assumption is made based on process values. The process values are as follows...20% plutonium for the master blend...2% to 6% for the final blend and pellets.

The CAR section also states, however, that "the plutonium content used for criticality calculations is as follows (actual design values are less than these)...22% for the master blend...6.3% for the final blend." The difference between the nominal values and bounding values used in the criticality calculations produces some additional conservatism in the calculated k_{eff}.

In addition to the lower plutonium content, there are other considerations that provide additional assurance of an adequate margin of subcriticality for MOX pellets, rods, and assemblies under AOA(2). The neutronic characteristics of finished MOX fuel have been well studied in order to enable it to be used as reactor fuel. This is possible because the final configuration is fixed and not easily subject to change. In addition, the chosen benchmarks used to validate AOA(2) are physically unusually, similar to anticipated design applications (i.e., they are both MOX fuel rod lattices). The benchmarks are also all well-moderated thermal systems for which the neutron cross sections are well behaved (i.e., smoothly varying), have been measured with a high degree of confidence, and have been tested in many applications over the years. For these systems, there is relatively little apparent margin in physical parameters; density, material form, and geometry are expected to be modeled fairly accurately. However, this is offset by the reduced sensitivity to the underlying parameters, the lower risk, and the smaller uncertainty in cross sections and the physical configuration.

As discussed in FSER Section 6.1.3.4.3, CAR Tables 6-3 and 6-4 also contain "typical, order-of-magnitude values" of volume, dimension, and mass limits characteristic of operations at the MFFF. Although the applicant stated that these values will not be referenced in NCSEs, they provide some information as to the degree of margin in proposed MFFF operations. As stated by the applicant, these values represent "'preliminary best estimate' nominal Pu values typical of those expected to be found in the AP and MP processes" (DCS, 2001a, RAI 104). The staff compared the values in CAR Tables 6-3 and 6-4 to corresponding published values in ANSI/ANS-8.1-1983 (R1988) (ANSI, 1988) and criticality handbooks TID-7028 and TID-7016. Mass values were compared to those in ANSI/ANS-8.12-1987 and TID-7016. The staff noted that the numbers included in Tables 6-3 and 6-4 are based on conservative assumptions of density, reflection (for the fully reflected cases), and geometry (for mass and volume limits). Based on this, the staff has reasonable assurance that these preliminary best-estimate values are satisfactorily conservative to provide for an adequate margin of subcriticality for safety.

Therefore, based on the proposed administrative margin of 0.05 and the margin in the system parameters defined by the commitments and preliminary process information above, the staff has reasonable assurance that there will be acceptable margin of subcriticality for safety for normal conditions at the MFFF.

<u>Credible Abnormal Conditions</u>

Section 6.4.3.3.4 of NUREG-1718 (NRC, 2000) states that the fact that a condition is abnormal may be used as "justification for having a lower margin of subcriticality than would be permissible for normal conditions." In such cases, "the abnormal conditions should meet the standard of being at least 'unlikely' from the standpoint of the double contingency principle" and, in addition, "the increased risk associated with the less conservative margin should be commensurate with and offset by the unlikelihood of achieving the condition." Based on the applicant's commitment to follow the DCP, the staff has reasonable assurance that achieving an

abnormal condition will be at least unlikely (FSER Section 6.1.4.2). In addition, an administrative margin of 0.05 has historically been found acceptable by the NRC for abnormal conditions at those facilities most similar to the MFFF. Therefore, the staff finds an administrative margin of 0.05 acceptable under abnormal conditions for proposed MFFF operations.

The staff has reasonable assurance that an administrative margin of 0.05, together with the commitments and preliminary process information described above, will provide an acceptable margin of subcriticality for safety, for both normal and credible abnormal conditions at the MFFF.

6.1.3.6 NCSE Criticality Controls

The applicant stated that the criticality controls credited in NCSEs will be identified and evaluated during development of the ISA and will later be identified as IROFS in its ISA summary, to be submitted as part of any later DCS application for a license to possess and use licensed material. Since the specific controls relied on as NCS IROFS are not design information, the identification of such controls is not needed for the review of the revised CAR; therefore, this approach is acceptable to the staff.

6.1.4 Design Bases of the PSSCs

6.1.4.1 Description of PSSCs

Pursuant to 10 CFR 70.23(b), before approving construction of the facility, the NRC must find that design bases of the PSSCs provide reasonable assurance of protection against natural phenomena and the consequences of potential accidents. Revised CAR Section 6.4, "Design Bases," states that the "[p]rincipal SSCs are described in Chapter 5 of this document. Specific IROFS associated with criticality safety will be identified in the ISA." Revised CAR Table 5.6-1, "MFFF Principal SSCs," identifies the PSSCs for criticality hazards as "criticality control," with specific controls to be determined during the design of the facility.

Because the applicant can choose any of the criticality control methods described in this revised CAR chapter to prevent criticality, the staff did not evaluate any specific controls during review of the revised CAR. However, based on the commitments in this chapter, the staff has reasonable assurance that the design bases of the PSSCs will provide adequate protection against the consequences of a criticality accident (as required by 10 CFR 70.23(b)), even though the specific IROFS have not yet been identified.

6.1.4.2 Commitment to DCP

The commitment to the DCP, including the corresponding definition of unlikely, is part of the design basis. The DCP requires (ANSI/ANS-8.1-1983 (R1988)) that "at least two unlikely, independent, and concurrent changes in process conditions" must occur before criticality is possible. Additionally, 10 CFR 70.64(a)(9) requires that the design of the proposed facility adhere to the DCP. As discussed below, the staff considers the commitment to the DCP acceptable and, in this regard, finds that the applicant has met the double contingency part of the 10 CFR 70.64(a)(9) requirement for construction authorization. The commitments contained in the remainder of Chapter 6 provide the additional measures necessary for criticality control, which satisfies the requirements of 10 CFR 70.64(a)(9) for construction authorization.

To provide reasonable assurance that the design will result in controls that are sufficiently reliable and available when needed to perform their safety function, the meaning of "unlikely" for each contingency must be unambiguously defined. As discussed in FSER Section 6.1.4.3, "unlikely" for meeting the DCP is defined as "not expected...during the facility lifetime." This is qualitatively consistent with a failure probability on the order of once in 100 years.

The applicant has stated that criticality will be made highly unlikely. An accidental criticality has the potential to exceed the threshold under 10 CFR 70.61, "Performance Requirements," for doses to facility workers. Dose mitigation in the form of shielding will not be credited in reducing the consequence level, as discussed in FSER Section 6.1.4.3. In revised CAR Section 5.4.3, the use of deterministic methods for demonstrating that criticality hazards are highly unlikely is described. This consisted of application of double contingency, Appendix B, "Quality Assurance Criteria for Nuclear Power Plants and Fuel Reprocessing Plants," to 10 CFR Part 50, "Domestic Licensing of Production and Utilization Facilities," industry codes and standards, and management measures for failure detection. The applicant, however, did not explain in the revised CAR how it will determine when an acceptable level of likelihood has been reached or how these generic criteria will be applied to specific controls. The applicant provided additional details on its method for ensuring that criticality hazards would be made highly unlikely (DCS, 2003b, 2003d). The description of the accident sequences leading to criticality and the controls relied on to prevent criticality will be described in NCSEs. The description of accident sequences would include the initiating event, methods of prevention, including at least two independent controls and their safety function, redundancy and diversity, the safety margin, failure modes, and failure detection and surveillance requirements. The description of controls would include the safety function, quality classification (QI-1a or QI-1b), operating range and limits, testing and maintenance requirements, environmental design factors, natural phenomena response, required instrumentation, and applicable codes and standards.

In accordance with the DCP, a least two independent controls whose failure is unlikely must be specified. For passively controlled units whose failure is not credible, it is considered sufficient to place them under the plant's configuration management program and assign the highest quality level (QL-1a), as done for sole IROFS. To show incredibility, it must be demonstrated that the system will remain subcritical under all credible process conditions. As discussed in FSER Section 6.1.3.4, accident sequences in which there are no credible means to achieve criticality do not require dual independent controls to comply with the DCP. Therefore, this approach is acceptable to the staff.

For systems with credible failure paths leading to criticality, active or passive engineered controls must be classified at least QL-1b, and administrative controls must be simple and unambiguous.

The likelihood determination will be based on consideration of all the applicable availability and reliability qualities, consistent with Section 5.4.3.2 of NUREG-1718 (NRC, 2000), which include those specified above. The staff considers this sufficient to ensure that all factors that affect accident sequence and control likelihood will be taken into consideration. In addition, for each independent and unlikely-to-fail control, there must be either a means to detect the failure of the control (on a period to be justified in the NCSEs) or safety margin (sufficient to ensure that repeated failure of the controls does not lead to criticality). Based on this, the staff has reasonable assurance that adherence to the DCP (in which each control failure is sufficiently reliable and available to be made unlikely), combined with failure detection or additional safety margin as described above, provides reasonable assurance that accident sequences leading to criticality will be highly unlikely.

For the first case, the requirement that a means to detect the control failure be provided can result in a significant reduction in the likelihood that both independent controls will be in a failed state at the same time, provided that this time period is sufficiently short compared to the expected time between failures of the control. This period will be specified and justified in the NCSEs. The staff has reasonable assurance that the combination of two independent and unlikely-to-fail controls (each of which is not expected during the facility lifetime) with adequately justified failure detection will ensure that criticality is highly unlikely.

For the second case, the applicant has stated that the safety margin must be sufficiently robust that multiple failures of the control will not lead to a loss of subcriticality. Specifically, there must be sufficient margin to accommodate at least three independent failures of the controls. In evaluating abnormal conditions, the most reactive credible change in the controlled parameter must be evaluated. For processes relying on dual-parameter control, this typically leads to abnormal conditions that have considerable margin, which would ensure considerable safety margin for the sequence. In the case of both systems relying on dual- and single-parameter control, however, the requirement that the system must accommodate multiple independent failures provides a significant reduction in the likelihood of criticality. The staff considers the occurrence of multiple independent failures, each of which is not expected during the facility lifetime, to be highly unlikely.

Instead of monthly failure detection or additional safety margin, the applicant stated that it may use other means to ensure that criticality is highly unlikely. In this case, the applicant will show that these alternate means provide comparable assurance to the two methods described above. A list of any systems utilizing methods other than failure detection or additional margin will be provided in an ISA summary. Regardless of the means of demonstrating that criticality is highly unlikely, the rationale will be provided in NCSEs supporting a license application and ISA summary.

Section 5.4.3.2 of NUREG-1718 contains the following acceptance criterion:

> A purely qualitative method of defining "unlikely" and "highly unlikely" is acceptable if it incorporates all of the applicable availability and reliability qualities to an appropriate degree...one acceptable definition of "highly unlikely" is a system of IROFS that possesses double contingency protection with each of the applicable qualities to an appropriate degree.

In addition, the Commission has adopted a strategic goal of no inadvertent criticality. Pursuant to 10 CFR 70.64(a)(9), the staff, therefore, finds that the design basis of the facility adheres to the DCP for construction authorization. Based on this and the other commitments contained in Chapter 6 of the CAR, the staff therefore finds that the criticality BDC set forth in 10 CFR 70.64(a)(9) have been met for construction authorization.

6.1.4.3 Commitment to ANSI Standards

The staff reviewed the applicant's commitment to the ANSI/ANS-8 series consensus standards in criticality safety, as described in revised CAR Sections 6.4 and 6.5. In revised CAR Section 6.4, the applicant committed to the use of the ANSI/ANS-8 series standards endorsed in RG 3.71 (NRC, 1972) in the design of the facility. Revised CAR Section 6.4 commits, in general, to "comply with the guidance (shall statements) and implement the recommendations (should statements)" of the applicable standard but identifies several clarifications to specific commitments within certain of the standards. Thus, in the context of this section, "guidance" is equivalent to the "requirements" of the ANSI standards. Moreover, RG 3.71 endorses several of these standards conditionally. In those cases in which the regulations or regulatory guidance disagreed with the standard, the applicant was requested to clarify or modify the commitment. Revised CAR Section 6.5 contains a discussion of commitments to ANSI/ANS-8.23-1997, which the applicant does not consider part of the design basis of the facility. Specific commitments to these standards are summarized below:

* ANSI/ANS-8.1-1983 (R1988) (ANSI, 1988)—The applicant committed to comply with the guidance (shall statements) of ANSI/ANS-8.1-1983 (R1988) and implement the recommendations (should statements), with clarification of three provisions. The first clarification is that, for Section 4.2.2 of the ANSI standard, the applicant committed to follow the DCP, which requires that at least two unlikely, independent, and concurrent changes in process conditions must occur before criticality is possible. For the purposes of meeting this commitment, "unlikely" is defined as "events or event sequences that are not expected during the facility lifetime, but are considered credible." This commitment will be met for those processes and areas in which criticality is determined to be credible. The staff notes that assessment of credibility will be determined during the performance of the ISA (covered in revised CAR Chapter 5). The staff further notes that a definition of unlikely that is qualitatively consistent with a probability of failure on the order of 10^{-2} per year is considered to be acceptable in NUREG-1718 (NRC, 2000). The phrase "not expected during the facility lifetime" is, therefore, acceptable provided the lifetime of the facility is assumed to be greater than approximately 100 years.[8] The applicant clarified that, while this is a qualitative determination, it is consistent with a failure probability on the order of once in 100 years (DCS, 2002c, 2002e). Therefore, this is acceptable to the staff. The second clarification is that, for Section 4.2.3 of the ANSI standard, the applicant committed to follow the standard (which lists several different types of control methods including engineered and administrative controls) but committed to relying on engineered features whenever practical and to justify the use of administrative controls. This is consistent with the preferred design approach and is, therefore, acceptable to the staff. The third clarification is that, for Section 4.3.2 of the ANSI standard, the applicant committed that, where an extension to the area(s) of applicability is required, the calculational method will be supplemented by other calculational methods to provide a better estimate of bias in the extended area(s) or through an increase in the margin of subcriticality. In making the extension, trends in the bias and any additional uncertainty must be considered in determining the appropriate amount of margin. Section 4.3.2 of the ANSI standard states that the AOA may be extended by making use of trends in the

---

[8]The term "unlikely" as used in the DCP should not be confused with the term "unlikely" in the performance requirements of 10 CFR 70.61(c) for intermediate-consequence events.

bias, and Section 4.3.3 states that the uncertainty in the bias shall contain allowances for extensions to the AOA. These commitments are consistent with the standard. The standard has been endorsed, and so this process is acceptable to the staff. However, this should not be construed as allowing redefinition of the applicant's AOA more broadly than that described in the NRC-reviewed validation reports. The proposed k_{eff} limits are acceptable to the staff with the AOAs as defined in the validation reports (and documented in this FSER). As stated in the authorization condition, the proposed k_{eff} limits and definition of the AOAs cannot be changed without NRC review and approval.

With regard to the subcritical limits in ANSI/ANS-8.1-1983 (R1988), the staff considers the use of subcritical limits from the standard instead of explicit calculation to be an acceptable practice. These results have been endorsed in RG 3.71 (NRC, 1972) and, as single parameter limits, are very conservative. In addition, since the standard was merely reaffirmed in the 1988 version, the staff considers the use of the 1988 version acceptable.

- ANSI/ANS-8.3-1997, "Criticality Accident Alarm System"—The applicant committed to comply with the guidance (shall statements) of ANSI/ANS-8.3-1997 and implement the recommendations (should statements), as modified by RG 3.71, with clarification of one provision. The clarification is that, for Section 4.1.3 of this standard, the applicant committed to evaluate the overall risk to personnel, specifically with respect to the risk from operational interruption and relocation following false alarms. The staff considers it appropriate to consider overall risk but notes that 10 CFR 70.24(a) requires a criticality alarm in all areas of the facility where more than the specified quantities of fissionable materials are handled or stored. A request for exemption from the requirements of 10 CFR 70.24(a) will be handled as described in revised CAR Section 6.3.2 and may include risk arguments. This is consistent with the rule and regulatory practices and, therefore, is acceptable to the staff.

- ANSI/ANS-8.7-1975, "Guide for Nuclear Criticality Safety in the Storage of Fissile Materials"—The applicant did not consider ANSI/ANS-8.7-1975 to be part of the design basis of the facility. The general commitment to ANSI/ANS-8.1-1983 (R1988) and technical practices as described in revised CAR Sections 6.3.3.2.8 and 6.3.4.3.2.8 are sufficient to ensure that criticality safety is appropriately provided for fissile material storage areas.

- ANSI/ANS-8.9-1975, "Guide for Nuclear Criticality Safety for Steel-Pipe Intersections Containing Aqueous Solutions of Fissile Materials"—The applicant stated that ANSI/ANS-8.9-1975 has been withdrawn by the ANS-8 working group and will not be used in the design of the facility. Piping configurations containing aqueous solutions of fissile material will be evaluated by calculation in accordance with ANSI/ANS-8.1-1983 (R1988). Because using validated methods to determine subcritical limits is an acceptable methodology, the staff determined that this approach is acceptable.

- ANSI/ANS-8.10-1983, "Criteria for Nuclear Criticality Safety Controls in Operations with Shielding and Confinement"—The applicant did not consider ANSI/ANS-8.10-1983 to be part of the design basis of the facility because the approach used for the facility is to prevent criticality in accordance with the DCP, rather than rely on shielding and confinement for dose mitigation. In addition, the applicant stated that events with the potential to exceed 10 CFR 70.61 requirements will be made highly unlikely, "including criticality events, without regard to actual dose consequences" (DCS, 2003b). Because shielding will not be credited in this way (it must still be considered for detector coverage), the staff considers it appropriate to exclude ANSI/ANS-8.10-1983 as a design basis for the facility.

- ANSI/ANS-8.12-1987, "Nuclear Criticality Control and Safety of Plutonium-Uranium Fuel Mixtures Outside Reactors"—The applicant did not consider ANSI/ANS-8.12-1987 to be part of the design basis of the facility. The staff notes that this standard does not contain any administrative requirements to which the applicant should commit. This standard only contains subcritical limits for certain plutonium-uranium mixtures. In the absence of a commitment to this standard, the commitment to ANSI/ANS-8.1-1983 (R1988) ensures the use of validated methods in computer calculations to demonstrate subcriticality. Therefore, the staff considers it appropriate to exclude ANSI/ANS-8.12-1987 as a design basis for the facility.

- ANSI/ANS-8.15-1981, "Nuclear Criticality Control of Special Actinide Elements"—The applicant did not consider ANSI/ANS-8.15-1981 to be part of the design basis of the facility. Criticality control of special actinide nuclides will be explicitly evaluated by calculation in accordance with ANSI/ANS-8.1-1983 (R1988). Because using validated methods to determine subcritical limits is an acceptable methodology, the staff determined that this approach is acceptable.

- ANSI/ANS-8.17-1984, "Criticality Safety Criteria for the Handling, Storage, and Transportation of LWR Fuel Outside Reactors"—The applicant committed to comply with the guidance (shall statements) of ANSI/ANS-8.17-1984 and implement the recommendations (should statements) with clarification of two provisions. The first clarification is that, for Section 4.11 of the standard, the applicant committed to use of the DCP for the handling, storage, and transportation of fuel units and rods. Compliance with the DCP is required for the facility by 10 CFR 70.64(a)(9), and, thus, this commitment is acceptable to the staff. The second clarification is that, for Section 5.1 of the standard, the applicant committed that, where an extension to the area(s) of applicability is required, the calculational method will be supplemented by other calculational methods to provide a better estimate of bias in the extended area(s) or through an increase in the margin of subcriticality. This is consistent with the endorsed standard and is, therefore, acceptable to the staff.

- ANSI/ANS-8.19-1996, "Administrative Practices for Nuclear Criticality Safety"—The applicant committed to comply with the guidance (shall statements) of ANSI/ANS-8.19-1996 and implement the recommendations (should statements), with the exception that no commitments are made to Section 10 of the standard regarding emergency response to criticality accidents. ANSI/ANS-8.19 is currently undergoing review by the working group and has the potential to be changed in the near term. However, the staff notes that the applicant has committed to ANSI/ANS-8.23-1997, which contains many of the same requirements.

Additionally, the staff notes that emergency response procedures are not at issue in deciding whether to approve the revised CAR. These procedures will be evaluated by the staff if DCS submits an application for a license to possess and use licensed material.

- ANSI/ANS-8.20-1991, "Nuclear Criticality Safety Training"—The applicant committed to comply with the guidance (shall statements) of ANSI/ANS-8.20-1991 and implement the recommendation (should statements) without exception or clarification. This is, therefore, acceptable to the staff.

- ANSI/ANS-8.21-1995, "Use of Fixed Neutron Absorbers in Nuclear Facilities Outside Reactors"—The applicant committed to comply with the guidance (shall statements) of ANSI/ANS-8.21-1995 (this standard contains no recommendations) without exception or clarification. This is, therefore, acceptable to the staff.

- ANSI/ANS-8.22-1997, "Nuclear Criticality Safety Based on Limiting and Controlling Moderators"—The applicant committed to comply with the guidance (shall statements) of ANSI/ANS-8.22-1997 and implement the recommendations (should statements), with clarification of one provision. This clarification is that, for Section 4.1.7 of the standard, the applicant committed to administrative controls to limit combustible loading for fire protection and to justify fire protection provisions in all fissile material processing, handling, or storage areas. This approach is acceptable to the criticality staff since this mainly affects fire protection, although the effects on criticality safety of a fire or initiation of fire protection measures (including both engineered systems and administrative responses) should be evaluated in NCSEs supporting a license application.

- ANSI/ANS-8.23-1997, "Nuclear Criticality Accident Emergency Planning and Response"—As stated in revised CAR Section 6.5, the applicant committed to comply with the guidance (shall statements) and implement the recommendations (should statements) of ANSI/ANS-8.23-1997 without exception or clarification. While the applicant did not consider this part of the design basis of the PSSCs, this commitment provides reasonable assurance that consideration will be given to emergency planning during design. This is, therefore, acceptable to the staff.

- In addition to those standards to which the applicant has committed above, the following criticality safety standards are referenced in NUREG-1718 (NRC, 2000):

  — ANSI/ANS-8.5-1996, "Use of Borosilicate-Glass Raschig Rings as a Neutron Absorber in Solutions of Fissile Material"

  — ANSI/ANS-8.6-1983 (R1987), "Safety in Conducting Subcritical Neutron-Multiplication Measurements In Situ"

Revised CAR Section 6.3.4.3.2.9 states that, wherever neutron absorber control is used, it will be part of the geometry and fixed by design. In addition, the applicant stated (DCS, 2001a, RAI 77; and DCS, 2002e) that it did not envision using raschig rings for criticality control in facility operations but will instead rely only on fixed neutron absorbers in accordance with ANSI/ANS-8.21-1995. The applicant also did not intend to conduct subcritical neutron multiplication measurements at the facility. Therefore, the staff concurs that commitments to these two standards are not applicable to the design of the facility and, therefore, are not part of the design basis of the facility.

Therefore, the following standards have been identified, in whole or in part, as part of the design basis of the facility—ANSI/ANS-8.1-1983 (R1988), ANSI/ANS-8.3-1997, ANSI/ANS-8.17-1984, ANSI/ANS-8.19-1996, ANSI/ANS-8.20-1991, ANSI/ANS-8.21-1995, and ANSI/ANS-8.22-1997. The applicant has also committed to ANSI/ANS-8.23-1997. The other standards are either not applicable to the design of the facility or are adequately covered by other commitments. This, therefore, represents an acceptable set of design bases of the facility. However, nothing precludes the use of any other standards endorsed in RG 3.71.

6.2 Evaluation Findings

In Section 6.4 of the revised CAR, DCS provided design basis information for NCS PSSCs that it identified for the facility. Based on the staff's review of the revised CAR and supporting information provided by the applicant relevant to NCS, the staff finds that, with the following conditions contained in this chapter of the FSER, DCS has met the 10 CFR 70.61(d) performance requirement and the BDC set forth in 10 CFR 70.64(a)(9) to the extent required for construction authorization. Therefore, pursuant to 10 CFR 70.23(b), the staff finds that the design bases of the PSSCs identified by the applicant will provide reasonable assurance of protection against natural phenomena and the consequences of potential accidents. The following construction authorization condition (S-1) applies:

Construction Authorization Condition

Notwithstanding commitments in the Construction Authorization Request, the facility will be designed so that k_{eff} shall not exceed an Upper Subcritical Limit of 0.9249 for normal and credible abnormal conditions covered by AOA(4). DCS shall not increase the k_{eff} limits or change the AOA boundaries beyond those contained in Chapter 6 of the Final Safety Evaluation Report (FSER) without prior NRC review and approval.

The following open items in the April 30, 2002, draft safety evaluation report have been closed—NCS-1, NCS-2, NCS-3, NCS-4, NCS-5, NCS-6, NCS-7, and NCS-8.

6.3 References

(ANSI, 1988) American National Standards Institute/American Nuclear Society. ANSI/ANS-8.1-1983 (R1988), "Nuclear Criticality Safety in Operations with Fissionable Materials Outside Reactors." 1988.

(ANSI, 1996) American National Standards Institute/American Nuclear Society. ANSI/ANS-8.19-1996, "Administrative Practices for Nuclear Criticality Safety." 1996.

(ANSI, 1997) American National Standards Institute/American Nuclear Society. ANSI/NAS-8.3-1997, "Criticality Accident Alarm System." 1997.

(DCS, 2001a) Hastings, P., Duke Cogema Stone & Webster. Letter to U.S. Nuclear Regulatory Commission, RE: Response to Request for Additional Information—Construction Authorization Request. August 31, 2001.

(DCS, 2001b) Hastings, P., Duke Cogema Stone & Webster. Letter to U.S. Nuclear Regulatory Commission, RE: Clarification of Responses to NRC Request for Additional Information. December 5, 2001.

(DCS, 2002a) Hastings, P., Duke Cogema Stone & Webster. Letter to U.S. Nuclear Regulatory Commission, RE: Clarification of Responses to NRC Request for Additional Information. January 7, 2002.

(DCS, 2002b) Hastings, P., Duke Cogema Stone & Webster. Letter to U.S. Nuclear Regulatory Commission, RE: Clarification of Responses to NRC Request for Additional Information. February 11, 2002.

(DCS, 2002c) Hastings, P., Duke Cogema Stone & Webster. Letter to U.S. Nuclear Regulatory Commission, RE: Clarification of Responses to NRC Request for Additional Information. March 8, 2002.

(DCS, 2002d) Ihde, R., Duke Cogema Stone & Webster. Letter to U.S. Nuclear Regulatory Commission, RE: Mixed Oxide Fuel Fabrication Facility Construction Authorization Request Revision. October 31, 2002 (including page changes through February 9, 2005).

(DCS, 2002e) Duke Cogema Stone & Webster. Letter to U.S. Nuclear Regulatory Commission, RE: Requests for Additional Information, Clarifications, and Open Item Mapping into the Construction Authorization Request Revision. November 22, 2002.

(DCS, 2002f) Duke Cogema Stone & Webster. Letter to U.S. Nuclear Regulatory Commission, RE: Mixed Oxide Fuel Fabrication Facility Construction Authorization Request Change Pages. December 20, 2002.

(DCS, 2003a) Hastings, P., Duke Cogema Stone & Webster. Letter to U.S. Nuclear Regulatory Commission, RE: Duke Cogema Stone & Webster Mixed Oxide Fuel Fabrication Facility Criticality Validation Report—Criticality Validation Report—Revision 1 of Part I and Original Issue of Part II. January 8, 2003.

(DCS, 2003b) Hastings, P., Duke Cogema Stone & Webster. Letter to U.S. Nuclear Regulatory Commission, RE: Responses to Site Description, Instrumentation, and Criticality Open Items/Additional NRC Questions on Construction Authorization Request (CAR) Revision. February 11, 2003.

(DCS, 2003c) Hastings, P., Duke Cogema Stone & Webster. Letter to U.S. Nuclear Regulatory Commission, RE: Mixed Oxide (MOX) Fuel Fabrication Facility Construction Authorization Request Change Pages. February 18, 2003.

(DCS, 2003d) Hastings, P., Duke Cogema Stone & Webster. Letter to U.S. Nuclear Regulatory Commission, RE: Mixed Oxide (MOX) Fuel Fabrication Facility Construction Authorization Request Change Pages. April 10, 2003.

(DCS, 2003e) Hastings, P., Duke Cogema Stone & Webster. Letter to U.S. Nuclear Regulatory Commission, RE: Response to DSER Open Item NCS-04. June 13, 2003.

(DCS, 2003f) Hastings, P., Duke Cogema Stone & Webster. Letter to U.S. Nuclear Regulatory Commission, RE: Criticality Validation Report–Revision 3 of Part I, Revision 2 of Part II, and Revision 1 of Part III, Response to DSER Open Item NCS-04. July 2, 2003.

(DCS, 2003g) Hastings, P., Duke Cogema Stone & Webster. Letter to U.S. Nuclear Regulatory Commission, RE: Response to Request for Additional Information—MFFF Criticality Validation Report (DSER Open Item NCS-04). July 29, 2003.

(DCS, 2003h) Hastings, P., Duke Cogema Stone & Webster. Letter to U.S. Nuclear Regulatory Commission, RE: Criticality Validation Report—Revision 3 of Part II: Response to DSER Open Item NCS-04. October 10, 2003.

(DCS, 2004a) Hastings, P., Duke Cogema Stone & Webster. Letter to U.S. Nuclear Regulatory Commission, RE: Mixed Oxide Fuel Fabrication Facility (MFFF), Response to Questions on the Criticality Validation Report—Revision 3 of Part II. February 12, 2004.

(DCS, 2005) Ashe, K., Duke Cogema Stone & Webster. Letter to U.S. Nuclear Regulatory Commission, RE: Criticality Review. February 28, 2005.

(Moseley, 1965) Moseley and Wing. "Properties of Plutonium Dioxide," RFP-503. 1965.

(NRC, 1972) U.S. Nuclear Regulatory Commission. Regulatory Guide 3.71, "Nuclear Criticality Safety Standards for Fuels and Material Facilities," 1972.

(NRC, 2000) U.S. Nuclear Regulatory Commission. NUREG-1718, "Standard Review Plan for the Review of an Application for a Mixed Oxide (MOX) Fuel Fabrication Facility." August 2000.

(NRC, 2001a) Giitter, J.G., U.S. Nuclear Regulatory Commission. Letter to P.S. Hastings, Duke Cogema Stone & Webster, RE: Mixed Oxide Fuel Fabrication Facility Construction Authorization—Request for Additional Information. June 21, 2001.

(NRC, 2001b) Persinko, A., U.S. Nuclear Regulatory Commission. Letter to P.S. Hastings, Duke Cogema Stone & Webster, RE: Duke Cogema Stone & Webster Quality Assurance Program for Construction of the Mixed Oxide Fuel Fabrication Facility. October 1, 2001.

(NRC, 2001c) Persinko, A., U.S. Nuclear Regulatory Commission. Letter to P.S. Hastings, Duke Cogema Stone & Webster, RE: Nuclear Criticality Safety Staff Qualifications and Administrative Margins for Fuel Fabrication Facilities. November 9, 2001.

(NRC, 2003a) Persinko, A., U.S. Nuclear Regulatory Commission. Letter to P.S. Hastings, Duke Cogema Stone & Webster, RE: Request for Additional Information—Mixed Oxide (MOX) Fuel Fabrication Facility Nuclear Criticality Safety. June 25, 2003.

(NRC, 2003b) Persinko, A., U.S. Nuclear Regulatory Commission. Memorandum to J.G. Giitter, U.S. Nuclear Regulatory Commission, RE: November 13, 2003, Meeting Summary: Meeting with Duke Cogema Stone & Webster to Discuss Nuclear Criticality Safety Related to Mixed Oxide Fuel Fabrication Facility Revised Construction Authorization Report. December 3, 2003.

(NRC, 2004a) Tripp, C.S., U.S. Nuclear Regulatory Commission. Memorandum to J.G. Giitter, U.S. Nuclear Regulatory Commission, RE: December 17–19, 2003, In-Office Review Summary: Duke Cogema Stone & Webster Construction Authorization Request Supporting Documentation for Criticality Code Validation. January 8, 2004.

(NRC, 2004b) Persinko, A., U.S. Nuclear Regulatory Commission. Letter to P.S. Hastings, Duke Cogema Stone & Webster, RE: Mixed Oxide (MOX) Fuel Fabrication Facility Nuclear Criticality Safety—Review of Validation Report. April 16, 2004.

(NRC, 2005) Brown, D.D., U.S. Nuclear Regulatory Commission. Memorandum to J.G. Giitter, U.S. Nuclear Regulatory Commission, RE: February 1–2, 2005, In-Office Review Summary: Duke Cogema Stone & Webster Construction Authorization Request Supporting Documentation for Criticality Code Validation. March 14, 2005.

7. FIRE PROTECTION

7.1 Conduct of Review

This chapter of the final safety evaluation report (FSER) contains the staff's review of fire protection issues described by the applicant, Duke Cogema Stone & Webster (DCS), in Chapter 7 of the revised construction authorization request (CAR). The objective of this review is to determine whether the applicant's commitments and goals related to fire protection are adequate to meet the regulatory acceptance criteria referenced below. The review will also determine whether the design of the proposed facility adequately protects against external and internal fires and whether the design bases of the principal structures, systems, and components (PSSCs) identified by the applicant adequately protect against natural phenomena and the consequences of potential accidents. The staff evaluated the information provided by the applicant for fire protection by reviewing Chapter 7 of the revised CAR, other sections of the revised CAR, and supplementary information provided by the applicant. The review of fire protection was closely coordinated with fire protection aspects of accident sequences described in the safety assessment of the design basis (see Chapter 5 of this FSER), the review of explosion protection aspects (see Chapter 8 of this FSER), and the review of other plant systems (see Chapter 11 of this FSER).

The staff reviewed how the fire protection information in the revised CAR addresses the following regulations:

- Title 10, Section 70.23(b), of the *Code of Federal Regulations* (10 CFR 70.23(b)) requires, as a prerequisite to construction approval, that the design bases of the PSSCs be found to provide reasonable assurance of protection against natural phenomena and the consequences of potential accidents.

- 10 CFR 70.64, "Requirements for New Facilities or New Processes at Existing Facilities," requires that the baseline design criteria (BDC) and defense-in-depth practices are incorporated into the design of new facilities. With respect to fire protection, 10 CFR 70.64(a)(3) requires that the facility design "provide for adequate protection against fires."

The review focused on the design basis of fire protection systems, their components, and other related information. For fire protection systems, the staff reviewed information provided by the applicant for the safety function, system description, and safety assessment. The review also encompassed proposed design basis considerations, such as redundancy, independence, reliability, and quality. The staff used Chapter 7 in NUREG-1718 (NRC, 2000) as guidance in performing the review.

Sections 7.1.1 through 7.1.4 of this FSER are the staff's evaluation of how the applicant addressed the fire protection acceptance criteria in NUREG-1718. Section 7.1.5 of this FSER is the staff's evaluation of the design bases PSSCs related to fire protection.

7.1.1 Organization and Conduct of Operations

The organization and conduct of operations are the management measures that assure that fire safety is administered appropriately at a licensed facility. Section 7.1 of the revised CAR describes the applicant's commitment to assure that PSSCs, as identified in the revised CAR, are available and reliable; fire protection organizational responsibilities are defined; transient ignition sources and combustibles are controlled; and the facility maintains a readiness to extinguish or limit the consequences of a fire. The applicant committed to develop a fire protection program and administrative controls in order to meet organizational and operational guidance of NUREG-1718. The conclusion of the review is that the applicant's commitment to provide fire protection organizational management, and to develop conduct of operations at the facility site, meets the regulatory acceptance criteria and fire protection BDC of 10 CFR Part 70.64(a)(3) for the construction approval stage. Therefore, it is acceptable. The following paragraphs provide more details on the applicant's plans for a fire protection program and administrative controls.

7.1.1.1 Fire Protection Program

In Section 7.1.1 of the revised CAR, the applicant committed to develop a fire protection program which establishes the fire protection policies for the facility site. The applicant stated that the objectives of the Fire Protection Program are to prevent fires from starting and to detect, control, and extinguish fires that do occur. The applicant will identify management and organizational responsibilities for the Fire Protection Program. The fire protection organization will develop prefire plans and fire safety training, procedures, and considerations. It will be responsible for maintenance, inspection, and quality assurance of fire protection hardware; design change control; recordkeeping; and administrative controls. An effective fire protection program extends the concept of defense-in-depth to fire protection in areas with PSSCs.

7.1.1.2 Administrative Controls

Administrative controls establish procedures primarily for the storage, handling, and use of combustible materials and ignition sources. In Section 7.1.2 of the revised CAR, the applicant committed to develop procedures that govern the handling of transient combustibles in buildings containing PSSCs, permit systems that control ignition sources, formal periodic surveillance inspections, compensatory measures for detection and suppression systems, and emergency action plans. The applicant committed to develop and implement testing, inspection, and maintenance procedures for facility fire protection systems and equipment. Systems that are designed and installed in accordance with National Fire Protection Association (NFPA) codes and standards are maintained according to NFPA requirements. The staff finds that the applicant's commitment to maintain all fire protection equipment, as recommended by NUREG-1718 (NRC, 2000), increases the overall availability and reliability of fire protection equipment and the performance of fire protection personnel. The applicant designated some administrative controls as fire safety PSSCs, which are discussed in Section 7.1.5 of the FSER.

7.1.2 Features and Systems

Plant fire protection features and systems include building construction, fire area determination, electrical installation, ventilation, detection and alarm, and suppression. Section 7.4 of

NUREG-1718 provides acceptance criteria for fire protection features and systems. For the reasons set forth below, the staff finds that the applicant's proposed fire protection features and systems meet the 10 CFR 70.64(a)(3) BDC for fire protection.

7.1.2.1 Construction

Section 7.2.2 of the revised CAR states that buildings where radioactive materials would be used, handled, or stored at the facility are Type I or Type II construction. Thus, the structural members, including walls, columns, beams, girders, trusses, arches, floors, and roofs are of approved noncombustible or limited-combustible materials and will have fire resistance ratings as specified by NFPA 220, "Standard on Types of Building Construction," issued 1995 (NFPA, 1995b). Buildings at the facility that contain PSSCs associated with hardware are Type I construction. Type I construction requires that exterior bearing walls are rated at least 3 hours and that interior bearing walls, trusses, beams, girders, and columns are rated at least 2 hours. This type of construction has the ability to remain structurally sound during anticipated fires. The exterior walls of the mixed oxide (MOX) fuel fabrication building (BMF) structure carry a minimum 2-hr fire rating in order to provide exposure protection. However, they would carry a minimum 2-hr fire rating in order to provide protection from exterior exposure fires. Because of the amount of depleted uranium it contains, the secured warehouse building (BSW) contains PSSCs—combustible loading controls and facility worker actions. The BSW construction would be classified as Type II. The staff concludes that the BSW does not need the fire resistance rating of a Type I building because it does not contain hardware PSSCs that need to be protected. In addition, buildings are protected from exterior fires by observing fire safety criteria recommended in NFPA 80A, "Recommended Practice for Protection of Buildings from Exterior Fire Exposures," issued 1996 (NFPA, 1996e). By following guidelines found in NFPA 80A, buildings and building components are able to withstand anticipated exterior fire events, such as from ruptured fuel tank, trucks, other buildings, and forestry or wildland. Thus, the staff finds that the preliminary construction features at the facility are adequate to meet the BDC of 10 CFR 70.64(a)(3) for fire safety.

7.1.2.2 Interior Surface

Section 7.2.2 of the revised CAR indicated that exposed interior walls or ceilings and any factory-installed facing material would have a Factory Mutual-approved or Underwriters Laboratories- (UL-) listed flame-spread rating of 25 or less, and a smoke-developed rating of 50 or less, in accordance with American Society for Testing and Materials (ASTM)-E-84, "Standard Test Method for Surface Burning Characteristics of Building Material," issued 1998 (ASTM, 1998). If carpets and rugs are used, they would be tested in accordance with NFPA 253, "Standard Test Method for Critical Radiant Flux of Floor Covering Systems Using a Radiant Heat Energy Source," issued 1995 (NFPA, 1995a). Interior finish materials with flame-spread and smoke-developed ratings, as stated, prevent fires from attaining full-room fire involvement. Thus, the staff finds that interior finish materials would not contribute to fire or smoke development.

7.1.2.3 Storage Racks

Revised CAR Section 7.2.3.6 states that racks for the storage of plutonium dioxide, uranium dioxide, or MOX in powder, pellet, or rod form would be noncombustible. The applicant will provide combustible loading controls to prevent the buildup of combustibles in areas where storage racks would be located. Limiting combustible materials in areas where special nuclear materials would be stored reduces the intensity of potential fires if they occur and minimizes structural damage to the storage racks. Noncombustible storage racks do not contribute to the fixed fire load; consequently, they provide stability to their contents during a fire. The staff finds that the provisions for storage racks are acceptable because they meet the guidance of NUREG-1718 (NRC, 2000).

7.1.2.4 Electrical Considerations

Section 11.5 of the revised CAR discusses the electrical systems proposed for use at the facility. In addition, the applicant provided details on fire and electrical safety interfaces during a briefing to the Advisory Committee on Reactor Safeguards (NRC, 2002). To prevent fires from initiating and to protect cables from fire exposure hazards, the electrical systems at the facility are designed with the following considerations:

- Cables in redundant electrical trains are separated in accordance with Institute of Electrical and Electronics Engineers (IEEE) 384-92, "Standard Criteria for Independence of Class 1E Equipment and Circuits," issued 1992 (IEEE, 1992a), electrical separation criteria.

- The main electrical trains (for ventilation and for the emergency control rooms (ECRs)) enter the BMF from different sides of the building (approximately 45.7 m (150 linear ft apart) but on the same floor. However, immediately after entering the building, they are routed to separate floors of the MOX process (MP) area (DCS, 2002).

- Other cables, backup power supplies, and equipment feeds are kept as far apart as practical and are routed separately to the extent practical.

- Cables supplying emergency electrical power to PSSCs are routed in conduit.

- Systems are designed so that electrical faults do not initiate fires. When the breaker trips or the fuse blows, the electrical circuit is interrupted, which prevents damage to electrical components and wiring or cable insulation. Faults that could result in the inability of the system to perform its intended safety function are detected and removed.

- Electrical characteristics of equipment and/or power systems are continuously monitored to assure adequate protection and operation.

- The number of trip devices that could shut down electrical systems that are designated PSSCs is kept to a minimum.

- Exposed electrical cable is qualified under IEEE 383, "Standard for Type Test of Class 1E Electric Cables, Field Splices and Connections for Nuclear Power Generation Stations," issued 1992 (IEEE, 1992b), (i.e., ignition resistant and self-extinguishing).

- The normal and standby alternating current (ac) power systems are designed according to NFPA 70, "National Electric Code," issued 1999 (NFPA, 1999b).

- Where electrical cables enter glovebox or decanning rooms, they are installed in solid stainless steel wireways or ducts (revised CAR Section 7.2.3.1).

The applicant's proposed electrical features follow the applicable provisions of NFPA 70-1999, IEEE 383-92, and IEEE 384-92 that provide reasonable assurance that electrical installations are free from fire hazards. This strategy allows for redundancy and independence of electrical wiring. As a result, automatic functioning of the detection, alarm, suppression, and barrier systems would be maintained; electrical cables would not pose a source of fire hazard to other systems; and cables are protected from anticipated fires. Thus, the staff finds that the protection of electrical wiring from fire is acceptable and adequate.

7.1.2.5 Fire Alarm and Detection Systems

Section 7.2.3.2 of the revised CAR states that the fire alarm systems are designed according to NFPA 72, "National Fire Alarm Code," issued 1996 (NFPA, 1996f). The BMF detection systems are tied into a central alarm panel that would be located in the aqueous polishing (AP) control room. The central fire alarm panel has a graphical display which would assist the BMF operators or the emergency response team in identifying and responding to alarms. Upon detection of a fire, audible and visual alarms would be heard or seen in the affected parts of the BMF.

The alarm systems are capable of annunciating and differentiating alarm or fire conditions, supervisory indicators, or trouble signals. Alarm signals are transmitted to the monitored alarm center at the U.S. Department of Energy Savannah River Site fire department and the AP control room. Initiating circuits are capable of transmitting an alarm under circuit fault conditions of single ground, open, or both.

Heat and smoke detectors supplement or can actuate fire-extinguishing systems, fire dampers, and door closure devices. Detection systems would be located throughout the facility in accordance with the principles of NFPA 72 (revised CAR Section 7.2.3.2). Automatic fire detection systems are selected on the basis of the fire hazard and their reliability in each area. Section 7.5.3 of the revised CAR states that detection systems in areas containing dispersable radioactive materials are identified as PSSCs for defense-in-depth reasons (see FSER Section 7.1.5.7). Each glove box is provided with a minimum of two detectors. Generally, smoke detectors would be deployed. However, heat detectors would be installed in gloveboxes that are prone to dusty conditions.

The primary power supply for the fire detection/alarm system is the normal power system, which has two sources of offsite ac power. In the event that both sources of normal power are lost, the detection/alarm system can be powered by the standby ac power systems and then by normal battery backup. The staff finds the design of the fire alarm and detection systems acceptable because it reduces malfunctions and assures prompt detection and notification of fires and trouble conditions.

7.1.2.6 Means of Egress Protection

The proposed facility layout largely complies with the 1997 version of NFPA 101, "Life Safety Code," issued 1997 (NFPA, 1997c). According to revised CAR Section 7.2.2, the means of egress are arranged and maintained to provide free and unobstructed illuminated egress. Buildings at the facility are designed to provide means of egress that are adequate in number, location, and capacity for emergencies.

[Text removed under 10 CFR 2.390]

[Text removed under 10 CFR 2.390]

Emergency lighting is provided for means of egress and for critical operations areas where manual operations must be performed during an outage of normal ac power sources. Standby generators support the emergency egress lighting. The proposed design calls for two standby generators, each of which could operate continuously for 24 hours. The staff finds that the preliminary emergency lighting provisions provide the necessary illumination in the event that power to normal lighting is interrupted. The provisions for means of egress and emergency lighting are adequate and acceptable.

7.1.2.7 Lightning Protection

The applicant designated the BMF and the emergency diesel generator (BEG) building as PSSC structures. Revised CAR Section 7.2.2 states that PSSC structures are provided with lightning protection in accordance with the applicable provisions of NFPA 780, "Lighting Protection Code," issued 1997 (NFPA, 1997a). Such lightning protection provides a means of directing a lightning discharge to the earth without damage to the property. The staff finds that the protection schemes adequately protect the PSSC structures from a lightning strike. Therefore, the provisions for lightning protection are acceptable.

7.1.2.8 Ventilation System

The areas of the BMF that contain PSSCs are provided with ventilation systems that minimize the spread of fire and the products of combustion. Section 11.4 of the revised CAR describes the four confinement zones that the BMF would contain; the staff's evaluation of this design feature is provided in FSER Section 11.4. Essentially, pressure gradients between the confinement zones ensure that leakage airflows from the zones of lowest contamination risk to zones of increasing contamination risk. During a fire, the main objective is to maintain differential pressure between the room of fire origin and the surrounding areas. The ventilation systems operate continuously to ensure that combustion products flow through the exhaust stacks of the gloveboxes, the process rooms, or the process cells.

The BMF design incorporates airlocks that offer access to the process rooms. The airlocks are separated from the process rooms by fire barriers and are exhausted by the secondary confinement system. The airlocks have minimum-leakage doors and are maintained at a positive pressure with respect to the process room, thereby reducing the spread of radioactive contaminants and combustion products from the process room. Stairwells are pressurized with respect to corridors, to minimize smoke infiltration during a fire in the MP or AP areas. The applicant's treatment of acceptance criteria for nuclear filter plenums is discussed below. The staff's evaluation of the PSSC confinement systems is provided in Sections 11.4.1.3 and 7.1.5.5 of the FSER.

7.1.2.8.1 Filter Plenum Construction and Protection

The final filter plenum enclosures are 2-hr fire-rated construction with appropriate doors and are located as far as practical from the postulated fires. The design calls for multiple separate filter housings and redundant trains of filter systems in separate fire areas. No ignition sources and only limited amounts of combustible materials would be permitted within the filter housings or filter housing rooms.

7.1.2.8.2 Ventilation Ductwork and Dampers

The proposed ductwork in the ventilation systems incorporates manual and automatic dampers and controls to distribute and regulate the movement of air. The ductwork would be welded stainless steel or welded galvanized pipe. Additional fire-rated protective features are provided when the ductwork passes through a fire barrier into another fire area. The fire resistance of the dampers is suitable for the fire barriers they penetrate. Dampers are manually or automatically operated depending on the stage of fire development and/or location of the fire. As a result, the operation of the exhaust and supply dampers would be pre-planned.

7.1.2.8.3 Filter Design and Protection

At the BMF, high efficiency particulate air (HEPA) filters would be used to prevent the release of radioactive materials. The HEPA filters meet the requirements of American Society of Mechanical Engineers (ASME)-AG-1, "Code on Nuclear Air and Gas Treatment," issued 1991 (ASME, 1991). The exhaust for the C3 and C4 confinement systems contains two sets of final filtration units. Each final filtration unit consists of a filter assembly housing, a two-stage spark arrestor, and two stages of HEPA filters. The spark arrestors are located upstream of the HEPA filter exhaust plenum. The first-stage spark arrestor is made of stainless steel wire mesh; it is designed to eliminate large, burning embers that may pass through the ductwork. The second-stage spark arrestor is noncombustible fiberglass and steel mesh designed to remove particles larger than 1.0 micron. The first-stage spark arrestor is composed of stainless steel fiber, and the second stage is composed of high-strength steel and glass fiber, while HEPA filters are fabricated of glass media with metallic frames and silicone gaskets. The filter housing assemblies are designed and fabricated to the same temperature ratings as the duct materials in which they are installed.

7.1.2.8.4 Protection of the Final HEPA Filters

Fire protection of the final HEPA filter systems would be provided by heat detectors, spark arrestors, and dilution of high-temperature exhaust streams. Exhaust systems are designed to operate continuously so that hot gases generated from one fire area will be diluted with air from unaffected fire areas in order to prevent prolonged exposure of the HEPA filters to temperatures above their maximum service temperature. The HEPA filters are rated to 232 °C (450 °F). Heat detectors located in the ductwork upstream of each final filtration unit would alarm in the event of high temperatures. The applicant performed analyses to demonstrate that maximum temperatures, soot, and pressures are not exceeded during credible fires (DCS, 2003b). In addition, the preliminary results indicated that the HEPA filters at the MOX fuel fabrication facility (MFFF) could perform their safety function under hypothetical severe heat and soot conditions. By the same letter, the applicant committed to provide demonstration (through testing) of the final design's function as part of the integrated safety analysis (ISA). Based on the applicant's design functions and the preliminary analyses and commitment, the staff concluded that the ventilation system is adequate to meet the BDC of 10 CFR 70.64(a)(3) for fire safety.

7.1.2.9 Fire Areas and Barriers

For facility design and operational purposes, the BMF is subdivided into fire areas. The applicant used guidance from NFPA 801, "Standards for Facilities Handling Radioactive Material," issued 1998 (NFPA, 1998a), to determine fire area boundaries. Section 7.2.3.1 of the revised CAR indicates that fire areas separate manufacturing from material storage areas, control or computer rooms from adjacent areas, emergency generators from each other, redundant trains of PSSCs as required, and electrical equipment or battery rooms from adjacent areas. There are more than 350 fire areas at the BMF, including 169 in the MP, 128 in the AP, and 56 in the shipping and receiving areas. Fire areas are separated by noncombustible, reinforced concrete walls with a minimum 2-hr fire rating; 3-hr fire barriers separate some hazardous areas. Hourly ratings are based on ASTM E-119, "Standard Test Methods for Fire Test of Building Construction and Materials," issued 1995 (ASTM, 1995), definitions. Construction details are in accordance with NFPA 221, "Fire Walls and Fire Barrier Walls," issued 1997 (NFPA, 1997b).

Openings in fire barriers would have appropriately rated self-closing fire doors, fire dampers, and fire-rated penetration seals. Fire doors are designed according to NFPA 80, "Standard for Fire Doors and Fire Windows," issued 1999 (NFPA, 1999a). The staff finds that the general selection of fire areas is appropriate for the hazards at the BMF. The fire area selection minimizes the potential size of a fire. Therefore, it potentially limits the fire exposure to PSSC equipment, material, and personnel. The applicant designated BMF and BEG fire barriers as PSSCs. Because of the quantity and combustibility of fuels found in the BMF, the applicant performed zone fire modeling and one-dimensional finite difference analysis to determine heat transfer to the fire barriers (NRC, 2003b; DCS, 2003c). The analyses indicate that, in most cases, the room temperatures would not exceed the ASTM E-119-95 temperature profile (to which the barriers would be rated). In scenarios where the predicted peak room temperature exceeded the ASTM E-119-95 peak temperature, the applicant demonstrated that the transient excessive temperatures would not lead to barrier failure caused by thermal shock, thermal stress, or spalling. The staff finds that the selection of the fire barriers provides an adequate margin of safety against potential flashover and/or rapid temperature rise. Thus, the design function of the barriers meets the BDC of 10 CFR 70.64(a)(3) for fire safety.

7.1.2.10 Water Supply and Drainage

The facility design incorporates a water-supply system in accordance with NFPA 801 (NFPA, 1998a) requirements. Section 7.2.3.4 of the revised CAR states that the fire protection water supply system consists of an underground loop around the facility site, fire hydrants, fire pumps, and two firewater storage tanks. One of the storage tanks is seismically qualified. The water supply system is designed to handle the largest sprinkler demand plus 500 gpm (1893 L/min) for hose streams. The distribution systems are hydraulically designed to meet NFPA 13, "Standard for the Installation of Sprinkler Systems," issued 1996 (NFPA, 1996k); NFPA 14, "Standard for the Installation of Standpipes and Hose Systems," issued 1996 (NFPA, 1996i); and NFPA 24, "Standard for the Installation of Private Service Mains and their Appurtenances," issued 1995 (NFPA, 1995d).

[Text removed under 10 CFR 2.390]

The staff determined that the facility water supply system accommodates the requirements for automatic and manual suppression activities at the facility. The provisions for water capacity are based on defense-in-depth and industry practices; therefore, they are adequate.

7.1.2.11 Fire Suppression

A combination of automatic suppression systems, fire hose stations, exterior hydrants, and manual extinguishing devices would provide fire suppression at the facility. Automatic suppression is provided in areas where potentially significant fire loading is present.

Because of criticality concerns, water-based fire suppression systems are not planned in areas where fissile material would be handled (revised CAR Section 7.2.3.3.1). Where fissile materials would normally be present, clean-agent suppression systems will be provided. In such cases, clean-agent suppression is provided as a defense-in-depth PSSC to the fire barriers. The staff evaluated clean-agent suppression as a PSSC in FSER Section 7.1.5.7. The clean-agent system will also be provided in areas of the facility that contain electrical and/or electronic equipment, such as computer rooms, motor control centers, and control rooms. Where needed, the clean-agent system would protect under floor areas as well. All clean-agent systems will be designed, installed, and maintained according to NFPA 2001, "Standard on Clean Agent Extinguishing Systems," issued 1996 (NFPA, 1996a).

Otherwise, water-based suppression is provided throughout the facility. To avoid possible ingress of water into areas where fissile material is handled, pre-action-type sprinklers are used. Pre-action systems reduce the chance of accidental discharge by requiring independent actions for water discharge. The secured warehouse and reagent processing buildings would deploy wet-pipe sprinkler systems (discharge water when elevated temperatures are detected). The truck bays of the shipping and receiving building would have an automatic deluge system (all sprinklers activate at once). Sprinklers are designed according to NFPA 13-1996. All water-based systems will be periodically inspected, tested, and maintained in accordance with NFPA

25, "Standard for the Inspection, Testing, and Maintenance of Water-Based Fire Protection Systems," issued 1998 (NFPA, 1998c).

Not all fire areas in the BMF would have automatic suppression. Suppression is not provided in some airlocks, solvent cells, plenums, and chases, as well as areas that are not normally occupied, have low combustible loading, or have no ignition sources. Plenums, chases, and solvent cells are difficult to inspect on a routine basis. Similarly, automatic suppression features are not provided in areas such as rod storage, where there are minimal ignition sources and combustible loading.

The standpipe and hose systems allow manual firefighting capabilities throughout the proposed facility. Because of criticality concerns, a dry-standpipe system, instead of the normally pressurized wet-standpipe system, is provided in the MP, AP, and shipping and receiving areas. The standpipe systems are designed for Class II service, so that both fire department and fire brigade personnel can use them. The standpipe system is designed per NFPA 14-1996.

Portable fire extinguishers would be present throughout the BMF and inside all buildings at the facility so occupants could extinguish small fires. Extinguishers are selected and located according to fire hazards and to their effectiveness. A combination of multipurpose dry chemical, metal-use, and carbon dioxide extinguishers would be used. The use of portable extinguishers would be in accordance with NFPA 10, "Standard for Portable Fire Extinguishers," issued 1998 (NFPA, 1998e). Specially configured portable carbon dioxide bottles would be present in rooms with gloveboxes. These portable bottles can be quickly connected to the glovebox to suppress fires within without overpressurizing the glovebox.

The design calls for packets of extinguishing powder to be present in gloveboxes and in the rooms where a zircaloy fire risk exists. In addition, zirconium swarfs are collected in covered metal containers and removed daily, stored, and disposed.

Based on the variety and redundancy of suppression features, the applicant did not place total reliance on a single fire-suppression feature at the BMF. Standpipe and hose systems (which both facility-trained personnel and the fire department can use) provide backup fire suppression to the automatic suppression systems. An additional backup suppression is the specially configured portable carbon dioxide bottles with quick disconnect fittings. According to revised CAR Section 7.3, the facility fire brigade would respond to firefighting emergencies. Development and training of the facility fire brigade would follow the requirements of NFPA 600, "Standard on Industrial Fire Brigades,"issued 1996 (NFPA, 1996b).

Based on information provided to date, the staff finds that the fire suppression strategy provides diversity and defense-in-depth and is therefore acceptable. Automatic and manual fire suppression controls the spread of fire, thereby reducing the challenge to fire barriers.

7.1.2.12 Combustible and Pyrophoric Metals

When plutonium dioxide is fully oxidized, it is not pyrophoric. Uranium dioxide does not oxidize in an inert atmosphere and oxidizes very slowly in air under process temperature conditions. DCS maintains that uranium oxide is not combustible under process conditions (DCS, 2000, p. 45). However, U.S. Nuclear Regulatory Commission (NRC) Information Notice 92-14, "Uranium

Oxide Fires at Fuel Cycle Facilities," issued February 1992 (NRC, 1992), indicates that normally stable uranium dioxide may be pyrophoric or oxidize rapidly even at room temperatures when in very fine powder form. Section 7.2.5 of the revised CAR indicates that, for process reasons, some gloveboxes containing plutonium and uranium oxides are inerted with nitrogen. Thus, the nitrogen systems provide additional fire protection as they limit the potential combustion of special nuclear materials and ordinary combustibles. The pyrophoricity of plutonium and uranium dioxides is discussed in FSER Section 11.3.

Zirconium is stored and handled in accordance with NFPA 482, "Production, Processing, Handling and Storage of Zirconium," issued 1996 (NFPA, 1996c). The applicant intends to use titanium for the electrolyzer circuit and associated equipment that could be exposed to silver(II) ions. Finely divided or powdered forms and thin metal sections of titanium exposed to hot sparks can ignite and burn. The staff's evaluation of titanium fires is discussed in FSER Section 11.2.

7.1.2.13 Glovebox Protection

Gloveboxes provide physical and visual access to internal equipment, processes, and material. Glovebox details are given in revised CAR Section 11.4.7.1 and the applicant's polycarbonate report (DCS, 2000). A typical glovebox is a large, stainless steel enclosure mounted on a structural stainless steel stand. Glovebox windows consist of rectangular polycarbonate panels (10-mm thick) that fit into frames in the glovebox walls and ceilings. Lead-impregnated polymer sheets or lead glass panels overlay windows where radiation shielding is required to reduce operator exposures.

Light fixtures are generally installed outside of the gloveboxes, and they provide illumination for the interior spaces through windows located in glovebox ceilings. The lenses of the fluorescent light fixtures are noncombustible.

The design of the proposed facility calls for each glovebox to be provided with a minimum of two detectors. Generally, smoke detectors would be deployed. However, if conditions such as dust limit their effectiveness, heat detectors would be installed. Actuation of the detectors transmits an alarm to the control room, operations utilizing hydraulic fluids are automatically shut down, the glovebox process fire doors are actuated to close, and operators can manually close the glovebox dampers if warranted.

The proposed facility design provides additional fire protection features in the process rooms containing gloveboxes, such as manual-injection carbon dioxide for each glovebox, inert atmospheres for some gloveboxes with dispersable materials, fire doors between interconnected process gloveboxes, and minimal ignition sources near glovebox windows. Clean-agent automatic suppression is also provided and designated a defense-in-depth PSSC (see FSER Section 7.1.5.7).

Fire areas are separated by fire barriers, which help to prevent fires from spreading throughout the proposed facility. The BMF fire barriers are rated a minimum of 2 hours and are designated PSSCs. All associated dampers and doors are fire rated for use in the barriers. To further reduce the fire risk, combustible loading controls are provided as a PSSC for areas with gloveboxes that store radiological materials. This PSSC limits the amount of fixed and transient combustibles by design and through the use of administrative controls.

The applicant's proposed use of polycarbonate is not in compliance with the requirements of NFPA 801 (NFPA, 1998a) requirements to use noncombustible materials. However, the applicant demonstrated that an acceptable level of fire protection is achievable with the use of polycarbonate. Compared to other plastic glovebox materials, polycarbonate is relatively difficult to sustain combustion or ignite (its reported flashpoint is 466 °C (871 °F), and the spontaneous ignition temperature is 577 °C (1071 °F), as obtained by the test method described in ASTM D-1929, "Standard Test Method for Determining Ignition Temperature of Plastics," issued 1986 (ASTM, 1986)).

The ISA and ISA summary to be prepared by DCS in support of its later application for a license to possess and use licensed material will evaluate whether the range of the reported properties for the polycarbonate are bounding for its expected use or conditions. In addition, the fire hazard analysis (FHA) will account for the polycarbonate fuel. Revised CAR Section 7.4 states that the design-basis fire scenarios will include an evaluation of the types of potential fires based on the combustible form and quantity, combustible type (including polycarbonate), and the fire severity, intensity, and duration. Based on the applicant's commitments to provide further evaluations in support of its later application and the information in the DCS polycarbonate report (DCS, 2000), the NRC considers polycarbonate to be a potential candidate material for use in glovebox window panels.

7.1.2.14 Flammable and Combustible Liquids and Gases

Chapter 8 of the revised CAR identifies facility fire areas with the potential for large spills of flammable or combustible liquids. Section 7.2.2 of the revised CAR states that flammable and combustible liquids would be stored and handled in accordance with NFPA 30, Flammable and Combustible Liquids Code," issued 1996 (NFPA, 1996h). Means for containing spills, such as dikes, and for drainage systems are provided for in accordance with NFPA 30. Most vessels containing flammable or combustible liquids would be placed in process cells. Process cells would have fire prevention features, such as eliminating ignition sources, earth grounding of vessels and pipes, and combustible loading controls (see FSER Section 7.1.5.3). The purification and solvent recovery cycles involve a solvent-diluent mixture, which is combustible. In addition to the PSSC process cell fire prevention features, the facility design incorporates the following additional measures to limit fires:

- Welded equipment and pipes are connected to the offgas treatment system to preclude overpressurization.

- Combustible chemicals are stored in welded tanks or equipment to avoid leakage.

- Chemicals are supplied through leak-tight welded lines.

- Vapor phase concentration of flammable constituents or mixtures below limits is maintained, as specified in NFPA 69, "Standard on Explosion Prevention Systems," issued 1997 (NFPA, 1997d).

- A minimum clearance is maintained from the tank bottoms for feed lines to preclude splashing or vapor formation.

- Offgases from process equipment are collected and adequately treated.

The NRC concludes that these additional protective features minimize the amount of fuel and consequently the potential heat output of a fire initiating in the process cell. Therefore, these features provide defense-in-depth for fire protection and reduce challenges to the process cell PSSC.

Flammable and combustible gases would be stored and handled in accordance with NFPA 50A, "Standard for Gaseous Hydrogen Systems at Consumer Sites," issued 1999 (NFPA, 1999c), and NFPA 55, "Standard for Portable Fire Extinguishers," issued 1998 (NFPA, 1998b). Flammable gas generation, delivery systems, and protection are evaluated in FSER Section 8.1.2.

The sintering furnaces in the MP areas would use hydrogen/argon atmospheres, while the calcining furnace in the AP area would use an air/oxygen atmosphere. The fire safety measures for the sintering furnaces, such as inert-gas purge and automatic shutoffs, are in accordance with NFPA 86C, "Industrial Furnaces Using a Special Processing Atmosphere," issued 1995 (NFPA, 1995c). Concerns regarding the adequacy of the overall safety provisions for the furnaces are discussed in FSER Section 11.3.

7.1.2.15 Special Hazards

The BMF battery rooms would be separated from other areas by 3-hr fire walls and have automatic suppression and provisions for ventilation and hydrogen gas detection. The ventilation rate is a minimum of two air changes per hour, and the potential hydrogen accumulation is limited to less than 2 percent of room volume. These requirements are adequate according to NFPA 111, "Stored Electrical Energy Emergency and Standby Power Systems," issued 1996 (NFPA, 1996d), and IEEE 484-1996, "Recommended Practice for Installation Design and Installation of Vented-Lead Acid Batteries for Stationary Applications," issued 1996 (IEEE, 1996).

There would be several laboratories in the BMF to provide physical and chemical analyses of samples from the MP or AP process. Fire protection for laboratories is designed to meet NFPA 45, "Standard for Fire Protection for Laboratories Using Chemicals," issued 1996 (NFPA, 1996g). The isolation of these special hazards reduces the potential fire damage, while escape routes are safeguarded. Therefore, they enhance safety by reducing challenges to the PSSCs.

7.1.3 Manual Firefighting Capability

The applicant performed a baseline needs assessment that determined the minimum necessary capabilities of the facility firefighting forces. The assessment evaluated the minimum staffing for the firefighting forces, organization and coordination of onsite and offsite resources, personal protective and firefighting equipment, and training and emergency planning. The assessment determined that an onsite emergency response team is required at the facility (revised CAR Section 7.3). The facility fire brigade will be developed in accordance with NFPA 600 (NFPA,

1996b). The staff finds that these plans for manual firefighting are adequate for approving construction of the facility.

7.1.4 Preliminary Fire Hazard Analysis

The preliminary FHA for the facility consists of a systematic analysis of the fire hazards, an identification of fire areas, and an evaluation of anticipated consequences given the features proposed to control the hazards.

The applicant provided a summary of the methodology and assumptions of the preliminary FHA in Section 7.4.1 of the revised CAR. The preliminary FHA concludes that potential fires are typically small and will not propagate beyond the fire area of origin, and that fire barriers could prevent facility-wide fire spread. Furthermore, programs to maintain the fire barriers and fire detection/suppression features increase the reliability of the fire barriers. The staff concludes that the preliminary FHA begins the process of documenting the adequacy of facility fire safety. It meets the regulatory acceptance criteria and fire protection BDC of 10 CFR Part 70.64(a)(3) for the construction authorization stage. Based on the applicant's commitment to perform an FHA, staff finds that the preliminary FHA is acceptable.

7.1.5 Design Bases of the PSSCs

In revised CAR Sections 5.5.2.2.6, 7.5.3, and 11.4.11, the applicant identified the design bases and safety functions of the PSSCs that are intended to provide adequate protection against fires. Pursuant to 10 CFR 70.23(b), this FSER documents the staff's determination whether those design bases provide reasonable assurance of protection against natural phenomena and the consequences of potential accidents.

7.1.5.1 Combustible Loading Controls

Revised CAR Tables 5.5-13b, 5.5-13a, and 5.5-14 identify combustible loading controls as a fire event group PSSC to protect the following:

- confinement barriers in the C1, C2, and C3 areas
- radiological material storage gloveboxes
- the pneumatic transfer system
- the secured warehouse

The specific confinement barriers include 3013 canisters, 3013 transport casks, fuel rods, MOX fuel transport casks, transfer containers, and the final C4 HEPA filters. Combustible loading controls are identified by the applicant as a PSSC for gloveboxes that store radiological material. For example, combustible loading controls would be provided for the sintered and green pellet glovebox stores and the plutonium dioxide buffer storage area.

[Text removed under 10 CFR 2.390]

The combustible-loading control PSSC limits the quantity of combustibles in specific fire areas in order to ensure that the various confinement barriers, the facility-wide system, and large stores of radiological material would not be adversely impacted by fire. In revised CAR Section 5.6.2.2, the applicant identified combustible loading controls to include control of the fixed combustible loads by design and control of transient combustibles by design and during operations. Transient combustible controls can be achieved through worker training, regular surveillance, and postings. Even when engineered features are relied on to mitigate potential fires, combustible loading controls are essential to limit the maximum fire growth rate. In addition to combustible loading controls, the applicant identified additional protective features for fire areas that contain the various plutonium confinement barriers. They are (1) ignition source controls, (2) training of facility workers in manual suppression of incipient stage fires, (3) automatic fire detection in all areas having confinement barriers, (4) automatic fire suppression in some areas, and (5) the passive boundary of the C2 confinement system.

The design bases for the combustible loading controls identified by the applicant as PSSCs are the guidelines within NFPA 801 (NFPA, 1998a). The administrative controls in NFPA 801 limit the unnecessary accumulation of combustibles in facilities handling radioactive materials. Based on industry experience, these controls restrict the occurrence and development of fires. Thus, credible fires do not significantly impact primary confinement barriers. The staff concludes that PSSC combustible loading controls and their associated design bases are adequate because they help to ensure that structures, systems, and components are not adversely impacted by fire.

7.1.5.2 Confinement Barriers

The applicant designated 3013 transport cask and MOX fuel transport cask confinement barriers as PSSCs. The function of transport cask confinement barriers is to withstand the design basis fire without breaching. The transportation casks are designed to meet the requirements of 10 CFR 71.73, "Hypothetical Accident Conditions."

[Text removed under 10 CFR 2.390]

However, because of the combustible-loading controls PSSC and the additional protection provided by automatic suppression

[Text removed under 10 CFR 2.390]

The staff concludes that the design bases for the confinement barriers of the MOX transport cask and the 3013 transport cask help to provide reasonable assurance of protection against the consequences of fires.

7.1.5.3 AP Process Cell Fire Prevention Features

One objective of a fire protection program is to prevent fires from starting. However, fires are postulated to occur in AP process cells because of the presence of solvents and other chemicals with low flash points. The applicant designated the AP process cell fire prevention features as PSSCs to ensure that fires in the process cells are highly unlikely. The safety strategy utilizes the following ignition source controls:

- elimination of ignition sources (including electrical equipment) within the AP process cells

- earth grounding of vessels and pipes to avoid static electricity

- the use of controls to ensure that potential chemical reactions that may result in a fire are made highly unlikely

- limiting combustible materials and the processes required combustible products

In addition, the following are provided:

- ventilation to prevent accumulations of vapors in the flammable range

- fire barriers rated at 2 hours or more (evaluated in FSER Section 7.1.5.6) to isolate process cells from each other and limit the effect of external fires

- all materials at risk contained in sealed vessels and pipes and isolated from any hazards inside the process cell

- lightning protection

Furthermore, the applicant identified procedures that apply to process cells:

- permit systems to control ignition sources
- procedures to prohibit leak testing using open flames or combustion-generated smoke.

The design bases for AP process cell fire prevention are practices promoted in NFPA 801 (NFPA, 1998a) and NFPA 30 (NFPA, 1996h), which are based on industry experience, and rely on the use of fundamental safety measures and controls regarding storage, exposure, ignition sources, combustibles, and separation. The staff concludes that the design bases for AP fire prevention features provide reasonable protection against fire initiation and development.

7.1.5.4 MOX Fuel Fabrication Structures

[Text removed under 10 CFR 2.390]

[Text removed under 10 CFR 2.390]

7.1.5.5 C3 and C4 Confinement Systems

The applicant identified the C4 (primary) and C3 (secondary) systems (revised CAR Section 11.4) as PSSCs that are designed to limit the release of radioactive materials to the environment. These systems are designed to remain operational during a design basis fire to ensure that any potential releases caused by a fire are filtered. Additionally, the design function of the C4 confinement system is to ensure that the final C4 HEPA filters are not impacted by fire. The design basis is Regulatory Guide 3.12, "General Design Guide for Ventilation Systems of Plutonium Processing and Fuel Fabrication Plants," issued August 1973 (NRC, 1973), which allows for the cooling of hot gas from a fire by mixing it with air from rooms that are not involved in the fire. However, if the fire is large enough to exceed set thresholds in the final filtration units or exhaust, the C3 and/or C4 systems may be isolated.

The HEPA filters are fabricated of glass media with metallic frames and silicone gaskets. The continuous service rating is 232 °C (450 °F). Preliminary analyses indicate that the exhaust flow dilution would reduce extremely high room temperatures of 1260 °C (2300 °F) to less than 204 °C (400 °F) in the final filter plenum. Thus, the final C3 and C4 HEPA filters are not expected to be affected by the maximum temperatures anticipated to result from credible fires within the BMF. Additionally, preliminary analyses indicated that the resulting soot and pressure would not exceed the capacity of the final HEPA filters.

The applicant committed to provide demonstration of the final design function (through testing) as part of the ISA. The ability of the final filtration units to withstand potential fires is described in FSER Section 11.4.1.3. The staff concludes that the ability of the final C4 and C3 HEPA filters to perform their safety function has been adequately demonstrated. Thus, the HEPA filters provide reasonable protection against the release of radioactive material during a fire.

7.1.5.6 Fire Barriers

The BMF fire barrier systems (including walls, doors, fire dampers, and penetration seals) are PSSC that prevent fires from spreading from one fire area to another. In addition, the applicant has identified fire barriers as PSSCs to ensure that facility-wide systems, such as pneumatic transfer tubes that contain or handle radioactive materials, do not spread fire. The design basis for fire barriers is NFPA 801 (NFPA, 1998a), NFPA 221 (NFPA, 1997b), NFPA 80 (NFPA, 1999a), and UL 555, "Standard for Fire Dampers and Ceiling Dampers," issued 1995 (UL, 1995).

To minimize the potential for fire to spread between fire areas, the applicant implemented the following measures:

- Fire barriers within the MOX BMF are designated PSSCs.

- Detection and suppression systems in areas with significant quantities of dispersable radioactive material are designated defense-in-depth PSSCs to provide a diverse means of mitigating risks.

- Buildings containing PSSCs are designed as Type I construction. In addition, fire barriers in those structures are rated a minimum of 2 hours.

- Fire barriers are noncombustible, reinforced concrete.

- Fire doors between fire areas are normally closed or self-closing.

- Barriers have adequate penetration seals; a penetration seal tracking program will record pertinent information regarding the emplacement and modification of fire barrier penetration seals.

- Airlocks are provided at the doors between secondary and tertiary confinement areas. The airlock is maintained at a positive pressure with respect to the secondary confinement area.

- Fire doors on gloveboxes are normally closed or close upon fire detection. Automatic closure is inhibited momentarily during operations when the collision zone is not clear.

- Transfer gloveboxes that pass through a fire barrier are equipped with fire doors that are rated at 2 hours.

- Fire dampers to the supply and exhaust systems are manually or automatically closed, as necessary to maintain the effectiveness of fire barriers.

- Combustible-loading controls are used as a PSSC to reduce the potential spread of hot gases through pneumatic transfer tubes (see FSER Section 7.1.5.1).

The applicant evaluated fire severity based on the maximum quantities of combustibles anticipated in each fire area. The quantities and types of combustibles and the ignition sources were based on those found in comparable spaces in the MELOX facility. The estimate of the combustible quantity included transient fuels (revised CAR Section 7.4.1.3). After accounting for all combustibles within a fire area and using reasonable estimates of fire growth, the applicant's calculations show that, in a few cases, the room temperatures could exceed temperatures that the barriers are qualified to withstand. Because of the type of fire growth rate assumed for the combustibles, the temperature profile results in steeper temperatures at the early stages of fire growth (DCS, 2003b). However, in no case is the peak temperature of the ASTM-E119 (ASTM, 1995) fire curve (to which the fire barriers would be rated) exceeded. The applicant reevaluated those scenarios with the early rapid temperature rise and determined (using finite difference analysis) that the first 2 inches of the concrete cover on the fire barriers would not exceed 93 °C (200 °F) when exposed to the ASTM E-119 time-temperature curve. The temperature necessary to cause spalling of concrete is 93 °C (200 °F). Thus, the analysis concludes that the fire barrier members would not fail because of thermal shock or stress (NRC, 2003b). In addition, the applicant will demonstrate fire barrier performance under credible fire conditions during the ISA. While fire barrier failure is not expected, if the ISA shows that such failure is credible, appropriate items relied on for safety (IROFS) would be identified (DCS, 2003a). Further, as part of its ISA, the applicant committed to evaluate the impact of hot gas transport within the transfer tubes and to identify isolation valves as IROFS if warranted by the ISA results.

For the reasons discussed above, the staff finds that designating the fire barriers as PSSCs as part of the MFFF design provides adequate protection against internal fires. The applicant's commitments to perform further analyses as part of its ISA and to identify appropriate IROFS if warranted constitute an acceptable fire safety strategy.

7.1.5.7 Suppression and Detection

Sections 7.4.2 and 7.5.3 of the revised CAR state that, where dispersable fissile material is present, fire detection and suppression is provided as a PSSC. The PSSC detection/suppression systems further ensure that a propagating fire cannot result in an unacceptable release of radioactive material to the environment and, therefore, provide defense-in-depth. Because of criticality concerns when using water, clean-agent suppression is provided. However, clean-agent suppression is not credited in the safety assessment or the preliminary FHA. According to revised CAR Section 7.5.3, the PSSC clean-agent suppression systems are designed to be operable after an earthquake. Therefore, if a postseismic fire were to occur in fire areas containing radioactive materials, postseismic firefighting capability will be available. All clean-agent systems will be designed, installed, and maintained according to NFPA 2001 (NRC, 1996a). To assure the reliability of clean-agent suppression systems, the effect of ventilation and agent soak time will be considered in their design. They will include 100-percent reserve. The reserve is not automatically connected. Instead, the facility fire brigade manually connects the reserve capacity when needed.

As discussed in FSER Section 7.1.2.5, smoke detection systems would be present in all areas of the BMF. Fire detection and alarm systems are designed according to NFPA 72 (NFPA, 1996f). Detection systems associated with automatic suppression in fire areas containing significant quantities of dispersable radioactive materials are designated as PSSCs. Detection and clean-agent suppression systems allow initiating fires to be quickly detected and controlled. As PSSCs, they reduce the impact of fire so that fire barriers can be more effective. Thus, the staff concludes that fire detection/suppression systems provide defense-in-depth to the fire barriers to help assure protection against fires.

7.1.5.8 Emergency Control Room Air-Conditioning System

To protect operators against external fires, the ECR air-conditioning system (ECRAS) is identified as a PSSC. The ECRAS is a collection of heating, ventilation, and air-conditioning components located within the shipping and receiving area. The safety function of the ECRAS is to ensure that conditions are habitable for the control room operators. According to revised CAR Section 11.4.2.7.2, the location of control room fresh air inlets minimizes the presence of contaminants. Section 11.4.1.3 of the FSER provides more detail on the ECRAS.

7.2 Evaluation Findings

In Section 7.5 of the revised CAR, the applicant provided design-basis information for the fire protection systems that were identified as PSSCs for the proposed MOX facility. Based on the staff's review of the revised CAR and supporting information provided by the applicant relevant to fire protection systems, the staff finds that DCS has met the BDC set forth in 10 CFR 70.64(a)(3) for fire safety. Further, the staff concludes, pursuant to 10 CFR 70.23(b), that the design bases of the PSSCs evaluated in this FSER will provide reasonable assurance of protection against natural phenomena and the consequences of potential accidents.

7.3 References

(ASME, 1991) American Society of Mechanical Engineers. ASME AG-1, "Code on Nuclear Air and Gas Treatment." 1991.

(ASTM, 1986) American Society for Testing and Materials. ASTM D-1929, "Standard Test Method for Determining Ignition Temperature of Plastics." 1986.

(ASTM, 1995) American Society for Testing and Materials. ASTM E-119, "Standard Test Methods for Fire Test of Building Construction and Materials." 1995.

(ASTM, 1998) American Society for Testing and Materials. ASTM E-84, "Standard Test Method for Surface Burning Characteristics of Building Material." 1998.

(DCS, 2000) Hastings P., Duke Cogema Stone & Webster. Letter to U.S. Nuclear Regulatory Commission, RE: Choice of MFFF Process Glovebox Window Material (DCS-NRC-000030). December 15, 2000.

(DCS, 2002) Hastings P., Duke Cogema Stone & Webster. Letter to U.S. Nuclear Regulatory Commission, RE: Evaluation of the Draft Safety Evaluation Report (DSER) on Construction of a Mixed Oxide Fuel Fabrication Facility. July 9, 2002.

(DCS, 2003a) Hastings, P., Duke Cogema Stone & Webster. Letter to U.S. Nuclear Regulatory Commission, RE: Responses to Financial Qualification, Fire Safety, Chemical Safety, Aqueous Processing, Material Processing and Ventilation Open Items/Additional NRC Questions on Construction Authorization Request (CAR) Revision. February 18, 2003.

(DCS, 2003b) Hastings, P., Duke Cogema Stone & Webster. Letter to U.S. Nuclear Regulatory Commission, RE: Construction Authorization Request Change Pages. April 10, 2003.

(DCS, 2003c) Hastings, P., Duke Cogema Stone & Webster. Letter to the U.S. Nuclear Regulatory Commission, RE: Response to DSER Open Item FS-2, DCS-NRC-000134. May 14, 2003.

(DOE, 1997) U.S. Department of Energy. DOE-STD-1066-97, "Fire Protection Design Criteria." Washington, DC, March 1997.

(IEEE, 1992a) Institute of Electrical and Electronics Engineers, Inc. Standard 384-92, "Standard Criteria for Independence of Class 1E Equipment and Circuits." 1992.

(IEEE, 1992b) Institute of Electrical and Electronics Engineers, Inc. IEEE Standard 383, "Standard for Type Test of Class 1E Electric Cables, Field Splices and Connections for Nuclear Power Generation Stations." 1992.

(IEEE, 1996) Institute of Electrical and Electronics Engineers, Inc. Standard 484-1996, "Recommended Practice for Installation Design and Installation of Vented-Lead Acid Batteries for Stationary Applications." 1996.

(NFPA, 1995a) National Fire Protection Association. Standard 253, "Standard Test Method for Critical Radiant Flux of Floor Covering Systems Using a Radiant Heat Energy Source." 1995.

(NFPA, 1995b) National Fire Protection Association. Standard 220, "Standard on Types of Building Construction." 1995.

(NFPA, 1995c) National Fire Protection Association. Standard 86C, "Industrial Furnaces Using a Special Processing Atmosphere." 1995.

(NFPA, 1995d) National Fire Protection Association. Standard 24, "Standard for the Installation of Private Service Mains and their Appurtenances." 1995.

(NFPA, 1996a) National Fire Protection Association. Standard 2001, "Standard on Clean Agent Extinguishing Systems." 1996.

(NFPA, 1996b) National Fire Protection Association. Standard 600, "Standard on Industrial Fire Brigades." 1996.

(NFPA, 1996c) National Fire Protection Association. Standard 482, "Production, Processing, Handling and Storage of Zirconium." 1996.

(NFPA, 1996d) National Fire Protection Association. Standard 111, "Stored Electrical Energy Emergency and Standby Power Systems." 1996.

(NFPA, 1996e) National Fire Protection Association. Standard 80A, "Recommended Practice for Protection of Buildings from Exterior Fire Exposures." 1996.

(NFPA, 1996f) National Fire Protection Association. Standard 72, "National Fire Alarm Code." 1996.

(NFPA, 1996g) National Fire Protection Association. Standard 45, "Standard for Fire Protection for Laboratories Using Chemicals." 1996.

(NFPA, 1996h) National Fire Protection Association. Standard 30, "Flammable and Combustible Liquids Code." 1996.

(NFPA, 1996i) National Fire Protection Association. Standard 14, "Standard for the Installation of Standpipes and Hose Systems." 1996.

(NFPA, 1996k) National Fire Protection Association. Standard 13, "Standard for the Installation of Sprinkler Systems." 1996.

(NFPA, 1997a) National Fire Protection Association. Standard 780, "Lightning Protection Code." 1997.

(NFPA, 1997b) National Fire Protection Association. Standard 221, "Fire Walls and Fire Barrier Walls." 1997.

(NFPA, 1997c) National Fire Protection Association. Standard 101, "Life Safety Code." 1997.

(NFPA, 1997d) National Fire Protection Association. Standard 69, "Standard on Explosion Prevention Systems." 1997.

(NFPA, 1998a) National Fire Protection Association. Standard 801, "Standards for Facilities Handling Radioactive Material." 1998.

(NFPA, 1998b) National Fire Protection Association. Standard 55, "Standard for Compressed and Liquefied Gases in Portable Cylinders." 1998.

(NFPA, 1998c) National Fire Protection Association. Standard 25, "Standard for the Inspection, Testing, and Maintenance of Water-Based Fire Protection Systems." 1998.

(NFPA, 1998e) National Fire Protection Association. Standard 10, "Standard for Portable Fire Extinguishers." 1998.

(NFPA, 1999a) National Fire Protection Association. Standard 80, "Standard for Fire Doors and Fire Windows." 1999.

(NFPA, 1999b) National Fire Protection Association. Standard 70, "National Electric Code." 1999.

(NFPA, 1999c) National Fire Protection Association. Standard 50A, "Standard for Gaseous Hydrogen Systems at Consumer Sites." 1999.

(NRC, 1973) U.S. Nuclear Regulatory Commission. Regulatory Guide 3.12, "General Design Guide for Ventilation Systems of Plutonium Processing and Fuel Fabrication Plants." August 1973.

(NRC, 1992) U.S. Nuclear Regulatory Commission. Information Notice 92-14, "Uranium Oxide Fires at Fuel Cycle Facilities." February 1992.

(NRC, 2000) U.S. Nuclear Regulatory Commission. NUREG-1718, "Standard Review Plan for the Review of an Application for a Mixed Oxide Fuel Fabrication Facility." August 2000.

(NRC, 2002) U.S. Nuclear Regulatory Commission. Briefing to the Advisory Committee on Reactor Safety on the MFFF, "Electrical and Instrument and Control Systems Overview." April 10, 2002.

(NRC, 2003a) Persinko, A., U.S. Nuclear Regulatory Commission. Meeting Summary: February 6–7 Meeting with Duke Cogema, Stone & Webster to Discuss the Mixed Oxide Fuel Fabrication Facility Revised Construction Authorization Report. March 5, 2003.

(NRC, 2003b) Persinko, A., U.S. Nuclear Regulatory Commission. July 29–August 1, 2003, Meeting Summary: Meeting with Duke Cogema, Stone & Webster to Discuss Open Items Related to the Mixed Oxide Fuel Fabrication Facility. August 29, 2003.

(UL, 1995) Underwriters Laboratories, Inc. Standard 555, "Standard for Fire Dampers and Ceiling Dampers." 1995.

This page intentionally left blank

8. CHEMICAL SAFETY

8.1 Conduct of Review

This chapter of the final safety evaluation report (FSER) contains the staff's review of chemical and process safety described by the applicant, Duke Cogema Stone & Webster (DCS), in Chapter 8 of the revised construction authorization request (revised CAR) (DCS, 2002e), with supporting process safety information from Chapters 5 and 11 of the revised CAR. The objective of this review is to determine whether the chemical process safety principal structures, systems, and components (PSSCs) and their design bases identified by the applicant provide reasonable assurance of protection against the consequences of potential accidents. The staff evaluated the information provided by the applicant for chemical process safety by reviewing the revised CAR, supplementary information provided by the applicant, and relevant documents available at the applicant's offices but not submitted by the applicant. The staff also reviewed technical literature as necessary to understand the process and safety requirements. The review of aqueous polishing (AP) and mixed oxide (MOX) process (MP) safety design bases and strategies was closely coordinated with the review of the radiation and chemical safety aspects of accident sequences described in the safety assessment of the design bases (see Chapter 5 of this FSER), the review of fire safety aspects (see Chapter 7 of this FSER), and the review of plant systems (see Chapter 11 of this FSER).

The staff reviewed the chemical process and safety information in the revised CAR and supporting documents against the following regulations:

- Title 10, Section 70.23(b), of the *Code of Federal Regulations* (10 CFR 70.23(b)) states, as a prerequisite to construction authorization, that the design bases of the PSSCs and the Quality Assurance Program must be found to provide reasonable assurance of protection against natural phenomena and the consequences of potential accidents.

- 10 CFR 70.64, "Requirements for New Facilities or New Processes at Existing Facilities," requires that baseline design criteria (BDC) and defense-in-depth practices be incorporated into the design of new facilities. With respect to chemical protection, 10 CFR 70.64(a)(5) requires that the MOX fuel fabrication facility (MFFF or the facility) design provide for adequate protection against chemical risks produced from licensed material, facility conditions which affect the safety of licensed material, and hazardous chemicals produced from licensed material. Related to chemical protection, 10 CFR 70.64(a)(3) requires that the facility design provide for adequate protection against fires and explosions, such as those that could be initiated by or involve chemicals at the proposed facility.

The review for this construction authorization focused on the design bases of chemical process safety systems and components. For each chemical process safety system, the staff reviewed information provided by the applicant for the safety function, system description, and safety analysis. The review also encompassed proposed design basis considerations, such as redundancy, independence, reliability, and quality. The staff used Chapter 8.0 of NUREG-1718, "Standard Review Plan for the Review of an Application for a Mixed Oxide (MOX) Fuel Fabrication Facility," issued August 2000 (NRC, 2000), as guidance in performing the review. As stated on page 8.0-2 of NUREG-1718, "information contained in the application should be of sufficient quality and detail to allow for an independent review, assessment, and verification by the NRC reviewers."

8.1.1 Background

As stated in the memorandum of understanding (MOU) between the U.S. Nuclear Regulatory Commission (NRC) and the Occupational Safety and Health Administration (OSHA), "Worker Protection at NRC-licensed Facilities," (NRC, 1998b), the NRC oversees chemical safety issues related to (1) radiation risk produced by radioactive materials, (2) chemical risk produced by radioactive materials, and (3) plant conditions that affect the safety and safe handling of radioactive materials, and, thus, represent an increased radiation risk to workers. The NRC does not oversee facility conditions that result in an occupational risk unless they affect the safe use of licensed radioactive material. The MOU provisions applicable to the proposed MOX facility are now codified in 10 CFR 70.64(a)(5).

The NRC staff reviewed the revised CAR submitted by the applicant (DCS, 2002e) for the following areas applicable to process safety at the construction authorization stage:

- chemistry description
- hazardous chemicals and potential interactions affecting licensed materials
- chemical accident sequences
- chemical accident consequences
- safety controls

Additional documentation from the applicant and the literature was reviewed as necessary to understand the process and safety requirements. DCS stated that the BDC of 10 CFR 70.64(a) are incorporated into the facility design and that applicable sections of the revised CAR are intended to demonstrate compliance with these BDCs (revised CAR, p. 5.5-67).

In performing its review of the revised CAR and supporting documents, the staff used the applicable guidance of Section 8.4 of NUREG-1718 for determining acceptance under 10 CFR Part 70, "Domestic Licensing of Special Nuclear Material," consistent with the construction authorization stage and the level of the design. The evaluation is summarized in the sections that follow.

8.1.2 Areas of Review and Evaluation Findings

8.1.2.1 Chemical Processes

The plutonium feedstock to the facility requires purification before it can be used in nuclear fuel. This purification process is termed AP. A solvent extraction process is used in the AP process to separate plutonium from gallium, americium, uranium, and other minor impurities. The solvent extraction process is similar to the plutonium-uranium recovery extraction (PUREX) process used domestically and worldwide in nuclear fuel processing (Bennedict, 1983). The resulting purified plutonium, sometimes referred to as polished plutonium, is converted into plutonium dioxide. Subsequently, powder processing operations combine the polished plutonium dioxide with uranium dioxide to form the nuclear fuel. This is often referred to as MP.

Chemical processes occur as part of the AP and MP operations, supported by reagent preparation in the reagent processing building.

The MOX fuel fabrication building and the reagent processing building form the core group of buildings for plutonium polishing (i.e., AP process) and MOX fuel fabrication (i.e., MP). These are described in Section 1.1.1.1.2 of this FSER.

The reagent processing building would be located adjacent to the main MFFF building. Most of the initial reagent dilution and mixing operations would occur here. The design has a below-grade collection tank room that would receive waste chemicals from the reagent and AP buildings. The AP waste chemicals could be slightly radioactive. In addition, the liquid solvent area will be located on the northwest side of the building for the collection of liquid waste solvent from the AP process and its subsequent transfer to the U.S. Department of Energy (DOE) Savannah River Site (SRS) for waste processing. The waste solvent would be slightly radioactive. The reagent processing building will be divided into discrete rooms/areas to segregate chemicals and the associated equipment and vessels to prevent inadvertent chemical interactions. Most reagents (e.g., nitric acid, hydrogen peroxide, hydroxylamine nitrate (HAN), hydrazine, oxalic acid, sodium carbonate, diluent (C10–C13 isoalkanes, hydrogenated tetrapropylene (HTP), or tetrapropylene hydrogenate (TPH)), and tributyl phosphate (TBP)) are stored and solutions would be prepared in the reagent processing building for use in the AP area. Sodium hydroxide and nitrates of silver and manganese are stored and prepared in the AP area. Liquid chemical containers would be located inside curbed areas or sumps would be provided to contain accidental spills. Chemicals are transferred to the AP area from the reagent processing building via piping located in a concrete, below-grade trench between the two buildings.

In addition, there would be a separate, secure warehouse building (referred to as BSW in Figure 1.1-2 of the revised CAR) that also stores depleted uranium dioxide, as a fine powder, in steel drums. This building would also contain several drums of depleted uranium nitrate solution. These drums would be stored in a room with a 2-hour fire resistance rating. The uranium dioxide powder is used to form the matrix for the MOX fuel pellets. The process adds depleted uranium nitrate solution as an isotopic diluent for enriched uranium within the process.
[Text removed under 10 CFR 2.390]

Section 8.1.2.3.3 provides a detailed discussion of depleted uranium.

The staff review identified the need for additional information regarding mass, energy, and radioactivity balances relative to what was provided in the revised CAR. The applicant has provided supplemental information in specific correspondence (DCS, 2001, 2002b, 2002d, 2002f). Mass balances have been found to generally agree to within 5–10 percent around process units. The staff notes that these values may change as the design progresses. The staff concludes that the additional information provided on the major balances and overall process meets the standard review plan (SRP) (NRC, 2000) acceptance criteria and is adequate and acceptable for the construction authorization stage.

Fluid transport system (FTS) categories and their applicable codes and standards are discussed further in Section 11.8 of this FSER. The FTS categories 1 and 2 represent PSSCs, with FTS-1 representing a higher qualification level because of higher anticipated quantities and concentrations of radionuclides (primarily plutonium and americium). Based upon the qualitative descriptions of FTS categories and radionuclide concentrations, the staff concludes that the majority of the AP equipment would have a PSSC and an FTS-1 or FTS-2 designation. Specific PSSC and design basis findings for chemical safety are discussed later in this chapter and in Sections 11.2 and 11.3 of this FSER.

8.1.2.1.1 AP Process

The AP process can be segmented into the following four operational areas (unit symbols are provided in parentheses and discussed in Section 11.2):

- plutonium purification process—includes the decanning unit, dissolution unit, dechlorination and dissolution unit, purification cycle, oxalic precipitation and oxidation unit, homogenization unit, milling unit, canning unit, and recanning unit

- recovery processes—includes the solvent recovery cycle, oxalic mother liquor recovery unit, and acid recovery unit

- waste storage—includes the liquid waste reception unit and waste organic solvent unit

- offgas treatment—includes the offgas treatment unit

The applicant will purchase a uranium nitrate solution as a reagent; consequently, the plutonium purification process design no longer includes a uranium oxide dissolution unit. Likewise, the applicant has eliminated the recycle of silver nitrate solutions from the recover processes. Thus, the design no longer includes a silver recovery unit. An automatic sampling system also exists for providing materials for radiochemical analyses.

Overviews of these operational areas are discussed in FSER Section 1.1. Figure 8.1-1 provides a process overview. Section 11.2 of this FSER provides a more detailed description and evaluation of the AP process.

8.1.2.1.2 MOX Process

The MP involves dry subprocesses (i.e., blending of uranium and plutonium dioxide powders, poreformer and lubricant additions, pelletizing, sintering, and rod and assembly processing). The MFFF uses the advanced micronized master blend (A-MIMAS) process for the manufacture of MOX fuel assemblies. The A-MIMAS process uses a two-step mixing procedure. In the first step, the plutonium dioxide powder is mixed with depleted uranium dioxide and recycled scrap powder to form a primary blend (master blend) with approximately 20-percent plutonium dioxide content of the total mass. This mix is then micronized (i.e., ball milled into a very fine powder). In the second step, the primary blend is forced through a sieve, poured into a jar, and mixed with depleted uranium dioxide and scrap powder to obtain the final blend with the specified plutonium content. The maximum plutonium dioxide content in the final blend is 6 percent of the total metal content (normally referred to as heavy metal). The two-step mixing process is used to ensure a consistent product. The MP areas include the following:

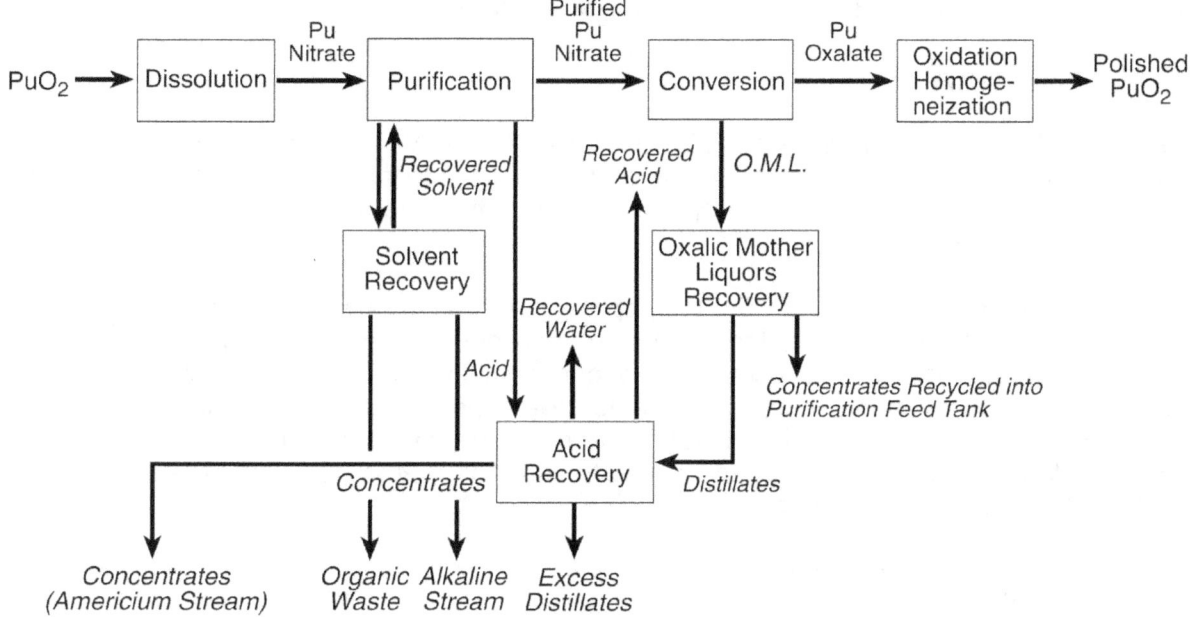

Figure 8.1-1 AP process overview

- The receiving area includes truck unloading, plutonium dioxide container handling, counting, and storage before and after transfer to the AP line. The function of the receiving area is to receive, unload, and store plutonium dioxide and uranium dioxide powder.

- The powder area has equipment for dosing MOX powder at the specified plutonium content in two steps for homogenizing and for pelletizing. The powder area receives uranium dioxide and plutonium dioxide powders and produces a mixture of specific plutonium content suitable for the production of MOX fuel pellets

- In the pellet process area, MOX pellets are sintered, ground, and sorted. The function of the pellet process area is to receive, store, process, and handle fuel pellets.

- In the fuel rod process area, pellets are loaded into rods, and the rods are inspected. The function of the fuel rod process area is to assemble, inspect, and store fuel rods.

- In the assembly area, rods are loaded into assemblies, and the assemblies are inspected and stored. The functions of the assembly area are to receive fuel rods and the required fuel assembly components and to assemble, inspect, and store completed MOX fuel assemblies.

- In a separate waste area, solid radioactive waste generated during the MP is processed, stored, and packaged for shipment.

The MP is reviewed in detail in Section 11.3 of this FSER.

8.1.2.1.3 Laboratory

Chemical and physical analyses of samples from the MP and AP areas will be conducted in the laboratory. The MFFF laboratory operations would also include laboratory liquid and solid waste management, temporary storage of scrap materials from the laboratory, MP adjustment in a test line, calibration, and document storage.

As shown in revised CAR Appendix 5A, the applicant's safety assessment notes that laboratory operations involve hazards associated with the major event types described in revised CAR Section 5.4.1.2.1. These events include loss of confinement/dispersal, fire, explosions, criticality, natural phenomena, external man-made events, external exposure, and chemical release. For this reason, the applicant stated that the following PSSCs are relied upon to reduce risks to the facility worker, the individual outside the controlled area (IOC), and the environment from potential hazards in the laboratory:

- C3/C4 confinement systems
- chemical safety controls
- laboratory material controls
- combustible loading controls
- criticality controls
- fire barriers
- fire detection and suppression
- fluid transfer systems
- gloveboxes
- glovebox pressure controls
- material handling controls
- material handling equipment
- Material Maintenance and Surveillance Programs
- process control safety subsystem
- facility worker action

As discussed in Section 5.6.2.7 of the revised CAR, laboratory material controls consist of administrative procedures that would be used to control the quantity of radiological and chemical materials in the laboratory. The safety function is to limit the extent of any potential explosion by limiting the quantity of hazardous chemicals that may be involved in any explosion and to limit the quantity of radiological/chemical material available for dispersion following a potential explosion. The applicant intended to develop procedures to establish limits on sample size, the number of samples that may be stored and used in the entire laboratory and in any one location, and the quantity of chemicals, reagents, or other hazardous materials that may be stored or used in a laboratory. The applicant will also develop procedures to ensure that laboratory operations are performed in accordance with safe laboratory operating practices. Procedures will be developed by the applicant and reviewed by the NRC staff as part of the license application phase.

As discussed in FSER Section 5, the staff independently evaluated the consequences of the controlling events within each event group for the laboratory. The staff's analysis confirms the applicant's assessment that the PSSCs listed above would be adequate to reduce the risk to all receptors to acceptable levels. Section 11.11 and Tables 11.3-34, 11.11-1, and 11.11-2 of the revised CAR provide information on the laboratory and the samples. The proposed safety strategies are similar to practices at many radiochemical facilities. The staff finds this approach acceptable for the construction authorization stage.

8.1.2.1.4 Chemical Reagents

Chemicals are received in various forms (solid, liquid, and gas) for use in the MFFF process. Most chemicals are stored in the reagent processing building, while some would be stored in the AP or MP areas. The main chemicals potentially stored in these areas have been identified by the applicant as the following:

- Solids

    — secured warehouse building—uranium dioxide

    — reagent processing building—oxalic acid and sodium carbonate

    — AP area—silver nitrate, sodium sulfite, plutonium dioxide, and manganese nitrate

    — MP area—azodicarbamide, zinc stearate, plutonium dioxide, and uranium dioxide

- Liquids

    — secured warehouse building—uranium (uranyl) nitrate solutions

    — reagent processing building—hydrazine hydrate, HAN, nitric acid, TBP, solvent/diluent (C10–C13 isoalkane), hydrogen peroxide, sodium hydroxide, hydrazine nitrate, silver nitrate, sodium carbonate, sodium sulfite, and zirconium nitrate

    — AP area—fresh and recovered nitric acid, aluminum nitrate solution, silver nitrate solution, manganese nitrate solution, solvent/diluent, HAN, hydrogen peroxide, oxalic acid, sodium carbonate, sodium hydroxide, sodium nitrate, sodium sulfite, TBP, uranyl nitrate, and zirconium nitrate solution

    — MP area—isopropanol

- Gases

    — reagent processing building—nitrogen tetraoxide (stored in a liquefied form in large cylinders and forced by pressurized air through an evaporator in the reagent processing building to form the gas, which is a mixture of nitrogen tetraoxide and nitrogen dioxide)

    — site—nitrogen, oxygen, hydrogen, argon, P10 (10 percent methane + 90 percent argon), helium, and 95 percent argon/5 percent hydrogen

— AP area—nitrogen, oxygen, P10 (10 percent methane + 90 percent argon), and 95 percent argon/5 percent hydrogen

— MP area—nitrogen, hydrogen, argon, P10 (10 percent methane + 90 percent argon), helium, and 95 percent argon/5 percent hydrogen

Storage facilities in the reagent processing building would contain the following:

- drums or tote tanks of reagents, including 13.6 N nitric acid, TBP, solvent/diluent, HAN (1.9 M), hydrazine hydrate ($N_2H_4.H_2O$; 22 percent hydrazine in water), hydrazine nitrate (30 percent solution in water), and sodium hydroxide (10N)

- cylinders of liquid nitrogen tetraoxide

- containers of hydrogen peroxide (30 wt%)

- storage of material for dissolving solid reagents, such as oxalic acid, silver nitrate, sodium sulfite, and sodium carbonate

A list of process chemicals can be found in Tables 8-1a to 8-1e and 8-2a to 8-2c of the revised CAR. Table 8.1-1 of this FSER provides a summary.

Chemicals and chemical mixtures in the process are usually used at lower concentrations than the reagent-grade chemicals potentially stored in the reagent processing building or AP area. The more dilute reagents usually pose reduced hazards.

The applicant provided information about the delivery of chemicals to the proposed facility and the potential impacts from spills and explosions. The applicant concluded that potential releases from the delivery of chemicals would not affect the IOC. The applicant stated that the emergency control room air conditioning system (ECRACS) is designated as a PSSC to ensure habitable conditions for all potential workers in the emergency control rooms (ECRs). Based upon completion of the chemical consequence analyses, measures will be provided, if necessary, to ensure that ECR workers are protected from potential chemical releases. Habitability is discussed further in FSER Section 8.1.2.6. The applicant stated that no chemical delivery scenarios have been identified that exceed the performance criteria of 10 CFR 70.61, "Performance Requirements." The applicant has identified an administrative PSSC entitled, "Hazardous Material Delivery Controls," which has the safety function of ensuring that the quantity of delivered hazardous material and its proximity to the MOX fuel fabrication building structure, the emergency generator building structure, and the waste transfer line are controlled to within the bounds shown in the safety analysis to produce acceptable results. Currently, this is expressed qualitatively because the designs are still evolving and numeric values may change; the specific values, detailed administrative controls, and procedures will be developed by the applicant and reviewed by the staff as part of its review of any later DCS application for a license to possess and use licensed material. The staff notes that controlling the deliveries of materials at nuclear facilities usually involves administrative controls. The staff finds this qualitative approach acceptable for the construction authorization stage because the potential

Table 8.1-1 Summary of Chemicals at the Proposed Facility

Chemical	Anticipated Annual Usage	Typical Estimated Onsite Quantity
	[Text removed under 10 CFR 2.390]	

hazard is identified and controlled by a method that has the capability to meet the regulatory requirements.

The staff review noted that sufficient general information on the chemical process description was provided by the revised CAR and the applicant's responses to Requests for Additional Information 111 and 113 (DCS, 2001, 2002f), recognizing that some of the information represents a snapshot in time as the design has continued to evolve. This includes information on mass balances, inventories, process physical parameters, chemical reactions, equipment, potential hazards, and proposed controls. The staff review noted several issues on the process chemistry that are discussed in Sections 8, 11.2, and 11.3 of this FSER. The staff finds that the overall process description itself is adequate and acceptable for the construction approval stage.

The applicant selected specific codes and standards that are typically used for similar purposes in the chemical and nuclear industries. The selected specific codes and standards are presented in general terms, and specific values are not mentioned and associated with specific structures, systems, and components (SSCs). FSER Sections 8.1.2.5, 8.1.2.6. 11.2, 11.3, and 11.8 discuss the equipment, SSCs, and codes and standards in more detail.

8.1.2.1.5 Chemical Process Inventories

The applicant provided chemical inventory information (revised CAR Sections 8.1, 8.2, 11.2, and 11.3). Common hazardous materials (e.g., vehicle fuel) and commonly used small quantities of solvents or gases are also used on site, and the applicant stated that specific inventories will be identified in the detailed design.

The staff reviewed the list of chemicals and their inventories. For construction, the chemical listing appears sufficiently complete, and the quantities appear consistent with the proposed activities and for preliminary safety assessments. The staff finds that this information is adequate and acceptable for the proposed activities and the construction authorization stage.

8.1.2.1.6 Chemical Process Ranges

The applicant provided information on ranges in Sections 8, 11.2, and 11.3 of the revised CAR and in supplementary submittals. In general, the focus is on design and steady-state conditions; ranges are not well addressed. Specific information on process ranges that might impact safety design bases are discussed in FSER Sections 8.1.2.5, 11.2, and 11.3. Additional information on the ranges will be provided by the applicant and reviewed by the staff as part of its review of any later DCS application for a license to possess and use licensed material.

8.1.2.1.7 Chemical Process Limits

The applicant provided information on chemical process limits in the revised CAR. In general, the focus is on design and steady-state conditions; limits are not well addressed. Specific information on process limits that might impact safety design bases are discussed in FSER Sections 8.1.2.5, 11.2, and 11.3. Additional information on the limits will be provided by the applicant and reviewed by the staff as part of its review of any later DCS application for a license to possess and use licensed material.

8.1.2.2 Hazardous Chemicals and Potential Interactions

This section identifies the major chemicals that would be stored and used at the facility, identifies potential interactions among these chemicals, and discusses potential unusual and unexpected reactions.

8.1.2.2.1 Chemicals

Plutonium would be received as impure plutonium dioxide in the AP area and transformed into various compounds in different solutions at various stages of the process to remove the impurities (e.g., gallium, americium, uranium). Purified plutonium will be converted back into plutonium dioxide before being transported to the MP plutonium dioxide buffer storage unit. During some AP operations, new chemical compounds would be produced. Table 8.1-2 identifies chemicals produced during normal operations. Several of these chemicals are potentially hazardous or are the precursors to potentially hazardous compounds, including hydrogen, nitric acid, nitrous acid, and nitrogen oxides. Table 8.1-3 in this FSER identifies hazardous and reactive chemicals potentially produced in significant quantities under off-normal conditions. These are hydrazoic acid, azides, and red oil. These compounds and identified controls (PSSCs and design bases) are discussed in detail in Section 8.1.2.5 of this FSER.

Nitrogen and helium gases at appropriate pressures would be used in the MP. Azodicarbamide (a poreformer) and zinc stearate are used in powder form in the pelletizing process. Other chemicals (liquids, solids, and gases) used in the MP area are present mainly in the laboratory for inspection and analysis purposes. Chemicals would be stored and used in the MFFF laboratory in small quantities and would thus be a smaller hazard than the chemicals referenced in the previous paragraph. In the revised CAR, the applicant stated that laboratory chemicals will be identified in the license application. The applicant provided a preliminary list of anticipated chemicals and quantities in the laboratory area (revised CAR Table 8-2c). Typical quantities are less than 10 kg (22 lb) of each reagent.

8.1.2.2.2 Chemical Interactions

Human error or equipment malfunction may result in inadvertent chemical interactions and initiate hazardous reactions. Hazardous chemical characteristics and incompatibilities with the associated materials/process conditions have been identified by the applicant and are summarized in Table 8.1-4. The staff concluded that this list was derived from standard guidelines. The applicant would conduct hazard and operability analyses (HAZOPs) as part of the integrated safety analysis (ISA) during detailed design and submit a complete chemical interaction matrix as part of the license application. As stated in the revised CAR, the applicant would control chemical preparation in accordance with operating procedures by trained personnel in order to minimize the potential for unexpected interactions. To minimize the risk to PSSCs associated with inadvertent chemical interactions, the applicant would prepare most chemical reagents for the AP process in the reagent processing building (a non-radiological building), with subsequent distribution to the AP area.

Sections 8.4.3.2 (B) and (C) of the SRP list the acceptance criteria for potential chemical interactions, with the understanding that some generalities are acceptable at the construction authorization stage. The staff notes that the applicant identified general precautions and several

Table 8.1-2 Reaction Products of the AP Process (Normal Operations)

Chemical	Formula	Comment
Alkaline Wastes (including dibutyl phosphate and monobutyl phosphate)	Various	Alkaline wastes generated in the solvent recovery unit as a result of washing solvent with sodium carbonate and sodium hydroxide solutions (Note 3)
Carbon Dioxide	CO_2	Reaction product when plutonium oxalate is transformed into plutonium dioxide in the oxalic precipitation and oxidation unit (Note 3)
Carbon Monoxide	CO	Reaction product when plutonium oxalate is transformed into plutonium dioxide in the oxalic precipitation and oxidation unit (trace quantities only) (Note 3)
Chlorine	Cl_2	Reaction product from dissolution of AFS material in the dechlorination dissolution unit (subsequently treated in the dechlorination scrubbing column) (Note 2)
Hydrogen	H_2	Produced from radiolysis and electrolysis reaction (Note 4)
Nitrogen Oxides	NO_x	Reaction product of dissolution units and excess reactant from purification cycle oxidation column (Note 2)
Nitrogen	N_2	Reaction product of several reactions in the purification cycle oxidation column; reaction product in dechlorination scrubbing column of the dechlorination dissolution unit (Note 3)
Nitric Acid	HNO_3	Reformed in nitric oxide scrubbing columns (offgas treatment unit) (Note 1)
Nitrous Acid	HNO_2	Always present in nitric acid solutions (Note 3)
Nitrous Oxide	N_2O	Reaction product of several reactions in the purification cycle oxidation column (Note 2)
Oxygen	O_2	Reaction product of hydrogen peroxide decomposition during plutonium dioxide dissolution in the dissolution unit (Note 3)
Plutonium Dioxide	PuO_2	Reformed in the calcining furnace of the oxalic precipitation and oxidation unit from the plutonium oxalate feed (Note 1)

Table 8.1-2 (cont.) Reaction Products of the AP Process (Normal Operations)

Chemical	Formula	Comment
Plutonium Oxalate	$Pu(C_2O_4)_2$	Precipitated in the oxalic precipitation and oxidation unit from the reaction of plutonium nitrate with oxalic acid (Note 1)
Plutonium (III, IV, VI) Nitrate	$Pu(NO_3)_3$, $Pu(NO_3)_4$, $PuO_2(NO_3)_2$	Plutonium (VI) nitrate formed from the dissolution of plutonium dioxide in the dissolution unit and in the evaporator of the oxalic mother liquor recovery unit (Note 1) Plutonium (IV) nitrate formed from the addition of hydrogen peroxide to the plutonium (VI) nitrate solution in the dissolution unit (Note 1) Plutonium (III) nitrate formed from the reduction of plutonium (IV) nitrate solution with HAN in the purification unit (Note 1)
Sodium Chloride	NaCl	Reaction product in the dechlorination scrubbing column of the dechlorination dissolution unit (Note 3)
Uranyl Nitrate	$UO_2(NO_3)_2$	Formed from the dissolution in nitric acid of the impure plutonium dioxide received at the proposed facility (Note 1)
Water	H_2O	Reaction product of several reactions in the purification cycle oxidation column; reaction product of hydrogen peroxide decomposition during plutonium dioxide dissolution in the dissolution unit; reaction product of uranium dioxide dissolution reaction in the uranium dioxide dissolution unit (Note 3)

Notes:
(1)   The applicant performed chemical consequence analyses for nitric acid, uranyl nitrate, and the plutonium compounds, using inventories from Table 8.2a of the revised CAR.
(2)   The applicant performed chemical consequence analyses for chlorine, nitric acid, and nitrogen oxides in selected areas of the facility based upon their projected quantities from reactions.
(3)   Because of the low rate of production and/or lack of toxicity, inventories are not quantified for the purposes of calculating chemical consequences to the site worker or the IOC from spills or releases.
(4)   The generation of hydrogen is considered in the design of the scavenging air system.

**Table 8.1-3 Potentially Hazardous Chemicals Produced in
AP under Off-Normal Conditions**

Chemical	Formula	Revised CAR Comment
Hydrazoic Acid	HN_3	Interaction of hydrazine nitrate and nitrous acid could initiate, under certain conditions, the formation of hydrazoic acid. However, any accumulation of hydrazoic acid, which occurs in the AP process, is 10 times lower than the explosive limits. In the gaseous phase, hydrazoic acid may decompose to nitrogen and hydrogen. (See FSER Section 8.1.2.5.3.2.)
Hydrazoic Salts (i.e., azides)	NaN_3 AgN_3	Interaction of hydrazine nitrate and nitrous acid could initiate, under certain conditions, the formation of hydrazoic salts (i.e., azides). However, since the solubility limits of these azides are not exceeded, precipitation and the potential for an explosion are prevented. (See FSER Section 8.1.2.5.3.3.)
Red Oil	(Various)	Red oil is an organic mixture, consisting of TBP and its complexes with plutonium nitrate and nitric acid, degradation products of TBP (e.g., dibutyl phosphate), and possibly various nitrated hydrocarbons. These are compounds that could react exothermically at temperatures higher than 130 °C (266 °F). Process unit design prevents the process fluid temperature from exceeding 130 °C (266 °F) by providing adequate margin between the maximum operating temperature and 130 °C (266 °F). (See FSER Section 8.1.2.5.5.)

specific hazards associated with the interaction of chemicals and also considered radiolysis effects. Specific hazards include red oil (solvent), HAN/hydrazine, azides, plutonium (VI) oxalate, hydrogen, and pyrophoric effects from certain metals and oxides. Thus, the staff finds that the overall process description itself is generally adequate and acceptable for the construction authorization stage. Significant specific chemical-related risks are discussed further in FSER Sections 8.1.2.5, 11.2, and 11.3.

8.1.2.2.3 Unusual and Unexpected Reactions

Additional chemical compounds are produced in some AP operations. The applicant stated that the behavior of these mixtures is well understood from experience at the La Hague facility in France and is included in the preliminary chemical process safety evaluation (e.g., Tables 8.1-2 to 8.1-4 of this FSER).

In general, for vapor and gaseous species, AP chemical interactions produce nitrogen oxides, carbon dioxide, carbon monoxide, and hydrogen, as well as possible plutonium, americium, and uranium mixtures entrained in nitric acid vapors. These chemicals would be generated in tanks and equipment and would be collected by the offgas treatment system. Solvent-diluent vapors would also be collected in a separate stream and treated by the offgas system before being

Table 8.1-4 Process Chemical Hazardous Characteristics and Incompatibilities
Revised CAR (DCS, 2002e)

Form	Chemical	Corrosive	Flammable	Explosive	Chemical Burn	Toxic	Incompatibilities
Liquid	Nitric Acid (13.6 N)	x			x	x	Organics, Hydrogen Peroxide, HAN, Hydrazine Monohydrate, Sodium Carbonate, Sodium Hydroxide
	Hydrogen Peroxide			x	x	x	Organics, Nirtic Acid, Manganese (metal), Hydrazine, Sodium Carbonate, Metallic Salts
	Tributyl Phosphate (solvent)		x	x	x	x	Ammonia, Nitric Acid, Oxidizing Agents, Strong Bases
	Diluent (C10 C13 isoalkane)		x	x		x	Oxidizing Agents, Oxygen
	Sodium Carbonate					x	Aluminum, Acids, Hydrogen Peroxide
	Hydroxylamine Nitrate	x		x	x	x	Bichromate and Permanganate of Potassium, Copper Sulfate, Zinc, Strong Oxiders, Strong Reducing Agents, Nitric Acid, Combustible Materials
	Hydrazine Monohydrate	x		x	x	x	Oxidizing Agents (Nitric Acid), Metals, Asbestos
	Sodium Hydroxide	x			x	x	Acids, Aluminum and Other Metals, Organic Halogens (especially Trichlorethylene), Sugars
	Aluminum Nitrate	x			x	x	Combustible Materials, Strong Reducing Agents, Metals, Water
	Hydrazine Nitrate	x			x	x	Acids, Strong Oxidizers, Metal Salts
	Isopropanol		x			x	Oxidizing Agents
	Zirconium Nitrate	x			x	x	Combustible Materials, Strong Reducing Agents, Metals
Gas	Dinitrogen Tetraoxide/Nitrogen Dioxide	x		x	x	x	Reducing Agents, Organics, Metals
	Hydrogen		x	x			
	Oxygen					x	Organics
Solid	Silver Nitrate (also present as liquid)	x		x	x	x	Ammonia, Carbonates, Chlorides
	Manganese Nitrate (also present as liquid)	x		x	x	x	Strong Reducing Agents, Combustible Materials
	Oxalic Acid (also present as liquid)				x	x	Silver, Sodium Chloride, Sodium Hypochlorite
	Azodicarbonamide						Strong Oxidizing Agents
	Zinc Stearate		x			x	Strong Oxidizing Agents, Acids

released to the stack.

The staff review notes that the list of chemicals produced in the process, as identified by the applicant, is reasonably comprehensive but that additional reactions and compounds may be added as the design progresses. The applicant's approach to identifying hazardous chemicals, interactions, and incompatibilities is based on experience and process understanding and appears capable of identifying any new potential hazards as the design evolves. The overall approach is reasonable and acceptable for the CAR stage.

FSER Section 8.1.2.5 discusses process safety information and specific concerns in more detail, and FSER Sections 11.2 and 11.3 discuss specific chemical concerns by process unit. The staff findings are presented in those sections.

8.1.2.3 Chemical Accident Sequences

The applicant provided information on chemical-related events in Sections 5.5 and 8 of the CAR. The applicant provided supplementary information in public meetings (NRC, 2003a, 2003b, 2003c), correspondence on the docket (DCS, 2003a), and in-office reviews of the preliminary hazards analysis, preliminary accident analysis, and chemical accident modeling reports (NRC, 2001c, 2002d). This section presents the staff's assessment of accident sequence bases, unmitigated sequences, estimated concentrations, and concentration limits for chemical safety.

This section also provides the methodology and results for the evaluation of chemical consequences that may be associated with a release of radiochemical materials.

As reflected in 10 CFR 70.64(a)(5), the design must provide adequate protection from the following:

(1)   chemical risks produced from licensed radioactive material

(2)   facility conditions which affect the safety of licensed material

(3)   hazardous chemicals produced from licensed materials

Staff evaluations and conclusions are presented in subsequent sections.

8.1.2.3.1 Chemical Consequence Limits

This section evaluates the chemical risks to workers and members of the public from events that release chemicals and radionuclides. The applicant stated that the chemical consequences are based on bounding analyses.

The 10 CFR 70.61 performance requirements define high- and intermediate-consequence events as including acute chemical exposures to individuals. The applicant primarily based the chemical consequence limits for accident categories on "Temporary Emergency Exposure Limits (TEELs)," Revision 18. The proposed limits are listed in Tables 8.1-5 and 8.1-6 of this FSER. The following are the TEEL definitions:

- TEEL—0 is the threshold concentration below which most people will experience no appreciable risk of health effects.

- TEEL—1 is the maximum concentration in air below which it is believed that nearly all individuals could be exposed without experiencing other than mild transient adverse health effects or perceiving a clearly defined objectionable odor.

- TEEL—2 is the maximum concentration in air below which it is believed that nearly all individuals could be exposed without experiencing or developing irreversible or other serious health effects or symptoms that could impair their abilities to take protective action.

- TEEL—3 is the maximum concentration in air below which it is believed that nearly all individuals could be exposed without experiencing or developing life-threatening health effects.

TEELs were developed by the DOE Subcommittee on Consequence Assessment and Protective Action (WSMS, 2000; DOE, 2004) to serve as temporary guidance until the American Industrial Hygiene Association (AIHA) publishes emergency response planning guideline (ERPG) concentrations. The TEEL values are used by DOE, the U.S. Department of Defense, and some other government agencies as ERPG approximations until peer-reviewed ERPGs become available. As noted in Table 8.1-5, the applicant used ERPG values for several chemicals. However, NRC regulations and guidance do not mention TEELs, TEEL methodology, or values derived from TEELs.

The TEEL definitions are similar to the ERPG definitions except for the 1-hour term because concentration-dependent chemical limits are based on a 15-minute exposure time and dose dependent chemicals are based on a 1-hour exposure time. The concentrations are then adjusted to the same 1-hour exposure timeframe by an exponential formula.

The ERPGs were developed by the AIHA ERPG committee (AIHA, 2004) as planning guidelines to anticipate adverse health effects from the exposure to toxic chemicals. The ERPGs do not contain safety factors that are usually incorporated into exposure guidelines.

The TEEL limits are subject to change over time. The TEEL limits are revised if new or additional toxicity data are found or new concentration limits are issued. For example, within the last 3 years, the TEEL-3 limits for nitric acid, hydrazine hydrate, and nitrogen tetraoxide, three important chemicals used in bulk quantities in the AP system, varied by more than 50 percent.

The NRC does not promulgate its own chemical consequence limits but relies on values from other government agencies and organizations that have a clear toxicological and regulatory basis. There are multiple limits available for assessing the impacts from potential chemical releases. The U.S. Environmental Protection Agency (EPA) is developing acute exposure guideline levels (AEGLs) as a government standard for emergency response (EPA, 2004). The AEGLs are defined as follows:

- AEGL-1 is the airborne concentration (expressed as ppm or mg/m^3) of a substance at or above which it is predicted that the general population, including susceptible but excluding hyper-susceptible individuals, could experience notable discomfort. Airborne concentrations below AEGL-1 represent exposure levels that could produce mild odor, taste, or other sensory irritations.

- AEGL-2 is the airborne concentration (expressed as ppm or mg/m^3) of a substance at or above which it is predicted that the general population, including susceptible but excluding hyper-susceptible individuals, could experience irreversible or other serious long-lasting effects or impaired ability to escape. Airborne levels below AEGL-2 but above AEGL-1 represent exposure levels which may cause notable discomfort.

- AEGL-3 is the airborne concentration (expressed as ppm or mg/m^3) of a substance at or above which it is predicted that the general population, including susceptible but excluding hyper-susceptible individuals, could experience life-threatening effects or death. Airborne levels below AEGL-3 but at or above AEGL-2 represent exposure levels that may cause irreversible or other serious, long-lasting effects or impaired ability to escape.

Currently, AEGLs have been established for several chemicals. Interim AEGLs are available for additional chemicals, including chlorine, hydrazine, nitric acid, and nitrogen dioxide, which are chemicals of concern in the AP process. There are some overlaps in the definitions of the AEGLs and TEELs. However, the high-consequence (Level 3 in Table 8.1-5) values proposed by the applicant, which come from TEEL Revision 18 and ERPGs, are generally higher than the AEGL values.

The staff comparison of the applicant's proposed values for selected chemicals to the limits contained in the National Institute for Occupational Safety and Health (NIOSH) immediately dangerous to life and health (IDLH) values (NIOSH, 1994, 2004; Casarett, 2001; Genium, 1999; Lewis, 1996) (of which 10 percent is sometimes used for members of the public to accommodate a wider range of population groups) found differences from the applicant values.

The chemical literature also includes other sources of limits. Those sources include military air guidelines for short-term chemical exposure to deployed military personnel, short-term public emergency guideline levels (a minimum-effects level), ceiling-limit values (also a minimum-effects level), and EPA toxic endpoints as part of its risk management program (USACHPPM, 1999; CFR, 2004). Many of the published values are different from those proposed by the applicant.

The staff has conducted significant reviews of chemical consequence limits because of their importance for meeting the performance requirements of 10 CFR 70.61 (NRC, 2004b). The SRPs for MOX and fuel cycle facilities (NRC, 2000; NRC, 2002b) mention AEGL and ERPG values as guidance. Other NRC guidance for control room habitability (discussed in Section 8.1.2.6 of this FSER) and NRC fuel cycle practice have used IDLH values. Licensing actions by the NRC on other facilities have used IDLH and TEEL values (e.g., NFS, 2002). In general, the ERPG/TEEL values tend to be higher when multiple guidance documents exist. There is a body of scientific information that is subject to data availability, interpretation, uncertainty, and discussion, and some values may have more conservatism and margin than others for chemical exposures. Chemical exposure limits do not correlate well with risk levels for different types of exposed populations (e.g., the general population as compared to a (healthy) worker population).

The staff notes that the differences are greater for the high-consequence limits (e.g., Level 3 values from Table 8.1-5, as compared to other values for similar purposes), and, for many chemicals listed in Table 8.1-5, there are significant differences between the Level 2 and Level 3 values.

The staff also considered the following:

- The NRC January meeting summary, dated March 5, 2003 (NRC, 2003b), states the DCS position as using TEEL-2 limits for site worker limits and that site worker impacts bound facility worker impacts.

- The May 30, 2003, letter from DCS (DCS, 2003c) discusses the offgas system and identifies a DCS commitment not to exceed a TEEL-2 limit for site workers.

- The June 5, 2003, letter from DCS (DCS, 2003d) discusses chemical spills in cells and states that, if necessary, existing MFFF features could be credited to ensure that chemical consequences are low. "Low" is understood to mean less than TEEL-2 for workers and less than TEEL-1 for the IOC.

- The applicant stated at a June 2003 meeting, as documented in the NRC June meeting summary, dated July 18, 2003 (NRC, 2003e), that, if low consequences are exceeded for any cell release, the applicant will upgrade existing devices to PSSCs.

In addition, page 8-12a of the revised CAR mentions comparing calculated concentrations of chemicals to TEEL-2 values for the site worker and to TEEL-1 values for the IOC. Thus, while the limits for determining chemical consequence categories are shown in Table 8.1-6, based upon the numerical values in Table 8.1-5, the applicant is committing to not exceeding Level 2 values for the site worker (the 100-meter receptor) and not exceeding Level 1 values for the IOC (the 160-meter receptor), as shown in Table 8.1-7. The staff finds that the Level 2 numerical values in Table 8.1-5 are below the corresponding IDLH values for each chemical, and the differences with other sets of values (e.g., AEGL-2s) are much smaller and would correspond to little or no health impact. Further, as stated above, the applicant committed to maintaining low chemical consequences to the site worker and IOC from events regulated by the NRC. Thus, the staff concludes that the use of the numerical values of Table 8.1-5 with the commitment to maintain low consequences to the site worker and IOC (Table 8.1-7) is a reasonable and consistent approach that provides for adequate protection of site worker and IOC health and safety and will result in a consistent designation of SSCs for site workers below that considered to be IDLH. Thus, the staff finds that the applicant's proposed approach is acceptable for addressing the applicable 10 CFR 70.61 performance requirements for chemical protection. This closes Open Item CS-05b, previously identified in the staff's revised draft safety evaluation report (DSER) (NRC, 2003d).

8.1.2.3.2 Chemical Accident Assumptions

The applicant considered a range of initial conditions, as well as the failure modes of storage

Table 8.1-5 Chemical Consequence Limits Proposed by the Applicant for the MFFF (mg/m³)

Name	Low-Consequence Level Level 1	Intermediate-Consequence Level Level 2	High-Consequence Level Level 3
Aluminum Nitrate (TEEL)	15	15	500
Azodicarbonamide (TEEL)	125	500	500
Chlorine (ERPG)	3	7.5	60
Chromic (III) Acid (TEEL)	1	2.5	25
Diluent (C_{10}–C_{13} Isoalkanes) (TEEL—see Note 1)	5	35	200
Decane (C_{10})	5	35	25000
Undecane (C_{11})	6	40	200
Dodecane (C_{12})	15	100	750
Tridecane (C_{13})	60	400	500
Ferrous Sulfate (Lab) (TEEL)	7.5	12.5	350
Fluorine (Lab) (ERPG)	0.75	7.5	30
Hydrazine Monohydrate (TEEL)	0.0075	0.06	50
Hydrazine Nitrate (TEEL)	3	5	5
Hydrofluoric Acid (ERPG)	1.5	15	40
Hydrochloric Acid (ERPG)	4	30	200
Hydrogen Peroxide (ERPG)	12.5	60	125
Hydroxylamine Nitrate (TEEL)	15	26	125
Iron (TEEL)	30	50	500
Isopropanol (TEEL)	1000	1000	5000
Manganese Nitrate (TEEL)	10	15	500
Manganous Sulfate (TEEL)	7.5	12.5	500
Nitric Acid (ERPG)	2.5	15	200
Nitric Oxide (TEEL)	30	30	125
Nitrogen Dioxide (TEEL)	7.5	7.5	35

Table 8.1-5 (cont.) Chemical Consequence Limits Proposed by the Applicant for the MFFF
(mg/m³)

Nitrogen Tetraoxide (TEEL)	15	15	75
Oxalic Acid (TEEL)	2	5	500
Potassium Permanganate (TEEL)	7.5	15	125
Silver Nitrate (TEEL)	0.03	0.05	10
Silver Oxide (TEEL)	30	50	75
Sodium (TEEL)	0.5	5	50
Sodium Carbonate (TEEL)	30	50	500
Sodium Hydroxide (ERPG)	0.5	5	50
Sodium Nitrite (TEEL)	0.125	1	60
Sulfuric Acid (ERPG)	2	10	30
Sulfamic Acid (TEEL)	40	250	500
Thenoyl Trifluoroacetone (TEEL)	3.5	25	125
Tributyl Phosphate (TEEL)	6	10	300
Uranium Dioxide (TEEL)	0.6	1	10
Uranyl Nitrate (TEEL)	1	1	10
Xylene (TEEL)	600	750	4000
Zinc Stearate (TEEL)	30	50	400
Zirconium Nitrate (TEEL)	35	35	50

(1) The TEEL value for the diluent represents the most conservative value in each category among the primary constituents n-decane, n-undedcane, n-dodecane, and n-tridecane.

containers and associated systems. The following release scenarios are stated in the revised CAR:

- leaks and ruptures involving equipment vessels and piping
- evaporating pools formed by spills and tank failures
- flashing and evaporating liquefied gases from pressurized storage

The applicant modeled releases using the total material at risk from the largest single tank or container. No credit was taken for process equipment installed to scrub and remove gases and vapors (i.e., installed for normal operations; assumed not to function during an accident involving a release). The applicant stated that more detailed accident sequences will be developed as necessary as part of the ISA during final design. The staff notes that the approach is consistent with similar facilities and the handling of similar chemical hazards (AIHA, 2004; Casarett, 2001;

DCS, 2002d; DOE, 2004; EPA, 1999, 2004; Genium, 1999; Lewis, 1996; NIOSH, 1994; NRC, 1998a, 2001b; Stumpf, 2000; USACHPPM, 1999; WSRC, 1994a, 1994b, 1995, 1999).

The distance of the possible receptors was also considered. A 100-m (328-ft) distance was used for calculations involving the site worker. The IOC receptor was assumed to be approximately 160 m (525 ft) from the point of release.

Table 8.1-6 Applicant's Use of Chemical Limits for Qualitatively Determining Chemical Consequence Categories

Consequence Category	Site or Facility Worker	IOC
High	Concentration $\geq$ Level 3	Concentration $\geq$ Level 2
Intermediate	Level 3 > Concentration $\geq$ Level 2	Level 2 > Concentration $\geq$ Level 1
Low	Level 2 > Concentration	Level 1 > Concentration

Table 8.1-7 Applicant's Commitments to Chemical Limits for Qualitative Chemical Consequences (DCS, 2003d, 2003e; NRC, 2003b, 2003e)

NRC Regulated Chemical Event Type	Site Worker*	IOC
All	Level 2 > Concentration	Level 1 > Concentration

* Site worker impacts bound facility worker impacts.

The applicant used the largest evaporation rate calculated from two models, as discussed in revised CAR Section 8.3.3, as input to the dispersion models for gases and vapors. The applicant used the dispersion coefficients derived from the ARCON96 model (NRC, 1999) (see FSER Sections 5.4 and 9) for estimating concentrations for the site worker (100-m (328-ft) receptor) and the IOC (160-m (525-ft) receptor). For chemicals that do not have appreciable vapor pressures and are released as fine particles (dust), the applicant used the five-factor formula for the release rate with dispersion coefficients calculated by the ARCON96 model (see FSER Section 9). The applicant no longer uses the EPA ALOHA code for chemical releases (EPA, 1999).

The staff notes that the ARCON96 code has been found acceptable (see FSER Section 9.1.1.4.3).

The analyses used the following input parameters and assumptions:

• release duration of 1 hour

• pool depth of 1 cm (0.4 in.)

- F atmospheric stability, with a 2.2 m/s (7.2 ft/s) windspeed, based on the 95th percentile χ/Q value from hourly measurements over a 5-year period (NRC, 1998a, 2001b)

- largest single-process vessel (common-mode failures that could affect adjacent or connected vessels were not considered)—477 L (126 gal) (each) for nitric acid, dodecane, TBP, and hydrazine hydrate; 681 L (180 gal) for HAN; and 908 L (240 gal) for nitrogen tetraoxide

[Text removed under 10 CFR 2.390]

- gas storage area gases will not reduce oxygen in personnel areas

- a temperature of 25 °C (77 °F)

- a ground-level release

- no mechanical or buoyancy plume rise

- a rural, flat-terrain topography

Section 5.5.2.10 of the revised CAR also indicates that direct chemical consequences with no direct radiological effects are not regulated by 10 CFR Part 70 because they do not impact or directly involve licensed material. In the case of a chemical release that has the potential to impact a facility worker and prevent the performance of a safety function, Section 5.5.2.10.6.1 of the revised CAR states that workers in the ECR perform a monitoring role during emergency conditions. Consequently, the ECRACS is a PSSC with a safety function of ensuring that habitable conditions are maintained in the ECR. The staff's evaluation of this issue is described in FSER Section 8.1.2.6.

Section 12.1 of the revised CAR states that the control of the facility would rely to a great extent on automated systems to ensure facility safety. It also mentions that the operations staff would be expected to monitor and confirm the status of confinement systems, fluid systems, and other facility systems and to recover from off-normal conditions. The applicant expected that few personnel actions would be relied upon for safety. At the present time, the applicant has identified the closing of dampers and the valving in of emergency scavenging air as two operator actions required outside of the ECR as part of a longer term response, and the applicant expected that the effects of a chemical release would have dissipated by the time these operator actions might become necessary (NRC, 2002d). The applicant stated that no safety-related operator or worker actions are necessary outside of the ECR that affect radiological safety, during or after a chemical release. Regarding PSSCs that are administrative controls, the applicant indicated in the revised CAR (Section 5.5.2.10.6.1) that these are either permissive in nature (e.g., no action until a sample is analyzed) or fail safe (i.e., crane stops).

8.1.2.3.3 Chemical Modeling Approach

The applicant has estimated hazardous chemical concentrations using techniques and models that are generally consistent with industry practice and generally follow the guidance on atmospheric and consequence modeling found in NUREG/CR-6410, "Nuclear Fuel Cycle Accident Analysis Handbook," issued 1998 (NRC, 1998a). The staff has reviewed the chemical modeling approach and finds that the basic models, approaches, and assumptions are consistent with NUREG/CR-6410 and are, therefore, acceptable.

As part of its review of any license application that the applicant may later submit, when more detailed design information is expected to be available, the staff will review the applicant's assumptions to ensure that they remain appropriate for the postulated application and physical phenomena and that input values lead to a conservative estimate of potential consequences. Items that the staff will evaluate include release quantities, spill temperatures, chemical concentrations, and other factors affecting potential hazardous chemical release rates.

8.1.2.4 Chemical Accident Consequences

This section presents the staff's assessment of accident consequences for chemical safety.

8.1.2.4.1 Analysis

The applicant evaluated potential chemical consequences for an IOC at the controlled area boundary, approximately 160 m (525 ft) from the point of release (i.e., the 160-m (525-ft) receptor).

For the site worker and facility worker, the applicant estimated the chemical consequences to be low. In the revised CAR, the applicant stated that the chemical consequence analyses for the individual at 100 m (328 ft) (site worker) bound the consequences for the facility worker. As stated in Section 11.2 of the FSER, the staff concurs with the applicant's qualitative distance analysis that, from the point of an in-cell release, the distance to a worker would be at least 100 m (328 ft). This closes a portion of the previously identified Open Item AP-13 from the revised DSER (NRC, 2003d).

The staff evaluation notes that the ECRACS is identified for chemical safety (protecting operators from chemical releases that could impact the safe handling of radioactive materials). As discussed in Section 8.1.2.6 of this FSER, the applicant provided design basis values for habitability in the ECR.

The applicant identified events involving the release of hazardous chemicals (primarily chlorine) produced from radioactive materials from the dechlorination and dissolution unit equipment and associated areas that can affect the facility worker (revised CAR Section 5.5.2.10.6.2). The applicant identified a mitigation strategy, and, as discussed in Section 8.1.2.3.1 above, the applicant must meet the chemical consequence limits set forth in Tables 8.1-5 and 8.1-7. The PSSCs are process entry controls for leaks occurring in cells, the C4 confinement system for leaks occurring in gloveboxes, and facility worker action for leaks occurring in C3 ventilated areas. The staff notes that administrative and ventilation control strategies similar to those proposed by the applicant are routinely used in similar hazardous situations in industry and should be capable of reducing consequences and/or frequencies as needed. Process entry controls would prevent worker access and hence exposure. The C4 confinement system would maintain negative pressure in gloveboxes and would direct any chemical release away from workers and, ultimately, would disperse it, thus preventing chemical exposure. Additionally, facility worker actions, such as evacuating affected areas or donning protective equipment, would mitigate chemical effects produced by site worker actions. Thus, the staff finds the applicant's approach for addressing chemical effects to be acceptable for the construction authorization stage. The applicant will be required to demonstrate the effectiveness of the safety strategies as part of any later application for a license to possess and use licensed material.

The applicant identified an event involving the release of hazardous chemicals and radioactive material (revised CAR Sections 5.5.2.10.6.3 and 8.5.1.10). This scenario involves the flow of NO_2/N_2O_4 at an abnormally high rate through the oxidation column. This has the potential for excessive chemical consequences for the worker, site worker (individual receptor at 100 m (328 ft)), and the IOC (160-m (525-ft) receptor). The applicant identified a safety strategy based upon mitigation. The PSSC is the process safety control subsystem (PSCS). The safety function is to ensure that the flow of NO_2/N_2O_4 is limited to under 44 kg/hr (97 lb/hr) (e.g., by active flow controls) to the oxidation column such that chemical consequences do not exceed the chemical consequence limits in Tables 8.1-5 and 8.1-7 (i.e., of low consequence) and, thus, are acceptable. The staff finds the applicant's fundamental approach of flow limitation for addressing these chemical effects will enable it to perform the required mitigative safety function with the necessary range of reliability; thus, it is acceptable for the construction authorization stage.

The staff had identified a concern with the airspeed and phenomenological models used to estimate evaporation rates from spills within the facility. Section 8.3 of the revised CAR states that the airspeed is calculated from the volumetric airflow rate through the room where the spill occurs divided by the minimal vertical cross-sectional area in that room; this has been identified by the applicant as an airspeed of approximately 0.1 m/s (0.33 ft/s) (NRC, 2003e). Similarly, Section 8.4.1 of the revised CAR states that the airspeed is calculated from the volumetric airflow rate through the room where the nitric acid spill occurs divided by the minimal vertical cross-sectional area in that room; this has been identified by the applicant as an airspeed of approximately 0.01 m/s (0.033 ft/s) (NRC, 2003e) for nitric acid spills or leaks in the AP area, with spill temperatures modeled up to the boiling point of nitric acid. The applicant used these air speeds as input into the evaporation models, which provide a release rate or source term. The staff notes that this use of such very low air velocities results in a significant reduction of the estimated consequences by factors of between 20 and 200 because the evaporation rate model used by the applicant is approximately linearly dependent with air velocity. Using these airspeeds for evaporation and source term estimation, combined with the 2.2-m/s (7.2-ft/s) windspeed for dispersion, the applicant concluded that the consequences would be low for the workers and IOC. The staff noted the following:

- The approach of using different airspeeds for evaporation and dispersion is an uncommon practice (e.g., NRC, 1998a).

- Use of the 2.2-m/s (7.2-ft/s) windspeed would likely result in exceeding the low-consequence level for nitric acid (Tables 8.1-5 and 8.1-7), currently identified by the applicant as 15 mg/m^3 for the workers and 2.5 mg/m^3 for the IOC. The applicant stated that the principal parameters which result in low consequences for all indoor releases are the spill/room size and the low indoor windspeed. The applicant did not consider these input airspeed parameters to be design basis values (DCS, 2003a). Thus, an appearance of a safety effect is credited without identification of PSSCs and design bases.

- Experience has shown that airflow velocities are usually uneven in cells, with higher velocities as the air exhaust inlets are approached. Most facility designs include exhaust inlets near the floor, where spills could likely occur. These real effects would likely result in much higher velocities near spills than the 0.01–0.1 m/s (0.033–0.33 ft/s) calculated on an average basis. For example, typical cell designs could produce air velocities of 0.5–1 m/s (1.6–3.3 ft/s) in the vicinity of the ventilation system inlet duct(s) based upon an average cell airspeed of 0.01 m/s (0.033 ft/s). Typical nuclear and chemical ventilation systems are designed for around 0.5–1 m/s (1.6–3.3 ft/s) at the inlet. The applicant's analysis does not account for these facts.

- There is no margin in the assumed air velocities of 0.01–0.1 m/s (0.033–0.33 ft/s).

- The selected models may no longer be suitable (e.g., laminar versus turbulent flow) or may not be supported by experimental data for the low velocities assumed by the applicant.

- Other phenomena may need to be considered and modeled by the applicant. These could include flashing and bulk convection effects. These would likely exceed the straight wind velocity effects upon evaporation. For example, bulk correction factors would increase release estimates by a factor of at least 3 at temperatures of approximately 110 °C (230 °F) and trend to very high values as the boiling point (approximately 118–120 °C (244–248 °F)) is approached (i.e., the flashing condition).

The applicant provided supplemental information (NRC, 2003e) indicating that there would be drip pans in the AP process cells (not currently credited for chemical consequence safety) that would reduce evaporation rates and keep the chemical consequence levels low, in accordance with the levels in Tables 8.1-5 and 8.1-7. The applicant also committed (DCS, 2003d, NRC, 2003e) to performing a detailed analysis of each applicable process cell as part of its ISA. If necessary, these drip pans could be credited at the license application stage to ensure that chemical consequences do not exceed the low level of concern (Level 2 for the site worker and Level 1 for the IOC, in accordance with Tables 8.1-5 and 8.1-7). The staff finds that the detailed design and as-built information to be developed as part of the ISA process are not required for CAR approval. The approach provides assurance of adequate safety and is found to be acceptable to the staff. This part of the previously identified Open Item CS-05b from the revised DSER (NRC, 2003d) is now considered closed.

8.1.2.4.2 Uncertainty

Estimates of risks are often accompanied by uncertainty because of the complexity of the postulated scenarios, the physical models used to describe them, and the design itself (particularly if it is preliminary). As stated in Section 8.4.3 of the revised CAR, the applicant believed that conservative modeling has been utilized for the chemical releases, and this bounds any uncertainty. The applicant will perform more detailed consequence analyses as part of any later application for a license to possess and use licensed material (DCS, 2003d; NRC, 2003e). As discussed in Section 8.1.2.4.1 above, the applicant has made a commitment to conduct a cell-by-cell analysis as part of the ISA and credit features, as necessary, to ensure that chemical consequences do not exceed the low level of concern. This should adequately address uncertainty.

8.1.2.5 Process Safety Controls

The applicant provided process safety information in Section 8.5 of the revised CAR. Additional information has also been obtained from the applicant in correspondence on the docket, in-office reviews of documentation, and public meetings (e.g., DCS, 2003a; NRC, 2002d, 2003a, 2003b, 2003c, 2003e, 2003f). This is reviewed and evaluated in the sections that follow.

Each of the MFFF facilities would have control requirements that would be incorporated into the overall design of the control system for process safety control. More information on the control system is available in FSER Section 11.6.

Reagents are stored and chemical mixtures are prepared in the reagent processing building and the reagent storage area of the AP area. The AP facility is broken down into process functional units, which are made up of one or more subassemblies performing consistent and elementary tasks. The applicant stated that the breakdown into control functional units allows each entity to be operated relatively independently in the given operating mode. The applicant intended to control process storage and operation conditions in order to prevent exothermic and potential autocatalytic reactions in the reagent processing building and the AP area. Autocatalytic and exothermic reactions of chemicals would be prevented through the control of the process parameters (e.g., reactant concentration, temperature, catalyst concentration in solution, and pressure) that affect the reactions.

The applicant stated that there is reasonable assurance that the PSSCs will be sufficiently reliable and available, and this assurance will be provided through the use of standard nuclear industry engineering practices. These practices are incorporated into the facility general design philosophy, design bases, system design, and commitments to applicable management measures. These practices ensure that applicable industry codes and standards are utilized, adequate safety margins are provided, engineering features are utilized to the extent practicable, the defense-in-depth philosophy is incorporated into the design, and PSSCs will be appropriately maintained. Significant chemical-related risks and associated process safety information are discussed in the following sections.

8.1.2.5.1 Safety Strategies for Events Involving Hydrogen

The staff notes that flammable liquids and gases would be delivered to and used in the proposed MOX facility. At the proposed facility, hydrogen is used as a reagent, generated by radiolysis, and generated electrolytically. The applicant's general strategies for the three respective cases are to limit the hydrogen concentration below its lower flammability limit (LFL) by dilution with ventilation air, dilute with scavenging air, and inhibit generation. The applicant identified the design basis for hydrogen as National Fire Protection Association (NFPA) 69, "Standard on Explosion Prevention Systems," issued 1997 (NFPA, 1997), with the application of

25 percent of the LFL (DCS, 2003c). The staff notes that a design-basis value of 25 percent of the LFL is endorsed by standards (e.g., NFPA 69 (NFPA, 1997), 801, "Standard for Fire Protection for Facilities Handling Radioactive Materials," issued 1998 (NFPA1998), 50A, "Gaseous Hydrogen Systems at Consumer Sites," issued 2000 (NFPA, 2000a), and 86C, "Industrial Furnaces Using a Special Processing Atmosphere," issued 2000 (2000b)), prior NRC activities (NRC, 1999), and DOE practices. The NRC inspects existing fuel cycle facilities using this criterion. The staff evaluation of this design basis is discussed in Section 11.2 of this FSER.

8.1.2.5.1.1 Argon-Hydrogen Mixture in Sintering Furnace and Hydrogen Storage. The sintering furnaces and the hydrogen storage unit/mixing station are two places associated with the use of an argon-hydrogen mixture (see FSER Section 11.3). The nominal mixture is nonexplosive with a minimum argon level of 95 percent and a maximum hydrogen concentration of 5 percent. Mixtures of argon and hydrogen containing more than about 5.8-percent hydrogen are flammable in air (DCS, 2001, 2002f). The applicant intended to minimize explosion risk by the following preventive measures:

•   Storage tanks containing argon and hydrogen and the mixing station are located approximately 60 m (197 ft) outside the MFFF building.

•   The composition of the argon-hydrogen mixture is monitored and controlled.

•   Welded supply tanks and feed lines are designed in accordance with the guidelines of the Compressed Gas Association (CGA).

•   Ignition sources (e.g., electricity, fire, and lightning) are either eliminated or precluded by design for the area containing the hydrogen storage system.

The applicant described its comprehensive strategy regarding potential explosions associated with the sintering furnace (DCS, 2002a, 2002c; NFPA, 2000a, 2000b; NRC, 2002a, 2003a, 2003b). This is discussed in FSER Section 11.3.1.2.4.

8.1.2.5.1.2 Radiolytic Hydrogen Production. Hydrogen production by radiolytic dissociation of hydrogenous molecules occurs because of the presence of radioactive material in solution. The applicant intended to preclude radiolytic-hydrogen explosion hazards by preventing the accumulation of hydrogen exceeding the LFL. This will be accomplished by adding dilution (scavenging) air and ventilation.

Hydrogen generation will occur in the AP process. For the air-hydrogen system under ambient conditions, the LFL is 4-percent hydrogen. The applicant stated in the revised CAR that hydrogen concentration in the free volume of AP equipment would be maintained below 1 vol% (i.e., less than 25 percent of the LFL) during normal operation by maintaining an adequate dilution airflow and ensuring that an exhaust path exists. Air for dilution would be provided by the scavenging air system. Emergency scavenging air would be supplied to those systems and vessels in which flammable gas concentrations could exceed the LFL within 7 days after a loss of ventilation. In such circumstances, emergency scavenging air would be supplied to maintain the hydrogen concentration below 25 percent of the LFL at all times. The applicant stated that the design basis for hydrogen is NFPA 69 (NFPA, 1997), with the application of 25 percent of the LFL (DCS, 2002e). The staff evaluation of this strategy is presented in Section 11.2.1.3.3.

Hydrogen production and accumulation may occur in the waste and byproducts, such as contaminated organic waste or organic-additive-bearing waste containing significant amounts of plutonium, scraps in transuranic waste containers, and other liquid waste. The applicant intended to use container filters to limit hydrogen accumulation while maintaining the confinement of radioactive materials. The waste containers have been identified as PSSCs with the safety function to ensure that hydrogen accumulation in excess of limits does not occur. The design basis is identified in Section 11.4.11.6 of the revised CAR and is a U.S. Department of Transportation standard for waste containers. This approach is consistent with regulated and successful nuclear industry experience and is acceptable to the staff at the construction authorization stage.

8.1.2.5.1.3 <u>Hydrogen Generated by Electrolysis</u>. In an efficient process and with relatively low voltages, hydrogen production by electrolysis is usually 5–10 percent of the passed current. At higher voltages, the cathode may blow hydrogen, and the majority of the current is involved with hydrogen generation. Using the applicant's stated electrical conditions, the rate could approach 10 cc/s or more, which can be an order of magnitude or more larger than that generated by radiolysis. The applicant identified an approach that inhibits hydrogen generation and, thus, prevents its accumulation. The applicant stated that nitric acid normality and voltage limits may be used to prevent hydrogen generation and keep the hydrogen concentration below the design basis of 25 percent of the LFL (DCS, 2002e, 2003f; NRC, 2003a). The applicant also indicated that acid concentration controls may be used. As discussed further in FSER Section 11.2.1.3.3, the applicant provided information that indicates that the proposed strategy has the ability to control hydrogen generated from electrolysis.

8.1.2.5.2 Solvents

The purification cycle and the solvent recovery cycle are two units in which solvent and diluent are used in processing. The purification cycle uses the solvent-diluent mixture for the extraction of plutonium. The solvent recovery cycle regenerates the solvent-diluent mixture by removing the degradation products and adjusting the TBP content and stores the treated solvent at a slightly acidic condition to prevent degradation by hydrolysis. The aqueous stream is washed with the diluent to remove traces of entrained solvent. The solvent-diluent utilized in these processes is combustible. According to NFPA 30 definitions, the diluent and solvent (TBP) are Class IIIa and IIIb combustible liquids, respectively. Combustible characteristics of these chemicals are presented in FSER Table 8.1-8.

The PSSCs and design bases associated with fire and explosion hazards within the process cells are addressed in Chapters 5 and 7 of this FSER, respectively. This discussion focuses on in-process areas, such as tank ullage spaces, piping, and ventilation systems (e.g., the offgas system).

With respect to potential explosions, the applicant identified a preventive strategy. A preventive strategy renders the potential event highly unlikely. This meets the intent of the performance requirements of 10 CFR 70.61 and is acceptable to the staff. Acceptable approaches include maintaining a temperature margin below the flashpoint, dilution of the flammable vapors by active ventilation, rendering the vapor nonflammable (i.e., above the upper flammability limit or inerting), or combinations thereof.

Table 8.1-8 Combustible Characteristics of Chemicals in the AP Area

Combustible	Location	Flashpoint, °C (°F)
Solvent—Diluent only	Pulsed column, diluent washing, and storage	> 57 (135)
Solvent—TBP only	Storage tanks and solvent recovery	> 120 (248)
Solvent—Diluent and TBP mixture	Purification cycle and solvent recovery	> 55 (131)
Solvent—Diluent, TBP, and aqueous	Purification cycle and solvent recovery	> 66 (151)

For each solvent or solvent mixture, there exists a relationship between the flashpoint of the liquid and the LFL of vapors above the liquid-vapor interface. To a close approximation, the flashpoint temperature of the liquid corresponds to 100 percent of the LFL in the vapor-air mixture near the surface of the liquid. Therefore, two acceptable control strategies for preventing solvent fires include (1) control of the solvent temperature below the liquid flashpoint and (2) prevention of flammable vapor-air mixtures above the surface of the liquid by maintaining vapor concentrations below some fraction of its LFL.

The applicant proposed to use the guidance provided in NFPA 69 (NFPA, 1997) as the design basis for the control of flammable mixture concentrations within the AP process vessels (DCS, 2003f). As noted by the applicant, NFPA 69 provides various options to accomplish this, including one of the following:

- The combustible concentration shall be maintained at or below 60 percent of the LFL when automatic instrumentation with safety interlocks is provided.

- The combustible concentration shall be maintained at or below 25 percent of the LFL.

The applicant identified six groups of flammable gas concentration limits and safety strategies, corresponding to six types of process units throughout the AP process. The limits, AP process areas of applicability, and corresponding safety strategy are described below.

Group No. 1—Solvent Extraction, Solvent Recovery, and Wastes

Inside process vessels in the solvent extraction, solvent recovery, and waste units, the applicant proposed a design-basis value of 60 percent of the LFL. This limit is in accordance with NFPA 69, as long as the system is equipped with automatic instrumentation with safety interlocks.

The PSSC, which performs the safety interlock function, is the PSCS. The PSCS will monitor the process temperature and terminate heat sources before exceeding a temperature limit. For example, under certain conditions, a temperature of 50 °C (122 °F) for HTP will ensure that flammable solvent vapors remain below 60 percent of the LFL. A second administrative PSSC, chemical safety controls, is also proposed by the applicant to control chemical concentrations and flow rates of reagents. The control of reagents is necessary to maintain process temperatures below those which would cause a temperature limit to be exceeded from exothermic chemical reactions.

Group No. 2—Oxalic Precipitation and Oxalic Mother Liquor Units

Inside process vessels in the oxalic precipitation and oxalic mother liquor units, the applicant proposed a design-basis value of 25 percent of the LFL. This limit is in accordance with NFPA 69 (NFPA, 1997).

The safety function of the PSCS is to prevent diluent from entering equipment at elevated temperatures. This will be implemented by online redundant measurements (e.g., density or conductivity).

Group No. 3—Equipment at Elevated Temperatures in the Acid Recovery Unit

Inside equipment at elevated temperatures in the acid recovery unit, the applicant proposed a design-basis value of 25 percent of the LFL. This limit is in accordance with NFPA 69.

The safety function of the PSCS is to prevent diluent from entering equipment at elevated temperatures. This will be implemented by online redundant measurements (e.g., density or conductivity).

Group No. 4—Equipment Not at Elevated Temperatures in the Acid Recovery Unit

Inside equipment not at elevated temperatures in the acid recovery unit, the applicant proposed a design-basis value of 60 percent of the LFL. This limit is in accordance with NFPA 69, as long as the system is equipped with automatic instrumentation with safety interlocks.

The PSSC, which performs the safety interlock function, is the PSCS. The PSCS will monitor the process temperature and terminate heat sources before exceeding a temperature limit. For example, under certain conditions, a temperature of 50 °C (122 °F) for HTP will ensure that flammable solvent vapors remain below 60 percent of the LFL. A second administrative PSSC, chemical safety controls, is also proposed by the applicant to control chemical concentrations and flowrates of reagents. The control of reagents is necessary to maintain process temperatures below those which would cause a temperature limit to be exceeded from exothermic chemical reactions.

Group No. 5—Hydrogen Concentration Associated with Radiolysis

Inside equipment that has the potential for hydrogen concentration associated with radiolysis, the applicant proposed a design-basis value of 25 percent of the LFL.

The safety function of the waste unit offgas treatment system is to provide an exhaust path for the removal of diluted hydrogen gas in process vessels. The instrument air system will also provide sufficient scavenging air to dilute the hydrogen that is generated during radiolysis. This is discussed and evaluated in Section 11.2.1.3.3 of the FSER.

Group No. 6—Hydrogen Generated by Electrolysis within the Dissolution Process

Inside equipment that has the potential for hydrogen gas concentration associated with electrolysis within the dissolution unit, the applicant proposed a design-basis value of 25 percent of the LFL.

The safety function of the PSCS is to limit the generation of hydrogen by ensuring that the normality of nitric acid is sufficiently high to ensure that the offgas is not flammable. This is discussed and evaluated in Section 11.2.1.3.3 of the FSER.

Staff Evaluation

The staff have evaluated the applicant's proposal to adopt the guidance of NFPA 69 as the design bases for AP process vessels containing flammable gases.

The guidance in NFPA 69 includes the following several key parts:

3-2 Basic Design Considerations: the following factors shall be considered in the design of a system to reduce the combustible concentration below the lower flammabilty limit:

(a) Required reduction in combustible concentration
(b) Variations in the process, process temperature and pressure, and materials being processed
(c) Operating controls
(d) Maintenance, inspection, and testing

3-3.1 The combustible concentration shall be maintained at or below 25% of the lower flammability limit. Exception: When automatic instrumentation with interlocks is provided, the combustible concentration shall be permitted to be maintained at or below 60% of the lower flammability limit.

3-4.1 Suitable instrumentation shall be provided to monitor the control of the concentration of combustible components.

The standard requirement of NFPA 69 is 25 percent of the LFL, and the use of automatic interlocks is an exception. The context of Section 3 is direct control of the combustible concentration. The code also requires instrumentation to monitor the control of the concentration of combustible components.

The staff review noted that the 25 percent limit is endorsed in the MOX SRP (NRC, 2000), other NRC guidance, DOE guidance, and NFPA codes for hydrogen and flammable vapors. The staff has drawn a conclusion regarding the acceptability of a flammable hydrogen concentration limit equivalent to 25 percent of the LFL in Section 11.2.1.3.3 of this FSER.

The staff has reviewed the applicant's proposed design basis of NFPA 69. This is a top-level, consensus standard often used in the chemical and nuclear industries. It has the ability to address the potential hazard of flammable solvent vapors. It is used as the basis for several acceptance criteria in Chapter 7 of the SRP. It provides for the determination of ranges of acceptable values. Consequently, the staff finds the general use of NFPA 69 by the applicant for solvents to be acceptable.

The staff concludes that the applicant has several different control strategies to pursue in the ISA. Some of these strategies have similarities to approaches used to control solvent flammabilities in other facilities and, in combination, may be capable of achieving very high reliabilities to render the flammability event highly unlikely. The applicant will need to demonstrate this high reliability as part of its later application for a license to possess and use licensed material.

If the applicant intends to pursue the use of interlocks in accordance with NFPA 69, then, in the ISA supporting the possession and use license application, the applicant must demonstrate that the proposed interlocks can perform their required safety function(s). Such a demonstration will likely require a clear calculational and experiential basis and setpoint analysis to meet the setpoint design basis. This may result in setpoint values below those allowed by NFPA 69. However, the staff concludes, on the basis of the foregoing evaluation, that the applicant has proposed a safety strategy and design basis for preventing solvent explosions in AP equipment which meet the requirements for construction authorization of the proposed MFFF. This closes Open Items CS-9 and AP-9 identified in the staff's revised DSER (NRC, 2003d).

8.1.2.5.3 Hydroxylamine Nitrate and Hydrazine in Nitrate Media

The AP process uses a mixture of HAN and nitric acid to strip (i.e., recover in the aqueous phase) plutonium from the solvent after removal of americium and gallium at the extraction step. The HAN reduces the plutonium (IV) to the plutonium (III) state. It possesses the proper plutonium (IV) to plutonium (III) reduction potentials and produces volatile reaction products—nitrogen, nitrous oxide, and water. These products do not contribute to the wastes generated by the process.

Hydrazine is used in conjunction with HAN to impede the HAN reaction on nitrous acid (always present in nitric acid solutions) and, thus, increase HAN availability for plutonium reduction. However, interaction of hydrazine nitrate and nitrous acid could initiate, under certain conditions, the formation of hydrazoic acid or azides (hydrazoic salts), which are hazardous compounds.

Besides limiting the plutonium losses in the process, hydrazine can enhance safety by keeping the nitrous acid concentration at a very low level. Nitrous acid is the main intermediate chemical that can lead to the formation of unstable compounds and conditions in the solvent extraction process.

The staff notes that energetic hydroxylamine-nitric acid reactions can occur under the right conditions, as evidenced by a DOE investigation of an accident at Hanford (DOE, 1997a, 1997b, 1998a). As a result of this explosion, DOE investigated the situation and concluded that the HAN phenomena involved the interdependence between at least the following five parameters:

(1)   chemical concentration of each reactant

(2)   molar ratio of nitric acid to HAN

(3)   temperature of the mixture

(4)   concentration of metal ions (as catalysts)

(5)   pressure of the system (appears to influence the severity of the reactions but not the initial autocatalytic initiation)

The DOE derived an instability index and a graph to link these parameters and generally account for the behavior of the system. The DOE also identified other precautions, such as frequent monitoring of HAN solutions, and mentioned that only dilute nitric acid should be added to HAN and that the addition should be performed slowly and in a well-ventilated tank. Dilute nitric acid was not specified further.

The material safety data sheet (MSDS) for hydroxylamine (which may be mixed with nitric acid to make HAN) also lists safety precautions that include temperature limits, venting/volume increase concerns, crystallization concerns, and hazardous reactions with oxidants, such as nitrates. The MSDS documentation identifies large volume increases (up to 2000 times the initial volume of the solution) during modest heating and violent decomposition at temperatures above 75 °C (167 °F), with the liberation of ammonia, nitrogen, and nitric oxide. Metallic impurities and surfaces, such as iron, can accelerate the reactions.

The use of HAN and hydrazine in nitric acid is a complex, multiparameter chemical system. This system involves the interrelationship of the chemical concentrations of each material, molar ratios (particularly for HAN and nitric acid), temperature, and concentration of metals (i.e., potential catalytic effect). These systems generally present three main hazards, including HAN autocatalytic decomposition, hydrazoic acid, and azides, which correspond to overpressurization and explosion events.

8.1.2.5.3.1 Hydroxylamine Nitrate Decomposition. The applicant provided its safety strategy in the revised CAR (Section 5.5.2.4.6.4). The staff evaluation and conclusions (NRC, 2004c) are summarized in this section of the FSER. In the revised CAR (DCS, 2002e), the applicant revised and explained its proposed safety strategy to prevent HAN explosions for the two categories—without and with nitrous oxide addition.

Category I—Process Vessels Containing HAN and Hydrazine Nitrate without Nitrous Oxide Addition

The applicant proposed a preventive safety strategy to reduce the risk to the worker, site worker, IOC, and environment, as summarized in Table 8.1-9. The PSSCs are the following:

•   The PSCS has the safety function of maintaining temperatures of these solutions to within safety limits.

•   Chemical safety controls have the safety function of ensuring that the concentrations of nitric acid, metal impurities, and HAN introduced into the process are within safety limits.

The design basis safe limits were developed by the applicant using a kinetic model comprised of five nonlinear ordinary differential equations and a number of rate constants and boundary conditions (NRC, 2003a; DCS, 2003c, 2003e, 2003g). The kinetic model includes generation and destruction terms for plutonium (IV), nitrous acid, HAN, hydrazine hydrate, and hydrazoic acid. The applicant used a commercial software package to solve the five-equation set. For a given set of initial conditions (i.e., starting temperature and initial concentrations of plutonium (IV), ionic strength, nitric acid, nitrous acid, HAN, hydrazine hydrate, and hydrazoic acid), the numerical solution to the problem shows the time-dependent changes in the concentrations of plutonium (IV), nitrous acid, HAN, hydrazine hydrate, and hydrazoic acid. The boundaries of safe operations are those values of limiting initial conditions for which the model calculates well-behaved reduction of plutonium (IV) and nearly stoichiometric consumption of HAN. In this mathematical model, the autocatalytic runaway condition is exhibited as a rapid consumption of HAN (usually in fractions of a second) and the generation of large amounts of nitrous acid.

In addition to the model, DCS has committed to conducting confirmatory testing to further substantiate the results of the model (DCS, 2003e). This testing will examine the reaction mechanism and kinetics of nitric and nitrous acid with nitrogen tetraoxide. DCS also plans to conduct testing to examine the reaction kinetics and mechanism for reactions between nitrous acid, hydrazine, and HAN. The investigations will be performed under a range of temperatures (0–70 °C (32–158 °F)), HAN concentrations (1 mM–2.5 M), and hydrazine concentration (0.1 mM–0.2 M). In addition, the plutonium redox behavior in nitric acid will be examined. Lastly, integral experiments that confirm the kinetic model predictions will be performed.

The staff reviewed the system of equations and rate constants. These have been derived from several different literature sources published over two decades. Rate constants were presented for conditions different from those anticipated at the proposed facility, and, consequently, the applicant used correlations to extrapolate the data. The staff used a different commercial software package to solve the series of differential equations and to evaluate the margin in the applicant's approach. The results are shown in Table 8.1-10. The staff evaluated the system of equations and results and found that they predicted a region of stability in process temperature, nitric acid concentration, and HAN concentration that is consistent with the experimental data used to develop the DOE instability index method (DOE, 1998a).

As a base case, the initial conditions were set equal to either the design basis values, where they had been proposed by the applicant, or the normal values expected for the purification process. Then, the value of each parameter was adjusted individually to find the region of instability with respect to that parameter. The staff assumed a nitrous acid concentration of 0.01 M. The system is insensitive to plutonium (IV) initial concentration, but a nominal value of 0.17 M (corresponding to 40 g Pu/L) is used. Ionic strength is adjusted automatically in the model as a function of plutonium ion and nitric acid concentrations.

Category II—Process Vessels Containing HAN and Hydrazine Nitrate with Nitrous Oxide Addition

In the revised CAR (DCS, 2002e) and by letters dated May 30, 2003, July 28, 2003, and October 6, 2003 (DCS, 2003c, 2003e, 2003g), the applicant revised and explained its proposed safety strategy to prevent HAN explosions in units with nitrous oxide addition. Nitrous oxide is added to some AP units to destroy excess HAN, hydrazine, and hydrazoic acid. In the revised strategy (summarized in Table 8.1-11), the overproduction of noncondensible gases resulting from HAN, hydrazine, and hydrazoic acid destruction by nitrous oxide would be prevented.

The purpose of the two PSSCs in Table 8.1-11 is to ensure that the pressure rating of process vessels is not exceeded. The applicable codes and standards for process vessels with fluids containing significant quantities of americium or plutonium are those for FTS-1, as described by the applicant in Section 11.8 of the revised CAR. The design basis values for pressure, temperature, flow, and volumetric capacities of process columns are provided in Table 11.8.2 of the revised CAR. Section 11.8 of this FSER contains the staff review and evaluation of these FTS codes and standards.

Table 8.1-9 PSSCs for Process Units with HAN and without Nitrous Oxide Addition

PSSC	Safety Function	Controlled Parameter	Design Basis
PSCS	Maintain temperature below safe limits	Temperature	< 50 °C (122 °F)
Chemical Safety Controls	Maintain maximum nitric acid concentration	[HNO_3]	< 6 M
	Maintain minimum hydrazine concentration	[N_2H_4]	≥ 0.1 M
	Maintain maximum HAN concentration	[HAN]	< 2.5 M
	Limiting residence time of nitric acid, HAN, hydrazine with nitric acid, and plutonium-bearing solution	time	as low as reasonable (probably several months)

Table 8.1-10 Design Basis Values for the Prevention of HAN Explosions

Controlled Parameter	Design Basis Value	Stable Value	Margin (%)
Temperature	< 50 °C (122 °F)	< 63 °C (145 °F)	13 °C (55 °F), (25%)
HNO_3	< 6.0 M	< 7.1 M	1.1 M, (18%)
HAN	< 2.5 M	*	*
N_2H_4	≥ 0.1 M	> 0.018 M	0.082 M, (550%)
HN_3	0.0 M	N/A	N/A

* The kinetic model with hydrazine = 0.1 M exhibits stability at all HAN concentrations below the design basis value.

Table 8.1-11 PSSCs for Process Units with HAN and Nitrous Oxide Addition

PSSC	Safety Function	Controlled Parameter	Design basis
Offgas Treatment System	Exhaust path for removal of offgases, which provides a means for heat transfer/pressure relief for affected process vessels	Heat transfer, pressure relief	Revised CAR Table 11.8-2
Chemical Safety Controls	Limit nitric acid concentration	$[HNO_3]$	Revised CAR Table 11.8-2
	Limit hydrazine concentration	$[N_2H_4]$	Revised CAR Table 11.8-2
	Limit HAN concentration	$[HAN]$	Revised CAR Table 11.8-2
	Limit hydrazoic acid concentration	$[HN_3]$	Revised CAR Table 11.8-2

The applicant also indicated that the tank in the reagent building would be inerted but has not determined whether the tank in the MOX building would be inerted. Inerting was intended for process control and not for safety purposes. The applicant did not anticipate either of these becoming PSSCs but planned HAZOPs at a later time to further investigate the hydrazine and HAN areas.

Staff Evaluation and Conclusions on HAN/Hydrazine

With regard to Category I HAN/hydrazine controls, the staff notes that its independent testing of the model equations indicates that significant margin is present in the safety strategy. In addition, the presence of multiple controls and measurements indicates the potential for high reliabilities. The applicant's commitments to confirmatory experimental testing to further substantiate the kinetic model and understand the HAN/hydrazine system provide additional confidence that the proposed strategy has the ability to address the safety concerns. The staff has also found literature analogues for HAN/hydrazine control that are consistent with the proposed approach (Biddle, 1968; WSRC, 2001b). Therefore, the staff concludes that the overall strategy has the ability to meet the performance requirements and is acceptable for construction authorization.

With regard to Category II HAN/hydrazine controls, the staff finds that the proposed codes and standards are consistent with industry approaches and provide for a value or range of values that protect the integrity of the vessels with a nitrous oxide addition from overpressurization. Therefore, the staff concludes that the overall strategy has the ability to meet the performance requirements and is acceptable for construction authorization.

Therefore, the applicant has provided sufficient information to close Open Item CS-02 from the revised DSER (NRC, 2003d).

Finally, the applicant intended to use hydrazine as a scavenger for nitrous acid and as a means to improve the efficacy of the plutonium stripping. Hydrazine can also react strongly with oxidants, such as nitric acid and peroxide. In the revised CAR, the applicant proposed the use of administrative controls to control hydrazine concentrations within safety limits. The applicant provided additional information indicating that the hydrazine would be received as hydrazine hydrate, at a certified concentration of 35 percent (22 percent as hydrazine). The applicant stated that solutions below 40-percent hydrazine (60 percent as the hydrate) possess no flash or fire point and are thus nonflammable. The applicant stated that the concentration would be confirmed by independent testing not to exceed 35-percent hydrazine (DCS, 2001). The applicant summarized the approach as follows:

- a certified or lot analysis by the manufacturer under a facility-approved quality assurance plan

- an analysis upon receipt at the facility

- an analysis after mixing/diluting for facility use

The staff evaluation concludes that the applicant's proposed approach using multiple independent administrative controls at different times to control hydrazine concentrations provides the potential for high reliability and, thus, reasonable assurance that the additions do not exceed 35-percent hydrazine. The staff finds this acceptable for the construction authorization stage.

8.1.2.5.3.2 Hydrazine and Hydrazoic Acid. Hydrazoic acid is also called hydrogen azide. It is formed when hydrazine reacts with the relatively small quantities of nitrous acid present in nitric acid solutions:

$$N_2H_5NO_3 + HNO_2 => HN_3 + 2 H_2O + HNO_3$$

Hydrazoic acid can undergo further oxidation to yield gases and water:

$$HN_3 + HNO_2 => N_2O + N_2 + H_2O$$

The first reaction generally proceeds faster than the second reaction under most of the conditions in the AP process. Thus, there is a net accumulation of hydrazoic acid, and, over time, an equilibrium concentration can be established. Hydrazoic acid exists in the aqueous phase and organic phases that have contacted the aqueous solutions. Thus, the process at the proposed facility is designed to destroy aqueous hydrazoic acid in the oxidation column (with nitrous/nitrous oxide fumes) and to remove the organic hydrazoic acid in the solvent recovery unit (as sodium azide, in the aqueous alkaline waste stream). The sodium azide is subsequently destroyed by sodium nitrite in the liquid waste reception unit.

Hydrazoic acid is a colorless liquid and extremely soluble in water. The multiple nitrogen bonds produce a molecule containing significant energy that can be released by decomposition reactions as follows:

$$2HN_3 => 3 N_2 + H_2 + 526KJ$$

Energetic decomposition of hydrazoic acid can occur if sufficient concentrations form in either the liquid or gaseous phases. The applicant has identified the following two type of hazards associated with hydrazoic acid:

(1)   an explosion related to a mixture of hydrazoic acid in air

(2)   an explosion related to the distillation and condensation of hydrazoic solutions

The applicant intended to relate the first type to the liquid phase concentration via the Henry's Law correlation.

The applicant adopted a preventive safety strategy to protect the worker, site worker, IOC, and environment. The PSSCs include the following:

- For the chemical safety controls, the first safety function is to assure that the proper concentration of hydrazine nitrate is introduced into the system, thus limiting the quantity of hydrazoic acid produced. The applicant has identified the design bases as a maximum hydrazine concentration of 0.14 mol/L and a hydrazine to hydrazoic acid yield of 39.3 percent or less.

  The second safety function is to ensure that hydrazoic acid does not accumulate or propagate into the acid recovery and oxalic mother liquor recovery units. This is accomplished by neutralization in the solvent recovery unit. The applicant intended to demonstrate the effectiveness of hydrazoic acid neutralization as part of the ISA at the time of the possession and use license application (NRC, 2003a). Sampling controls have been identified as a PSSC to confirm the destruction of the hydrazoic acid (revised CAR Section 8.5.1.8, p. 8-33).

- For the PSCS, the safety function is to limit the temperature of the solution, thereby limiting the vapor pressure and evaporation of hydrazoic acid. The applicant identified a design basis temperature not exceeding 60 °C.

The staff reviewed information on hydrazoic acid and concluded that it is a potential explosion hazard. The staff concurs with a prevention strategy for this hazard. The revised CAR does not provide adequate justification for the design bases. Supplemental information from the literature and the applicant (WSRC, 2001b; DCS, 2003b) indicates the following:

- The Henry's Law coefficient is sometimes used to estimate vapor concentrations of hydrazoic acid in HAN/hydrazine systems.

- A typical hydrazoic yield is around 33 percent.

The staff finds that limiting the hydrazine concentration and hydrazoic acid yield directly limits the hydrazoic acid concentrations in the liquid and gaseous phases, in a manner consistent with the literature. The use of neutralization prevents recycling of the hydrazoic acid into other parts of the process. In addition, vapor pressure is a strong function of temperature, and, thus, temperature control provides another direct control that can be very effective. The staff concludes that the three safety functions are reasonably independent and, in combination, have the ability to make the explosion event highly unlikely, in accordance with the performance requirements.

The staff concludes that the selected strategy, PSSCs, and design bases have the ability to meet the performance requirements and are acceptable for the construction authorization stage. This closes Open Item CS-03 from the revised DSER (NRC, 2003d).

8.1.2.5.3.3 <u>Hydrazine and Azides</u>. Hydrazoic acid forms metal azides with cations in near-neutral and alkaline solutions. These are slightly soluble or insoluble, depending on the cation. Some metal azides can become unstable under certain conditions and present a potential explosion hazard.

The applicant has adopted a preventive safety strategy to protect the worker, site worker, IOC, and environment. The PSSCs include the following:

- For the chemical safety controls, the first safety function is to assure that azides are not added to high-temperature process equipment (e.g., the calciner). The applicant identified the design bases as no addition of azides to high-temperature process equipment.

  The second safety function is to ensure that azides have been destroyed before acidification of the alkaline wastes and their transfer to the high alpha waste of the liquid waste reception unit. Sodium nitrite is added as the scavenging agent for the azides. Sampling controls have been identified as a PSSC to confirm the destruction of the azides (revised CAR Section 8.5.1.9, p. 8-33) before acidification.

  The third safety function is to ensure that tanks potentially containing azides are not left dry (revised CAR Section 8.5.1.9, p. 8-32).

- For the PSCS, the first safety function is to ensure that metal azides are not exposed to temperatures that would supply sufficient energy to initiate energetic azide decomposition. The temperature design basis has been identified as not exceeding 140 °C (284 °F).

  The second safety function is to limit and control conditions under which dryout can occur.

Silver azide can also form upstream in the process according to the reaction:

$$HN_3 + AgNO_3 => AgN_3 + HNO_3$$

However, the process separates silver from hydrazoic acid containing vessels and does not recycle hydrazoic acid. In addition, silver concentrations are estimated to be very low in vessels containing hydrazoic acid. Finally, the temperature design basis of 140 °C is below the decomposition temperature of silver azide.

The staff review found that the applicant identified the potential hazards of metal azides, the process includes steps to reduce these hazards, and PSSCs and design bases have been proposed to reduce potential hazards and risks even further. Two PSSCs with multiple safety functions have sufficient independence and, in combination, should have the ability to render the potential event highly unlikely. Consequently, the staff concludes that the approach provides adequate assurances of safety for construction authorization.

8.1.2.5.4 Hydrogen Peroxide

A 35-wt% hydrogen peroxide solution is received and diluted to 10 wt% for use in the dissolution units and the oxalic mother liquor recovery unit. Explosive vapors can be produced if a concentration of hydrogen peroxide solution exceeding 75 wt% is boiled. The applicant stated that administrative controls would be used to specify and dilute the hydrogen peroxide concentration before use in the process in order to prevent explosions. The PSSC is identified as chemical safety controls.

The applicant provided supplemental information (DCS, 2002f) stating that the peroxide would be received as 35-wt% solution and diluted to form a 10-wt% solution. The applicant summarized the approach as follows:

- a certified or lot analysis by the manufacturer under a facility approved quality assurance plan

- an analysis upon receipt at the facility

- an analysis after mixing/diluting for facility use

The staff evaluation concludes that the applicant's proposal to use multiple independent administrative controls at different times to control peroxide concentrations is acceptable.

8.1.2.5.5 TBP—Nitrate (Red Oil)

Tributyl phosphate is a phosphate ester containing three butyl groups that is added to the solvent phase as the extractant for uranium and plutonium nitrate complexes. A TBP-solvent mixture preferentially moves uranium and plutonium from the aqueous (acid) phase to the organic phase, leaving most of the impurities in the acid stream. In a separate stripping column, the solvent-TBP-uranium-plutonium mixture is contacted with nitric acid containing a reductant (HAN). The HAN changes the oxidation state of the plutonium to a form that is only soluble in the nitric acid, thus allowing its recovery and separation from uranium. The organic phase containing the uranium is contacted with dilute nitric acid in a separate step that reverses the equilibria, and the uranium nitrate moves back into the aqueous phase. In the absence of TBP, no uranium and plutonium would be removed from the aqueous phase, and purification could not be accomplished.

In the presence of a nitrate of heavy metals or in nitric acid solutions, the TBP will form nitrate complex compounds that could react exothermically. Exothermic TBP-nitrate reactions are frequently referred to as red oil reactions because of the reddish color that has been observed in nitrated TBP/diluent mixtures and residues found during experiments and after the incidents.

Red oil is an organic mixture, consisting of TBP and its complexes with plutonium nitrate and nitric acid, degradation products of TBP (e.g., dibutyl phosphate), and possibly various nitrated hydrocarbons. The applicant analyzed the risk associated with red oil in units where the aqueous phase is likely to contain traces of these products above ambient temperatures or come into contact with a heating source (e.g., acid recovery unit, oxalic mother liquor recovery unit).

8.1.2.5.5.1 Applicant's Proposed Approach for Red Oil. The applicant adopted a preventive safety strategy to protect the worker, site worker, IOC, and environment (revised CAR Section 5.5.2.4.6.7; NRC 2003a, 2003c). Three sets of PSSCs to prevent the overpressurization event are identified by the applicant in Chapter 5 of the revised CAR. The PSSCs include (1) the offgas treatment system, (2) the PSCS, and (3) chemical safety controls. These PSSCs apply to open and closed systems as described below:

The applicant defined a system's ability to accommodate a red oil reaction in terms of the system being open or closed. An open system is defined as one in which the vent area associated with the offgas treatment system is sufficient to prevent overpressurization, if the runaway reaction occurs. The offgas treatment system design assumes that the system is filled with100-percent solvent. A closed system is one in which the vent area is not sufficient to prevent overpressurization, if the runaway reaction occurs. A closed system will have substantial quantities of solvent present (estimated at about 40–60 percent; the exact values would be determined for specific vessels at the ISA stage) and must use an evaporative cooling correlation.

Offgas Treatment System

For closed systems, the safety function of the offgas treatment system is to provide an exhaust path for aqueous phase evaporative cooling in process vessels. The following provides the design basis:

> evaporative cooling = 1.2 (steam energy input at 133 °C (271 °F) + energy generated chemically)
>
> where 1.2 is a safety factor

For open systems, an additional function of the offgas treatment system in open systems is to provide venting of vessels and equipment that potentially contain TBP and its associated byproducts to prevent overpressurization, should the runaway reaction occur. The design-basis value for the vent size is 12.5 kg/cm^2 (0.008 mm^2/g) of organic material present in the vessel.

Process Safety Control Subsystem

For open and closed systems, there are three safety functions of the PSCS, including to (1) control the energy generation rate by restricting the bulk temperature of solutions that may contain degraded organic material to within safety limits, (2) limit the residence time of organic compounds in the presence of oxidizers (e.g., nitric acid) and radiation fields, and (3) for closed systems only, ensure an adequate aqueous phase inventory to provide evaporative cooling.

The design basis for the first safety function is to limit the temperature of steam used to heat process vessels to 133 °C (271 °F). The proposal to limit steam temperature to 133 °C (271 °F) should not be interpreted as a small operational margin from the red oil runaway initiation temperature of 137 °C (279 °F). Steam at this temperature is required to efficiently heat the nitric acid-water azeotrope to its boiling point of 120 °C (248 °F). The boiling point temperature should not be exceeded under normal conditions (vented). Therefore, the effective temperature margin is 17 °C (63 °F).

The residence time will be a controlled parameter of one or more items relied on for safety (IROFS) to be identified by the applicant in its ISA summary submitted as part of any later application for a license to possess and use licensed material.

The design basis for the third safety function is a bulk fluid maximum temperature of 125 °C (257 °F) with a maximum heatup rate of 2 °C (3.6 °F)/min after startup. Aqueous phase availability would be assured through continuous feed or injection if either the temperature limit or ramp rate is exceeded. Detailed evaluation of startup operations will be conducted as part of the ISA to ensure that an appropriate margin of safety exists to account for regimes where the temperature rate of change during startup may exceed 2 °C (3.6 °F)/min.

Chemical Safety Controls

The safety function of the chemical safety controls is to ensure that the diluent does not contain cyclic chain hydrocarbons. The exclusion of cyclic chain hydrocarbons minimizes the involvement of the diluent in the red oil reactions. The exclusion of degraded organic compounds prevents the phenomena from initiating at lower temperatures.

DCS has committed to further evaluate the red oil phenomena, including continuing analyses and experiments which could result in an increase or decrease of the temperature at which action is required to remain below the design basis value. DCS is also evaluating the effect of impurities on the initiation temperature in closed systems (NRC, 2003a).

8.1.2.5.5.2 <u>NRC Review of Red Oil</u>. The staff has reviewed the proposed use of evaporators for waste processing and reagent recovery at the planned facility. Contact between TBP with aqueous solutions of nitric acid and acidic solutions of metal nitrates is intrinsic to the PUREX and similar solvent extraction processes, and evaporator treatment of PUREX streams has been performed safely since the 1950s. However, several events have occurred involving a range of conditions. Extended contact between heated mixtures of TBP and nitric acid, nitrates, and/or heavy metal nitrate salts can form degradation products and intermediates (including nitrated esters and adducts) that may lead to violent exothermic reactions under certain conditions (Hou, 1996; ORNL, 1988; SRS, 1976; WSRC, 1994a, 1994b, 1994c, 1995).

Several explosive incidents have occurred in the United States (Savannah River in 1953 and 1975, Hanford in 1953) (ORNL, 1988; WSRC, 1994a, 1994b) and the former Soviet Union (Tomsk-7 in 1993) (DOE, 1993; IAEA, 1998; OECD, 1994). The Hanford and first Savannah River (1953) event involved evaporators undergoing heating with low-pressure steam; temperature control was not precise but ranged from 135–140 °C (275–284 °F). The second Savannah River incident (1975) involved a denitrator operating at greater than 150 °C (302 °F). The pretreatment step for the denitrator is an evaporator operated with controls and a design basis temperature not exceeding 135 °C (275 °F). Thus, the TBP and solvent mixture safely traveled through the evaporator before the event in the denitrator. The incident at Tomsk did not involve heated equipment; the reactions started in a tank at a nominal temperature of 50 °C (122 °F) with an organic layer estimated at 90 °C (194 °F). Other, less serious incidents have also occurred.

The staff notes from its review that these events occurred because of the unanticipated presence (either carryover or accumulation) of the TBP, solvent, and degradation products. The applicant indicated that solvent carryover can be considered as an anticipated event in the facility.

Westinghouse Savannah River Company has summarized much of the experimental information that is available to understand the phenomena and establish reasonable controls (WSRC, 1994a, 1994b). Experimental work has replicated portions of the observed phenomena. The important specific reactions are summarized as follows:

- hydrolysis reactions (occur slowly—rates increase with temperatures)

- dealkylation and nitration reactions (occur slowly—rates increase with temperature)

- pyrolysis (becomes significant at about 150 °C (302 °F) with little or no water present)

- butyl alcohol and butyl nitrate nitration/oxidation reactions (become significant at 90–100 °C (194–212 °F))

- TBP nitration/oxidation (becomes significant around 135 °C (275 °F))

- uranium and plutonium adduct reactions (become significant at 15–175 °C (59–347 °F))

These reactions evolve large quantities of non-condensible gases, including the reddish-brown nitrous oxide fumes observed in several red oil events and experiments. The chemical reactions are outlined in revised CAR Section 8.5.1.5.2, with the degradation mechanisms in revised CAR Figure 8.5-2 and the degradation products in revised CAR Table 8-8.

In general, these reactions can occur in aqueous or organic phases. Organic compounds have a greater affinity for and tend to concentrate in the organic phase, if it is present. Organic phases also tend to have poorer heat transfer characteristics than aqueous phases. Radiolysis also produces similar reactions and products, albeit at slower rates. Finally, the staff review found three basic routes to the red oil phenomena. If volatile species do not evaporate, the reactions become significant with heating at 90–100 °C (194–212 °F) and reaction runaway occurs at 135 °C (275 °F). Pyrolysis dominates at higher temperatures if little water is present.

8.1.2.5.5.3 NRC Review of Controls, Limits, and Safety Design Bases Proposed in the Literature. After a review of the Tomsk accident and additional experimental studies in the early 1990s, DOE researchers recommended conditions for controlling red oil effects and preventing related explosions at DOE facilities. Paddleford and Fauske (WSRC, 1995) conclude that a runaway reaction is possible in an open system (i.e., at atmospheric pressure), and self-heating was observed at temperatures in the vicinity of 130 °C (266 °F). They mention the following preventive measures:

- prevent accumulation of significant masses of TBP
- prevent significant interaction with strong nitric acid
- stir (mix) during additions and sampling
- maintain low temperatures
- maintain cooling system effectiveness
- keep TBP/nitric acid mixed with or thermally coupled to aqueous layers
- implement practical vent sizing (0.005 in.^{2}/kg of TBP in nitric acid)

The derived venting relationship is shown as Figure 8.1-2 and relates vent cross-sectional area to the mass of organic material present and the internal pressure generated from the red oil reactions. The recommended safe range is below a value of 32 kg organic compounds/cm^2 of vent area, which essentially prevents pressurization of the vessel containing the red oil/nitric acid mixture. For comparison, the applicant's design basis for the offgas treatment system in open systems is 12.5 kg/cm^2 (0.008 mm^2/g) of organic material present in the vessel. This compares to a 32-kg/cm^2 limit for rapid pressurization, as shown in Figure 8.1-2. Thus, the applicant's design basis for open systems is in the recommended safe range and provides for a safety margin of about 60 percent below the initiation conditions.

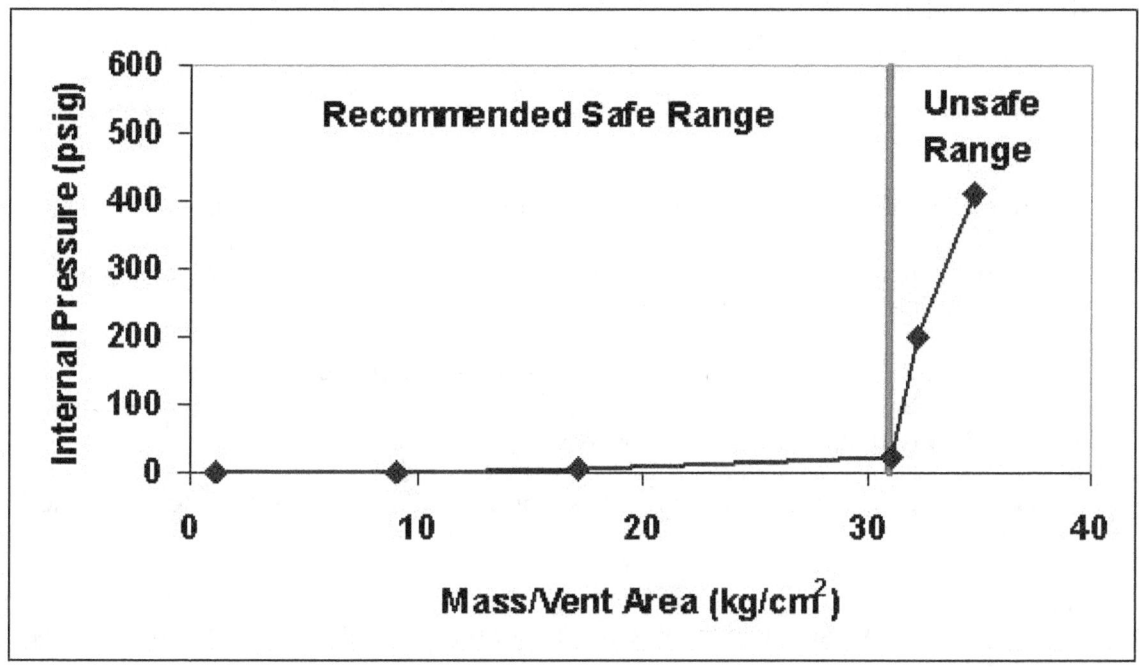

Figure 8.1-2 Empirical venting relationship

The DOE team recommendations include the following (DOE, 1993):

- adequate venting and relief disks on vessels

- adequate mixing

- vessel containment

- controls with interlocks and independent verification

- minimization of organic compound accumulations in vessels containing high concentrations of nitric acid

- presence of alarms on temperature and pressure controls

- controls and limits on strong oxidants and reductants

The Tomsk engineers mentioned the following lessons learned (DOE, 1993):

- use tanks capable of withstanding high pressures (20 atm (284 psi))
- limit the concentration of added nitric acid (e.g., for pH adjustments) to 6 M or less
- limit the total nitric acid concentration to 3 M or less
- perform continual air sparging (for mixing)
- maintain a temperature rise of no more than 7 °C (13 °F) in any organic layers
- keep the tank contents below 70 °C (158 °F)
- provide adequate relief valves and rupture disks

Hyder (WSRC, 1994a) mentioned the following lessons learned from the incidents and the experimental studies:

- Natural processes provide adequate cooling at temperatures below 80 °C (176 °F).

- Above this temperature, care should be taken to ensure that adequate cooling is available for the amount of TBP present.

- In well-ventilated systems, n-butanol evaporation and convective cooling are sufficient to control the reactions up to about 120 °C (248 °F) and the decomposition reactions become very rapid in the range of 130–150 °C (266–302 °F).

- Confined systems and circumstances that can give rise to confined systems should be avoided.

- Limits on TBP mass, concentration, or mass per unit surface area should be established. The SRS currently uses a concentration limit of 0.5 percent in the agitated (total) feed to an evaporator.

Hyder (WSRC, 1994b) further elaborated upon potential limits:

- Tanks in which TBP-nitrate complexes are or may be present should be adequately vented. A value of 0.032 cm^2/kg (0.005 $in.^2/kg$ or 31 kg/cm^2) of TBP is mentioned. This is based upon Figure 8.1-2.

- Chemically degraded TBP, or TBP that has sat for a long time in the presence of acids or radiation, should be purified before use in solvent extraction.

- Evaporators in which TBP might be introduced should be operated at a controlled temperature and their TBP content should be limited. Local high temperatures should be avoided. No portion or surface of the evaporator should exceed 130 °C (266 °F), and operating temperatures substantially lower than this are desirable. If evaporators or heating units must operate at higher temperatures, the means for preventing the introduction of TBP must be correspondingly rigorous.

- Evaporator bottoms and other solutions that may contain TBP should be cooled under conditions that ensure heat removal. Sparging and agitation may also be appropriate.

- Process design and operations should consider the potential for red oil reactions, particularly for nonroutine operations and situations.

The SRS currently operates evaporators in H Canyon (Jones, 2003). These evaporators are subject to DOE safety class (i.e., for the public) and safety significant (i.e., for the workers) controls to prevent a potential red oil runaway reaction and explosion (i.e., frequency under 1.0×10^{-6}/yr). The controls are generally divided into two categories, those that prevent excessive amounts of TBP entering the evaporators and those that prevent overheating. The latter include a 120 °C (248 °F) temperature safety limit (used in an analogous manner to the design basis), a high steam coil pressure interlock, and an alarm for the operator to manually check that the steamflow has been terminated. The staff notes that the 120 °C (248 °F) temperature limit corresponds to the normal boiling point of the water-nitric acid azeotrope (Lange, 1999).

The Defense Nuclear Facilities Safety Board (DNFSB) issued a technical report on the control of red oil explosions (DNFSB, 2003). This report emphasizes controls on temperature, pressure, mass, and concentration (acid), and the use of effective defense-in-depth practices for the prevention of a red oil explosion. The following paragraphs summarize the discussions in the report regarding the controls.

Temperature Control

The report identifies an initiation temperature of 130 °C (266 °F) for the runaway red oil reactions. Lower temperatures are needed for operationally protected temperatures and setpoints. H Canyon has an operational protected temperature (technical safety requirement, similar to the design basis) of 120 °C (248 °F) and an overtemperature safety setpoint of 117 °C (243 °F) (i.e., 3 °C is allowed for instrument errors and biases). Steam interlocks are set not to exceed 25 psig, which corresponds to 132 °C (269 °F).

Pressure Control

The report mentions passive vents in accordance with the Fauske correlation (> 0.063 mm^2/g of TBP (< 15.9 g TBP/mm^2 or < 1.59 kg TBP/cm^2)). It also mentions 312 g (11 oz) of red oil/mm^2 (31.2 kg/cm^2). With some safety margin, it recommends no more than 208 g/mm^2 (20.8 kg/cm^2) of red oil (DNFSB, 2003, p. 4-3). Page 5-2 of the DNFSB report mentions 41.5 cm^2 (6.44 in.^{2}) as the minimum vent for 1361 kg (3000 lb) of TBP. This corresponds to less than 32.7 kg TBP/cm^2

(465 lb TBP/in.^{2}) of vent. The applicant used 0.008 mm^2/g of organic compound (12.5 kg/cm^2).

Mass Control

Mass control is applied to the organic phase to keep it from entering heated equipment (prevention) or by omitting TBP (or equivalent) mass/concentration (mitigation).

The applicant used 30 percent TBP in branched dodecane (HPT). There are no organic mass limits needed for open systems. The applicant stated that organic mass limits for closed systems will be developed in the license application stage and will be a substantial fraction of the volume (e.g., 40–60 percent). Organic material entering an open or closed system is considered an anticipated event.

Concentration Control

Concentration control applies primarily to nitric acid (less than 10 M) and, to a lesser extent, nitrate salts/uranyl nitrate (less than 20 percent). The applicant planned to reconcentrate nitric acid to over 13.6 N. There are no limits on acid concentration.

8.1.2.5.5.4 <u>NRC Staff Findings and Conclusions on Red Oil</u>. The NRC staff has evaluated the applicant's proposed controls for the red oil phenomena (NRC, 2004a). For open systems, the applicant provided PSSCs and design bases. Significantly, the applicant's design basis for the offgas treatment system in open systems is 12.5 kg/cm^2 (0.008 mm^2/g) of organic material present in the vessel. By comparison to Figure 8.1-2, this is well within the safe range. Thus, the applicant's design basis for venting open systems provides for a safety margin of 60 percent from the initiation conditions. In addition, since the system cannot pressurize easily, the temperature will not exceed the normal boiling point of the azeotrope (120.4 °C (249 °F)), which is below the initiation temperature of the runaway red oil reactions. Therefore, the staff finds the applicant's approach for open systems acceptable. The remainder of the discussion focuses on the evaluation of the PSSCs and design bases for closed systems.

Staff Evaluation of Temperature Limit

The staff evaluated whether the average solution temperature limit of 125 °C (257 °F) is adequate. The staff evaluated additional literature from SRS which supports a red oil initiation limit of no greater than 130 °C (266 °F) for evaporators containing nitric acid and TBP. The SRS H-Canyon Safety Analysis Report (WSRC, 2001c) recommends an "always safe" limit of less than 130 °C (266 °F) but cites a temperature of 135 °C (275 °F) at which the initiation of a runaway reaction between TBP and concentrated (70 wt%) nitric acid was observed to occur. Using contemporary methods but similar experimental conditions, others (WSRC, 2001d) have concluded that some of the Colven data from 1956, which support initiation temperatures as low as 129 °C (264 °F), are outliers. The minimum initiation temperature found by these authors is 137 °C (279 °F) for 15 M nitric acid.

The applicant has committed to design the acid recovery evaporator system such that the bulk fluid temperature will not exceed 125 °C (257 °F). This will be accomplished by shutting off the steam and injecting aqueous phase material into the system. As long as the pressure is maintained at a pressure of approximately 1 atmosphere, the bulk fluid cannot exceed the nitric acid/water azeotrope boiling point, which is 120.4 °C (249 °F). Establishing a vent path that can provide sufficient mass flow to remove 1.2 times the heat input from the steam and the heat generated from hydrolysis and an adequate aqueous phase inventory provides the physical conditions necessary to limit the bulk temperature to approximately 120.4 °C (249 °F). Therefore, the applicant has proposed a shutdown temperature margin of 12 °C (137–125 °C) for the evaporators, plus an additional 20-percent margin for heat removal capacity by the offgas treatment system.

The 125 °C (257 °F) bulk fluid limit provides operational flexibility. The only way to increase the system temperature to 130 °C (266 °F) or above is to (1) boil off all of the aqueous phase and/or (2) pressurize the evaporator by blocking the vent path. With respect to system pressurization, staff evaluated the change in system pressure which would raise the azeotrope boiling point of 120.4 °C (249 °F) to 125 °C (257 °F). The pressure increase required to raise the temperature 4.6 °C (8.3 °F) is about 10 percent of the ambient offgas treatment system pressure. However, the availability and reliability of the vent path is not in question, as it will be addressed during the license application phase. With the steam temperature limited to 133 °C (271 °F) and the bulk fluid temperature at 125 °C (257 °F), the total heat input from the steam will only be approximately 3 percent of the heat generated by the chemical reaction because of the small delta-T driving force. Cool aqueous phase will be injected into the system should the bulk temperature or heatup rates be exceeded.

Past DOE experiments (WSRC, 1998) have analyzed steam heating of TBP/nitrogen-paraffin/nitric acid mixtures to determine whether a mixture of steam and air from an empty tank could heat the organic layer until an exothermic, runaway reaction could take place. A jet of superheated steam (90 percent) and air (10 percent) at 139 °C (282 °F) was used. The jet was cooled to the organic layer temperature, which reached a maximum temperature of 128 °C (262 °F). The organic layer was effectively limited to this temperature through evaporative cooling. This work demonstrates that red oil reactions are not susceptible to source point initiation caused by minor temperature variations that may be encountered during operations.

The staff inquired about the likelihood for preventing organic compounds inadvertently entering nitric acid recovery systems (e.g., Nomura, 1994). Since the applicant has not committed to PSSCs to prevent the introduction of organic mass into the evaporators, the staff requested information concerning the potential for foaming, two-phase flow and, ultimately, pressurization of the evaporator by blockage of the offgas treatment system vent (NRC, 2003a). The applicant committed to developing a fundamental understanding of the system by evaluating the mechanism and behavior of such events through modeling and experimentation, as needed (NRC 2003a). This fundamental understanding is intended to allow a determination of the appropriateness of the relationship of the vent area-to-mass organic ratio, including the potential for two-phase flow.

The staff considered the safety approach used for the SRS H-Canyon evaporators. The DOE identified the red oil initiation temperature for 14–15 M nitric acid in closed systems as 137 °C (279 °F) and has identified an always safe solution temperature limit of 130 °C (266 °F) for its systems (WSRC, 2001c). An always safe value of 130 °C (266 °F) ensures that the red oil initiation temperature lies above this value for all foreseeable acid concentrations in either open or closed systems. To protect this always safe value, operational limits (safety class set points, as specified in the SRS technical safety requirements) of 120°C (248 °F) are placed on solutions in H Canyon evaporators. However, the nitric acid solution concentrations are typically less than 50 wt%. This is lower than the 67 wt% azeotrope concentration that applicant proposed as part of its nitric acid recovery system. As a result, the 120 °C (248 °F) limit is sufficiently above the actual boiling point of the H Canyon nonazeotropic solutions, which allows the evaporators to both operate efficiently and maintain an adequate safety margin. Since the applicant proposed to recover the azeotrope, an always safe (not to exceed) value of 125 °C (257 °F) allows the recovery of the azeotrope (B.P. = 120.4 °C (249 °F)) while maintaining a 12 °C (22 °F) margin from the 137 °C (279 °F) initiation temperature for 14–15 M nitric acid in closed systems. The applicant will establish safety set points, with margin, less than the 125 °C (257 °F) design-basis value as part of any later application it may file for a license to possess and use licensed material.

The Shape of the Evaporator

The staff evaluated the shape of the proposed acid recovery evaporators. Specifically, the staff evaluated the hypothesis that a tall and thin (high aspect ratio) thermosiphon evaporator would have a high likelihood of inducing the separation of organic and aqueous phases. This potential would be highest during startup and shutdown modes of operation. However, the condition of two separate phases is already assumed in the safety assessment (i.e., the red oil phenomena would occur in a separate organic phase after its contact with nitric acid). This organic phase could be lighter than the aqueous phase and float on the surface. Alternatively, the organic phase may be heavier than the aqueous phase (because of complexation of the TBP with heavy metals, such as uranium or plutonium) and be located on the bottom, which is a condition referred to as phase inversion. In either case, the safety functions of the proposed red oil controls do not rely on the miscibility of the organic phase with aqueous solutions. The applicant assumed that a distinct and separate organic phase may be present that requires temperature controls, venting, active cooling, and antifoaming controls. In its November 2003 report, the DNFSB (DNFSB, 2003) noted that "if solution temperature sensors are used it is important that they be located such that the organic phase temperature can be measured with or without phase inversion." The staff agrees in principle with this recommendation as important to the final design of temperature measurement IROFS. Information on the placement of temperature measurement devices is not required at the CAR stage.

November 2003 DNFSB Technical Report

On November 13, 2003, the DNFSB issued a technical report, "Control of Red Oil Explosions in Defense Nuclear Facilities, 2003" (DNFSB, 2003). In the report, DNFSB recommends that two or more of four controls should be used to prevent a runaway red oil reaction and explosion of the detonable gases produced by the reaction, as described above in Section 8.1.2.5.5.3.

In describing the temperature control of the evaporators (Section 4.1 of the report), the DNFSB staff provided the following specific recommendation:

> ...to be assured that red oil conditions are not present in an evaporator, controls for temperature, pressure, and concentration should all be utilized. (p. 4-3)

Of the four controls listed above, the applicant identified only temperature controls for the closed acid recovery evaporators at the MOX facility. Under the definition of a closed system, the pressure controls in these vessels are not adequate to prevent an overpressurization event. In addition, mass controls, though present in the design, are not credited in the safety assessment as PSSCs, and there are no limits on nitric acid concentration.

However, the staff concludes that the applicant provided sufficient defense-in-depth provisions by committing to a multitiered approach that includes (1) a combination of multiple, independent temperature controls, (2) adequate aqueous phase evaporative cooling provided by the offgas treatment system, (3) the exclusion of cyclic chain hydrocarbons, and (4) the commitment to perform additional research on the runaway initiation temperature and the effect of impurities on the initiation temperature.

For the CAR stage, the applicant provided sufficient controls and margin such that the bulk temperature will not exceed 125 °C (257 °F). This temperature is 9–15 °C (16–27 °F) below the 134–140 °C (273–284 °F) range of experimentally measured runaway initiation temperature data (which average 137 °C (279 °F)). The applicant's proposed aqueous injection system extends beyond the safety requirements at DOE facilities and the operating French MOX facility. Other possible operational concerns related to the evaporator startup, shutdown, and possible abnormal conditions are best addressed in the ISA when more specific design information is expected to be available.

Consequently, the staff finds the applicant's approach acceptable for the CAR stage. This closes Open Item CS-01 from the revised DSER (NRC, 2003d).

8.1.2.5.6 Plutonium (VI) Oxalate

Plutonium can exist in several different oxidation states in aqueous solutions. The most common are the +3, +4, and +6 states. The applicant intended to precipitate plutonium (IV) oxalate and expeditiously calcine it to the plutonium dioxide powder, in alignment with good handling practices for plutonium (DOE, 1998b), for subsequent MOX fuel manufacture. Plutonium (VI) forms slowly in nitric acid solutions and may also be present if the valence adjustments are not performed correctly. Upon heating, plutonium (VI) oxalate ($PuO_2C_2O_4$) exhibits an endothermic peak around 142 °C (288 °F) because of dehydration (it is usually present as the trihydrate) and a subsequent rapid exothermic peak at approximately 219 °C (426 °F) associated with oxidation of the oxalate. Section 8.5.1.7 of the revised CAR mentions a potential explosive hazard with plutonium (VI) oxalate introduction into the calcining furnace and its subsequent rapid decomposition. As described in revised CAR Section 5.5.2.4.6.12, the safety strategy is prevention using the PSSC of chemical safety control. The safety function is to perform a measurement of the valency of the plutonium before adding oxalic acid to it in the oxalic precipitation and oxidation unit. If plutonium (VI) is detected, feed would be terminated. The revised CAR also indicates that the design basis is to prevent the introduction of plutonium (VI) oxalate into heated equipment where temperatures in excess of 219 °C (426 °F) are credible. In addition, controls will be in place to ensure that temperatures do not exceed 219 °C (426 °F) where plutonium (VI) oxalate might be present (e.g., the oxalic mother liquor recovery unit and the oxalic precipitation and oxidation unit).

The applicant will develop a measurement system for plutonium (VI) as part of the final design process. Analytical methods under consideration include ultraviolet-visible spectrum and alpha spectrum spectroscopy, with detection/resolution limits of 1×10^{-5} and 1×10^{-9} M for plutonium (VI), respectively (NRC, 2003a, 2003c).

The applicant also provided information on the design basis for the calciner (NRC, 2003a, 2003c). According to Section 11.8 of the revised CAR, the design basis pressure criteria for storage tanks, process columns, and exchangers are the greater of maximum pressure acting at the top of the vessel in normal operating conditions or the maximum pressure acting at the top of the vessel during transient or accident conditions, plus 10 percent. The applicant committed to the same design-basis criteria for the calciner furnace.

The staff notes the detection limits of plutonium (VI) are very low regardless of the method used and effectively prevent significant quantities from entering the calciner (i.e., the safety strategy of prevention). In addition, the design basis of the calciner (the maximum pressure plus 10 percent) assures that pressure increases from the rapid decomposition of any plutonium (VI) oxalate that remains after detection will not challenge the vessel's integrity. The staff review also found that the highest temperature in any piece of AP equipment other than the calciner is 135 °C (275 °F); this provides a significant margin vis-a-vis the exotherm temperature of approximately 219 °C (426 °F). The staff finds the approach acceptable for the construction authorization stage.

8.1.2.6 Habitability Issues

The purpose of this section is to review the criterion used by the applicant to ensure the habitability of the facility during routine operations and during an accident or event.

8.1.2.6.1 Design Approach

The applicant stated that facility workers mainly perform a monitoring role during emergency conditions. The applicant identified the ECRACS as the PSSC to allow the performance of the monitoring safety role. The description of the ECRACS is given in Section 11.4.2.7 of the revised CAR. The following safety functions are attributed to the ECRACS:

- maintain a habitable environment in each of the two ECRs for the facility personnel
- provide cooling to the emergency electrical rooms

The system will include two 100-percent capacity air filter trains, one for each ECR. Each unit consists of a filtration unit and a booster fan for each intake. Each filter housing contains a hazardous gas removal cartridge and/or an organic vapor cartridge, and high-efficiency particulate air (HEPA) filter cartridges. The filter housings are manufactured from stainless steel.

The control strategy is provided as follows:

- Each ECR intake is continuously monitored for hazardous chemicals.

- Upon detection of a hazardous chemical above allowable limits, the intake is automatically isolated and switched to the recirculation mode using a filtration unit with HEPA filtration and hazardous gas removal elements.

- An alarm sounds if hazardous chemical concentrations are detected at both intakes.

- The alarm alerts operators to don emergency self-contained breathing apparatuses (SCBAs).

The applicant stated that monitoring would be performed for those chemicals for which unmitigated release could result in control room concentrations exceeding the limits in Table 8.1-12. Specific setpoints for actuation of air inlet isolation valves and ventilation recirculation would be determined during final design. Initial calculations by the applicant have identified releases of hydrazine or nitrogen tetraoxide as having the potential for resulting in concentrations at or above these limits. Calculations would be made during final design to verify the list of chemicals monitored.

The table contains limits from two different sources. The preferred limit is the IDLH value from NIOSH/OSHA. If a TEEL-3 value is lower than the IDLH value for a given chemical, the applicant intended to apply the TEEL-3 limit for that chemical. The applicant will use TEEL-2 limits where no IDLH value exists.

The applicant has also proposed standards for breathing air (revised CAR Section 11.9.1.11) that would be used as design information for the chemical processes that involve chemicals which could displace oxygen (i.e., the chemicals are not toxic or hazardous), thus decreasing its concentration in the air. For operating procedures to avoid asphyxiating atmospheres, the applicant specified that the high ventilation rates preclude the creation of an asphyxiating atmosphere (DCS, 2001). The applicant referenced the CGA publication, "Accident Prevention in Oxygen Rich and Oxygen-Deficient Atmospheres," issued 1992 (CGA, 1992b), as guidance.

The applicant proposed two standards to assure the quality of breathable air:

- CGA G-7.1, "Commodity Specification for Air," issued 1992 (CGA, 1992a)

- American National Standards Institute (ANSI) Z88.2-92, "American National Standard for Respiratory Protection," issued 1992 (ANSI, 1992)

8.1.2.6.2 Asphyxiating and Hazardous Atmospheres

The excess or lack of oxygen may become the initiators of additional accidents. For example, nonflammable materials can become flammable with more oxygen present, while operating personnel can suffer anoxia and experience difficulties controlling the plant with less oxygen present. If the atmospheric concentration of oxygen exceeds 23 percent, it is called an oxygen-rich atmosphere (CGA, 1992b, CGA P-14, Section 2.1). An oxygen-deficient atmosphere is one that contains less than 19.5 percent of oxygen (CGA, 1992b, CGA P-14, Section 2.2). An atmosphere containing less than 19.5-percent oxygen can cause problems in the cognitive abilities and impair the operational capacity of personnel. Moreover, the exposure to insufficient concentrations of oxygen for a prolonged time may cause convulsions, unconsciousness, and death. It is important to consider that the exposure effects may vary depending on the individual (CGA, 1992b, 2001).

With regard to design basis values for chemicals that can cause a hazardous atmosphere, the applicant will use the values in Table 8.1-12 based upon IDLHs, TEEL-3s, and TEEL-2s. The applicant committed (DCS, 2002e) to actively monitor the ECRACS air intakes for hazardous chemicals that could exceed Table 8.1-12 values at the intakes.

8.1.2.6.3 Habitability Standards

Regulatory guidance on control room habitability and ventilation systems is summarized as follows:

- Regulatory Guide (RG) 1.78, "Assumptions in Evaluating the Habitability of a Nuclear Power Plant Control Room During Postulated Hazardous Release," issued 2001 (NRC, 2001a), provides general guidance for the evaluation of control room habitability. This document has information about hazardous chemicals and the toxic limits of some components. Although the information included in this document is more qualitative than quantitative, it depicts some information that would be acceptable to the NRC as control room design bases. The document concludes that a chemical exposure period not exceeding 2 minutes at the IDLH value provides sufficient margin (i.e., as compared to the maximum 30-minute exposure and different functions (donning SCBAs as compared to evacuation) assumed in the derivation of IDLH values).

- ANSI/ASHRAE 62-1989, "Ventilation for Acceptable Indoor Air Quality," issued 1990 (ASHRAE, 1990), includes tables with information about outdoor air requirements to achieve the stipulated indoor air quality. These requirements only apply to some parts of the facility (e.g., offices, lobbies, reception areas, cafeterias, conference rooms, restrooms, lockers and dressing rooms, and elevators). Appendix C includes some standards and dose limits that can be used as bases for air pollutants (e.g., carbon monoxide, nitrogen dioxide, and particulates). The dose limits are depicted in terms of short-term exposure limits, threshold limit values, and weighted average concentrations. Although this information could be used for air ventilation design, it is important to remember that the document specifies that these tables "are not part of the standard," but are "included in the standard only for information purposes."

- The National Ski Patrol document, "Components and Hazardous Atmospheres (Prehospital Emergency Care & Crisis Intervention)," issued 1992 (NSP, 1992), provides guidance for constituents in air considered dangerous for human beings, based upon a 30-minute or longer exposure.

Most of the limits shown in ANSI/ASHRAE 62-1989 are included in 29 CFR 1910.1000, "Air Contaminants."

8.1.2.6.4 Staff Conclusions on Habitability

The staff notes that the applicant identified a safety function for workers in the ECRs. To protect workers in the ECRs, the applicant identified a safety function for the ECRACS to maintain habitable conditions. The staff concludes that it is reasonable and prudent to staff the ECRs and maintain habitable conditions, such that the facility can be monitored and, if necessary, emergency actions can be taken (NRC, 2001a, 2003g). Such approaches are often used in the nuclear and process industries. The staff finds the top-level safety strategy acceptable.

The revised CAR states that the safety function is to maintain habitable conditions for operators. The ECRACS is identified as the PSSC. The revised CAR, page 8-11, states, "Monitoring will be performed for those chemicals whose unmitigated release could result in control room concentrations above the limits specified in Table 8-12." Table 8.1-12 contains concentrations that NRC RG 1.78 indicates are only suitable for 2-minute exposures while the control room staff dons protective clothing and SCBA.

Consequently, these limits (Table 8.1-12, the design basis for ECR habitability) do not represent habitable concentrations. However, the applicant also identified chemical concentration levels for accident consequence categorization (FSER Table 8.1-5). Level 1 on this table represents "[t]he maximum concentration in air below which it is believed nearly all individuals could be exposed without experiencing other than mild transient adverse health effects or perceiving a clearly defined objectionable odor." The staff notes that Level 1 would not impair the activities of operators in the ECR and approximates a habitable condition. The staff concludes the applicant should maintain hazardous chemical concentrations below Level 1 of FSER Table 8.1-5 for the duration of potential hazardous chemical events and proposes the following condition:

Construction Authorization Condition

In addition to the safety functions and design bases specified in the revised CAR, the facility will be designed so that a safety function of the Emergency Control Room (ECR) Air Conditioning System shall be to maintain hazardous chemical concentrations in each ECR below CAR Table 8-5 Level 1 concentrations for the duration of credible hazardous chemical release events.

The CAR Table 8-5 Level 1 values use an exposure duration basis of 1 hour. As noted in Section 8.1.6.1.2 of this FSER, longer ECR durations might require lower chemical concentrations. The staff review of the revised CAR (Section 11.4) found that the ECRACS will include specific components for removing hazardous chemicals. The staff found that the current state-of-the-art technology of scrubbers and absorbers (e.g., Perry, 1997) likely has the ability to design and fabricate these components to meet the desired chemical concentration levels corresponding to habitability. In addition, technology that is considered to be state-of-the-art implies the ability for these systems to be built with redundant and diverse components and independent trains and, thus, the ability to meet high reliabilities corresponding to highly unlikely failure rates. The staff expects the specific details of meeting the FSER Table 8.1-5 Level 1 values (or lower values for longer durations, if appropriate) to be demonstrated by the applicant in the ISA summary, which would be submitted as part of any later application for a license to possess and use licensed material. The staff concludes that the applicant has now provided suitable habitability design bases (for oxygen levels, impurities, and hazardous materials) that are consistent with its proposed approach and provide sufficient margin. This closes Open Item CS-10 from the staff's revised DSER (NRC, 2003d).

Table 8.1-12 Chemical Limits Proposed by the Applicant for Control Room Consequence Calculations at the MFFF (mg/m³)

Chemical Name	Applicant's Proposed Concentration Limit (mg/m³)	Source of Limit	NRC Staff Observation
Aluminum Nitrate	15	TEEL-2	No IDLH
Azodicarbonamide	500	TEEL-2	Rev 18; Rev 19 TEEL-2 is 200
Chromic (III) Acid	25 (as CrIII)	IDLH	IDLH
Chlorine	29	IDLH	IDLH
Diluent (C10–C13 Isoalkanes)	35	TEEL-2	No IDLH; dodecane TEEL-2 = 80
Ferrous Sulfate	12.5	TEEL-2	No IDLH
Fluorine	30	TEEL-3	IDLH = 40
Hydrazine Monohydrate	0.06	TEEL-2	No IDLH
Hydrazine Nitrate	5	TEEL-2	No IDLH
Hydrofluoric Acid	25	IDLH	IDLH
Hydrochloric Acid	75	IDLH	IDLH
Hydrogen Peroxide	106	IDLH	IDLH
Hydroxylamine Nitrate	26	TEEL-2	No IDLH
Iron	500	TEEL-3	IDLH = 2,500
Isopropanol	5000	IDLH	IDLH (also 10% of LEL)
Manganese Nitrate	500 (as Mn)	IDLH	IDLH
Manganous Sulfate	500 (as Mn)	IDLH	IDLH
Nitric Acid	66	IDLH	IDLH
Nitric Oxide	125	IDLH	IDLH
Nitrogen Dioxide	35	TEEL-3	IDLH = 40
Nitrogen Tetraoxide	15	TEEL-2	No IDLH
Oxalic Acid	500	IDLH	IDLH
Potassium Permanganate	15	TEEL-2	No IDLH
Silver Nitrate	10 (as Ag)	IDLH	IDLH
Silver Oxide	10 (as Ag)	IDLH	IDLH
Sodium	5	TEEL-2	No IDLH
Sodium Carbonate	50	TEEL-2	No IDLH
Sodium Hydroxide	10	IDLH	IDLH
Sodium Nitrite	1	TEEL-2	No IDLH
Sodium Sulfite	50	TEEL-2	No IDLH
Sulfuric Acid	15	IDLH	IDLH
Sulfamic Acid	250	TEEL-2	No IDLH
Thenoyl Trifluoroacetone	25	TEEL-2	No IDLH
Tributyl Phosphate	300	TEEL-3	IDLH = 330
Uranium Dioxide	10 (as U)	IDLH	IDLH
Uranyl Nitrate	10 (as U)	IDLH	IDLH
Xylene	3900	IDLH	IDLH
Zinc Stearate	50	TEEL-2	No IDLH
Zirconium Nitrate	10 (as Zr)	IDLH	IDLH

In the license application, the applicant must demonstrate that habitable conditions can be maintained in the ECRs (i.e., the CAR Table 8-5 Level 1 values) or propose and demonstrate an alternative.

8.1.2.7 Baseline Design Criteria

The applicant stated that the BDC specified in 10 CFR 70.64(a) are incorporated into the design and operation of the proposed facility (revised CAR Section 5.5.5.4). The applicant stated that information demonstrating compliance with these criteria is provided in the applicable chapters of the revised CAR. For chemical protection, 10 CFR 70.64(a)(5) states the following:

> Chemical protection. The design must provide for adequate protection against chemical risks produced from licensed material, facility conditions which affect the safety of licensed material, and hazardous chemicals produced from licensed material.

Chapter 8 of the SRP (NRC, 2000) contains guidance and references to other peer-reviewed work on the subject of chemical safety. The applicant indicated that reagents are stored and chemical mixtures are prepared in the reagent processing building and the reagent storage part of the AP area. They are generally separated from each other and radioactive materials. The applicant will avoid mixing incompatible materials by design, controls, and procedures. The AP and MP facilities are broken down into process functional units, which are made up of one or more subassemblies performing consistent and elementary tasks. The applicant stated that the breakdown into control functional units allows each entity to be operated relatively independently in the given operating mode. The staff review notes that this separation and independence is consistent with accepted practices for safe operations.

The applicant will control process storage and operation conditions in order to prevent exothermic and potential autocatalytic reactions in the reagent processing building and the AP and MP areas. Autocatalytic and exothermic reactions of chemicals would be prevented through control of the process parameters (e.g., reactant concentration, temperature, catalyst concentration in solution, and pressure) that affect the reactions. Many of these controls have been identified as PSSCs with design bases. Codes and standards have been applied on a specific level, often as design bases of PSSCs.

The applicant stated that there is reasonable assurance that the PSSCs will be sufficiently reliable and available. This assurance will be provided through the use of standard nuclear industry engineering practices (e.g., reasonably and generally accepted good engineering practices (RAGAGEP)). These practices are incorporated into the facility general design philosophy, design bases, system design, and commitments to applicable management measures. These practices ensure that applicable industry codes and standards are utilized, adequate safety margins are provided, engineering features are utilized to the extent practicable, the defense-in-depth philosophy is incorporated into the design, and PSSCs will be appropriately maintained.

The staff review finds that the applicant has provided sufficient information to meet the requirements of 10 CFR 70.64(a)(5). The staff review using the SRP notes that the previously identified open items in MP have been closed by additional safety strategies and information from the applicant, and the staff concludes that the applicant has satisfied this BDC.

Related to chemical protection, the explosion protection BDC is stated as part of the fire protection BDC in 10 CFR 70.64(a)(3):

> Fire protection. The design must provide for adequate protection against fires and explosions.

Chapters 7 and 8 of the SRP (NRC, 2000) describe the fire protection/explosion BDC and include guidance and references to other peer-reviewed work on the subject. Guidance, codes, and standards have been applied on a specific level, often as design bases of PSSCs.

The applicant stated that there is reasonable assurance that the PSSCs will be sufficiently reliable and available. This assurance will be provided through the use of standard nuclear industry engineering practices (e.g., RAGAGEP). These practices are incorporated into the facility general design philosophy, design bases, system design, and commitments to applicable management measures. These practices ensure that applicable industry codes and standards are utilized, adequate safety margins are provided, engineering features are utilized to the extent practicable, the defense-in-depth philosophy is incorporated into the design, and PSSCs will be appropriately maintained.

The staff review finds that the applicant has provided sufficient information to meet the requirements of 10 CFR 70.64(a)(3). The staff review using the SRP notes that the previously identified open items related to MP have been closed by additional safety strategies and information from the applicant, and the staff concludes that the applicant has satisfied this BDC.

8.1.3 Chemical Process Safety Interfaces

The applicant will make controls established for chemical safety consistent with those established for radiological safety and criticality safety, along with the associated management measures. Accordingly, the Chemical Safety Program will be conducted under the same elements of programmatic infrastructure described in Chapters 4, 12, and 14 of the revised CAR and interfaces with the management measures discussed in Chapter 15 of the revised CAR.

8.2 Evaluation Findings

In revised CAR Chapters 5, 8, and 11, DCS provided design basis information for chemical process safety PSSCs identified for the proposed facility. Based on the staff's review of these revised CAR chapters and supporting information provided by the applicant relevant to chemical process safety, the staff finds, for the reasons discussed above, that DCS has met the BDC set forth in 10 CFR 70.64(a)(3) for explosions and 10 CFR 70.64(a)(5) for chemical safety. Further, the staff concludes, pursuant to 10 CFR 70.23(b), that the design bases of the PSSCs identified by the applicant will provide reasonable assurance of protection against natural phenomena and the consequences of potential accidents.

The following previously identified open items in the April 30, 2002, and April 30, 2003, DSERs (NRC, 2003d) have now been closed in a satisfactory manner, including CS-01, CS-02, CS-03, CS-05b, CS-9, and CS-10.

8.3 <u>References</u>

(ACGIH, 1994) Proceedings of the American Conference of Governmental Industrial Hygienists. "Threshold Limit Values for Chemical Substances and Physical Agents and Biological Exposure Indices." 1994.

(AIHA, 2004) American Industrial Hygiene Association. "AIHA's 2004 ERPGs/WEELs Update Sets." March 24, 2004. http://www.aiha.org/publicationsadvertising/html/poerpgweels.htm (November 15, 2004)

(ASHRAE, 1990) American National Standards Institute/American Society of Heating, Refrigerating and Air-Conditioning Engineers. ANSI/ASHRAE 62-1989, "Ventilation for Acceptable Indoor Air Quality." Atlanta, Georgia, 1990.

(ANSI, 1992) American National Standards Institute. ANSI-Z88.2-92, "American National Standard for Respiratory Protection." New York, 1992.

(Bennedict, 1983) Bennedict, Pigford, and Levi. *Nuclear Chemical Engineering*. McGraw Hill: New York, 1983.

(Biddle, 1968) Biddle, P. and J.H. Miles. "Rate of Reaction of Nitrous Acid with Hydrazine and with Sulphamic Acid: Its Application to Nitrous Acid Control in Two-Phase Industrial Systems," Journal of Inorganic and Nuclear Chemistry. Vol. 30, pp. 1291–1297. 1968.

(Casarett, 2001) Casarett and Douyll. *TOXICOLOGY: The Basic Science of Poisons*, Second Edition. McGraw Hill: New York, 2001.

(CFR, 2004) Code of Federal Regulations, Title 40, *Protection of Environment*, Part 68, "Chemical Accident Prevention Provisions," Appendix A.

(CGA, 1992a) Compressed Gas Association, Inc. G-7.1, "Commodity Specification for Air. 1992." Arlington, Virginia, 1992.

(CGA, 1992b) Compressed Gas Association, Inc. P-14, "Accident Prevention in Oxygen Rich and Oxygen-Deficient Atmospheres." Arlington, Virginia, 1992.

(CGA, 2001) Compressed Gas Association, Inc. SB-2-2001, "Oxygen-Deficient Atmospheres." Chantilly, Virginia, 2001.

(DCS, 2001) Hastings, P., Duke Cogema Stone & Webster. Letter to U.S. Nuclear Regulatory Commission, RE: Response to Request for Additional Information. August 31, 2001.

(DCS, 2002a) Hastings, P., Duke Cogema Stone & Webster. Letter to U.S. Nuclear Regulatory Commission, RE: Clarification of Responses to NRC Request for Additional Information. January 7, 2002.

(DCS, 2002b) Hastings, P., Duke Cogema Stone & Webster. Letter to U.S. Nuclear Regulatory Commission RE: Clarification of Responses to NRC Request for Additional Information, DCS-

NRC-000083. February 11, 2002.

(DCS, 2002c) Hastings, P., Duke Cogema Stone & Webster. Letter to U.S. Nuclear Regulatory Commission, RE: Clarification of Responses to NRC Request for Additional Information. March 8, 2002.

(DCS, 2002d) Hastings, P., Duke Cogema Stone & Webster. Letter to U.S. Nuclear Regulatory Commission, RE: Clarification of Responses to NRC Request for Additional Information. April 23, 2002.

(DCS, 2002e) Hastings, P., Duke Cogema Stone & Webster. Letter to U.S. Nuclear Regulatory Commission, "Response to DSER Open Item CS-2." July 28, 2002.

(DCS, 2002f) Hastings, P., Duke Cogema Stone & Webster. Letter to U.S. Nuclear Regulatory Commission, RE: Requests for Additional Information, Clarifications, and Open Item Mapping into the Construction Authorization Request Revision. November 22, 2002.

(DCS, 2003a) Hastings, P., Duke Cogema Stone & Webster. Letter to U.S. Nuclear Regulatory Commission, RE: Responses to Financial Qualification, Fire Safety, Chemical Safety, Aqueous Processing, Material Processing and Ventilation Open Items/Additional NRC Questions on Construction Authorization Request (CAR) Revision. February 18, 2003.

(DCS, 2003b) Hastings, P., Duke Cogema Stone & Webster. Letter to U.S. Nuclear Regulatory Commission, RE: Response to DSER Open Items AP-3 and CS-3. May 23, 2003.

(DCS, 2003c) Hastings, P., Duke Cogema Stone & Webster. Letter to U.S. Nuclear Regulatory Commission, RE: Response to DSER Open Items AP-10 and CS-2. May 30, 2003.

(DCS, 2003d) Hastings, P., Duke Cogema Stone & Webster. Letter to U.S. Nuclear Regulatory Commission, RE: Meeting Minutes from the 02–04 June 2003 Public Meeting on Chemical Safety Open Items. June 5, 2003.

(DCS, 2003e) Hastings, P., Duke Cogema Stone & Webster. Letter to U.S. Nuclear Regulatory Commission, RE: Response to DSER Open Item CS-2. July 28, 2003.

(DCS, 2003f) Hastings, P., Duke Cogema Stone & Webster. Letter to U.S. Nuclear Regulatory Commission, RE: Response to Request for Additional Information—Chemical Safety Open Items CS-09, AP-02, AP-08, and AP-09. September 29, 2003.

(DCS, 2003g) Hastings, P., Duke Cogema Stone & Webster. Letter to U.S. Nuclear Regulatory Commission, RE: Response to Request for Additional Information—Chemical Safety Open Items CS-01 and CS-02. October 6, 2003.

(DNFSB, 2003) Conway, J., Defense Nuclear Facilities Safety Board. Letter to S. Abraham, U.S. Department of Energy, RE: DNFSB/TECH-33, Control of Red Oil Explosions in Defense Nuclear Facilities. November 13, 2003.

(DOE, 1993) U.S. Department of Energy. DOE-DP-0120, "Trip Report: Moscow and Tomsk,

Russia, June 19–29, 1993." Washington, DC, September 1993.

(DOE, 1997a) U.S. Department of Energy. DOE/RL-97-63, "Summary Report—Accident Investigation Board Report on the May 14, 1997 Chemical Explosion at the Plutonium Reclamation Facility, Hanford, Washington." Washington, DC, July 26, 1997.

(DOE, 1997b) U.S. Department of Energy. DOE/RL-97-59, "Final Report—Accident Investigation Board Report on the May 14, 1997 Chemical Explosion at the Plutonium Reclamation Facility, Hanford, Washington." Washington, DC, July 26, 1997.

(DOE, 1998a) U.S. Department of Energy. DOE/EH-0555, "Technical Report on Hydroxyloamine Nitrates." Washington, DC, February 1998.

(DOE, 1998b) U.S. Department of Energy. DOE Std 1128-98, "Guide of Good Practices for Occupational Radiological Protection in Plutonium Facilities." Washington, DC, June 1, 1998.

(DOE, 2004) U.S. Department of Energy. "SCAPA." http://www.orau.gov/emi/scapa/ (November 15, 2004)

(EPA, 1999) U.S. Environmental Protection Agency. "ALOHA: Areal Locations of Hazardous Atmospheres, User's Manual." Washington, DC, January 1999.

(EPA, 2004) U.S. Environmental Protection Agency. "Acute Exposure Guideline Levels." July 7, 2004. http://www.epa.gov/oppt/aegl/chemlist.htm (November 15, 2004)

(Genium, 1999) *Genium's Handbook of Safety, Health, and Environmental Data for Common Hazardous Substances*. McGraw Hill: New York, 1999.

(Hou, 1996) Hou, Y., E.K. Barefield, D.W. Tedder, and S.I. Abdel-Khalik. "Thermal Decomposition of Nitrated Tributyl Nitrate," *Nuclear Technology*. Vol. 113, pp. 304–315. March 31, 1996.

(IAEA, 1998) International Atomic Energy Agency. "The Radiological Accident in the Reprocessing Plant at Tomsk." Vienna, Austria, 1998.

(Jones, 2003) Jones, R.P. "A Potential Inadequacy Related to Operator Response Time Potentially Inadequate t Effectively Respond to an Analyzed Condition." U.S. Department of Energy, Occurrence Report Number SR-WSRC-HCAN-2003-0006. March 21, 2003.

(Lange, 1999) *Lange's Handbook of Chemistry*, 15th Edition. McGraw Hill: New York, 1999.

(Lewis, 1996) Lewis, R.J., Sr. *Sax's Dangerous Properties of Industrial Materials,* 9th Edition. Van Nostrand Reinhold: New York, 1996.

(Malachowski, 1999) Malachowski, M.J. *Health Effects of Toxic Substances*, 2nd Edition. Government Institutes, a Division of ABS Group, Inc. Rockville, Maryland, 1999.

(NFPA, 1997) National Fire Protection Association. NFPA Standard 69, "Standard on Explosion

Prevention Systems." 1997.

(NFPA, 1998) National Fire Protection Association. NFPA Standard 801, "Standard for Fire Protection for Facilities Handling Radioactive Materials." 1998.

(NFPA, 2000a) National Fire Protection Association. NFPA Standard 50A, "Gaseous Hydrogen Systems at Consumer Sites." 2000.

(NFPA, 2000b) National Fire Protection Association. NFPA Standard 86C, "Industrial Furnaces Using a Special Processing Atmosphere." 2000.

(NFS, 2002) Moore, B.M., Nuclear Fuel Services, Inc. Letter to G. Janosko, U.S. Nuclear Regulatory Commission, RE: ISA Summary for the BLEU Preparation Facility Processes. October 14, 2002.

(NIOSH, 1994) National Institute for Occupational Safety and Health. "NIOSH Pocket Guide to Chemical Hazards." June 1994.

(NIOSH, 2004) National Institute for Occupational Safety and Health. http://www.cdc.gov/niosh/homepage.html (November 15, 2004)

(Nomura, 1994) Nomura, Y., R. Leicht, and P. Ashton. "Fault Tree Analysis of System Anomaly Leading to Red Oil Explosion in Plutonium Evaporator," Journal of Nuclear Science and Technology. Vol. 31, No. 8, pp. 850–860. August 1994.

(NRC, 1992) U.S. Nuclear Regulatory Commission. Information Notice 92-14, "Uranium Oxide Fires at Fuel Cycle Facilities." Washington, DC, February 21, 1992.

(NRC, 1998a) U.S. Nuclear Regulatory Commission. NUREG/CR-6410, "Nuclear Fuel Cycle Facility Accident Analysis Handbook." Washington, DC, March 1998.

(NRC, 1998b) U.S. Nuclear Regulatory Commission. "Memorandum of Understanding between the Nuclear Regulatory Commission and the Occupational Safety and Health Administration: Worker Protection at NRC-licensed Facilities." Federal Register, Vol. 53, No. 210, pp. 43950–43951. October 31, 1998.

(NRC, 1999) Pierson, R., U.S. Nuclear Regulatory Commission. Memorandum to D. Clark Gibbs, U.S. Department of Energy, RE: Estimates of Hydrogen Generation From Wastes at the Proposed, TWRS-P Facility. April 21, 1999.

(NRC, 2000) U.S. Nuclear Regulatory Commission. NUREG-1718, "Standard Review Plan for the Review of an Application for a Mixed Oxide (MOX) Fuel Fabrication Facility." Washington, DC, August 2000.

(NRC, 2001a) U.S. Nuclear Regulatory Commission. Regulatory Guide 1.78, "Assumptions in Evaluating the Habitability of a Nuclear Power Plant Control Room During Postulated Hazardous Release." Washington, DC, 2001.

(NRC, 2001b) Brown, D., U.S. Nuclear Regulatory Commission. Memorandum to E.J. Leeds, U.S. Nuclear Regulatory Commission, RE: 2/21–22/02 In-Office Review Summary of DCS Construction Authorization Request Supporting Documents for the MFFF. March 11, 2001.

(NRC, 2001c) Persinko, A., U.S. Nuclear Regulatory Commission. Memorandum to E.J. Leeds, U.S. Nuclear Regulatory Commission, RE: 11/27–29/01 In-Office Review Summary of DCS Construction Authorization Request Supporting Documents for the MFFF. December 18, 2001.

(NRC, 2002a) Persinko, A., U.S. Nuclear Regulatory Commission. Memorandum to E.J. Leeds, U.S. Nuclear Regulatory Commission, RE: 2/22/02 Phone Call Summary: DCS Construction Authorization Request Supporting Documents for the MFFF. February 28, 2002.

(NRC, 2002b) U.S. Nuclear Regulatory Commission. NUREG-1520, "Standard Review Plan for the Review of a License Application for a Fuel Cycle Facility." Washington, DC, March 2002.

(NRC, 2002c) U.S. Nuclear Regulatory Commission. Draft Regulatory Guide 1114, "Control Room Habitability at Nuclear Power Reactors." Washington, DC, March 2002.

(NRC, 2002d) Persinko, A., U.S. Nuclear Regulatory Commission. Memorandum to M.N. Leach, U.S. Nuclear Regulatory Commission, RE: August 28–30, 2002, In-Office Review Summary of DCS Construction Authorization Request Supporting Documents for the MFFF. December 31, 2002.

(NRC, 2003a) Persinko, A., U.S. Nuclear Regulatory Commission. Memorandum to M.N. Leach, U.S. Nuclear Regulatory Commission, RE: December 10–12 Meeting Summary: Meeting with DCS to Discuss Mixed Oxide Fuel Fabrication Facility Revised Construction Authorization Report. January 31, 2003.

(NRC, 2003b) Persinko, A., U.S. Nuclear Regulatory Commission. Memorandum to M.N. Leach, U.S. Nuclear Regulatory Commission, RE: January 15–16 Meeting Summary: Meeting with DCS to Discuss Mixed Oxide Fuel Fabrication Facility Revised Construction Authorization Report. March 5, 2003.

(NRC, 2003c) Persinko, A., U.S. Nuclear Regulatory Commission. Memorandum to M.N. Leach, U.S. Nuclear Regulatory Commission, RE: February 6–7 Meeting Summary: Meeting with DCS to Discuss Mixed Oxide Fuel Fabrication Facility Revised Construction Authorization Report. March 5, 2003.

(NRC, 2003d) Pierson, R., U.S. Nuclear Regulatory Commission. Letter to R. Idhe, Duke Cogema Stone & Webster, RE: Draft Safety Evaluation Report on Construction of Proposed Mixed Oxide Fuel Fabrication Facility, Revision 1. April 30, 2003.

(NRC, 2003e) Persinko, A., U.S. Nuclear Regulatory Commission. Memorandum to K. Gibson, U.S. Nuclear Regulatory Commission, RE: June 2–4, 2003, Meeting Summary: Meeting with Duke Cogema Stone & Webster to Discuss Chemical Safety Related to Mixed Oxide Fuel Fabrication Facility. July 18, 2003.

(NRC, 2003f) Persinko, A., U.S. Nuclear Regulatory Commission. Memorandum to K. Gibson,

U.S. Nuclear Regulatory Commission, RE: July 29–August 1, 2003, Meeting Summary: Meeting with Duke Cogema Stone & Webster to Open Items Related to Mixed Oxide Fuel Fabrication Facility. August 29, 2003.

(NRC, 2003g) U.S. Nuclear Regulatory Commission. Advisory Committee on Reactor Safeguards Full Committee Meeting Transcripts. http://www.nrc.gov/reading-rm/doc-collections/acrs/tr/fullcommittee/2003/ November 6, 2003

(NRC, 2004a) Giitter, J., et al, U.S. Nuclear Regulatory Commission. Memorandum to H. Peterson, U.S. Nuclear Regulatory Commission, RE: Determination on Position for Closure of Chemical Safety Open Item CS-1 Pertaining to Red Oil Explosions. April 12, 2004.

(NRC, 2004b) Giitter, J., et al, U.S. Nuclear Regulatory Commission. Memorandum to H. Peterson, U.S. Nuclear Regulatory Commission, RE: Safety Evaluation and Staff Positions on the Closure of Remaining Chemical Safety Open Items CS-5b Pertaining to the MOX Fuel Fabrication Facility Applicant's Use of Temporary Emergency Exposure Limits. April 16, 2004.

(NRC, 2004c) Giitter, J., et al, U.S. Nuclear Regulatory Commission. Memorandum to S. Magruder, U.S. Nuclear Regulatory Commission, RE: Determination on Position for Closure of Chemical Safety Item CS-2 Pertaining to HAN Explosions. August 23, 2004.

(NSP, 1992) National Ski Patrol. "Components and Hazardous Atmospheres (Prehospital Emergency Care & Crisis Intervention)," 4th Edition, p. 349. Brady Morton Series. 1992. http://www.patrol.org/resource/petri/airgas.html

(OECD, 1994) Organization for Economic Cooperation and Development/Nuclear Energy Agency. Proceedings at a Topical Meeting on Safety of the Nuclear Fuel, "Tomsk-7 Nuclear Event: Causes, Consequences, and Lessons Learned." September 20–21, 1994.

(ORNL, 1988) Oak Ridge National Laboratory. ORNL/TM-10798, "The Red Oil Problem and Its Impact on Purex Safety." Oak Ridge, Tennessee, June 16, 1988.

(Perry, 1997) Perry's Chemical Engineers' Handbook, 7th Edition. McGraw Hill: New York, 1997.

(Sitting, 1994) Sitting, Marshall. World-wide Limits for Toxic and Hazardous Chemicals in Air, Water, and Soil. Noyes Publications: New Jersey, 1994.

(SRS, 1976) Savannah River Site. DP-1418, "Behavior of Tributyl Phosphate in A-Line Processes." August 1976.

(Stumpf, 2000) Stumpf, D.K. TITAN II: A History of a Cold War Missile Program. The University of Arkansas Press: Fayetteville, 2000.

(USACHPPM, 1999) U.S. Army Center for Health Promotion and Preventive Medicine. USACHPPM TG 230A, "Short-Term Chemical Exposure Guideline for Deployed Military Personnel." May 1999.

(WSMS, 2000) Westinghouse Safety Management Solutions. WSMS-SAE-00-0266, "ERPGs and TEELs for Chemicals of Concern: Rev. 17." Aiken, South Carolina, August 20, 2000.

(WSRC, 1994a) Westinghouse Savannah River Company. WSRC-TR-94-059, "Safe Conditions for Contacting Nitric Acid or Nitrates with Tri-n-Butyl Phosphate (TBP)." Aiken, South Carolina, January 1994.

(WSRC, 1994b) Westinghouse Savannah River Company. WSRC-TR-94-0372, "Safe Handling of TBP and Nitrates in the Nuclear Process Industry." Aiken, South Carolina, July 1994.

(WSRC, 1994c) Westinghouse Savannah River Company. WSRC-MS-94-0649, "Safe Venting of "Red Oil" Runaway Reactions." Aiken, South Carolina, December 1994.

(WSRC, 1995) Westinghouse Savannah River Company. WSRC-RP-95-259, "Thermal Decomposition of Nitrated Tributyl Phosphate." Aiken, South Carolina, January 1995.

(WSRC, 1998) Laurinat, J.E., N.M. Hassan, T.S. Rudisill, N.M. Askew, Westinghouse Savannah River Company. WSRC-TR-98-00171, "Analysis of Steam Heating of a Two-Layer TBP/N-Paraffin/Nitric Acid Mixtures." Aiken, South Carolina, July 22, 1998.

(WSRC, 1999) Westinghouse Savannah River Company. WSRC-MS-98-00899, "Inventory or Consequence-Based Evaluation of Hazardous Chemicals: Recommendations for DOE Facility Safety Analysis." Aiken, South Carolina, June 9, 1999.

(WSRC, 2001a) Westinghouse Savannah River Company. WSRC-SA-2001-00004, Revision 3, "F Canyon SAR." Aiken, South Carolina, 2001.

(WSRC, 2001b) Hallman, D.F., Westinghouse Savannah River Company. WSRC-TR-2000-00443, Revision 0, "Hydrozoic Acid Controls and Risks when Processing Plutonium Solutions in HB-Line Phase II." Aiken, South Carolina, January 31, 2001.

(WSRC, 2001c) Westinghouse Savannah River Company. WSRC-SA-2001-00008, Revision 4, H-Canyon Safety Analysis Report, Section 8.3.2.2.1, "Explosion–TBP-Nitric Acid ("Red Oil") Runaway Reactions." Aiken, South Carolina, 2001.

(WSRC, 2001d) Rudisill, T.S. and W.J. Crooks III, Westinghouse Savannah River Company. WSRC-MS-2001-00214, "Initiation Temperature for Runaway Tri-n-Butyl Phosphate/Nitric Acid Reaction." Aiken, South Carolina, September 14, 2001.

This page intentionally left blank

9. RADIATION SAFETY

9.1 Conduct of Review

This chapter of the final safety evaluation report (FSER) contains the U.S. Nuclear Regulatory Commission (NRC) staff's review of radiation safety features described by Duke Cogema Stone & Webster (DCS or the applicant) in Chapter 9 of the revised construction authorization request (CAR) (DCS, 2002e). The objective of this review is to determine whether the applicant's principle structures, systems, and components (PSSCs) and their design bases provide reasonable assurance of worker protection against natural phenomena and the consequences of potential accidents.

The radiological safety of workers is the subject of this FSER chapter. Further, for the purposes of this FSER, the staff has adopted the applicant's usage for names of individuals for whom accident risks must be limited. That is, facility workers are workers in the restricted area inside a room of the mixed oxide (MOX) fuel fabrication facility (MFFF) near a potential accident release point. Site workers are workers considered to be outside the MFFF located 100 meters from the ventilation exhaust stack. An individual outside the controlled area (IOC) is an individual at or beyond the proposed MFFF controlled area boundary. A fourth receptor protected under the provisions of Title 10, Section 70.61, "Performance Requirements," of the *Code of Federal Regulations* (10 CFR 70.61) is the environment, which is the area outside the proposed restricted area. The staff's evaluation of radiation safety protection of the IOC and the environment is provided in Chapter 10 of this FSER.

The staff evaluated the information provided by the applicant for radiation safety by reviewing Chapter 9 of the revised CAR, other sections of the revised CAR, supplementary information provided by the applicant, and relevant documents available at the applicant's offices but not submitted by the applicant. The review of the radiation safety design bases and strategies was closely coordinated with the review of accident sequences described in the safety assessment of the design bases (see Chapter 5 of this FSER).

The staff reviewed how the radiation safety information in the revised CAR addresses or relates to the following regulations:

*   10 CFR 70.23(b) states that, as a prerequisite to construction approval, the design bases of the PSSCs and the Quality Assurance Program must be found to provide reasonable assurance of protection against natural phenomena and the consequences of potential accidents.

*   Pursuant to 10 CFR 70.61(b)(1), the risk of credible high-consequence events to workers must be limited. For workers, a high-consequence event is an internally or externally initiated event that results in an acute 1 sievert (Sv) (100 rem) total effective dose equivalent (TEDE). Controls must be used which either make the occurrence of such events highly unlikely or make their consequences less severe than an acute 1 Sv (100 rem) TEDE.

*   Pursuant to 10 CFR 70.61(c)(1), the risk of credible intermediate-consequence events to workers must be limited. For workers, an intermediate-consequence event is an

internally or externally initiated event that is not a high-consequence event that results in an acute 0.25 Sv (25 rem) TEDE. Controls must be used which either make the occurrence of such events unlikely or make their consequences less severe than an acute 0.25 Sv (25 rem) TEDE.

The review of this revised CAR focused on the design basis of radiation safety systems, components, and other related information. For radiation safety systems, the staff reviewed information provided by the applicant for the safety function, system description, and safety analysis. The staff used Chapter 9.0 in NUREG-1718, "Standard Review Plan for the Review of an Application for a Mixed Oxide (MOX) Fuel Fabrication Facility," issued 2000, as guidance in performing the review (NRC, 2000).

9.1.1 Radiation Safety Design Features

The objective of the staff's review under this section was to determine whether the applicant's design for construction and operation of the proposed facility is adequate to protect the radiological health and safety of workers (including site workers) from accidents and credible natural phenomena hazards (NPHs). Specifically, the staff's review of the applicant's proposed radiation safety design features of the facility focused on the PSSCs required to reduce the risk from intermediate-consequence events (those producing an acute 0.25 Sv (25 rem) TEDE to a worker) and high-consequence events (those producing an acute 1 Sv (100 rem) TEDE to a worker).

9.1.1.1 ALARA Design Considerations

The applicant has committed to designing the MFFF in accordance with as low as reasonably achievable (ALARA) considerations. Below, the staff discusses the applicant's commitment to ALARA design considerations that might affect the design of radiation safety PSSCs.

The staff evaluated organizational relationships and responsibilities with respect to the performance of reviews of radiation safety design features, application of ALARA into design-stage collective dose estimates, descriptions and elements of the design review process for radiation safety design features, and the applicant's plan to use experience from past designs and from operating plants to develop improved radiation safety design features.

The applicant identified the organizational relationships and responsibilities for performing radiological design and design reviews in Section 9.1.1.1 of the CAR. These relationships and responsibilities are consistent with those described in the previously approved MOX Project Quality Assurance Plan, Revision 3 (DCS, 2002c; NRC, 2003a), and are thus acceptable to the staff.

The applicant's approach to design-stage collective dose estimates includes performing dose assessments in accordance with NRC Regulatory Guides (RGs) 8.19, "Occupational Radiation Dose Assessment in Light Water Reactor Power Plants Design State Man-Rem Estimates," issued 1979, and 8.34, "Monitoring Criteria and Methods to Calculate Occupational Radiation Doses," issued 1992 (NRC, 1979; NRC, 1992), using the ABAQUES method and radiation zoning criteria for minimizing the direct (external) component of collective radiation dose, and using reviews of experience at the MELOX facility and considerations of facility design to evaluate and minimize collective internal doses.

The staff evaluated the applicant's approach to design-stage collective dose estimates. The staff finds that this approach will reduce the time spent in radiation areas, improve accessibility to components requiring periodic maintenance or inservice inspection, reduce the retention of radioactivity throughout plant systems, reduce contamination, facilitate decommissioning and minimize secondary waste production, and incorporate experience from operating plants and past designs. This approach is consistent with the acceptance criteria in Section 9.1.4.1.3 of NUREG-1718 (NRC, 2000) and is, therefore, acceptable.

With respect to external doses, the applicant did not identify sources of direct radiation that pose an unacceptable accident or natural phenomena risk. Therefore, the ABAQUES method and radiation zoning criteria are not design bases for PSSCs identified in the safety assessment. The applicant's ALARA design philosophy to prevent or mitigate the hazard of radioactive material intake will rely primarily on experience at the MELOX facility in Marcoule, France. General measures and design philosophies for reducing the risk posed by a loss of material confinement that are described in the revised CAR include remote operations and automation, use of gloveboxes with negative pressure relative to occupied spaces, preventive maintenance, and airborne radioactivity monitoring. The NRC staff has reviewed and evaluated the sources of radioactive material that the applicant proposes to use, namely weapons-grade plutonium and depleted uranium. The staff agrees that direct radiation from these sources will not pose an unacceptable accident or natural phenomena risk.

9.1.1.2 Facility Design Features

The information required to support the staff's review of the applicant's safety assessment includes facility and process drawings and descriptions with radiation safety design features that are PSSCs relied upon in the safety assessment. In the revised CAR, the applicant provided facility and process drawings and descriptions, including ventilation system drawings, for the locations and access control points for restricted areas, the controlled area, and passive and active material confinement boundaries. This information facilitated a clear understanding of the intended function of radiation safety PSSCs, commensurate with the expected level of design detail available at the construction authorization stage. The level of information provided by the applicant is, therefore, acceptable to the staff.

9.1.1.3 Source Identification

The applicant proposed that the plutonium oxide that would be shipped to an operating facility would contain mostly plutonium isotopes (e.g., ^{238}Pu, ^{239}Pu, ^{240}Pu) and small quantities of other impurities. The impurities will include elements and radionuclides such as gallium, isotopes of uranium and americium, and radioactive decay products of uranium, plutonium, and americium. The mass distribution of the radionuclides used by the applicant in its safety assessment is provided in FSER Table 9.1-1 (DCS, 2002e, Table 9-3).

The staff confirmed by calculation that the radioactive decay progeny of the plutonium that the applicant proposed to use would not make a significant contribution to the dose to workers resulting from accidents at the proposed facility. The staff calculated that the radionuclides listed in FSER Table 9.1-1 would account for greater than 99.99 percent of the dose resulting from inhalation. Therefore, for the purposes of the safety assessment, the staff considers the applicant's proposed list of radionuclides and mass fractions to be complete.

The aqueous polishing (AP) process would remove the impurities from the plutonium oxide as the first step in the fuel fabrication process. The purified form of plutonium oxide would be transferred to the powder blending and fuel fabrication process. The impurities would be temporarily stored in tanks in solution form before being transferred to the proposed waste solidification building for further treatment.

The fuel fabrication process would use depleted uranium, which is less than 0.7 percent by mass ^{235}U. A depleted uranium nitrate solution $(DU(NO_3)_2)$ would be used to dilute high-enriched uranium that would be separated from plutonium oxide feedstock in the AP process. All tanks containing depleted uranium nitrate would be in the AP process.

Depleted uranium is also the major component of fuel that would be manufactured at the facility. Depleted uranium dioxide (DUO_2) powder would be used directly in the powder blend and pellet manufacturing process. The powder would comprise approximately 94 percent of the mass of each fuel pellet.

Table 9.1-1 Mass Fractions of Isotopes in Aged Weapons-Grade Plutonium

Isotope	0-yr-old WGPu, gm/gm WGPu	40-yr-old WGPu, gm/gm WGPu	70-yr-old WGPu, gm/gm WGPu
^{236}Pu	1.00×10^{-9}	5.98×10^{-14}	4.06×10^{-17}
^{238}Pu	6.86×10^{-4}	5.00×10^{-4}	3.95×10^{-4}
^{239}Pu	9.21×10^{-1}	9.20×10^{-1}	9.19×10^{-1}
^{240}Pu	6.18×10^{-2}	6.15×10^{-2}	6.13×10^{-2}
^{241}Pu	1.00×10^{-2}	1.00×10^{-2} (1)	1.00×10^{-2} (1)
^{242}Pu	1.00×10^{-3}	1.00×10^{-3}	1.00×10^{-3}
^{235}U	$0.00 \times 10^{+00}$	1.83×10^{-2} (2)	1.90×10^{-2} (2)
^{238}U	$0.00 \times 10^{+00}$	1.26×10^{-3} (2)	1.26×10^{-3} (2)
^{241}Am	$0.00 \times 10^{+00}$	7.00×10^{-3}	7.92×10^{-3} (1)

1. The applicant has made a conservative assumption that ^{241}Pu, which has a 14.4-year half-life, does not decay.

2. The mass fraction of uranium isotopes represents ingrowth from the radioactive decay of ^{239}Pu and ^{242}Pu and isotopes that would be present in MFFF feedstock for other reasons.

Given a facility design production rate of 70 metric tons heavy metal of fuel per year, this corresponds to an annual demand for 65.8 tons of depleted uranium dioxide feedstock.

In summary, the applicant identified six categories of radioactive sources at the facility that would be present in quantities that could pose unacceptable risks to workers from accidents and NPHs. These are the impure plutonium oxide ($PuO_2(U)$), the purified plutonium oxide ($PuO_2(P)$), the AP high alpha activity waste stream containing mostly americium and other radioactive decay products, depleted uranium oxide, the master blend, and the final blend. The purified plutonium oxide source, master blend, and final blend categories represent three distinct concentrations of plutonium oxide (in decreasing concentration of the plutonium oxide)—(1) a 100-percent plutonium oxide powder, (2) a 20-percent plutonium oxide-depleted uranium oxide (MOX) mixture, and (3) a 6-percent MOX final blend. The isotopic mass distribution of radionuclides in each source category is provided in FSER Table 9.1-2.

The applicant provided an estimate of the amount of licensed material from each source category that could be present in each processing unit throughout an operating facility. For units in which multiple sources may be present, only the source likely to result in the bounding accident consequences was identified.

Table 9.1-2 Mass Fractions of Isotopes in Each Source Category for the Safety Assessment

Radio-nuclide	$PuO_2(U)$ (1)	$PuO_2(P)$ (2)	High Alpha Activity Waste (Liquid)	Depleted Uranium (3)	20% MOX	6% MOX
^{236}Pu	4.07×10^{-17}	4.10×10^{-17}	-----	-----	8.20×10^{-18}	2.46×10^{-18}
^{238}Pu	3.95×10^{-4}	3.98×10^{-4}	-----	-----	7.96×10^{-5}	2.39×10^{-5}
^{239}Pu	9.20×10^{-1}	9.27×10^{-1}	-----	-----	1.85×10^{-1}	5.56×10^{-2}
^{240}Pu	6.14×10^{-2}	6.18×10^{-2}	-----	-----	1.24×10^{-2}	3.71×10^{-3}
^{241}Pu	1.00×10^{-2}	1.01×10^{-2}	-----	-----	2.02×10^{-3}	6.06×10^{-4}
^{242}Pu	1.00×10^{-3}	1.00×10^{-3}	-----	-----	2.00×10^{-4}	6.00×10^{-5}
^{235}U	(4)	-----		2.500×10^{-3}	2.00×10^{-3}	2.35×10^{-3}
^{238}U	(4)	-----		9.975×10^{-1}	7.98×10^{-1}	9.38×10^{-1}
^{241}Am	7.00×10^{-3}	-----	0.78 g/L	-----	-----	-----

1. $PuO_2(U)$ is unpolished plutonium oxide feed.
2. $PuO_2(P)$ is polished plutonium oxide.
3. Depleted uranium is both a feed material and a waste stream. These values are for feed.
4. Uranium isotopes do not contribute significantly to dose from inhalation events as compared to isotopes of plutonium and americium.

The applicant also provided the photon and neutron energy spectra and intensities per unit mass of licensed material associated with both purified and impure plutonium oxide (FSER Tables 9.1-3 and 9.1-4). The photon energy spectra are based on the decay of the isotopes present in freshly separated weapons-grade plutonium that would occur over a 70-year period. This accounts for the ingrowth of beta- and gamma-emitting radionuclides that would contribute to direct gamma radiation exposures. The applicant also included in these spectra the contribution from (α,n) interactions. The staff independently verified the intensity and energy distribution of these spectra using simple spreadsheet models and the mass distributions of isotopes provided by the applicant. The staff finds that these spectra provide an acceptable basis for the proposed shielding design.

Table 9.1-3 Photon Energy Spectrum for MFFF MOX

Photon Energy (MeV)	Intensity (gamma/sec/gram)	
	20% MOX	5% MOX
0.015	4.79×10^7	1.20×10^7
0.025	7.03×10^1	1.76×10^1
0.038	4.94×10^4	1.23×10^4
0.058	1.27×10^5	3.17×10^4
0.085	1.36×10^4	3.40×10^3
0.125	4.21×10^4	1.06×10^4
0.225	5.93×10^3	1.57×10^3
0.375	2.72×10^4	6.80×10^3
0.575	1.02×10^3	2.55×10^2
0.85	1.39×10^2	3.47×10^1
1.25	1.25×10^1	3.13×10^0
1.75	5.67×10^0	1.42×10^0
2.25	3.24×10^0	8.14×10^{-1}
2.75	1.86×10^0	4.67×10^{-1}
3.5	1.65×10^0	4.13×10^{-1}
5	6.94×10^{-1}	1.74×10^{-1}
7	7.84×10^{-2}	1.97×10^{-2}
9.5	8.91×10^{-3}	2.24×10^{-3}
Total	4.82×10^7	1.20×10^7

Table 9.1-4 Neutron Energy Distribution for MFFF MOX

Neutron Source	Intensity (neutrons/sec/gram)	
	20% MOX	5% MOX
Spontaneous Fission	11.8	2.96
(α, n)	12.0	3.0
Total	23.8	5.96

Energy spectra and source intensities per unit mass of depleted uranium were not provided by the applicant. However, given the low specific activity of this material, the staff does not anticipate that depleted uranium compounds would pose a direct radiation hazard to workers that would challenge the performance requirements in 10 CFR Part 70, "Domestic Licensing of Special Nuclear Material."

Another consideration in the hazard evaluation of radioactive sources is the solubility of inhaled radioactive compounds. In general, inhalation of soluble compounds of plutonium results in a committed dose approximately twice as high as the inhalation of insoluble compounds, namely plutonium oxide. However, soluble plutonium compounds, such as plutonium nitrate, tend to have higher molecular weights, which result in lower radioactivity per gram of molecular weight of the compound. To simplify the safety assessment, the applicant demonstrated a conservative assumption that the licensed nuclear material in all process units of the proposed facility would be insoluble plutonium oxide.

The applicant proposed the following relation:

$$[(\text{specific activity}) \times (DCF_Y)]_{\text{applicant}} \geq [(\text{specific activity}) \times (DCF_W)]_{\text{nitrate, oxalate, et al.}}$$

In this relation, the specific activity refers to the quantity of radioactivity per gram of molecular weight of the compound. DCF_Y and DCF_W are the dose conversion factors (DCFs) for transportability class Y and class W, respectively, from Federal Guidance Report No. 11, "Limiting Values of Radionuclide Intake and Air Concentration and Dose Conversion Factors for Inhalation, Submersion and Ingestion," issued 1988 (EPA, 1988). This relation demonstrates that, for worker radiation dose calculations in the safety assessment, the applicant's assumption that all plutonium compounds are plutonium dioxide is conservative. This is because a mass unit of plutonium in the insoluble applicant form (class Y) results in a higher dose than a mass unit of plutonium in higher molecular weight soluble compounds (class W). The staff evaluated this assumption and find that it holds for the types of licensed nuclear materials proposed for use in the facility. Therefore, for the purposes of the safety assessment, the applicant's use of DCFs for the least soluble compounds is acceptable to the staff.

The depleted uranium compounds also pose an inhalation hazard to workers. However, the limiting hazard in the case of depleted uranium dioxide is chemical toxicity, not radiological dose.

For this reason, the staff's evaluation of the applicant's safety assessment for this and other hazardous chemicals is provided in Chapter 8 of this FSER.

9.1.1.4 Safety Assessment of the Design Bases for PSSCs

The methodology presented by the applicant for its safety assessment of the design bases is consistent with the methodology presented in the NUREG/CR-6410, "Nuclear Fuel Cycle Facility Accident Analysis Handbook," issued 1998 (NRC, 1998). The following sections describe the staff's evaluation of the applicant's methodology for radiological consequence assessment in support of the safety assessment of the design bases.

With the exception of the postulated criticality accident, which involves a significant direct radiation hazard to facility workers, the radiation hazard of most concern in the safety assessment is the release of airborne radioactive material to occupied spaces. Therefore, the staff focused its review on the applicant's estimates of potential accidental releases of radioactive material from engineered confinement systems.

9.1.1.4.1 Facility Worker Consequence Assessment

To reduce the risk to the facility worker, the applicant proposed a methodology in the safety assessment, described in Section 5.4.4.2 of the revised CAR, in which event consequences to facility workers are qualitatively estimated. Since the applicant assumed that the facility worker would be at the location of the release, the risks from events involving plutonium and americium are deemed to be qualitatively unacceptable and PSSCs are deterministically applied. Further, the applicant assumed that unmitigated consequences from the release of uranium are low and PSSCs are not applied.

However, the applicant's safety assessment includes several events in which the proposed PSSC for protection of the facility worker is facility worker action to mitigate the consequences. For example, in the event of a loss of radioactive material confinement involving a fuel assembly drop, waste container drop, or a small loss of confinement at a glovebox, the worker is expected to recognize the event and leave the affected area. For these events, the staff evaluated the applicant's quantitative dose consequence estimates in order to evaluate the adequacy of this safety strategy. The staff's evaluation of this approach is discussed further in FSER Section 9.1.2.3.

9.1.1.4.2 Accident Release Estimates for Site Worker Radiation Safety Features

The applicant used a five-factor approach to estimate accidental releases of radioactivity to the atmosphere that would result in a site worker receiving a dose. The total quantity released, or the release rate over time, is referred to as the source term.

The source terms derived in the applicant's assessment are the products of five parameters. The first parameter is the quantity of radioactive material that is postulated to be involved in the event. The remaining four parameters are reduction factors that are applied to account for different physical phenomena that would limit the amount of licensed nuclear material to which a facility or site worker would be exposed. The following is the general form of the source term formula:

Source Term, kilograms = MAR × DR × ARF × RF × LPF   Eq. 9-1

where MAR is the quantity of material at risk during the event, in kilograms,

DR is the unitless damage ratio of material actually impacted by the event (0<DR<1),

ARF is the atmospheric release fraction, which is the unitless fraction of impacted material (MAR × DR) that can become airborne (0< ARF<1),

RF is the respirable fraction, which is the unitless fraction of airborne material that can be inhaled into the human respiratory system (0<RF<1), and

LPF is the unitless fraction of airborne material that breaches the confinement barrier.

The material at risk was provided by the applicant for more than 200 individual process units throughout the proposed facility. These include the six major types of radioactive sources as described in FSER Section 9.1.1.3. The applicant's hazard analysis considered many different types of events that could cause an adverse human health or environmental effect as a result of accidental exposure to these radioactive sources. These types of accidents are categorized into the major events of fires, explosions, loss of confinement, load drops, and nuclear criticality. The staff evaluation of the applicant's hazard assessment, including event types that were screened from further consideration, is provided in Chapter 5 of this FSER.

The applicant set the damage ratio (DR) equal to 1, with the exception that the DRs assumed for pellets exposed to overpressurization gas flows and pressurized rods that have been breached are 0.01 and 0.001, respectively. If rods are dropped, the DR is assumed by the applicant to be equal to 0.02. The applicant also set the DR equal to 0.1 for a fire in the secured warehouse building. The staff has verified that the applicant's assumptions are consistent with staff's guidance provided in NUREG/CR-6410 (NRC, 1998). Therefore, these assumptions are acceptable to the staff.

The applicant chose values for the atmospheric release fraction (ARF) and respirable fraction (RF) based on values in NUREG/CR-6410 (NRC, 1998). The values chosen by the applicant for these parameters are provided in FSER Table 9.1-5.

The staff reviewed the applicant's bases for choosing ARFs and RFs and found them to be consistent with those in NUREG/CR-6410. The applicant selected a value of 3.0×10^{-5} for the ARF for dropped fuel rods. The basis for this value is a study performed by Sandia National Laboratory (SNL, 1991). These assumptions are acceptable to the staff because they are based either on data previously used and accepted by the staff for safety assessments or on data derived from experiments related to the actual hazard.

Table 9.1-5 ARFs and RFs

Release Form	Release Fraction	Explosive Detonation	Explosive Over-pressurization	Fire/Boil	Drop	Entrainment
Solution	ARF	1.0	5.0×10^{-5}	2×10^{-3}	2.0×10^{-5}	4.0×10^{-7}
	RF	0.01	0.8	1.0	1.0	1.0
	ARF × RF	0.01	4.0×10^{-5}	2×10^{-3}	2.0×10^{-5}	4.0×10^{-7}
Powder	ARF	1.0	5.0×10^{-3}	6.0×10^{-3}	2.0×10^{-3}	4.0×10^{-5}
	RF	0.2	0.3	0.1	0.3	1.0
	ARF × RF	0.2	1.5×10^{-3}	6.0×10^{-4}	6.0×10^{-4}	4.0×10^{-5}
Pellet	ARF	0.01	5.0×10^{-3}	5.0×10^{-4}	1.0	NA
	RF	1.0	0.3	0.5	1.1×10^{-5}	
	ARF × RF	0.01	1.5×10^{-3}	2.5×10^{-4}	1.1×10^{-5}	
Rod	ARF	0.01	3.0×10^{-5}	0.0	3.0×10^{-5}	NA
	RF	1.0	1.0	1.0	1.0	
	ARF × RF	0.01	3.0×10^{-5}	0.0	3.0×10^{-5}	
Filter (unencased)	ARF	2.0×10^{-6}	0.01	1.0×10^{-4}	1.0×10^{-2}	NA
	RF	1.0	1.0	1.0	1.0	
	ARF × RF	2.0×10^{-6}	0.01	1.0×10^{-4}	1.0×10^{-2}	

As used by the applicant, the leak path factor generally accounts for the particulate matter removal efficiency of the high-efficiency particulate air (HEPA) filters used in the ventilation confinement systems. When relied upon to mitigate the effects of an accident, the applicant assumed a 99-percent removal efficiency (i.e., 1-percent leak path factor) per stage. Each HEPA system relied upon for safety includes two banks or stages of HEPA filters in series. Assuming that the effective leak path factor for a system of staged HEPA filters is the product of the individual leak path factors for successive filter stages, the applicant applied a leak path factor of 10^{-4} for systems relied upon in its safety assessment. The combination of efficiencies in this manner is acceptable to the staff because it is consistent with the guidance in Section F.2.1.3 of the staff's accident analysis handbook (NRC, 1998).

When prevention alone, rather than mitigation, was the applicant's preferred safety strategy, the applicant applied a leak path factor equal to 0. The applicant used a leak path factor equal to 1 when the HEPA filters were either unlikely to function as needed or not required to mitigate the event consequences (see FSER Section 9.1.1.4.4).

NRC RGs 3.71, "Nuclear Criticality Safety Standards for Fuels and material Facilities," issued

1972, and 3.35, "Assumptions Used for Evaluating the Potential Radiological Consequences of Accidental Nuclear Criticality in a Plutonium Processing and Fuel Fabrication Plant," issued 1972 (NRC,1972c; NRC,1972b), were used by the applicant to develop source terms for direct radiation and airborne releases resulting from a criticality accident. However, since the NRC has withdrawn these guides, the staff used the current guidance in NUREG/CR-6410 (NRC, 1998) to estimate the downwind consequences to a site worker of a criticality accident. By so doing, the staff independently evaluated the applicant's source terms. The staff finds that the applicant's analysis is consistent with the current guidance and is, therefore, acceptable.

9.1.1.4.3 Dose Assessment for the Site Worker

The applicant's methodology for dose assessment relies on an assumption that the principle human health hazard posed by releases of radioactive material from the proposed MOX facility is inhalation of radioactive material downwind of the facility. Other pathways of exposure would include direct radiation from the passing plume and exposure to ground surfaces contaminated by material depositing on the ground as the plume passes. However, the staff confirmed by calculation that, with the exception of the postulated criticality event, the direct radiation and ground contamination pathways are negligible as compared to the inhalation pathway.

To calculate the 50-year committed effective dose equivalent (CEDE) from inhalation doses from passing plumes, the applicant applied a simple formula involving the source term (Eq. 9.1), the atmospheric dispersion factor (χ/Q), a human receptor's breathing rate (B.R.), and the DCF (from EPA, 1988):

$$CEDE_i \text{ [rem]} = \text{Source Term}_i \text{ [kg]} \times \chi/Q \text{ [s m}^{-3}] \times \text{B.R. [m}^3 \text{ s}^{-1}] \times DCF_i \text{ [rem } \mu Ci^{-1}] \times C_i \text{ [}\mu Ci \text{ kg}^{-1}]$$

where $CEDE_i$ is the committed dose from the ith radionuclide, and
C_i is the specific activity of the ith radionuclide.

Atmospheric dispersion factors were calculated by the applicant using site-specific meteorological data from the U.S. Department of Energy's Savannah River Site (SRS) H-Area meteorological tower collected from 1987 through 1996. The ARCON96 model (NRC, 1999) was used to estimate factors for the site worker located 100 meters from the plant stack. The value calculated by the applicant is 6.1×10^{-4} s m^{-3}.

The use of ARCON96 to estimate near-field atmospheric dispersion in the vicinity of buildings and structures is described in NUREG/CR-6410 (NRC, 1999) and RGs 1.78, "Evaluating the Habitability of a Nuclear Power Plant Control Room During a Postulated Hazardous Chemical Release," issued December 2001, and 1.194, "Atmospheric Relative Concentrations for Control Room Radiological Habitability Assessments at Nuclear Power Plants," issued June 2003 (NRC, 2001b; NRC, 2003b).

The staff verified by independent calculations that the meteorological data used by the applicant in its safety assessment is consistent with data published by the SRS for the H-Area meteorological tower (DOE, 1999). The staff also performed independent calculations for the site worker atmospheric dispersion factor and calculated a value of 6.1×10^{-4} s m^{-3}.

The breathing rate of 3.47×10^{-4} m^3 s^{-1} assumed by the applicant is consistent with guidance provided by the NRC in RG 1.25 (NRC 1972a) and is equivalent to a volume of 10 cubic meters inhaled during an 8-hour workday. This assumption is based on NRC guidance applicable to fuel handling and is, therefore, acceptable to the staff for use in the applicant's safety assessment.

The U.S. Environmental Protection Agency DCFs used by the applicant (EPA, 1988) are based on the recommendations of the International Commission on Radiation Protection. These are the same recommendations that form the basis for NRC radiation protection standards in 10 CFR Part 20. Therefore, these factors are acceptable to the staff.

9.1.1.4.4 Verification of Low-Consequence Events

Unmitigated event consequences result from an accident sequence when mitigative controls either fail or do not exist. Unmitigated event consequences are those consequences calculated by the applicant before determining and taking credit for PSSCs that would reduce the risk of the event. However, in some cases, the unmitigated event consequence is so low that it falls below the intermediate-consequence threshold values for workers specified in 10 CFR 70.61(c)(1). These events, referred to as low-consequence events, require no PSSCs to lower the risk. The applicant identified 22 hazard assessment events as low-consequence events.

The staff performed independent calculations to verify the applicant's assertion that some events would be low-consequence events and would not require PSSCs to further reduce the accident risk. Based on the staff's confirmatory analysis, the staff accepts the applicant's categorization.

9.1.2 Radiation Protection Program

The purpose of this review is to determine whether the applicant's Radiation Protection Program is adequate to protect the radiological health and safety of the workers and to comply with the regulatory requirements of 10 CFR Part 19, "Notices, Instructions and Reports to Workers: Inspection and Investigations"; 10 CFR Part 20, "Standards for Protection Against Radiation"; and 10 CFR Part 70, to the extent that such programmatic elements are relied upon to provide reasonable assurance of protection against natural phenomena and the consequences of potential accidents.

The staff reviewed the revised CAR using the guidance in NUREG-1718, Section 9.2 (NRC,2000).

9.1.2.1 ALARA

The applicant's commitment to an ALARA program at the construction authorization stage includes a management commitment to this policy, an ALARA committee, administrative control levels and dose limits for design, and internal audits and assessments.

The staff has reviewed the applicant's brief description of the ALARA program and its related safety assessment of the design bases, in which the applicant did not identify any PSSCs or management measures within the purview of its ALARA program. The staff finds that, at the construction authorization stage, no such PSSCs or management measures are required. However, the regulations in 10 CFR Part 20 and 10 CFR Part 70 contain specific requirements for an ALARA program that would have to be fully and adequately addressed as part of any later application for a license to possess and use licensed material.

9.1.2.2 Radiation Safety Procedures and Radiation Work Permits

The applicant identified seven major components of its Radiation Safety Procedures and Work Control Program in the CAR. These are work planning, entry and exit control, radiological work controls, posting and labeling, release of materials and equipment, sealed radioactive source accountability and control, and radioactive material receipt.

Of these controls, the entry and exit controls, specifically process cell entry controls, are relied on in the safety assessment of the design bases as a PSSC to protect the facility worker, as shown in FSER Table 9.1-6. The applicant proposes several controlling parameters in the design basis for the process cell entry controls. To prevent access during normal operations, the applicant proposed radiation work permits, signs and postings, and barricades. The reliance on a combination of administrative and engineered controls for this safety function provides adequate assurance that it will be highly unlikely for facility workers to inadvertently enter process cells upon failure of an administrative control. Therefore, this safety strategy is acceptable to the staff.

The applicant proposed to further control radiation dose during any maintenance activities at the proposed facility by committing to meet the requirements of 10 CFR 20.1601 and 10 CFR 20.1602 for high and very high radiation areas, respectively. The applicant's commitment to meet these regulatory requirements is acceptable to the staff.

9.1.2.3 Training

In the CAR, the applicant identified facility worker action and facility worker controls as PSSCs for several internal hazards at the proposed facility. As shown in Table 9.1-6, design basis controlling parameters for facility worker actions would include evacuation of the affected area. Facility worker controls are those actions taken by the facility worker before commencing an activity that could result in an event with unacceptable dose consequences. In any subsequent license application, in accordance with 10 CFR 70.62(d), the applicant would be required to specify the management measures (such as training and procedures) which would adequately ensure that facility worker actions and controls are reliable to serve their intended safety function (i.e., the facility worker understands what to do and when to do it).

The applicant's calculations demonstrate that dose consequences would be adequately mitigated by the facility worker's response, which is to evacuate the affected area within 30 seconds. The applicant's calculation demonstrates an adequate margin of safety in that the dose to the facility worker would be at least 10 times lower than the 0.25 Sv (25 rem) intermediate consequence threshold for workers specified in 10 CFR 70.61(c)(1). Based on the staff's review of the applicant's calculations, the staff finds that self-protective actions, including escape from the affected area, would be sufficient to mitigate the consequences for those events for which this action is credited.

9.1.2.4 Air Sampling

The applicant has committed to air sampling for facility worker exposure monitoring and control. The applicant planned to use measurements of airborne radioactivity concentrations to estimate internal dose in accordance with 10 CFR 20.1204. However, the Airborne Radioactivity Monitoring Program was not identified as a PSSC by the applicant in its safety assessment.

The staff has reviewed the applicant's brief description of the Airborne Radioactivity Monitoring Program and its related safety assessment of the design bases, in which the applicant did not identify any PSSCs or management measures within the purview of this program. The staff finds that, at the construction authorization stage, no such PSSCs or management measures are required. However, the regulations in 10 CFR Part 20 and 10 CFR Part 70 contain specific requirements for such a program that would have to be fully and adequately addressed in any subsequent license application.

9.1.2.5 Contamination Control

The staff has reviewed the applicant's brief description of the Contamination Control Program and its related safety assessment of the design bases, in which the applicant did not identify PSSCs or management measures within the purview of this program. The staff finds that, at the construction authorization stage, no such PSSCs or management measures are required. However, the regulations in 10 CFR Part 20 and 10 CFR Part 70 contain specific requirements for such a program that would have to be fully and adequately addressed in any subsequent license application.

9.1.2.6 External Exposure, Internal Exposure, and Summing Internal and External Exposures

The applicant provided a brief description of its Direct Exposure Control Program, Internal Exposure Control Program, and procedures for summing dose contributions from both direct exposure and the intake of radioactivity in the revised CAR. These programs are not identified as either PSSCs or management measures in its safety assessment.

Though not applicable to safety assessments for potential accidents, the applicant provided estimates of total radiation doses to facility workers during normal operations based on experience during operation of a similar facility, the MELOX plant in Marcoule, France.

The applicant's estimates of external exposure hazards at the proposed MFFF involve an adjustment of the external dose rates at the MELOX plant to account for the different isotopic distribution of plutonium isotopes in the reactor-grade plutonium used at MELOX, as compared to the weapons-grade plutonium proposed for use at the MFFF. This adjustment includes both a 20:1 MELOX:MFFF adjustment of the photon intensity and an 11:1 adjustment in the neutron intensity per unit mass of plutonium. With these adjustments, the external collective dose is estimated to be 12 person-rem per year, including a 10 person-rem contribution from the MOX process (MP) units and 2 person-rem from the AP area.

For internal dose estimates, the applicant surveyed recent records at the MELOX MFFF in Marcoule, France. From 1996 through July 2001, 41 individuals received an internal dose. Thirty workers received less than 10 percent of the allowable limit on intake (ALI) (based on a 0.02 Sv (2 rem) annual limit (ICRP, 1991)), 10 workers received between 10 and 33.3 percent

Table 9.1-6 Radiation Protection PSSCs

PSSC	Safety Function	Controlling Parameters
Process cell entry controls (administrative)	Prevent the entry of personnel into process cells during normal operations.	Radiation work permits Signs and postings Barricades
	Ensure that facility workers do not receive a dose in excess of limits while performing maintenance in the AP process cells.	Radiation work permits Signs and postings Commitment to 10 CFR 20.1602 for very high radiation areas
Facility worker action (administrative)	Ensure that facility workers take proper actions to limit chemical and radiological exposure.	Facility worker response to exit the affected area
Facility worker controls (administrative)	Ensure that facility workers take proper actions before bagout operations to limit radiological exposure. Ensure that facility workers take proper actions during maintenance activities to limit radiological exposure.	Facility worker prejob preparation to prevent and/or limit dose during tasks involving transient primary confinements or maintenance in AP/MP C3 areas

of the ALI, and 1 worker received between 33.3 and 100 percent of the ALI. Using these data, the staff calculated a collective dose of approximately 13 person-rem. Given the 5-year period over which these intakes occurred, the average annual internal dose is approximately 3 person-rem for workers at the MELOX facility. The applicant estimated a value of 4.5 person-rem per year. The MELOX facility is larger than the proposed MFFF and has a higher throughput of material. In addition, MELOX processes reactor-grade plutonium, which delivers about 10 times more dose from inhalation than the weapons-grade plutonium proposed for use in the MFFF. Table 9.1-7 presents a comparison of the inhalation doses per unit mass of plutonium dioxide for reactor-grade and weapons-grade plutonium. Therefore, the staff concurs that this collective dose is a reasonable upper bound of doses expected at the MFFF.

Therefore, the applicant's estimate that the TEDE among facility workers at the proposed MFFF would be less than 20 person-rem is reasonable to the staff. The applicant further assumed that the total workforce will number approximately 400 facility workers. Therefore, the average TEDE per facility worker is less than 0.5 mSv (50 mrem) per facility worker per year, well below the occupational dose limits of 50 mSv (5000 mrem) set forth in 10 CFR 20.1201(a).

Table 9.1-7 Comparison of the Radiotoxicity of Reactor-Grade Plutonium and Weapons-Grade Plutonium; TEDE Per Gram of Plutonium Inhaled

Pu Isotope	DCF (rem/μCi)	Specific Activity of Pu isotope (μCi/g)	Reactor-Grade Isotopic Mass Fraction	Reactor-Grade Dose Fraction (rem/g Pu)	Weapons-Grade Isotopic Mass Fraction	Weapons-Grade Dose Fraction (rem/g Pu)
^{238}Pu	4×10^{-3}	1.71×10^{7}	0.0149	1.0×10^{3}	0.000395	2.7×10^{1}
^{239}Pu	4×10^{-3}	6.13×10^{4}	0.595	1.5×10^{2}	0.92	2.3×10^{2}
^{240}Pu	4×10^{-3}	2.28×10^{5}	0.240	2.2×10^{2}	0.0614	5.6×10^{1}
^{241}Pu	2×10^{-1}	1.03×10^{8}	0.103	2.1×10^{6}	0.01	2.1×10^{5}
^{242}Pu	4×10^{-3}	3.93×10^{3}	0.04	6.3×10^{-1}	0.001	1.6×10^{-2}
TOTAL				**2.1×10^{6}**		**2.1×10^{5}**

9.1.2.7 Respiratory Protection

As discussed above in Section 9.1.2.4, the applicant proposed that facility worker actions be relied upon to mitigate the consequences of some facility hazards (DCS, 2002b, page 16). The applicant did not explicitly identify design basis controlled parameters and values associated with facility worker respiratory protection equipment for this purpose. Rather, the applicant stated that facility workers are assumed to be in the impacted area for no more than 30 seconds and that they do not use their personnel protective equipment. The staff has reviewed the applicant's calculations and agrees that facility worker action, an administrative PSSC that requires facility workers to escape the affected area, provides an adequate level of protection for the facility worker for those events where this PSSC is applied.

9.1.2.8 Instrumentation

The applicant has committed to implement its Instrumentation Calibration and Maintenance Program to ensure that radiation protection instrumentation is calibrated and maintained in accordance with American National Standards Institute N323-1978 (R1993), "Radiation Protection Instrumentation Test and Calibration."

The staff has reviewed the applicant's brief description of the Instrumentation Calibration and Maintenance Program and its related safety assessment of the design bases, in which the applicant did not identify PSSCs or management measures within the purview of this program. The staff finds that, at the construction authorization stage, no such PSSCs or management measures are required. However, the regulations in 10 CFR Part 20 and 10 CFR Part 70 contain specific requirements for such a program that would have to be fully and adequately addressed in any subsequent license application.

9.1.3 Radiation Safety Design Bases

The PSSCs for facility worker radiological protection, which are elements of the applicant's proposed Radiation Protection Program and were identified in the applicant's safety assessment of the design bases, are listed in FSER Table 9.1-6. This table lists the radiation safety PSSCs, functions, and controlled parameters for the design described in the revised CAR. The staff has reviewed whether these functions and parameters provide reasonable assurance that the applicant's design for construction and operation of the facility would be adequate to protect the radiological health and safety of workers and to comply with the regulatory requirements of 10 CFR Part 20 and 10 CFR Part 70 during routine and nonroutine operations, including anticipated events. The staff's evaluation findings are provided in FSER Section 9.2.

9.2 Evaluation Findings

In Chapter 5 of the revised CAR, DCS provided design basis information for radiation protection elements that it identified as PSSCs for the proposed facility. Based on the staff's review of the revised CAR and supporting information provided by the applicant relevant to radiation protection, the staff finds that the design bases of the PSSCs identified by the applicant will provide reasonable assurance of protection against natural phenomena and the consequences of potential accidents, in accordance with 10 CFR 70.23(b).

As discussed above in FSER Section 9.1, pursuant to 10 CFR 70.61(b)(1) and 10 CFR 70.61(c)(1), the risk of credible high-consequence and intermediate-consequence events must be limited. For workers, a high-consequence event is an internally or externally initiated event that results in an acute 1 Sv (100 rem) TEDE, and an intermediate-consequence event is one which results in an acute 0.25 Sv (25 rem) TEDE. For such high-consequence events, controls must be used which either make the occurrence of these events highly unlikely or make their consequences less severe than an acute 1 Sv (100 rem) TEDE. For such intermediate-consequence events, controls must be used which either make the occurrence of these events unlikely or make their consequences less severe than an acute 0.25 Sv (25 rem) TEDE. Based on staff's review of the safety assessment methodology presented in the revised CAR, the staff finds that the design bases of the PSSCs identified by the applicant will reduce facility worker dose consequences to acceptable levels for both high-consequence and intermediate-consequence events. Accordingly, the staff concludes that, for worker doses, the applicant has met the 10 CFR 70.61(b)(1) and 10 CFR 70.61(c)(1) performance requirements.

9.3 References

(DCS, 2002a) Hastings, P.S., Duke Cogema Stone & Webster. Letter to U.S. Nuclear Regulatory Commission, RE: DCS-NRC-000081, "Clarification of Responses to NRC Request for Additional Information." January 7, 2002.

(DCS, 2002b) Hastings, P.S., Duke Cogema Stone & Webster. Letter to U.S. Nuclear Regulatory Commission, RE: DCS-NRC-000083, "Clarification of Responses to NRC Request for Additional Information." February 11, 2002.

(DCS, 2002c) Hastings, P.S., Duke Cogema Stone & Webster. Letter to U.S. Nuclear Regulatory Commission, RE: MOX Project Quality Assurance Program Plan. March 26, 2002.

(DCS, 2002d) Hastings, P.S., Duke Cogema Stone & Webster. Letter to U.S. Nuclear Regulatory Commission, RE: Clarification of Responses to NRC Request for Additional Information. April 23, 2002.

(DCS, 2002e) Ihde, R., Duke Cogema Stone & Webster. U.S. Nuclear Regulatory Commission, RE: MOX Fuel Fabrication Facility Construction Authorization Request. October 31, 2002 (including page changes through February 9, 2005).

(DOE, 2000) U.S. Department of Energy. DOE-STD-3009-2000, "DOE Standard: Stabilization, Packaging, and Storage of Plutonium-Bearing Materials." Washington, DC, 2000.

(EPA, 1988) U.S. Environmental Protection Agency. EPA-520/1-88-020, Federal Guidance Report No. 11, "Limiting Values of Radionuclide Intake and Air Concentration and Dose Conversion Factors for Inhalation, Submersion and Ingestion." Washington, DC, 1988.

(ICRP, 1983) International Commission on Radiation Protection. ICRP Publication 38, "Radionuclide Transformations: Energy and Intensity of Emissions." 1983.

(ICRP, 1991) International Commission on Radiation Protection. ICRP Publication 60, "1990 Recommendations of the International Commission on Radiological Protection." 1991.

(NRC, 1972a) U.S. Nuclear Regulatory Commission. Regulatory Guide 1.25, "Assumptions Used for Evaluating the Potential Radiological Consequences of a Fuel Handling Accident in the Fuel Handling and Storage Facility for Boiling and Pressurized Water Reactors (Safety Guide 25)." Washington, DC, 1972.

(NRC, 1972b) U.S. Nuclear Regulatory Commission. Regulatory Guide 3.35, "Assumptions Used for Evaluating the Potential Radiological Consequences of Accidental Nuclear Criticality in a Plutonium Processing and Fuel Fabrication Plant." Washington, DC, 1972.

(NRC, 1972c) U.S. Nuclear Regulatory Commission. Regulatory Guide 3.71, "Nuclear Criticality Safety Standards for Fuels and Material Facilities." Washington, DC, 1972.

(NRC, 1979) U.S. Nuclear Regulatory Commission. Regulatory Guide 8.19, "Occupational Radiation Dose Assessment in Light Water Reactor Power Plants Design Stage Man-Rem Estimates." Washington, DC, 1979.

(NRC, 1992) U.S. Nuclear Regulatory Commission. Regulatory Guide 8.34, "Monitoring Criteria and Methods to Calculate Occupational Radiation Doses." Washington, DC, 1992.

(NRC, 1998) U.S. Nuclear Regulatory Commission. NUREG/CR-6410, "Nuclear Fuel Cycle Facility Accident Analysis Handbook." Washington, DC, March 1998.

(NRC, 1999) U.S. Nuclear Regulatory Commission. "ARCON96 Code System to Calculate Atmospheric Relative Concentrations in Building Wakes." CCC-664 Radiation Safety Information Computational Center. Washington, DC, 1999.

(NRC, 2000) U.S. Nuclear Regulatory Commission. NUREG-1718, "Standard Review Plan for the Review of an Application for a Mixed Oxide (MOX) Fuel Fabrication Facility." Washington, DC, 2000.

(NRC, 2001a) U.S. Nuclear Regulatory Commission. NUREG-1747, "Overview and Summary of NRC Involvement with DOE in the Tank Waste Remediation System-Privatization (TWRS-P) Program." Washington, DC, 2001.

(NRC, 2001b) U.S. Nuclear Regulatory Commission. Regulatory Guide 1.78, "Evaluating the Habitability of a Nuclear Power Plant Control Room During a Postulated Hazardous Chemical Release." Washington, DC, December 2001.

(NRC, 2003a) Persinko, A., U.S. Nuclear Regulatory Commission. Letter to P. Hastings, Duke COGEMA Stone & Webster, RE: Safety Evaluation Report: Quality Assurance Program for Construction of MOX Fuel Fabrication Facility. January 10, 2003.

(NRC, 2003b) U.S. Nuclear Regulatory Commission. Regulatory Guide 1.194, "Atmospheric Relative Concentrations for Control Room Radiological Habitability Assessments at Nuclear Power Plants." Washington, DC, June 2003.

(SNL, 1991) Sandia National Laboratories. SAND87-7082, "Effects of a Potential Drop of a Shipping Cask, a Waste Container, and a Bare Fuel Assembly During Waste-Handling Operations; Yucca Mountain Site Characterization Project." 1991.

This page intentionally left blank

10. ENVIRONMENTAL PROTECTION

10.1 Conduct of Review

This chapter of the final safety evaluation report (FSER) contains the staff's review of environmental protection measures described by Duke Cogema Stone & Webster (DCS or the applicant) in Chapter 10 of the revised construction authorization request (CAR) (DCS, 2002e). The objective of this review is to determine whether principle structures, systems, and components (PSSCs) and their design bases identified by the applicant provide reasonable assurance of protection of an individual outside the controlled area (IOC) and the environment against natural phenomena and the consequences of potential accidents. The staff evaluated the information provided by the applicant for environmental protection by reviewing Chapter 10 of the revised CAR, other sections of the revised CAR, supplementary information provided by the applicant, and relevant documents available at the applicant's offices but not submitted by the applicant. In some cases, the staff also performed independent calculations. The review of environmental protection design bases and strategies was closely coordinated with the review of the accident sequences described in the safety assessment of the design bases (see Chapter 5 of this FSER).

As previously discussed in FSER Chapter 1, the set of U.S. Nuclear Regulatory Commission (NRC) radiation safety requirements applicable to an individual depends on whether that individual is a worker exposed to radiation as part of his assigned employment duties or is a member of the public. The radiological safety of workers is discussed in FSER Chapter 9, and the radiation protection standards applicable to the public, or IOC, are covered in this FSER chapter.

The staff reviewed how the environmental protection information in the revised CAR addresses or relates to the following regulations:

- Title 10, Section 20.1101(b), of the *Code of Federal Regulations* (10 CFR 20.1101(b)) states that a licensee shall use, to the extent practical, procedures and engineering controls based upon sound radiation protection principles to achieve doses to members of the public that are as low as reasonably achievable (ALARA).

- 10 CFR 20.1301(b) states, "If the licensee permits members of the public to have access to controlled areas, the limits for members of the public continue to apply to those individuals."

- 10 CFR 20.1302(c) states the following:

  Upon approval from the Commission, the licensee may adjust the effluent concentration values in appendix B to part 20, Table 2, for members of the public, to take into account the actual physical and chemical characteristics of the effluents (e.g., aerosol size distribution, solubility, density, radioactive decay equilibrium, chemical form).

- 10 CFR 70.23(b) states that, as a prerequisite to construction approval, the design bases of the PSSCs and the Quality Assurance Program must be found to provide reasonable assurance of protection against natural phenomena and the consequences of potential accidents.

- Pursuant to 10 CFR 70.61(b)(2), the risk of credible high-consequence events to any IOC must be limited. For such individuals, a high-consequence event is an internally or externally initiated event that results in an acute 0.25 sievert (Sv) (25 rem) total effective dose equivalent (TEDE). Controls must be used which either make the occurrence of such events highly unlikely or make their consequences less severe than an acute 0.25 Sv (25 rem) TEDE.

- Pursuant to 10 CFR 70.61(c)(2), the risk of credible intermediate-consequence events to any IOC must be limited. For such individuals, an intermediate-consequence event is an internally or externally initiated event that is not a high-consequence event and that results in an acute 0.05 Sv (5 rem) TEDE. Controls must be used which either make the occurrence of such intermediate-consequence events unlikely or make their consequences less severe than an acute 0.05 Sv (5 rem) TEDE.

- For environmental protection purposes, an intermediate-consequence event is one which produces a 24-hour averaged release of radioactive material outside the restricted area in concentrations exceeding 5000 times the values in Table 2 of Appendix B to 10 CFR Part 20, "Standards for Protection Against Radiation." Pursuant to 10 CFR 70.61(c)(3), controls must be used which either make the occurrence of such intermediate-consequence events unlikely or make their environmental consequences less severe than those described above.

In the revised CAR, the applicant described its commitment to environmental protection in the areas of (1) radiation safety (ALARA) goals for effluent control and waste minimization, (2) design of effluent and environmental monitoring for normal and off-normal operations, and (3) design bases for PSSCs derived from the safety assessment, as necessary to ensure environmental protection. The staff focused its review on the applicant's safety assessment of the design bases for environmental protection PSSCs, which is discussed below in FSER Section 10.1.3. The staff also evaluated preliminary information provided by the applicant on ALARA goals and the Effluent and Environmental Monitoring Program. With regard to the design bases for environmental and IOC protection, this FSER chapter addresses the applicant's consequence methodology and results used to identify PSSCs that are relied upon to meet the IOC and environmental performance requirements of 10 CFR 70.61(b)(2) and 10 CFR 70.61(c)(2–3). The staff used Chapter 10.0 in NUREG-1718, "Standard Review Plan for the Review of an Application for a Mixed Oxide (MOX) Fuel Fabrication Facility," issued August 2000, as guidance in performing the review (NRC, 2000).

10.1.1 Radiation Safety

The staff evaluated the applicant's radiation safety measures for environmental protection based on the applicant's methods to maintain public doses at ALARA levels in accordance with 10 CFR 20.1101 and the applicant's waste minimization practices.

10.1.1.1 ALARA Design Goals for Effluent Control

The applicant defined ALARA design goals for effluent control in revised CAR Section 10.1.1.

The ALARA design goal is for airborne radioactive effluents released from the proposed mixed oxide (MOX) fuel fabrication facility (MFFF). This goal is 20 percent of the effluent concentration limits in column 1 of Table 2 in Appendix B to 10 CFR Part 20, as determined at the restricted area boundary. The human dose corresponding to this goal, assuming an individual were present continuously over 1 year at the restricted area boundary, is 0.1 mSv (10 mrem) TEDE. Therefore, this goal affords an initial level of protection for members of the public in the controlled area of 10percent of the 1 mSv (100 mrem) TEDE limit described in 10 CFR 20.1301(b). This fraction is consistent with staff expectations that an initial goal of 10 to 20 percent of Appendix B values or less can be achieved by almost all materials facility licensees, as stated in Regulatory Guide (RG) 8.37, "ALARA Levels for Effluents from Materials Facilities," issued 1973 (NRC, 1973b); therefore, it is acceptable to the staff.

The applicant has not defined liquid effluent ALARA goals because the proposed facility will not discharge liquid effluent directly to the environment. This is acceptable because the applicant's proposal is to transfer low-level waste containing NRC-licensed material to the U.S. Department of Energy (DOE) at the Savannah River Site (SRS). The DOE would perform additional treatment before discharging this material. Therefore, these discharges would be regulated by DOE and would be subject to DOE ALARA considerations.

10.1.1.2 Air Effluent Controls to Maintain Public Doses at ALARA Levels

The confinement ventilation systems are described in FSER Section 11.4. In summary, there are five major sources of ventilation exhaust that would contribute to air emissions from the facility stack. The following ventilation and air-conditioning systems are PSSCs or have individual components that are PSSCs:

- process cell off-gas treatment system
  — final filters
  — pressure boundary downstream of the final filters

- process cell ventilation system
  — final filters
  — pressure boundary downstream of the final filters
  — tornado dampers

- medium depressurization exhaust system (which maintains the C2 confinement zone)
  — final filters
  — pressure boundary downstream of the final filters
  — tornado dampers

- high depressurization exhaust system (C3 confinement zone)

- very high depressurization exhaust system (C4 confinement zone)

The applicant's proposed design bases for these systems relies on NRC RG 3.12, "General Design Guide for Ventilation Systems of Plutonium Processing and Fuel Fabrication Plants," issued 1973 (NRC, 1973a), and the American Society of Heating, Refrigerating and Air-Conditioning Engineers report, "Design Guide for Department of Energy Nuclear Facilities" (ASHRAE, 1993), for guidance. Additional design bases are described in FSER Section 11.4.

The staff concludes that a commitment to these codes and standards should ensure that engineered effluent controls will be designed and constructed to meet the requirements for capacity, compartmentalization, safe shutdown, corrosion resistance, and efficiency required during normal and likely facility conditions to maintain public doses at ALARA levels. The staff will review the applicant's process controls and procedures that will augment these engineered controls and form the basis for a complete ALARA program during review of any license application that DCS may later submit.

10.1.1.3 Liquid Effluent Controls to Maintain Public Doses at ALARA Levels

The proposed facility would not have liquid effluents that discharge directly to the environment. There are, however, five categories of liquid waste that must be managed. These waste streams would be transferred to SRS before final treatment and disposal. The five waste streams are the high alpha activity waste, stripped uranium waste, solvent wastes, low-level liquid waste (including chlorine wastes), and nonhazardous liquid waste. Estimated volumes of each waste type are provided in FSER Table 10.1-1. Waste minimization practices identified by the applicant are discussed in FSER Section 10.1.1.5. The PSSCs required to safety handle, store, and transfer liquid wastes are discussed in FSER Chapter 5 and FSER Section 11.8.

10.1.1.4 ALARA Review and Reports to Management

The applicant has committed to a program of measuring trends in environmental monitoring and surveillance data against the effluent ALARA goals on a quarterly basis. The goals will be reevaluated and new goals will be established for the following year.

The staff has reviewed the applicant's brief description of the operational ALARA program and its related safety assessment of the design bases, in which the applicant did not identify PSSCs or management measures within the purview of this program. The staff finds that, at the construction authorization stage, no such PSSCs or management measures are required. However, the regulations in 10 CFR Part 20 and 10 CFR Part 70 contain specific requirements for such a program that would have to be fully and adequately addressed in any subsequent license application.

10.1.1.5 Waste Minimization

The applicant provided an overview of its commitment to waste minimization practices in Section 10.1.4 of the revised CAR. The applicant's proposal for incorporating waste minimization practices into the design process focuses on recycling and reuse of materials. During operations, the applicant proposed to rely on waste management procedures to separate and segregate solid and liquid wastes and remove packaging and shipping materials before entry into contaminated areas.

The applicant would use active and passive confinement systems and vacuum systems inside gloveboxes. These systems are designed to allow recycling of materials from the secondary waste streams in the aqueous polishing (AP) process and MOX process (MP) scraps back to the main processes. Specific AP process waste minimization steps include acid recovery and solvent regeneration.

Table 10.1-1 Potential for Intermediate-Consequence Events Involving Liquid Waste Streams

Waste Category	Maximum Volume (liters/ yr) (gallons/ yr)	Radionuclide	Maximum Concentration (µCi/mL)	5000 times Table 2, Appendix B, to 10 CFR Part 20 (µCi/mL)	Potential to Exceed Environmental Performance Requirement?	Disposition (5)
High Alpha Activity Waste	39,000 (10,300)	^{241}Am	1,550 (1)	1×10^{-4}	Yes (6)	WSB
Stripped Uranium	167,000 (44,000)	^{238}U	0.01 (2)	1.5×10^{-3}	Yes (7)	WSB
Solvent LLW	10,600 (2,800)	^{239}Pu	1.2×10^{-4} (3)	1×10^{-4}	Yes (8)	SRS Solvent Recovery
Low-Level Aqueous Waste	1,080,000 (285,000)	alpha	3×10^{-6} (4)	1×10^{-4}	No	WSB
Nonhazardous Liquid Waste	33,300,000 (8,800,000)	*Radionuclide concentrations in nonhazardous liquid waste are far below levels that could cause an intermediate consequence event.*				Approved NPDES Facilities at SRS

(1)   Based on 3.1×10^{15} Bq (84,000 Ci) of ^{241}Am per year (DCS, 2004, Table 3-3) and normal volume
(2)   Based on 5,000 kg (11,000 lb) of uranium per year (DCS, 2004, Table 3-3) and normal volume
(3)   Based on 17.2 mg (3.8×10^{-5} lb) per year of plutonium (DCS, 2004, Table 3-3) and normal volume
(4)   Based on 1.12×10^{8} Bq (3.03×10^{-3} Ci) alpha-emitting radioactivity per year (DCS, 2004, Table 3-3) and normal volume
(5)   WSB = waste solidification building
(6)   Waste transfer line is a PSSC for this waste
(7)   Waste transfer line is a PSSC for this waste
(8)   PSSCs are applied to ensure safe transfer of this waste to DOE without spillage

The applicant's safety assessment of the design bases considered hazards in these areas. However, specific waste minimization practices and program commitments are not relied upon in the safety assessment to reduce the risk of these hazards. The PSSCs relied upon to reduce the risks of hazards in the acid recovery and solvent regeneration systems are described in Chapters 5, 8, and 11 of this FSER.

10.1.2 Effluent and Environmental Monitoring

The staff has reviewed the applicant's brief description of the Effluent and Environmental Monitoring Programs and its related safety assessment of the design bases, in which the applicant did not identify PSSCs or management measures within the purview of these programs. The staff finds that, at the construction authorization stage, no such PSSCs or management measures are required. However, the regulations in 10 CFR Part 20 contain specific requirements for such programs that would have to be fully and adequately addressed in any subsequent license application.

The following sections describe the staff's evaluation of the Effluent and Environmental Monitoring Programs, to the extent that such information was provided by the applicant in the environmental report and revised CAR (DCS, 2002b; DCS, 2002e).

10.1.2.1 Concentrations of Radionuclides in Air Effluents and Public Doses

The applicant provided an estimate of maximum controlled area radionuclide concentrations, which is based on annual releases described in its environmental report (DCS, 2004, Table D-7), a 50-percent atmospheric dispersion parameter value (χ/Q) of 2.5×10^{-4} s/m^3, a distance to a receptor from the plant stack of 52 meters (171 feet), and the assumption that releases occur from ground level. This calculation demonstrates that the average controlled area concentration immediately outside the restricted area would be less than 34 percent of the ALARA goal. The staff performed an independent calculation using the methodology described in the National Council on Radiation Protection and Measurements Report 123, as described in NUREG-1718. The staff assumed a 28-meter (92-foot) stack height, no plume rise, and a site-specific 3.6-m/s (8 mph) annual average windspeed and that the wind blows downwind toward a receptor 100 percent of the time. The staff's estimate of the χ/Q is 5×10^{-5} s m^{-3} at a distance of 400 meters (1312 feet). The staff's calculation demonstrates that the concentration outside the controlled area would be less than 7 percent of the applicant's ALARA goal. The staff's analysis is summarized in FSER Table 10.1-2.

The applicant's estimate of the maximum potential dose to an individual member of the public from SRS is 3.3×10^{-5} mSv (3.3×10^{-3} mrem) per year. The staff performed an independent analyses using GENII, the Hanford environmental radiation dosimetry software system. The staff's result is 5.2×10^{-6} mSv (5.2×10^{-4} mrem) per year, which is in good agreement with the applicant's value, given the uncertainties inherent in the assumptions related to environmental dosimetry analyses. The staff's value is 0.00052 percent of the NRC's 10 CFR Part 20 1 mSv (100 mrem) standard for members of the public.

Based on the staff's independent calculation, the known or expected concentrations of radioactive material in airborne effluents from the proposed facility would be below the limits in Table 2 of Appendix B to 10 CFR Part 20. This forms an acceptable basis for future ALARA evaluations as required by 10 CFR 20.1101(b).

10.1.2.2 Physical and Chemical Characteristics of Radionuclides in Discharges

With regard to the provisions of 10 CFR 20.1302(c), the applicant did not propose to adjust the effluent concentration values that appear in Table 2 of Appendix B to 10 CFR Part 20 for members of the public by taking into account the actual physical and chemical characteristics of the effluents (e.g., aerosol size distribution, solubility, density, radioactive decay equilibrium, chemical form). Instead, the applicant intended to demonstrate compliance with the annual dose limit of 10 CFR 20.1301 using the dose methodology provided for in 10 CFR 20.1302(b)(1) and not the concentration-based methodology provided for in 10 CFR 20.1302(b)(2).

This approach is consistent with the requirements of 10 CFR Part 20 and, therefore, is acceptable to the staff.

10.1.2.3 Air Effluent Discharge Location and Effluent Monitoring

Revised CAR Section 10.2.1.4 indicates that the discharge location for radioactive air effluents from the proposed facility would be the facility stack located on the roof of the MP process building. This stack would be 12.2 m (40 ft) tall and would be located on the top of the MOX fuel fabrication building, which would be 22.3 m (73 ft) tall. The stack would discharge 5,419 m³/min (191,360 cubic feet per minute) during normal operations. The applicant has committed to maintaining an operable continuous sample collection system in accordance with RG 4.16, Revision 1, "Monitoring and Reporting Radioactivity in Releases of Radioactive Materials in Liquid and Gaseous Effluents from Nuclear Fuel Processing and Fabrication Plants and Uranium Hexafluoride Production Plants," issued 1985 (NRC, 1985).

The applicant will be expected to provide additional information (including, for example, sample collection and analysis procedures, a description of action levels, pathway analyses for public doses, and recording and reporting procedures) in any subsequent license application.

10.1.2.4 Environmental Monitoring Program

The applicant performed preconstruction environmental monitoring measurements, as described in revised CAR Section 10.3. These measurements were conducted in accordance with DOE Order 5400.1 and established a baseline of existing radiological, chemical, physical, and biological conditions in the area of the site. These measurements were used to evaluate whether contaminants could pose a potential safety concern for construction personnel. The data may also be used by the applicant and/or DOE to apply for environmental permits.

To accomplish these goals, the applicant made use of the data provided from the extensive SRS Environmental Monitoring Program and augmented the SRS environmental studies with additional sample collections.

With regard to expected impacts, the nonradiological impacts to the environment from the construction of the facility, if authorized, are expected to be minimal. The applicant expected that nonradiological monitoring prescribed through various environmental permits will be sufficient through construction and operation of the facility.

The staff reviewed the Plutonium Disposition Program Preconstruction Environmental Monitoring Report (Fledderman, 2002) and finds that the report is consistent with the guidance in the Standard Review Plan (NRC, 2000); therefore, it is acceptable to the staff.

10.1.3 Safety Assessment of the Design Bases

The staff's evaluation of the applicant's environmental protection measures at the construction authorization stage focuses on the potential accident sequences that result in radiological releases to the environment and the PSSCs relied upon for safety that are specified by the applicant to reduce the risk of these accidents.

Table 10.1-2 Air Effluent Concentrations from the MFFF

Radionuclide	Annual Releases [1]	NRC Average RABC [2]	DCS Average RABC [3]	20% of Appendix B to 10 CFR Part 20	NRC Ratio	DCS Ratio
^{236}Pu	1.3×10^{-8}	2.1×10^{-26}	1.0×10^{-25}	1×10^{-14}	1.7×10^{-12}	1.0×10^{-11}
^{238}Pu	8.5×10^{0}	1.3×10^{-17}	6.8×10^{-17}	4×10^{-15}	3.4×10^{-3}	1.7×10^{-2}
^{239}Pu	9.1×10^{1}	1.4×10^{-16}	7.6×10^{-16}	4×10^{-15}	3.6×10^{-2}	1.8×10^{-1}
^{240}Pu	2.3×10^{1}	3.6×10^{-17}	1.8×10^{-16}	4×10^{-15}	9.1×10^{-3}	4.6×10^{-2}
^{241}Pu	1.01×10^{2}	1.6×10^{-16}	8.1×10^{-16}	2×10^{-13}	8.0×10^{-4}	5.0×10^{-3}
^{242}Pu	6.1×10^{-3}	9.7×10^{-21}	4.9×10^{-20}	4×10^{-15}	2.4×10^{-6}	1.2×10^{-5}
^{241}Am	4.8×10^{1}	7.6×10^{-17}	3.8×10^{-16}	4×10^{-15}	1.9×10^{-2}	9.6×10^{-2}
^{234}U	5.1×10^{-3}	8.1×10^{-21}	4.1×10^{-20}	1×10^{-14}	8.1×10^{-7}	4.1×10^{-6}
^{235}U	2.1×10^{-4}	3.3×10^{-22}	1.7×10^{-21}	1×10^{-14}	2.8×10^{-8}	1.4×10^{-7}
^{238}U	1.2×10^{-2}	1.9×10^{-20}	9.6×10^{-20}	1×10^{-14}	1.6×10^{-6}	8.0×10^{-6}
				Totals	**0.07**	**0.34**

(1)   DCS, 2002b

(2)   RABC = restricted area boundary concentration. The atmospheric dispersion factor (χ/Q) estimated by the staff for this table is based on a 28-m (92-ft) stack height, no plume rise, and a site-specific 3.6 m/s (8 mph) annual average windspeed (ASHRAE, 1993). The maximum value is 5×10^{-5} s/m^3 at a distance of 400 m (1312 ft).

(3)   The atmospheric dispersion factor (χ/Q) estimated by the applicant is based on 50 percent meteorology and a distance to the receptor of 52 m (171 ft) (DCS, 2004). The maximum value is 2.5×10^{-4} s/m^3.

10.1.3.1 Consequence Assessment Methodologies

In its safety assessment, the applicant calculated committed doses to an IOC and concentrations of radioactive material in the environment outside the restricted area from each accident to demonstrate that risks from event consequences were reduced to acceptable levels. The consequence assessment methodology used by the applicant for dose consequences at the controlled area boundary is the same methodology used for the site worker, as described in Chapter 9 of this FSER, with the exception of the value of the atmospheric dispersion factor. The atmospheric dispersion factor that the applicant derived for the distance from the facility to the controlled area boundary is 2.5×10^{-4} s/m^3.

The consequence assessment methodology used to comply with the 10 CFR 70.61(c)(3) performance requirement is also similar to the methodology presented for site workers in Chapter 9 of this FSER. The principal difference is that the applicant did not use the respirable fraction (RF) as a reduction factor in calculations demonstrating that concentrations fall below the 10 CFR 70.61(c)(3) intermediate-consequence threshold. This is because the 10 CFR 70.61(c)(3) concentration pertains to total concentrations in the environment and not to doses to human receptors resulting from intake of respirable particles. This approach is acceptable to the staff.

The proposed restricted area boundary for the facility would be approximately 52 meters from the facility discharge stack (DCS, 2004). The atmospheric dispersion factor that the applicant derived for this location is 2.79×10^{-4} s/m^3.

As a result, the following equation is used to calculate environmental consequences:

$$[EC]_x = \{[\text{Source Term} / RF] \times [\chi/Q]^{RA} \times [f]_x\} / (3600 \text{ s hr}^{-1} \times 24 \text{ hr})$$

The source term is the same as that described in Chapter 9 of this FSER, the RF is divided back into the source term to negate the reduction applied for human consequence source terms, f_x is the specific activity and the fraction of the total quantity of the material at risk that is the radionuclide X, and $[\chi/Q]^{RA}$ is the value of the atmospheric dispersion factor for the facility stack, as described above.

The use of this equation is consistent with the staff's guidance in NUREG/CR-6410, "Nuclear Fuel Cycle Facility Accident Analysis Handbook," issued 1998 (NRC, 1998), and the regulations in 10 CFR 70.61(c)(3); therefore, it is acceptable to the staff.

10.1.3.2 Radiation Doses to Individuals Outside the Controlled Area from Accidents

In Chapter 5 of the revised CAR, the applicant presented the mitigated bounding event consequences for the five major categories of events (i.e., fire, explosion, loss of confinement, load handling events, and criticality). The methodology that the applicant used to estimate radiation doses to the IOC from potential accidents is substantially similar to that described in FSER Section 9.1.1.4, with the exception that the postulated distance to the IOC is 160 meters, as compared to the 100-meter distance to the site worker. As stated in FSER Chapter 9, the staff finds that the applicant used acceptable methods for estimating the consequences from accident sequences that result in radiological releases that could result in doses to the IOC. The applicant's estimates are bounding and its assumptions are appropriately conservative, given the available level of design at the construction authorization stage.

The mitigated consequences fall below the performance requirements of 10 CFR 70.61, "Performance Requirements." Therefore, the staff finds that the applicant's estimates of IOC accident consequences are acceptable.

10.1.3.3 Environmental Consequences

The staff reviewed the applicant's estimates of environmental consequences of bounding events provided in the revised CAR. The staff finds that the consequences are below the intermediate-consequence level described in 10 CFR 70.61(c)(3); therefore, they are acceptable.

10.1.3.4 PSSCs for Protection of the IOC and the Environment

The PSSCs required for protection of the IOC and the environment for each of the controlling events are discussed in Chapter 5 of this FSER and in other sections as referenced in FSER Table 5-1. As discussed above, the applicant used acceptable methods for estimating consequences from accident sequences that result in radiological releases to the environment.

10.2 Evaluation Findings

In Section 10 of the revised CAR, DCS provided information on public and environmental protection ALARA goals and methodologies to demonstrate compliance with NRC standards for radiation safety applicable to members of the public in accordance with 10 CFR 20.1101(b), 10 CFR 20.1301(b), and 10 CFR 20.1302(c). Based on the staff's review of this information, it finds that the applicant has established acceptable ALARA goals and methodologies to ensure public and environmental radiation safety.

In Section 10.5 of the revised CAR, DCS provided design basis information for the IOC and environmental protection PSSCs that it identified for the proposed facility. Based on the staff's review of the revised CAR and supporting information provided by the applicant relevant to environmental protection, the staff finds that the design bases of the PSSCs identified by the applicant will provide reasonable assurance of protection against natural phenomena and the consequences of potential accidents, in accordance with 10 CFR 70.23(b).

Further, pursuant to 10 CFR 70.61(b)(2), 10 CFR 70.61(c)(2), and 10 CFR 70.61(c)(3), and based on staff's review of the safety assessment methodology presented in the revised CAR, the staff finds that the design bases of the PSSCs identified by the applicant will reduce IOC dose and environmental contamination consequences to acceptable levels.

10.3 References

(ASHRAE, 1993) American Society of Heating, Refrigeration and Air-Conditioning Engineers. "Design Guide for Department of Energy Nuclear Facilities." 1993.

(DCS, 2002a) Hastings, P.S., Duke Cogema Stone & Webster. Letter to U.S. Nuclear Regulatory Commission, RE: Clarification of Responses to NRC Request for Additional Information. 2002.

(DCS, 2002b) Hastings, P.S., Duke Cogema Stone & Webster. Letter to U.S. Nuclear Regulatory Commission, MFFF Environmental Report. 2002.

(DCS, 2002c) Hastings, P.S., Duke Cogema Stone & Webster. Letter to U.S. Nuclear Regulatory Commission, DCS-NRC-000083, Clarification of Responses to NRC Request for Additional Information. 2002.

(DCS, 2002d) Hastings, P.S., Duke Cogema Stone & Webster. Letter to U.S. Nuclear Regulatory Commission, Clarification of Responses to NRC Request for Additional Information. 2002.

(DCS, 2002e) Idhe, R.H., Duke Cogema Stone & Webster. Letter to U.S. Nuclear Regulatory Commission, RE: Construction Authorization Request Revision. October 31, 2002 (including page changes through February 9, 2004).

(Fledderman, 2002) Fledderman, P.D. Plutonium Disposition Program Preconstruction Environmental Monitoring Report, Westinghouse Savannah River Company, ESH-EMS-2002-1141, Revision 0. 2002.

(Napier, 1988) Napier, B.A. et al. "GENII—The Hanford Environmental Radiation Dosimetry Software System." PNL-6584. Prepared by Pacific Northwest Laboratory, Richland, WA, for U.S. Department of Energy. 1988.

(NCRP, 1996) National Council on Radiation Protection and Measurements. NCRP No. 123, "Screening Models for Releases of Radionuclides to Atmosphere, Surface Water and Ground, Recommendations of the National Council on Radiation Protection and Measurements." 1996.

(NRC, 1973a) U.S. Nuclear Regulatory Commission. Regulatory Guide 3.12, "General Design Guide for Ventilation Systems of Plutonium Processing and Fuel Fabrication Plants." Washington, DC, 1973.

(NRC, 1973b) U.S. Nuclear Regulatory Commission. Regulatory Guide 8.37, "ALARA Levels for Effluents from Materials Facilities." Washington, DC, 1973.

(NRC, 1985) U.S. Nuclear Regulatory Commission. Regulatory Guide 4.16, Revision 1, "Monitoring and Reporting Radioactivity in Releases of Radioactive Materials in Liquid and Gaseous Effluents from Nuclear Fuel Processing and Fabrication Plants and Uranium Hexafluoride Production Plants." Washington, DC, 1985.

(NRC, 1990) U.S. Nuclear Regulatory Commission. NUREG/CR-4691, "MELCOR Accident Consequence Code System (MACCS)." Washington, DC, 1990.

(NRC, 1998) U.S. Nuclear Regulatory Commission. NUREG/CR-6410, "Nuclear Fuel Cycle Facility Accident Analysis Handbook." Washington, DC, 1998.

(NRC, 2000) U.S. Nuclear Regulatory Commission. NUREG-1718, "Standard Review Plan for the Review of an Application for a Mixed Oxide (MOX) Fuel Fabrication Facility." Washington, DC, August 2000.

This page intentionally left blank

11. PLANT SYSTEMS

11.1 Civil Structural Systems

11.1.1 Conduct of Review

This chapter of the final safety evaluation report (FSER) contains the U.S. Nuclear Regulatory Commission (NRC) staff's review of the civil structural systems described by the applicant, Duke Cogema Stone & Webster (DCS) in Chapter 11 of the revised construction authorization request (CAR) (DCS, 2002b). The objective of this review is to determine whether the principal structures, systems, and components (PSSCs) and their design bases identified by the applicant provide reasonable assurance of protection against natural phenomena and the consequences of potential accidents. The staff evaluated the information provided by the applicant for the civil structural systems by reviewing Chapter 11 of the revised CAR, other sections of the revised CAR, supplementary information provided by the applicant, and relevant documents available at the applicant's offices but not submitted by the applicant. The review of the civil structural systems design bases and strategies was closely coordinated with the review of the civil structural aspects of accident sequences described in the safety assessment of the design bases (see Chapter 5 of this FSER) and the review of other plant systems.

The staff reviewed how the information in the revised CAR addresses the following regulations:

- Title 10, Section 70.23(b), of the *Code of Federal Regulations* (10 CFR 70.23(b)) states, as a prerequisite to construction approval, that the design bases of the PSSCs and the Quality Assurance Program be found to provide reasonable assurance of protection against natural phenomena and the consequences of potential accidents.

- 10 CFR 70.64, "Requirements for New Facilities or New Processes at Existing Facilities," requires that baseline design criteria (BDC) and defense-in-depth practices be incorporated into the design of new facilities. With respect to structural systems, 10 CFR 70.64(a)(2) requires that the mixed oxide (MOX) fuel fabrication facility (MFFF, or the facility) design "provide for adequate protection against natural phenomena with consideration of the most severe documented historical events for the site."

The review for this construction approval focused on the design basis of the civil structural systems and other related information. For each civil structural system, the staff reviewed information provided by the applicant for the safety function, system description, and safety analysis. The review also encompassed proposed design basis considerations such as redundancy, independence, reliability, and quality. The staff used Chapter 11.0 in NUREG-1718, "Standard Review Plan for the Review of an Application for a Mixed Oxide (MOX) Fuel Fabrication Facility" (NRC, 2000), as guidance in performing the review.

The following three general areas of Section 11.1, "Civil Structural Systems," of the revised CAR are reviewed for this section of the FSER:

- classification of civil structural systems
- codes and standards
- structural design criteria

11.1.1.1 System Description

The civil structural systems for the proposed facility include the buildings; support structures; and facilities that house, support, confine, or contain various plant systems, components, and equipment associated with licensed nuclear materials, or hazardous chemicals associated with licensed nuclear materials, as well as support buildings.

11.1.1.1.1 Function

As described in Sections 11.1.1 and 11.1.7 of the revised CAR, the safety functions for the civil structural systems would accomplish the following:

- Support PSSCs during normal, severe, and extreme environmental conditions.

- Provide confinement functions as part of secondary and tertiary confinement systems.

- Protect PSSCs from the effects of normal, severe, and extreme environmental loads.

- Protect PSSCs from the effects of temperature extremes, including design basis internal and external fires.

- Protect PSSCs from the effects of design basis man-induced events, including potential load drops.

11.1.1.1.2 Major Components

The civil structural systems considered to be seismic Categories (SCs) I and II or conventional seismic structures were identified in Section 11.1.3 of the revised CAR. The major components for each of the seismic categories are as follows:

- SC-I structures include the MOX fuel fabrication building (which includes the MOX processing area, aqueous polishing (AP) area, and the shipping and receiving area), emergency fuel storage vault, and the emergency generator building.

- SC-II structures include the safe haven buildings.

- Conventional seismic (CS) structures include the reagent process building, administration building, secured warehouse building, technical support building, standby generator building, receiving warehouse building, and miscellaneous site structures (e.g., the bulk gas storage pad; the heating, ventilation, and air conditioning (HVAC) and process chiller pads; and the electric transformer pads) specified in Section 11.1.3.11 of the revised CAR.

A detailed description for each major component is provided in Section 11.1.3 of the revised CAR.

11.1.1.1.3 Control Concepts

This section is not applicable to civil structural systems.

11.1.1.1.4 System Interfaces

Civil structural systems interface with the site and all plants systems because they provide protection and support for structures, systems, and components (SSCs).

11.1.1.2 Classification of Civil Structural Systems

The classification outlined in Section 11.1, "Civil Structural Systems," of the revised CAR consists of three levels—(1) SC-I, (2) SC-II, and (3) CS structures. The design loadings considered for the civil structures in each category are as follows:

- SC-I—normal, severe, and extreme environmental loads, including the design basis earthquake and tornado

- SC-II—normal, severe, and design basis earthquake

- CS—normal, severe, and conventional seismic loads as specified by the Uniform Building Code

11.1.1.3 Design Basis of the PSSCs

11.1.1.3.1 Codes and Standards

This section contains a review of Sections 11.1.6.1.3, "Codes and Standards for SC-II Structures," and 11.1.7.3, "Codes and Standards for SC–I Structures," of the revised CAR.

The designs of the SC-1 civil structural systems and the associated steel and concrete components conform to standard engineering practice. A comprehensive list of the applicable codes and standards is provided in Section 11.1.7.3 of the revised CAR. The American Concrete Institute (ACI) report, ACI 349-97, "Code Requirements for Nuclear Safety-Related Concrete Structures," issued 1997 (ACI, 1997), and American National Standards Institute/American Institute of Steel Construction (ANSI/AISC) N690-1994, "Specification for the Design, Fabrication, and Erection of Steel Safety-Related Structures for Nuclear Facilities," issued 1994 (AISC, 1994), including Supplement 1, would be supplemented by specific provisions. In Section 11.1.6.1.3 of the revised CAR, the applicant referenced the same codes and standards as those for the SC-I structures for the design of the SC-II civil structural systems, except for ACI 349.1R–91, "Reinforced Concrete Design for Thermal Effects on Nuclear Power Plant Structures, issued 1996 (ACI, 1996); ACI 349-97; ACI 349.2R–97, "Embedment Design Examples," issued 1998 (ACI, 1998); and ANSI/AISC N690-1994, including Supplement 1. Additionally, ACI 318-99, "Building Code Requirements for Structural Concrete," issued 1999 (ACI, 1999), was included for SC-II civil structures.

The staff reviewed the codes and standards for the designs of SC-I and SC-II civil structural systems and concluded that the cited codes and standards are consensus standards that provide reasonable guidance consistent with the categorization assigned to the buildings.

11.1.1.3.2 Structural Design Criteria

This section contains a review of the structural design criteria and load combinations for the civil structural systems discussed in the revised CAR.

[Text removed under 10 CFR 2.390]

Structural design loads for SC-I and SC-II structures were discussed in Sections 11.1.7.4, "Loads and Loading Combinations for SC-I Structures," and 11.1.6.1.4, "Values for SC-II Structures," of the revised CAR. Design criteria and loads anticipated for the structures were divided into (1) normal loads, (2) severe environmental loads, and (3) extreme environmental loads. The normal loads include dead, live, hydrostatic fluid pressure, lateral soil pressure, thermal, and component reaction loads. The severe environmental loads include wind and flood loads. The extreme environmental loads include seismic, tornado, and explosive loads, and post-earthquake settlements. The only extreme environmental load considered in the design of the SC-II structures was the seismic load. The staff finds that the applicant's evaluation satisfies the requirements of 10 CFR 70.64(a)(2), which states that structures must be designed to provide for adequate protection against natural phenomena with consideration of the most severe documented historical events for the site.

11.1.1.3.2.1 Normal Loads.

- Dead Loads

  Dead loads are gravity loads induced by the mass of the structure, permanent equipments, and any permanent hydrostatic loads with constant fluid levels. This definition for dead loads is consistent with the American Society of Civil Engineers (ASCE) Standard 7-98, "Minimum Design Loads for Buildings and Other Structures," issued 1998 (ASCE, 1998), and is acceptable to the staff. The CAR also indicates that effects of differential settlement would be considered in determining dead loads.

- Live Loads

  Live loads are loads produced by the use and occupancy of the building. The live loads considered for the civil structures of the proposed MFFF include floor, rain, snow and ice, transportation vehicle and heavy floor, as well as crane, monorail, hoist, and elevator loads.

  — Floor Live Loads

    The minimum uniformly distributed live loads for the civil structures were established by the applicant in accordance with the ASCE Standard 7-98 (ASCE, 1998). Specifically, the floor live loads identified include the following:

platform and work area	6.0 kN/m² (125 psf)
light storage	6.0 kN/m² (125 psf)
heavy storage	12.0 kN/m² (250 psf)
heavy operation	12.0 kN/m² (250 psf)
office	4.79 kN/m² (100 psf)
computer room	7.18 kN/m² (150 psf)
dining/meeting rooms	4.79 kN/m² (100 psf)
laboratory	9.58 kN/m² (200 psf)
toilet areas	4.79 kN/m² (100 psf)
mechanical (utility) rooms	7.18 kN/m² (150 psf)
electrical rooms	7.18 kN/m² (150 psf)
stairs, fire escapes, and corridors	4.79 kN/m² (100 psf)
roof	2.4 kN/m² (50 psf)
transportation vehicle loads	14.37 kN/m² (300 psf) or forklift truck of 26.69 kN (6 kips) capacity

The staff reviewed the design basis floor live loads discussed in Section 11.1.7.4.1.1 of the revised CAR and additional information provided in the applicant's March 8, 2002, letter (DCS, 2002a) and determined that the floor live loads and the roof loads are appropriate and acceptable for the design of the facility civil structures.

— Rain Loads

The design basis rain loads for the civil structures were determined by the applicant in accordance with the requirements of ASCE Standard 7-98 (ASCE, 1998). The design rain load for the roof system of the SC-I structures is 2.4 kN/m² (50 psf), which is equivalent to more than 24.4 cm (9.6 in.) of standing water on the roof because of deflection of the roof or blockage of the primary roof drains. The CAR further states that "parapets or other structures, which could potentially contribute to significant ponding, are not used on the roofs of SC-1 structures," and that the rain load does not combine with the roof live load in the load combinations. The staff reviewed the design basis rain load and determined that it is appropriate and acceptable.

— Snow and Ice Loads

The design basis snow and ice loads value was determined by the applicant to be 0.48 kN/m² (10 psf). This value was estimated based on the 100-year maximum ground snow and ice loads. The importance factor for the snow and ice loads value is 1.2. The staff reviewed the design basis maximum ground snow and ice loads value and found that it was based on acceptable methods.

— Transportation Vehicle Loads and Heavy Floor Loads

The design basis load for transportation vehicular truck traffic in designated building areas was determined in accordance with the standard loadings defined by the American Association of State Highway and Transportation Officials in 1996 (AASHO, 1996). The minimum truck loading of HS 20-44 is used for wheel-loading design. The heavy floor loading considered in areas used for transportation, transfer, and storage of finished fuel assemblies is 14.37 kN/m^2 (300 psf), or forklift truck of 26.69 kN (6 kips) capacity. The staff reviewed the design basis transportation vehicle loads and heavy floor loads and determined that they are appropriate and acceptable.

— Crane, Elevator, and Hoist Loads

These design loads apply to structural members and components to support permanently installed cranes, monorails, hoists, and elevators. Section 11.1.7.4.1.1 of the revised CAR states that the design basis crane, monorail, hoist, and elevator loads would envelop the full-rated capacity of the cranes, monorails, hoists, and elevators, including impact loads and test load requirements. The effects of a crane load drop were evaluated in accordance with the guidance provided in NUREG-0612, "Control of Heavy Loads at Nuclear Plants," issued 1980 (NRC, 1980), for SC-I structures. The staff reviewed the design basis crane, monorail, hoist, and elevator loads and found that these design basis loads are appropriate and acceptable.

— Hydrostatic Fluid Pressure Loads

The revised CAR indicates that no hydrostatic fluid pressure loads caused by fluid held inside the buildings of the facility are currently identified. The staff agrees with this statement. However, if the applicant later determines that there is fluid inside the building of the facility, hydrostatic fluid pressure loads should be included in the design.

— Lateral Soil Pressure Loads

Section 11.1.7.4.1.1 of the revised CAR indicates that a determination of the lateral soil pressure loads on structures, elements of structures, or both because of retaining soil would be based on the density of the soil and any surcharge load, plus the hydrostatic pressure caused by ground water or soil saturation.

The minimum lateral soil pressure loads on structures and/or elements of structures resulting from retaining soil are defined in ASCE Standard 7-98 (ASCE, 1998). Earthquake-induced soil pressure on structures or embedded wall design would be developed in accordance with ASCE Standard 4-98, "Seismic Analysis of Safety-Related Nuclear Structures and Commentary," issued 1999 (ASCE, 1999). No hydrostatic pressure is expected because the ground water table at the site is below the MFFF.

The staff reviewed the approach presented by the applicant for determining lateral soil pressure loads and found that the approach is based on acceptable standards and engineering practices.

—   Thermal Loads

Thermal loads consist of thermally induced forces and moments on the structural components of buildings. These loads would result from operating and environmental conditions. The thermally induced loads would be design dependent. Consequently, determination and consideration of these thermally induced loads would be an integral part of a design. For the design of civil structural systems, the applicant would consider the effects of thermal expansion loads caused by axial restraint of the structural components, as well as loads resulting from thermal gradients. The applicant also indicated that it would determine these thermally induced loads based on the most critical transient or steady-state condition. The staff reviewed the information that the applicant presented in the revised CAR and found that its consideration of thermal expansion loads and thermal gradient loads for the design of civil structural systems is acceptable.

—   Equipment Reaction Loads

The equipment reaction loads included those from pipes, HVAC ducts, conduits, and cable trays. These loads would be design dependent and need to be assessed during design. The applicant stated that the equipment reaction loads would be determined based on the most critical transient or steady-state condition. The applicant further indicated that it would make sure that the final designs envelop the actual equipment reaction loads. The staff reviewed the information presented regarding consideration of equipment reaction loads in the design and found that the bounding/enveloping approach is acceptable.

11.1.1.3.2.2 Severe Environmental Loads.

•   Wind Loads

Figure 6-1b in ASCE Standard 7-98 (ASCE, 1998) identifies a design basis wind speed of 160.93 km/h (100 mph) for the region. Information provided in Table 11.1-2 of the revised CAR for the Savannah River Site (SRS) identifies a design basis wind speed of 209.22 km/h (130 mph), which is higher than the value provided in ASCE 7-98. The revised CAR also indicates that the wind loads calculated based on the design basis wind would be determined by the procedures provided in ASCE 7-98. The approach for determining the wind loads is acceptable to the staff because the approach used is the same or similar to consensus standards approved by the staff.

The applicant also included the consideration of windborne missiles in the design of the civil structural systems.

[Text removed under 10 CFR 2.390]

11-7

Considering the effects of windborne missiles in the design is consistent with ASCE Standard 7-98 (ASCE, 1998), which requires the inclusion of windborne debris in areas where the basic windspeed is equal to or greater than 193.12 km/h (120 mph). The inclusion of windborne missiles in the design is acceptable to the staff because it meets the guidance provided in ASCE 7-98.

- Flood Loads

The maximum probable flood level for the site is at elevation 68.43 m (224.5 ft) above mean sea level. The corresponding site grade level is approximately 82.91 m (272 ft) above mean sea level. Because the site grade level is much higher than the maximum probable flood level at the site, the proposed facility is considered to be a flood-dry site and will be free from the adverse effects of the maximum probable flood. Consequently, the loads resulting from flood water do not need to be considered in the design of civil structural systems. Analysis of the maximum probable flood level was based on the surface hydrology of the region and potential dam failure resulting from seismic events given in Section 1.3.4.2, "Floods," of the revised CAR.

The staff reviewed the discussion regarding the flood loads and concludes that the facility design is consistent with the design criteria of Regulatory Guide (RG) 3.40, "Design Basis Floods for Fuel Reprocessing Plants and for Plutonium Processing and Fuel Fabrication Plants," issued 1977 (NRC, 1977), and that the approach used for conducting the flood analysis is consistent with that outlined in Section 5.1.3 of ANSI/American Nuclear Society (ANS) Standard 2.8-1992, "Determining Design Basis Flooding at Power Reactor Sites," issued 1992 (ANS, 1992), which determined design basis flooding at power reactor sites. Therefore, the information provided is acceptable to the staff.

11.1.1.3.2.3 Extreme Environmental Loads.

- Seismic Loads

Table 11.1-2 and Section 1.3.6, "Seismology," of the revised CAR list the design basis earthquake ground motions based on probabilistic seismological studies specific to SRS. The applicant's design basis horizontal and vertical response spectra for the facility were developed based on RG 1.60, "Design Response Spectra for Seismic Design of Nuclear Power Plants," issued 1973 (NRC, 1973a), horizontal and vertical spectrum shapes anchored at a peak ground acceleration of 0.20 g. A more detailed discussion on the applicant's design spectra is provided in Section 1.3.1.5.5 of this FSER.

According to the revised CAR, the design seismic loads for the SC-I structures of the facility would be determined by considering the design basis earthquake accelerations in the three orthogonal directions (two horizontal and one vertical).

To support soil-structure interaction analysis, synthetic time histories for the three components of the design basis ground motion would be generated. The response spectra of the synthetic time histories would envelop the design response spectra and meet the minimum power spectral density requirement in accordance with the criteria in Section 3.7.1, "Seismic System Analysis," of NUREG-0800, "Standard Review Plan for the Review of Safety Analysis Reports for Nuclear Power Plants," issued 1987 (NRC, 1987), according to the revised CAR.

A simplified three-dimensional finite-element model that ignored the embedment and simulated intact slabs would be used for the soil-structure interaction analysis of the MOX fuel fabrication building. A three-dimensional lumped mass stick model representing the building characteristics and the assumption that the roofs and floors are rigid diaphragms would be used for the soil-structure interaction analysis of the emergency generator building.

According to the revised CAR, the soil model for the soil-structure interaction analysis would include a sufficient number of idealized soil layers from the ground surface to the bedrock. This soil model was developed using the information from the soil exploration and site-response analysis. The structural damping values used in the analysis would be in accordance with RG 1.61, "Damping Values for Seismic Design of Nuclear Power Plants," (NRC, 1973b) for a safe-shutdown earthquake and are acceptable to the staff. From the soil-structure interaction analysis, the response spectra at the foundation and each floor and roof level would be obtained in order to develop acceleration profiles for building design.

Three soil conditions (lower, best, and upper bound) accounting for material property uncertainties would be considered in the soil-structure interaction analysis to develop the in-structure response spectra at each direction and for a given structural level. A broadening of in-structure spectrum peaks would be applied to the spectrum envelope to account for structural material properties. Specific extents of spectrum-peak broadening for the MOX fuel fabrication and emergency generator buildings were discussed in Section 11.1.7.4.1.3, "Extreme Environmental Loads," of the safety analysis report. The staff determined that the approach for consideration of uncertainties associated with structural and soil material properties is consistent with the design guidance in RG 1.122, "Development of Floor Design Response Spectra for Seismic Design of Floor-Supported Equipment or Components," issued 1978 (NRC, 1978).

For the MOX fuel fabrication and emergency generator buildings, a three-dimensional finite-element model would be used to conduct a static analysis using the equivalent accelerations of each level developed from the three-dimensional soil-structure analysis. The acceleration profiles that envelop the acceleration profiles from the three soil conditions would be used for the finite-element analysis. Results from the equivalent static three-dimensional finite-element analysis based on the acceleration profiles applied in the three directions would be combined using the 100-40-40 percent rule as described in Section 3.2.7.1.2 of ASCE Standard 4-98 (ASCE, 1999).

For other structures, the approach of applying the applicable seismic response to the base of the finite element models may be used. Guidance provided by RG 1.92, "Combining Modal Responses and Spatial Components in Seismic Response Analysis," issued 1976 (NRC, 1976), and ASCE 4-98 would be used for combining modal responses and collinear responses from the individual earthquake components.

The staff reviewed the approach presented by the applicant for determining the seismic loads and found the approach to be acceptable.

- Tornado Loads for SC-I Structures

  Table 11.1-2 of the revised CAR provides the design basis tornado windspeed, atmospheric pressure change, and rate of pressure drop. In determining design tornado loads, the procedure provided in ASCE Standard 7-98 (ASCE, 1998) would be used. The staff reviewed the approach presented by the applicant and found the approach to be acceptable based on the guidance provided in ASCE 7-98. The three types of tornado loads on the facility structures are as follows:

  — Tornado Wind Pressure Loads

  The tornado wind pressure loads are defined in Table 11.1-2 of the revised CAR. These pressure loads are based on Section 6 of ASCE 7-98, which is an industry consensus standard that has been accepted by the staff. Therefore, the information presented is acceptable to the staff.

  — Tornado-Created Differential Pressure Loads

  The tornado-created differential pressure loads are defined in Table 11.1-2 of the revised CAR. These pressure loads are based on guidance provided in Section 3.3.2 of NUREG-0800 (NRC, 1987), and, therefore, it is acceptable to the staff.

  — Tornado-Generated Missile Loads

  The design basis for tornado-generated missiles is presented in Table 11.1-2 of the CAR. In Section 11.1.7.2.1.2 of the CAR, the applicant stated that the analysis for tornado-generated missiles is complete. However, in its March 8, 2002, letter, the applicant stated that the CAR incorrectly implies that the analysis for tornado-generated missiles has been completed. The applicant revised the CAR to say that the analysis will be provided as part of its license application. This deferral is acceptable to the staff based on the guidance provided in (1) Section 11.4.6.2 of NUREG-1718 (NRC, 2000), which states that the adequacy of civil structural system designs pertaining to tornado missiles may be determined based on the integrated safety analysis (ISA), the summary of which need not be submitted as part of the revised CAR (see 10 CFR 70.65(b)), and (2) Sections 3.5.2 and 3.5.3 of NUREG-0800 (NRC, 1987).

  [Text removed under 10 CFR 2.390]

- Site Proximity Missile (Except Aircraft) Load

  The applicant indicated that the distance from the nearest transportation route to the facility and the emergency diesel generator (EDG) buildings is greater than the safe distance established based on the available inventories of hazardous materials.

11-10

Therefore, the effects of explosion-generated missiles from the transportation route are not a concern for design. Furthermore, the applicant stated that analyses to identify site proximity missiles that might be generated from events within the safe distance will be submitted as part of its license application (DCS, 2002a, Enclosure A, p. 49). This deferral is acceptable to the staff based on the guidance provided in Section 11.4.6.2 of NUREG-1718 (NRC, 2000) stating that the adequacy of civil structural system designs may be determined based on the ISA, the summary of which need not be submitted as part of the revised CAR (see 10 CFR 70.65(b)).

- Foundation Design Criteria

The foundation design criteria are discussed in MFFF site geotechnical reports (DCS, 2001; DCS, 2003) supporting the revised CAR. The MOX fuel fabrication building has two foundation levels. The base level of the MOX processing area is at grade. The base level of the AP area is approximately 5.3 m (17.5 ft) below grade. A portion of the base level of the shipping and receiving area is at grade, and the remainder is 5.3 m (14 ft) below grade. The base mat for the MOX fuel fabrication building will be 2 m (6.5 ft) thick, and it will be 0.91 m (3 ft) thick for the emergency generator building. Some soils at the foundation locations will be removed and replaced with engineered structural fill. Approximately 3.05 m (10 ft) of the engineered fill material will be placed beneath the main foundation level of the MOX fuel fabrication building and approximately 1.54 m (5 ft) beneath the emergency generator building. The engineered fill will be extended at least 3.05 m (10 ft) beyond the edges of the foundations and will be compacted to at least 95 percent of the maximum dry density determined in accordance with American Society for Testing and Materials (ASTM) DI557.

The ultimate bearing capacity is approximately 3.4 MPa (71.2 ksf) for the MOX fuel fabrication building and approximately 0.65 MPa (13.5 ksf) for the EDG building. The third MFFF site geotechnical report (DCS, 2003) indicates that the ultimate bearing capacities were determined based on conservative soil strength properties. For the final design, a minimum factor of safety of more than 6 will be maintained for both buildings. The staff reviewed the estimated bearing capacities and found that the values provided are acceptable.

- Settlements

The first MFFF site geotechnical report (DCS, 2001) correctly states that the critical foundation load condition that limits the allowable foundation bearing capacity will be settlement. The settlements at the site come from two potential sources—(1) compaction of soft soil materials, including soft zones, and (2) localized liquefaction.

Two methods (DCS, 2003) were used to estimate settlements for the SC-I structures—recompression indices and numerical modeling. The estimated settlement based on the recompression indices for the MOX fuel fabrication building is approximately 4.32 cm (1.7 in.) at the edge of the foundation and 7.62 cm (3.0 in.) at the center of the foundation. The numerical modeling incorporating the presence of soft zones estimates the primary settlements ranging from 6.86–8.89 cm (2.7–3.5 in.). Including the secondary consolidation, the total estimated settlements of the MOX fuel fabrication building are approximately 8.13–10.16 cm (3.2–4.0 in.). These primary

11-11

settlement estimates were determined assuming that the in situ soils were preconsolidated to pressures in excess of the sum of the initial overburden pressures and the pressure induced by structure loads, but that the soft zone soils were normally consolidated under the existing overburden pressures (DCS, 2003, Table 7-1).

The appropriateness of the use of this preconsolidation pressure was justified by the third MFFF site geotechnical report (DCS, 2003) based on the measured settlements for Building 221-S (vitrification building) at the SRS site. The soil conditions at the Building 221-S site are similar to those at the MFFF site except that no soft zones are present at the Building 221-S site (DCS, 2003). The average foundation pressure for Building 221-S (239.4–263.3 kPa (5–5.5 ksf)) is approximately 10–20 percent smaller than that of the MOX fuel fabrication building (292.5 kPa (6.11 ksf)). The dimensions of Building 221-S (110.3×35.6 m (362×117 ft)) are relatively smaller than those of the MOX fuel fabrication building (123.1 m and 140.2×91.4 m (408 ft and 460×300 ft)). Measured primary settlements at Building 221-S range from 3.1–6.6 cm (1.22–2.6 in.). These settlements are comparable to the 6.86–7.34 cm (2.7–2.9 in.) settlements estimated for the MOX fuel fabrication building if no soft zones are included in the analysis. The difference may be attributed to relatively larger structural dimensions and expected foundation pressures for the MOX fuel fabrication building than those for Building 221-S (DCS, 2003).

The staff reviewed the information presented in the site geotechnical reports regarding settlement analysis and determined that the approach for estimating structural settlements is acceptable because it is based on current industry guidance and analysis techniques.

The postearthquake dynamic settlement of the potentially liquefiable soil was estimated in the third MFFF site geotechnical report (DCS, 2003) based on the PC-3+ design basis ground motion and the 1886 Charleston motion. The settlement ranges from 0.66–3.73 cm (0.26–1.47 in.) for the design basis ground motion and 1.55–5.64 cm (0.61–2.22 in.) for the 1886 Charleston motion. The applicant indicated that these dynamic settlements might occur in loose or soft strata below the ground water level, at a depth of 18.29 m (60 ft) or more. There are two significantly stiffer soil layers more than 12.19 m (40 ft) thick between the potentially liquefiable zones and the foundations. According to the MFFF site geotechnical reports (DCS, 2001; DCS, 2003), these two soil layers would tend to redistribute the estimated dynamic settlement such that no significant differential settlement would occur at the foundation level.

The staff reviewed the information presented in the site geotechnical report regarding the dynamic settlements and concur that the postearthquake-induced dynamic settlements resulting from localized liquefaction will not create stability problems for the foundations for the SC-I structures.

- Aircraft Crash Hazard

As discussed in Section 1.1.1.9 of this FSER, the estimated total aircraft crash probability per year, including those induced by the federal airways and SRS helicopters for the MFFF building, is $2.99×10^{8}$ and $6.67×10^{8}$ for the emergency generator building, based on the applicant's calculation. Each of the estimated probabilities is smaller than the 10^{7} annual probability of unacceptable radiological consequences indicated in Section 3.5.1.6

of NUREG-0800 (NRC, 1987). As a result, the applicant concluded that aircraft crash hazards at the facility are not a design concern.

In its review of the aircraft hazard analysis, the staff also included an in-office review of the applicant's calculation of projected future flight information (Section 5.1.5.2 of this FSER). The staff found these calculations to be acceptable.

- Load Combinations

   The load combinations for both SC-I and SC-II civil structures were determined using Section 3.8.4, "Other Seismic Category I Structures," of NUREG-0800 (NRC, 1987) as a guide. Tornado, tornado missile, and explosion loads will not be considered in the load combinations of SC-II structures. The staff reviewed the various load combinations presented in Sections 11.1.7.4.2.1, "Loading Combinations for SC-I Concrete Structures," and 11.1.7.4.2.2, "Loading Combinations for SC-I Steel Structures," of the revised CAR and determined that these load combinations are the same as those suggested in NUREG-0800 for the design of structures. Therefore, the load combinations are acceptable to the staff.

11.1.2 Evaluation Findings

In Section 11.1 of the revised CAR, DCS provided design basis information for civil structural systems that it identified as PSSCs for the proposed facility. Based on the staff's review of the revised CAR and supporting information provided by the applicant relevant to civil structural systems, the staff finds that DCS has met the BDC set forth in 10 CFR 70.64(a)(2). Further, the staff concludes, pursuant to 10 CFR 70.23(b), that the design bases of the PSSCs identified by the applicant will provide reasonable assurance of protection against natural phenomena and the consequences of potential accidents.

11.1.3 References

(AASHO, 1996) American Association of State Highway and Transportation Officials. "Standard Specifications for Highway Bridges," 16th Edition. Washington, DC, 1996.

(ACI, 1996) American Concrete Institute. ACI 349.1R-91, "Reinforced Concrete Design for Thermal Effects on Nuclear Power Plant Structures." Reapproved, 1996. Detroit, Michigan, 1996.

(ACI, 1997) American Concrete Institute. ACI 349-97, "Code Requirements for Nuclear Safety-Related Concrete Structures." Detroit, Michigan, 1997.

(ACI, 1998) American Concrete Institute. ACI 349.2R-97, "Embedment Design Examples." Detroit, Michigan, 1998.

(ACI, 1999) American Concrete Institute. ACI 318-99, "Building Code Requirements for Structural Concrete." Detroit, Michigan, 1999.

(AISC, 1994) American National Standards Institute/American Institute of Steel Construction. ANSI/AISC N690–1994, "Specification for the Design, Fabrication, and Erection of Steel Safety-Related Structures for Nuclear Facilities." Chicago, Illinois, 1994.

(ANS, 1992) American National Standards Institute/American Nuclear Society. ANSI/ANS-2.8-1992, "Determining Design Basis Flooding at Power Reactor Sites." La Grange Park, Illinois, 1992.

(ASCE, 1998) American Society of Civil Engineers. ASCE 7-98, "Minimum Design Loads for Buildings and Other Structures." Reston, Virginia, 1998.

(ASCE, 1999) American Society of Civil Engineers. ASCE 4-98, "Seismic Analysis of Safety-Related Nuclear Structures and Commentary." Reston, Virginia, 1999.

(DCS, 2001) Hastings, P., Duke Cogema Stone & Webster. Letter to U.S. Nuclear Regulatory Commission, RE: MOX Fuel Fabrication Facility Site Geotechnical Report (DCS01–WRS–DS–NTE–G–00005–C). August 10, 2001.

(DCS, 2002a) Hastings, P., Duke Cogema Stone & Webster. Letter to U.S. Nuclear Regulatory Commission, RE: Clarification of Responses to NRC Request for Additional Information (DCS–NRC–000085). March 8, 2002.

(DCS, 2002b) Ihde, R., Duke Cogema Stone & Webster. Letter to U.S. Nuclear Regulatory Commission, Mixed Oxide Fuel Fabrication Facility Construction Authorization Request Revision. October 31, 2002.

(DCS, 2002c) Hastings, P., Duke Cogema Stone & Webster. Letter to U.S. Nuclear Regulatory Commission, Requests for Additional Information, Clarification, and Open Item Mapping into the Construction Authorization Request Revision (DCS01–NRC–000120). November 22, 2002.

(DCS, 2003) Duke Cogema Stone & Webster. MOX Fuel Fabrication Facility Site Geotechnical Report (DCS01-WRS-DS-NTE-G-00005-E), report prepared for the Department of Energy, Chicago Operations Office. June 30, 2003.

(DCS, 2004) Ihde, R, Duke Cogema Stone and Webster. Letter to W. Kane, U.S. Nuclear Regulatory Commission, RE: Mixed Oxide Fuel Fabrication Facility—Construction Authorization Request. 2004.

(NRC, 1973a) U.S. Nuclear Regulatory Commission. Regulatory Guide 1.60, "Design Response Spectra for Seismic Design of Nuclear Power Plants." Washington, DC, 1973.

(NRC, 1973b) U.S. Nuclear Regulatory Commission. Regulatory Guide 1.61, "Damping Values for Seismic Design of Nuclear Power Plants." Washington, DC, 1973.

(NRC, 1976) U.S. Nuclear Regulatory Commission. Regulatory Guide 1.92, "Combining Modal Responses and Spatial Components in Seismic Response Analysis." Washington, DC, 1976.

(NRC, 1977) U.S. Nuclear Regulatory Commission. Regulatory Guide 3.40, "Design Basis Floods for Fuel Reprocessing Plants and for Plutonium Processing and Fuel Fabrication Plants." Washington, DC, 1977.

(NRC, 1978) U.S. Nuclear Regulatory Commission. Regulatory Guide 1.122, "Development of Floor Design Response Spectra for Seismic Design of Floor-Supported Equipment or Components." Washington, DC, 1978.

(NRC, 1980) U.S. Nuclear Regulatory Commission. NUREG-0612, "Control of Heavy Loads at Nuclear Power Plants." Washington, DC, 1980.

(NRC, 1987) U.S. Nuclear Regulatory Commission. NUREG-0800, "Standard Review Plan for the Review of Safety Analysis Reports for Nuclear Power Plants." Washington, DC, 1987.

(NRC, 2000) U.S. Nuclear Regulatory Commission. NUREG-1718, "Standard Review Plan for the Review of an Application for a Mixed Oxide (MOX) Fuel Fabrication Facility." Washington, DC, 2000.

(NRC, 2001) Giitter, J.G., U. S. Nuclear Regulatory Commission. Letter to P. Hastings, Duke Cogema Stone & Webster, RE: Mixed Oxide Fuel Fabrication Facility Construction Authorization—Request for Additional Information. June 21, 2001.

11.2 <u>Aqueous Polishing Process and Chemistry</u>

11.2.1 Conduct of Review

This section of the FSER contains the staff's review of the AP process safety described by the applicant in Section 11.3 of the revised CAR (DCS, 2002e), with supporting process safety information from Chapters 5, 8, and 11 of the revised CAR, supplementary information provided by the applicant, and relevant documents available at the applicant's offices but not submitted by the applicant. The staff also reviewed technical literature as necessary to understand the process and safety requirements. The objective of this review is to determine whether the chemical process safety PSSCs and their design bases provide reasonable assurance of protection against natural phenomena and the consequences of potential accidents. The review of AP safety design bases and strategies was closely coordinated with the review of the radiation and chemical safety aspects of accident sequences described in the safety assessment of the design bases (see Chapter 5 of this FSER), the review of fire safety aspects (see Chapter 7 of this FSER), and the review of plant systems (see Chapter 11 of this FSER).

The staff evaluated the AP process and chemistry information in the revised CAR against the following regulations:

- 10 CFR 70.23(a) states, as a prerequisite to construction approval, that the design bases of the PSSCs and the Quality Assurance Program be found to provide reasonable assurance of protection against natural phenomena and the consequences of potential accidents.

- 10 CFR 70.64 requires that BDC and defense-in-depth practices be incorporated into the design of new facilities or new processes at existing facilities. With respect to chemical protection, 10 CFR 70.64(a)(5) requires that the MFFF design provide for adequate protection against chemical risks produced from licensed material, facility conditions which affect the safety of licensed material, and hazardous chemicals produced from licensed material. Related to chemical protection, 10 CFR 70.64(a)(3) requires that the facility design provide for adequate protection against fires and explosions, such as those that could be initiated by or involve chemicals at the proposed facility.

The review of the revised CAR focused on the design basis of chemical process safety systems, their components, and other related information. For each chemical process safety system, the staff reviewed information provided by the applicant for the safety function, system description, and safety analysis. The review also encompassed proposed design basis considerations such as redundancy, independence, reliability, and quality. The staff used Chapter 8.0 of NUREG-1718 as guidance in performing the review (NRC, 2000).

As stated in the memorandum of understanding (MOU) between the NRC and the Occupational Safety and Health Administration, "Worker Protection at NRC-licensed Facilities," (*Federal Register*, Vol. 53, No. 210, October 31, 1998, pp. 43950–43951), the NRC oversees chemical safety issues related to (1) radiation risk produced by radioactive materials, (2) chemical risk produced by radioactive materials, and (3) plant conditions that affect the safety and safe handling of radioactive materials. These chemical safety issues are important because they represent an increased radiation risk to workers. The NRC does not oversee facility conditions that result in an occupational risk but do not affect the safe use of licensed radioactive material. The MOU provisions applicable to the proposed MOX facility are now codified in 10 CFR 70.64(a)(5).

The NRC staff reviewed the following areas of the revised CAR applicable to process safety at the construction approval stage and consistent with the level of design (NRC, 2000, p. 8.0-8):

- AP description
- hazardous chemicals and potential interactions affecting licensed materials
- AP chemical accident sequences
- AP chemical accident consequences
- AP safety controls

Additional documentation from the applicant and the literature was reviewed as necessary to understand the process and safety requirements. In addition, the revised CAR incorporates the BDC of 10 CFR 70.64(a) into the design and operations of the proposed facility (see revised CAR, p. 5.5-53), and applicable sections of the revised CAR are intended to demonstrate compliance with these BDCs. The staff's detailed evaluation of the proposed AP process is presented in the sections that follow.

11.2.1.1 System Description of the AP Process

This section provides a description and overview of the AP process, including design, operational, and process flow information. A summary of the major components and functions is discussed in Section 11.2.1.2 of this FSER.

The AP process is designed to receive weapons-grade plutonium from the proposed pit disassembly and conversion facility (PDCF) and alternate feedstock (AFS) at SRS and to remove the impurities from the feed plutonium from the PDCF and AFS for use in the MOX process (MP). There are four major steps in AP:

(1)   The first step is the dissolution of plutonium dioxide powder by electrically generated silver (II) ions. The dissolution step involves the use of silver as a catalyst in a nitric acid medium. This electrolytically assisted route is independent of the powder characteristics. Dissolution is complete, with the rate determined by the generation of the silver (II) ions, at relatively mild temperatures (20–40 °C (68–104 °F); nominal temperature 30 °C (86 °F)). Dechlorination is required before dissolution for those powders containing chlorides (i.e., for AFS).

(2)   The second step involves plutonium purification by solvent extraction. Purification extracts and recovers plutonium, regenerates the solvent, and recycles nitric acid. Solvent extraction removes impurities such as gallium, uranium, and americium (CAR Section 11.3). The proposed facility would use a modified plutonium uranium reduction extraction (PUREX) process (e.g., Bennedict, 1981).

(3)   Third, the process includes conversion into plutonium dioxide by continuous oxalate calcination. Precipitation and calcination of the plutonium oxalate is a standard process used to prepare plutonium dioxide fuels.

(4)   Finally, recovery and recycle efforts are completed. Process side streams are treated to recover and recycle chemical reagents and reduce waste generation, as appropriate.

The proposed design of the AP process is similar to the design currently employed at La Hague's plutonium finishing facilities. Departures from the La Hague design result from U.S. regulatory requirements, lessons learned at La Hague, or manufacturing and throughput requirements specific to the facility.

11.2.1.2 Major Components and Functions

The AP process can be segmented into the following four operational areas (unit symbols are provided in parentheses):

(1)   plutonium purification process—includes the decanning unit (KDA), dissolution unit (KDB), dechlorination and dissolution unit (KDD), purification cycle (KPA), oxalic precipitation and oxidation unit (KCA), homogenization unit (KCB), milling unit (KDM), canning unit (KCC), and recanning unit (KDR), but no longer includes a uranium oxide dissolution unit

(2)   recovery processes—includes the solvent recovery cycle (KPB), oxalic mother liquor recovery unit (KCD), and acid recovery unit (KPC), but no longer includes a silver recovery unit.

(3)   waste storage—includes the liquid waste reception unit (KWD) and waste organic solvent unit (KWS)

(4)   offgas treatment—includes the offgas treatment unit (KWG)

11.2.1.2.1 Decanning Unit

The decanning unit consists of a series of workstations and gloveboxes distributed between the MP and AP areas. All cans entering the facility via this unit are initially packaged in 3013 standard containers and meet the 3013 standard. For this reason, the staff notes that such powders have been stabilized and are not pyrophoric and do not contain significant volatile species.

The three main functions of the decanning unit include the following:

(1)   density measurement (PDCF powders only)

(2)   opening of outer, inner, and convenience cans

(3)   transfer of powder to a dissolution dosing hopper (PDCF powder) or to a reusable can for the milling unit (AFS powder)

Density is determined from an x-ray level measurement on the entire, sealed 3013 container. Cans containing PDCF powders with a density less than 7 g/cc are sent to the dissolver for processing. Any PDCF powders with densities near or greater than 7 g/cc require ball milling and a subsequent, confirmatory density measurement to meet the less-than-7-g/cc limit. All AFS powders are ball milled. In addition, AFS powders are analyzed as follows:

•   If chlorinated species are detected, the reusable can is emptied into the dosing hopper of one of the electrolyzers in the dechlorination and dissolution unit.

•   If the powder does not contain chlorinated species, the reusable can is fed into the dosing hopper of the dissolution unit's electrolyzer.

•   If the powder contains chemical species not compatible with the AP process, the powder is transferred to the recanning unit for repackaging.

The plutonium feed design basis is discussed in Section 11.2.1.4 of this FSER.

11.2.1.2.2 Milling Unit

The milling unit consists of a series of gloveboxes. The functions of the milling unit are as follows:

•   Mill AFS powders to homogenize the plutonium dioxide and decrease the particle size to a level compatible with the dissolution process.

•   Perform density measurements on the milled powders.

•   Sample the powder for chemical analysis.

•   Store powders while awaiting analytical results.

As described in Section 11.2.1.2.1, the AFS powders are analyzed in the laboratory. Depending on the results of those analyses, the powder is transferred to different processes. If the chloride concentration is found to exceed 500 micrograms/gram of plutonium (500 ppm), the powder is directed to the feeding hopper for one of the two dechlorination electrolyzers (dechlorination and dissolution unit). If the chloride level exceeds the design basis values discussed in Section 11.2.1.4, the reusable can is sent to the recanning unit for repackaging in a 3013 container.

11.2.1.2.3 Recanning Unit

The recanning unit consists of two gloveboxes and two workstations designed to repackage powders that are not compatible with the AP process into 3013 containers. The powder is analyzed before transfer out of any unit in the facility. The operations include welding, tightness, and contamination checks.

11.2.1.2.4 Dissolution Unit

The function of the dissolution unit is to dissolve plutonium dioxide powders containing less than 500 ppm chlorides. The plutonium dioxide is electrolytically dissolved in the dissolution unit as a precursor for separating impurities (specifically americium, gallium, and uranium) in the purification cycle. Samples from the dilution and sampling tank are analyzed to determine the fissile material content and the required degree of dilution before being sent to the purification cycle feed tank.

The dissolution kinetics are improved by augmenting the reaction with a strong oxidizing agent (in this case, by electrolytic dissolution with silver ions (Ag[II])). Silver ions (Ag[II]) are electrolytically produced in a cylindrical compartment. The electrolytic dissolution takes place in a 6 N nitric acid solution at nominally 30 °C (86 °F). The general dissolution process may be described as follows:

<u>Electrolytic production of Ag^{2+} (at the anode)</u>:
$$Ag^+ = Ag^{2+} + e^- \qquad\qquad (11.2\text{-}1)$$

<u>Dissolution of PuO_2 powder</u>:
$$PuO_2 \text{ (solid)} + Ag^{2+} = PuO_2^+ \text{ (solid)} + Ag^+ \qquad\qquad (11.2\text{-}2)$$
$$PuO_2^+ \text{ (solid)} + HNO_3 = PuO_2^+ \text{ (solution)} \qquad\qquad (11.2\text{-}3)$$
$$PuO_2^+ \text{ (solution)} + Ag^{2+} = PuO_2^{2+} \text{ (solution)} + Ag^+ \qquad\qquad (11.2\text{-}4)$$

This gives the following general reaction:

$$PuO_2 \text{ (solid)} + 2Ag^{2+} = (PuO_2)^{2+} \text{ (solution)} + 2Ag^+ \qquad\qquad (11.2\text{-}5)$$

Ag^+ ions are oxidized at the anode. The staff review notes that electrolytic dissolution of the plutonium dioxide is indirect; electrolysis produces silver (II), which affects the actual dissolution of the plutonium dioxide. If a sufficient concentration of silver is not available, other anode (and near-anode) reactions might occur, such as the following:

$$2\, H_2O = O_2 \text{ (gas)} + 4H^+ + 4e^- \text{ (water electrolysis/oxidation)} \qquad\qquad (11.2\text{-}6)$$

$$2\, Ag^{2+} + H_2O = 2\, Ag^+ + 2\, H^+ + (\tfrac{1}{2})\, O_2 \text{ (gas)} \qquad\qquad (11.2\text{-}7)$$

(water oxidation by silver)

Water oxidation by silver (II) is promoted at temperatures of 40 °C (104 °F) and above (Bourges, 1985). The production of oxygen may have safety concerns. For example, the presence of oxygen can lead to explosions with hydrogen.

The applicant stated that the following reduction reaction takes place at the cathode:

$$NO_3^- + 3H^+ + 2e^- = HNO_2 + H_2O \qquad\qquad (11.2\text{-}8)$$

The literature also identifies the following cathode reactions (Bray, 1987, 1992; Bourges, 1985):

$$2\,H^+ + NO_3^- + e^- = NO_2\ (gas) + H_2O \qquad\qquad (11.2\text{-}9)$$

$$4\,H^+ + NO_3^- + 3e^- = NO\ (gas) + 2\,H_2O \qquad\qquad (11.2\text{-}10)$$

Hydrogen generation can also occur at the cathode:

$$2\,H^+ + 2\,e^- = H_2\ (gas) \qquad\qquad (11.2\text{-}11)$$

The staff review notes hydrogen generation usually occurs at a low rate at all times. Under off-normal conditions, such as overvoltage or reduced acid concentrations, hydrogen generation has the potential to increase substantially and become the dominant cathode reaction.

Dissolution occurs when a current is applied. The joule heating effect of the electrical current supplied is reduced by cooling the anolyte and the catholyte. The staff review notes that electrolytic processes usually operate at 90–95 percent efficiency (i.e., 90–95 percent of the current goes towards the intended reaction) under the best conditions. The literature indicates that the silver-assisted plutonium dissolution reaction occurs with 30–70 percent efficiencies (Bray, 1987, 1992; Bourges, 1985). Side reactions almost invariably occur and likely involve the evolution of gases, such as hydrogen, oxygen, and nitrogen oxides. At higher and lower electrode voltages, the electrolyzer would operate in a different regime, and a higher percentage of the current could produce gas evolution.

The dissolution unit is operated in batches. The dissolution unit and the companion dechlorination and dissolution unit are designed to treat 22 kg/day (48.51 lb/day) of plutonium dioxide. The operating ranges depend on the quantity of AFS powder processed. During the processing of AFS powder (which may occur over the first 3 years of operation), the dissolution units are anticipated to process between 4 kg/week (8.8 lbs/week) of powder (i.e., one batch from the dechlorination and dissolution unit) up to a maximum of 129 kg/week (284 lbs/week) of powder. This correlates to six dissolutions per week for both lines of the dechlorination and dissolution unit and six batches of the single line in the dissolution unit (13.5 kg/batch (29.8 lb/batch)). During the processing of PDCF powders, the minimum processed is expected to be 13.54 kg/week (29.8 lbs/week) of powder (i.e., one batch through the single electrolyzer of the dissolution unit). The maximum processed is estimated as 243 kg/week (536 lbs/week) of powder, corresponding to six dissolutions per week for all three electrolyzers. The nominal flow rate to the purification cycle is approximately 14.4 L/h (3.8 gal/h).

Hydrogen peroxide is added to the receiving tank used for interim storage in order to adjust the oxidation state of the plutonium from (VI) to (IV). A plutonium (IV) oxidation state allows for better extraction and separations processing. The peroxide also reduces any excess silver (II) to silver (I). The uranium impurity exists as the ^{235}U isotope (from radioactive decay of ^{239}Pu and nonseparable portions of the original pit—essentially more than 93 percent uranium enrichment or assay). Consequently, an isotopic dilution to around 30 percent assay is made by feeding the appropriate quantity of depleted uranium nitrate solution (0.25 percent ^{235}U) to the dilution and sampling tank. Other adjustments (e.g., acidity) may also be made to the solution in the dilution and sampling tank to optimize subsequent purification of the plutonium.

11.2.1.2.5 Dechlorination and Dissolution Unit

The dechlorination and dissolution unit includes two electrolyzers, each the same size as the single electrolyzer in the dissolution unit. However, as noted previously, batch sizes are different depending on the feed material.

The functions of the dechlorination and dissolution unit are to remove chloride ions from the plutonium dioxide powder and then dissolve the material. The unit processes powders with chloride levels exceeding 500 ppm. The unit can also be used to process nonchlorinated feed materials (e.g., the PDCF feed).

Each electrolyzer line includes a scrubbing column for the removal of chlorine. The electrolyzers are operated in two modes. The first mode is dechlorination—electricity is applied and the chlorine gas is evolved at the anode as follows:

$$2 \; Cl^- = Cl_2 \; (g) + 2 \; e^- \qquad\qquad\qquad (11.2\text{-}12)$$

The applicant indicated that the process removes greater than 99 percent of the initial chloride in the powder, based on operation with 6 N nitric acid and a temperature of 60 °C (140 °F).

In the second mode of operation (i.e., after chloride removal), the electrolyzer is operated in a manner completely analogous to the dissolution unit electrolyzer. Silver ions are fed into the electrolyzer, and the powder is dissolved.

11.2.1.2.6 Purification Cycle

The purification cycle uses a modification of the standard PUREX process (Bennedict, 1981). Plutonium nitrate from the dissolution unit is received, and plutonium is solvent extracted and scrubbed for impurities. The plutonium with uranium left in the stream is stripped after adjustment of the plutonium valence to the trivalent state. The purification cycle includes plutonium reception, recycle, and transfer to the oxalic precipitation and oxidation unit. The purification cycle also sends the solvent/diluent stream to the solvent recovery cycle and the raffinate stream to the acid recovery unit.

The main goal of the purification cycle is to separate plutonium from impurities contained in the solution coming from the dissolution unit. In the revised CAR, the applicant identified the main functions of the purification cycle as follows:

- Receive plutonium nitrate from the dissolution unit and the dechlorination and dissolution unit.

- Receive recycled plutonium nitrate from the oxalic mother liquor recovery unit.

- Receive solutions with high plutonium content from the laboratories.

- Perform plutonium extraction and impurities scrubbing.

- Perform plutonium stripping and uranium scrubbing.

- Perform further plutonium purification by additional plutonium stripping and diluent washing.

- Adjust plutonium to the tetravalent state.

- Perform plutonium stripping in a plutonium barrier.

- Perform scrub bed uranium stripping and diluent washing.

- Control purified plutonium and transfer to the oxalic precipitation and oxidation unit or recycle to the beginning of the purification cycle.

- Control and dilute scrubbed uranium and prepare uranium for transfer to the liquid waste reception unit.

- Wash, control, and transfer raffinates to the acid recovery unit.

- Wash, control, and transfer solvent/diluent to the solvent recovery cycle.

- Destroy residual hydroxylamine nitrate (HAN)/hydrazine and hydrazoic acid in the aqueous phase transferred to the oxalic precipitation and oxidation unit.

The extraction process is continuous, but the feed solutions from the dissolution unit are received in batches. The raffinate and the plutonium nitrate solutions are transferred continuously to the acid recovery unit inlet buffer storage and to the oxalic precipitation and oxidation unit inlet buffer storage, respectively.

Plutonium nitrate solution is batched to the feed tank for plutonium extraction and impurities scrubbing. Plutonium (IV) in the aqueous solution (4.5 N HNO_3) is extracted by the solvent (30 percent tributyl phosphate (TBP) in a branched dodecane—the latter is sometimes referred to as hydrogenated propylene tetramer (HPT)) in a pulsed extraction column. The impurities remain primarily in the aqueous phase. The solvent stream is scrubbed by 1.5 N nitric acid in a pulsed scrubbing column to ensure good decontamination. The aqueous raffinates are washed by the diluent in a pulsed column and transferred to the raffinate reception tank.

Plutonium (IV) is reduced to plutonium (III) by HAN ($[NH_3OH^+][NO_3^-]$), and plutonium (III) is stripped in a pulsed stripping column (Reactions 11.2-13 and 11.2-14). Hydrazine nitrate is introduced to prevent parasitic reoxidation of plutonium (III) back to plutonium (IV) (Reactions

11.2-15 and 11.2-16). The stripped plutonium is washed with diluent in a pulsed diluent washing column before the final valence adjustment. Unstripped plutonium is extracted in the plutonium barrier mixer-settler bank. Hydrazine nitrate and HAN are introduced in the last stage of the plutonium barrier. The solvent from the plutonium barrier flows to the uranium-stripping mixer-settler bank. These reactions are written as follows:

Plutonium reduction by HAN ($NH_3OH[NO_3]$):
$$2[NH_3OH]^+ + 4Pu^{+4} = 4Pu^{+3} + N_2O(g) + H_2O + 6H^+ \qquad (11.2\text{-}13)$$
$$2[NH_3OH]^+ + 2Pu^{+4} = 2Pu^{+3} + N_2(g) + 2H_2O + 4H^+ \qquad (11.2\text{-}14)$$

Plutonium reduction by hydrazine:
$$4Pu^{4+} + N_2H_5^+ + H_2O = 4\,Pu^{3+} + N_2O + 5\,H^+ \qquad (11.2\text{-}15)$$

Parasitic reoxidation of Pu(III) to Pu(IV):
$$2Pu^{3+} + 2HNO_2 + 3H^+ + NO3^- = 2Pu^{4+} + 3HNO_2 + H_2 \qquad (11.2\text{-}16)$$

Uranium is stripped (recovered from the organic phase) in a slightly acidic 0.02 N nitric acid solution in a uranium-stripping, mixer-settler bank. The stripped uranium stream is washed with diluent in a three-stage, diluent-washing, mixer-settler bank. Isotopic dilution with depleted uranium nitrate solution is performed as necessary. The stripped solvent from the uranium-stripping mixer-settler is directed to the solvent recovery cycle. The aqueous phase from the uranium diluent washing is directed to the liquid waste reception unit.

The final valence adjustment of plutonium (III) to plutonium (IV) is achieved by oxidizing the plutonium (III) solution with nitrous fumes (essentially a nitrogen dioxide/nitrogen tetraoxide mixture). In this process, excess HAN and hydrazine are eliminated, and air stripping of the plutonium solution in an air-stripping column destroys the nitrous acid. The plutonium nitrate solution is received in the plutonium reception tank from where it is transferred to the batch constitution tanks of the oxalic precipitation and oxidation unit.

The selected aqueous-to-organic ratios in the plutonium extraction and plutonium stripping operations enable the process to obtain a plutonium concentration close to 40 g/L (0.34 lb/gal) at the outlet of the purification cycle.

The purification cycle operates continuously. The feeding solutions from the dissolution unit are received in batches. Thus, the nominal flow rate to purification is approximately 15.1 L/h (4 gal/h). This cycle is designed to process 14.5 kg/day (32 lb/day) of plutonium. The operating range of the purification cycle is 11–19 kg/h (24.3–42 lb/h) of plutonium.

11.2.1.2.7 Solvent Recovery Cycle

The solvent recovery cycle operates continuously in conjunction with the purification cycle. Standard PUREX methods are used to wash the solvent and remove the degradation products.

In the revised CAR, the applicant identified the functions of the solvent recovery cycle as follows:

•   Recover the used solvent from the purification cycle to prevent the accumulation of degradation products.

- Renew the solvent and adjust its TBP content.

- Store the treated solvent and continuously feed the purification cycle.

- Perform a diluent wash operation on the aqueous effluents produced by this operation to remove traces of entrained solvent. Effluent in this section refers to effluent from individual process units to other process units; the facility discharges no radioactive liquid effluent directly into the environment.

Washing removes undesirable byproducts, such as hydrazoic acid and TBP degradation products, from the solvent to prevent their accumulation and associated deleterious impact upon process and safety performance. The solvent is washed by contact with an aqueous solution of sodium carbonate and sodium hydroxide in mixer-settler banks. Washing removes hydrazoic acid (as sodium azide) and TBP degradation products (mainly monobutyl phosphate and dibutyl phosphate) from the solvent to the aqueous stream. The washed solvent is collected in a buffer tank where it is cooled. The purification cycle is continuously fed at a controlled flow rate using a dosing pump. The excess solvent, generated by the diluent wash and the content adjustment TBP wash, is transferred to the liquid waste reception unit. The aqueous effluents generated by washing undergo a diluent wash in a mixer-settler battery (one stage) at ambient temperature to remove traces of entrained solvent. The aqueous-to-organic phase ratio for this operation is around 100:1.

The diluent is recycled in the mixer-settler with a specific system including an airlift and two pots. The recycling flow rate equals the incoming aqueous flow rate from the mixer-settler bank. The aqueous-to-organic ratio is close to 1 when recycling is in operation.

The unit is designed to treat solvents at a flow rate of 17.4 L/h (4.6 gal/h), which corresponds to 14.2 kg/day (31.7 lb/day) of plutonium.

11.2.1.2.8 Oxalic Precipitation and Oxidation Unit

In the revised CAR, the applicant identified the functions of the oxalic precipitation and oxidation unit as follows:

- Receive purified plutonium nitrate concentrated to approximately 40 g/L (0.34 lb/gal) from the purification cycle and prepare uniform batches.

- Precipitate out the plutonium nitrate as oxalate.

- Produce plutonium dioxide after filtering, drying, and calcining the oxalate. The filtering operation includes drawing off the mother liquors, washing, and dewatering the plutonium oxalate cake.

- Transfer the plutonium dioxide to the homogenization unit and transfer the mother liquors and the filter washing solutions to the oxalic mother liquor recovery unit. Solutions are transferred by air lifts and flow control valves as described in Section 11.3.2.8.2 of the revised CAR.

- Ensure reducing agents, hydrazoic acid, and plutonium (VI) do not propagate into downstream processing units (e.g., the oxalic mother liquor recovery unit).

The following depicts the precipitation reaction:

$$Pu(NO_3)_4 + 2\ H_2C_2O_4 = Pu(C_2O_4)_2\ (s) + 4\ HNO_3 \hspace{3em} (11.2\text{-}17)$$

The conversion line is rated for the processing of 1.1 kg/h (2.3 lb/h) (25.7 kg/day (56.6 lb/day)) of plutonium. The operating range of the oxalic precipitation and oxidation unit is 0–1.25 kg/h (2.75 lb/h) of plutonium. Plutonium nitrate solutions arrive from the purification cycle where acidity and valency are adjusted. They are received in alternate batches in two annular tanks to form a batch with a volume of 0.6 m³ (21.2 ft³ or about 158.5 gal). Solutions are transferred by a pump to two dosing wheels, which supply one precipitator each. The solutions flow by gravity to the precipitators.

Precipitation takes place in two precipitators which are connected in parallel. The oxalate reagents are injected into each precipitator. The plutonium oxalate precipitate carried by the mother liquors escapes via the precipitator overflows and flows by gravity to a rotary filter. Rotation ensures that dewatering and cake removal are continuously performed. The filter removes the plutonium oxalate cake, plate by plate, with a scraper. The removed cake is collected by a screw mechanism and falls into a chute. It enters the next processing operation (the calcination furnace) by gravity.

The furnace consists of two main parts—a drying zone where the plutonium oxalate is dried, and a calcining zone where the oxalate is transformed into plutonium dioxide in an oxidizing atmosphere of oxygen. The following shows the overall reaction:

$$Pu(C_2O_4)_2 + O_2 = PuO_2 + 4\ CO_2 \hspace{3em} (11.2\text{-}18)$$

The staff notes that this is plutonium (IV) oxalate, which generally requires a higher temperature (400–500 °C) for conversion to the oxide than either plutonium (III) or plutonium (VI) oxalates (around 200 °C). Potential concerns and controls for plutonium (VI) oxalate are discussed further in Section 8.1.2.5.1.6 of this FSER. Potential concerns and controls with plutonium (III) oxalate are discussed in Section 11.3 under plutonium pyrophoricity.

The oxalic mother liquors, which are collected in separator pots, flow by gravity to the oxalic mother liquor recovery unit. The filtered mother liquors are adjusted to approximately 3.3 N with recovered 13.6 N nitric acid to avoid any risk of precipitation of plutonium oxalate caused by residual oxalic acid.

The gases produced during drying and calcination of the plutonium oxalate (e.g., carbon dioxide and steam, nitric oxides, and trace organics), the excess of oxygen, and the air from upstream and downstream of the process are removed by a negative-pressure circuit. This system comprises a filter, a condenser, a demister, an electric heater, two high-efficiency particulate air (HEPA) filters, and two fans.

11.2.1.2.9 Homogenization Area

The plutonium dioxide produced in the oxalic precipitation and oxidation unit is continuously fed by gravity from the calcination furnace into one of the two separating hoppers installed in parallel. The plutonium material balance is determined by weighing the filled cans (canning unit) and by determining the plutonium content of the hopper by powder sampling. Sampling ensures that all the finished product specifications are met in each batch of plutonium dioxide in each hopper and checks the isotopic composition of the plutonium dioxide for the finished product of each batch in each hopper.

In the revised CAR, the applicant identified the functions of the homogenization unit as follows:

•   Receive and homogenize the plutonium dioxide produced in the oxalic precipitation and oxidation unit.

•   Fill cans with plutonium dioxide in such a manner that the mass of plutonium per can is constant.

•   Prepare samples for laboratory analysis to characterize the batch.

•   Perform sample-based residual moisture measurement.

•   Perform thermogravimetry analysis.

•   Store reference samples.

11.2.1.2.10 Canning Unit

The canning unit is designed to package plutonium dioxide powder in reusable stainless steel cans and transfer them one by one to the MP plutonium dioxide buffer storage unit to prepare the MOX powder. It is also used to establish the plutonium dioxide powder material balance. The nominal capacity is about 10 cans of plutonium dioxide per day, each filled with approximately 2.4 kg (5.3 lb) of plutonium dioxide. The plutonium dioxide powder is fed by gravity from the homogenizer at a temperature not exceeding 150 °C (302 °F). Full plutonium dioxide cans are transferred pneumatically in a shuttle to the MP plutonium dioxide buffer storage unit. Cans that are rejected because of overfilling (as indicated by weighing) or unsatisfactory laboratory results are transferred and recycled to the appropriate upstream process. The nominal flow rates are as follows:

•   plutonium dioxide inlet from the homogenization unit—1.2 kg/h (2.6 lb/h)
•   plutonium dioxide outlet—approximately 10 full reusable cans per day

11.2.1.2.11 Oxalic Mother Liquor Recovery Unit

In the revised CAR, the applicant identified the functions of the oxalic mother liquor recovery unit as follows:

•   Continuously receive oxalic acid mother liquors adjusted to 3.3 N with nitric acid from the oxalic precipitation and oxidation unit.

- Continuously receive ventilation effluent droplets from the oxidation and degassing columns.

- Concentrate the oxalic mother liquors in a subcritical evaporator to destroy the oxalic acid and remove plutonium from the distillates.

- Check and transfer the distillates to the acid recovery unit.

- Monitor and recycle, batchwise, the concentrates to the top of the purification cycle.

The nominal capacity corresponds to the processing of the materials generated by the precipitation of 24 kg (HM)/day (52.9 lb/day) of plutonium. This translates into a liquor flow rate of approximately 250 L/h (66 gal/h). The oxalic acid mother liquor recovery unit operates continuously, unlike the oxalic acid precipitation and oxidation unit, which produces the oxalic mother liquors. Consequently, the design includes buffer tanks with more than 3 days capacity. This allows for the independent operation of the two units. The mother liquor solution flows by gravity into the buffer tanks (two tanks, about 1 m³ (264 gal) each). After sampling for plutonium concentration, an airlift transfers the solution into a feed tank (also of 1 m³ (264 gal) capacity). These tanks have a geometrically safe, annular design. A double airlift transfers the solution from the feed tank into an evaporator. The evaporator concentrates the liquor and generates a relatively clean overhead product (distillate). In the evaporator, residual plutonium oxalate is converted into plutonium nitrate and oxalic acid. In the presence of manganese (II) ions (added as a catalyst) and excess nitric acid, the resulting free oxalic acid decomposes into carbon dioxide, water, and nitric oxide. The following equations depict these reactions:

$$Pu(C_2O_4)_2 + 4\ HNO_3 = Pu(NO_3)_4 + 2\ H_2C_2O_4 \text{ (Mn(II) as catalyst)} \quad (11.2\text{-}17)$$

$$H_2C_2O_4 + 2\ HNO_3 = 2\ CO_2 + 2\ NO_2 + 2\ H_2O \text{ (Mn(II) as catalyst)} \quad (11.2\text{-}18)$$

The evaporator exposes the plutonium nitrate to prolonged boiling (100-135 °C (212-275 °F)) in a highly acidic and oxidizing environment. Consequently, plutonium (IV) and plutonium (III) are oxidized to plutonium (VI) nitrate by reactions such as the following:

$$3\ Pu^{+3} + HNO_3 + 3H^+ = 3\ Pu^{+4} + NO + 2\ H_2O^- \quad (11.2\text{-}19)$$

$$3\ Pu^{+4} + 2\ H_2O = 2\ Pu^{+3} + PuO_2^{+2} + 4\ H^+ \quad \text{(Mn(II) as catalyst)} \quad (11.2\text{-}20)$$

The distillate is condensed and cooled, and a small percentage is returned to the evaporator/column system to supply reflux via a pot. The net distillate product is analyzed for the plutonium concentration by an online neutron counter. In the revised CAR, the applicant stated that, if the concentration is sufficiently low, the distillate is routed to the acid recovery unit. If the plutonium specification is exceeded, the distillate is transferred to the buffer tanks for recycling and retreatment.

The concentrates are removed from the evaporator by an airlift and placed in small buffer tanks. Because of the oxidation reactions, the plutonium is present in the hexavalent oxidation state (as PuO_2^{+2}). The applicant noted in the revised CAR that, if the residual oxalate concentration meets requirements, then the concentrates are returned to the purification unit via an airlift.

11.2.1.2.12 Acid Recovery Unit

In the revised CAR, the applicant identified the following functions for the acid recovery unit:

- Receive extraction raffinates from the purification cycle, oxalic mother liquor distillates from the oxalic mother liquor recovery unit, and effluents from laboratories in batches and continuously receive active liquid effluents from the offgas treatment unit equipment ventilation.

- Concentrate the radioactivity contained in the effluents and send it to the liquid waste reception unit.

- Recover concentrated acid for recycling in the process.

- Recover distillates from the rectification column for use in the offgas treatment unit and the purification cycle, with excess liquid to the liquid waste reception unit.

The acid recovery unit uses evaporation as the principal treatment method. The feed tank receives the following materials:

- raffinates from the purification cycle in 1500 L (396 gal) batches

- oxalic mother liquor distillates (oxalic mother liquor recovery unit evaporator 3000) in 2500 L (660 gal) batches

- recombined acid from the offgas treatment unit (continuously)

- effluents from laboratories in batches

The solution is transferred by double-stage airlifts from the feed tank to the boiler of the first evaporator, a natural recirculation, thermosiphon design. The heating power is kept constant by regulating the steam pressure in the boiler. The concentrates are drained off into a 500 L (132 gal) tank discontinuously, several times a day. The concentrates contain the americium and gallium impurities and are sent to the liquid waste reception unit. The overhead product is fed to the second evaporator, which has a similar design and includes a rectification column on the overhead product. The reflux system at the top of the column can be used to spray the upper trays and improve decontamination. The recovered acid is drawn off as a concentrate product by airlift and cooled. The acid drawoff flow rate is regulated to maintain the desired acidity. The acid is recycled within the facility. The distillate product is continuously transferred by pump for AP water recycling. Any excess recycle water is analyzed and temporarily stored before transfer by pump to the liquid waste reception unit.

11.2.1.2.13 Silver Recovery Unit

The applicant decided to eliminate the silver recovery unit for economic reasons. The silver nitrate used in the dissolution unit is part of the high alpha waste stream that will be sent to the acid recovery unit.

11.2.1.2.14 Offgas Treatment Unit

The offgas treatment unit ventilation system is dedicated to maintaining a system negative pressure for AP process confinement and for removal vapors and gases from processing equipment. In the revised CAR, the applicant identified the following functions of this unit:

- Remove plutonium from offgases collected from the dissolution unit, the dechlorination and dissolution unit (during the dissolution step), the oxalic precipitation and oxidation unit, the oxalic mother liquors recovery unit, the acid recovery unit, the liquid waste reception unit, and the oxidation and degassing columns (purification cycle).

- Recombine the nitrous fumes in a specific nitric oxide scrubbing column.

- Clean, by water scrubbing, the offgases collected from all AP units.

- Treat the offgas flow by HEPA filtration before releasing to the stack.

- Maintain negative pressure in the tanks and equipment connected to the process ventilation system.

A specific offgas treatment unit extraction system is dedicated to the pulsed purification columns, with similar functions:

- Treat offgases by HEPA filtration before releasing to the stack.
- Maintain negative pressure in the pulsation system and the pulsed columns legs.

The offgases containing nitric oxide (from dissolution/oxidation and degassing columns) are gathered downstream of a cap impactor to remove droplets. The collected effluent stream is recycled, by gravity, to the oxalic mother liquor recovery. Offgases are then scrubbed with recycled effluents and with recovered distillates from the acid recovery unit. The offgases pass through a demister, using an air ejector to provide the motive force. The extraction rate is regulated based on the pressure in the scrubbing column.

Normal ventilation gases (i.e., process vents) are combined with the treated nitric oxide gas streams. These gases are scrubbed with recycled effluents and then with water. The washed gases successively pass through a cooler, a demister, an electric heater, a double HEPA filtering line (2×2), and an exhauster before release through the stack.

The pulsation air from solvent extraction is treated in a separate (extraction) line. The air successively passes through an electric heater, a HEPA filtering line (2×2), and an exhauster before release through the stack.

The offgas treatment unit operates continuously. The nitric oxide scrubbing column is designed to treat approximately 62 N m^3/h (36 cubic feet per minute (cfm)). The main ventilation line (offgas scrubbing and filters) is designed to process approximately 600 N m^3/h (350 cfm). The designed capacity of the column pulsation air extraction is 150 N m^3/h (88 cfm).

The applicant stated that specific operating limits and the associated items relied on for safety (IROFS) will be provided in the ISA.

11.2.1.2.15 Liquid Waste Reception Unit

The liquid waste reception unit receives liquid waste from the AP process for temporary storage before sending it to SRS for treatment and processing (see Sections 11.3.2.14 and 10.1.4 of the revised CAR). Waste stream descriptions and quantities are summarized in Table 11.2-1. The functions of this unit are to treat the following liquid waste streams:

•   The low-level liquid waste stream, which comprises of (1) the room HVAC condensate, rinse water from laboratories, and washing water from sanitaries which are potentially noncontaminated and are collected as low-low-level liquid waste, (2) the distillate stream from the acid recovery unit which is contaminated and slightly acidic, (3) miscellaneous floor washes from C2/C3 rooms and overflows or drip tray material from some of the reagent tanks in the AP building, and (4) the chloride stream from the scrubbers used during the dechlorination step for AFS feeds (i.e., from the dechlorination and dissolution unit).
                                [Text removed under 10 CFR 2.390]

•   The high alpha waste stream is a combination of three waste streams—americium, alkaline waste, and excess acid. The americium stream collects americium and gallium nitrates and all of the silver used in the dissolution unit, along with traces of plutonium. The alkaline waste stream from the solvent recovery area contains dilute caustic soda (NaOH), sodium carbonate, sodium azide, and traces of uranium and plutonium. The excess acid stream from the acid recovery unit contains high alpha activity excess acid. The high alpha storage tank along with the high alpha buffer storage tank are a holding point for high alpha wastes and provide 90 days of storage.
                                [Text removed under 10 CFR 2.390]

•   The stripped uranium (less than 1 percent ^{235}U) waste stream receives the contents of the uranium dilution tanks in the purification cycle.
                                [Text removed under 10 CFR 2.390]

11.2.1.2.16 Waste Organic Solvent Unit

The function of the waste organic solvent unit is to handle the organic solvent generated by the process and transfer it to SRS for handling and disposal.

The unit consists of an intermediate solvent tank, a final solvent tank, and a carboy filling station. It is designed to collect waste solvent for sampling in order to assure compliance with the SRS waste acceptance criteria (WAC).

The intermediate solvent tank is located in a closed process cell and has mixing and automatic sampling capabilities. Sampling is conducted to determine compliance with the SRS WAC. If the waste solvent meets the WAC, it is transferred to the final waste solvent tank.

The final waste solvent tank receives solvent from the intermediate tank and from the solvent recovery cycle unit solvent tank. This final tank is located in an open process cell and has mixing and manual sampling capabilities. Waste solvent is transferred from this tank to a

Table 11.2-1 Waste Stream Descriptions and Quantities in the Waste Reception Unit

Waste Stream Designation	Annual Volume, Liters (gal)	Main Chemical or Isotope	Disposition, Liters (gal)
Liquid americium stream (concentrated stream from acid recovery)	11,400 (3,000) (PDCF) 16,700 (4,400) (AFS)	^{241}Am < 24.5 kg/y (54 lb/y) (3.1×10^{15} Bq (84,000 Ci)) Pu <150 g/y (0.33 lb/y) Hydrogen ions 9.4 N Nitrate salts 2200 kg/y (4850 lb/y) from silver nitrate Silver < 300 kg/y (661 lb/y) Trace quantities of thallium, lead, and mercury	High alpha waste to WSB 39,000 (10,300) (PDCF) 31,900 (8,421) (AFS)
Excess acid Stream	5,000 (1,321) (AFS) 15,500 (4,100) (PDCF)	Am < 90 mg/y (2×10^4 lb/y) (rectification step after two evaporation steps) Hydrogen ions 13.6 N	
Alkaline stream	12,100 (3,200) (PDCF) 10,200 (2,700) (AFS)	Pu < 16 g/y (0.035 lb/y) U < 13 g/y (0.029 lb/y) Na < 110 kg/y (243 lb/y)	
Stripped uranium stream	167,000 (44,000) (PDCF) 151,000 (40,000) (AFS)	Pu < 0.1 mg/L U < 3100 kg/y (6834 lb/y)(1% assay) Hydrogen ions 0.14 N	Stripped uranium to WSB 167,000 (44,000) (PDCF) 151,000 (40,000) (AFS)
Excess low level radioactive solvent wastes	10,600 (2,800) (PDCF) 9,800 (2,600) (AFS)	Solvent 30% TBP in dodecane Pu < 17.2 mg/y (3.8×10^5 lb/y)	LLW solvent to SRS solvent recovery 10,600 (2,800) (PDCF) 9,800 (2,600) (AFS)
Distillate waste	386,000 (102,000) (PDCF) 265,000 (70,000) (AFS)	^{241}Am 241 < 0.85 mg/y (1.9×10^6 lb/y) Activity 2.39×10^8 Bq/y Hydrogen ions 0.02 N	Liquid LLW 1,080,000 (285,000) (PDCF) 1,030,000 (273,000) (AFS)
Chloride removal waste	75,700 (20,000) (AFS)	Only when processing AFS materials Chloride < 5 g/L (0.04 lb/gal)	
Rinsing water	598,000 (158,000) (PDCF) 598,000 (158,000) (AFS)	Alpha activity < 4 Bq/L Hydrogen ions 0 0.6 N	
Internal HVAC condensate	94,600 (25,000)	Trace contamination	

carboy. The slightly contaminated solvent is anticipated to be a low-level waste (LLW). This waste is stored in a 300-gallon carboy or other suitable vessel and transferred by truck to SRS for disposition.

11.2.1.2.17 Uranium Dissolution Unit

The applicant decided to purchase depleted uranium nitrate solution as a reagent from an outside supplier. Consequently, the uranium dissolving unit has been eliminated from the proposed facility (DCS, 2002e).

11.2.1.2.18 Sampling System

The sampling system is used for radioactive and chemical solutions. Section 11.3.2.16 of the revised CAR discusses the following three liquid sampling system approaches that the applicant intends to use at the facility:

(1)   In direct sampling, the solution is directly extracted from the process equipment by gravity flow or with a recycling pump into a vial. Direct sampling is limited to nonaggressive reagents or effluents of suspect origin. A large sample volume provides a lower detection limit.

(2)   In suction sampling, a vial is filled by suction through a needle by the vacuum in the vial. Aggressive reagents can be sampled manually but with vacuum vial filling. Particles are not expected in the sampling system because all AP process solutions are expected to be clear. A moving cask is used for suction filling of active liquids.

(3)   With remote sampling, the solution is lifted up by an airlift head from which direct vacuum sampling is carried out. For concentrated radioactive liquid waste, remote sampling under a box is required.

The applicant stated that all sampling systems will be qualified using engineering studies and/or evaluations.

11.2.1.3 Staff Review of the AP Process System

11.2.1.3.1 Chloride Concentration of AFS Powder (KDM)

The staff evaluated the risk to the site worker and the individual outside the controlled area (IOC) from chlorine emissions resulting from AP processes into the environment. Alternate plutonium dioxide feedstock that contains greater than 500 ppm chloride salts would be processed in the dechlorination electrolyzers. The maximum concentration of chloride salts in alternate feedstock is expected to be a mass fraction of 20–33 percent chloride. If such a high chloride-containing stream is directed to the normal (nondechlorination) dissolution electrolyzer, which does not have a chlorine scrubber, by mistake, up to 3–5 kg (6.6–11 lbs) of chlorine gas could be released. This would result in an intermediate-level consequence to the IOC if the release were to occur in less than 1 hour. The staff identified this as Open Item AP-07 in the revised draft safety evaluation report (DSER) (NRC, 2003c).

The staff has analyzed this event and concludes that the electrical current necessary to evolve this rate of material would be significantly larger than the steady-state load of the electrolyzer, and that the electrical current would have to exist for a relatively long time (15–60 minutes or more) and would likely exceed standard protection devices, such as circuit breakers, and/or the capacity of the electrolyzer itself. The controls would likely terminate the current before significant quantities of chlorine could be evolved. Thus, the staff concludes that this is an unlikely or highly unlikely event, and that no additional controls are necessary to meet the performance requirements in 10 CFR 70.61. "Performance Requirements." This closes AP-07 from the revised DSER (NRC, 2003c).

11.2.1.3.2 Electrolyzers (KDB and KDD)

The applicant provided information (DCS, 2001) that discussed a loss of confinement scenario for the electrolyzer, based on an overtemperature situation caused by a control system failure, electric isolation failure, or a loss of cooling

[Text removed under 10 CFR 2.390]

The applicant concluded that the event must be either prevented or mitigated and selected a prevention strategy to reduce the risk to the facility worker and the environment, based on shutdown of the electrolyzer and natural cooling. The applicant identified the safety design basis as the detection of the high temperature (identified as greater than 70 °C (158 °F)) and shutdown of the electrolyzer and related processes without exceeding any design limits or chemical control limits, using assigned channels on the emergency control system. Shutdown was understood to be a termination of the electrical current. The PSSCs identified by the applicant are the temperature and shutdown controls, as well as the process safety instrumentation and control (I&C) system.

The applicant's response further noted that the electrolyzer is geometrically safe to preclude potential criticality events. The applicant mentioned isolation of the anode and cathode and an isolation monitoring system; these are not identified as PSSCs. The applicant also stated that the scavenging and emergency air systems would be used to preclude the possibility of explosions, based on the rate of hydrogen generated by radiolysis. Consequently, the applicant indicated that the voltage to the electrolyzer would be limited.

The safety strategy for the site worker and the IOC relies upon mitigation features. The PSSC is the C3 confinement system, which has the safety function of filtering radioactive materials from the C3 air before release into the atmosphere.

In subsequent discussions (DCS, 2002d; NRC, 2002c), the applicant indicated that there will be multiple, redundant temperature sensors located at the highest temperature location(s). These would provide temperature signals to the process safety I&C system (a PSSC), which would terminate electrical power. The applicant has also stated the temperature design basis of 70 °C (125 °F) incorporates the effects of potential reaction exotherms. The power is actually terminated before the temperature reaches 70 °C (125 °F).

The NRC staff conducted a literature review on the electrical dissolution of plutonium dioxide (Bray, 1987, 1992; Bourges, 1985). Key points from this review are as follows:

- Plutonium dioxide from different sources and calcination conditions (e.g., high-fired) can be effectively and rapidly dissolved.

- Multiple reactions are involved.

- Current efficiencies are in the 30–70 percent range.

- Nitric oxide is formed at the cathode; ammonia/ammonium ions can also be produced.

- Hydrogen is produced over a wide range of acidities and cathodic materials; at catholyte concentrations below about 2 N nitric acid, measured hydrogen gas concentrations exceeded 1 percent in the evolved gases. Potential reactions between evolved hydrogen and nitric oxide were not discussed.

- Hydrogen is a concern because of potential flammability in the gas space and hydrogen embrittlement in certain metals used in construction (e.g., titanium).

- Polarization curves were not available.

Several safety or operational limits are mentioned in the literature, which would require shutting off the power if exceeded. These limits include catholyte density (e.g., an indirect measure of acid concentration), catholyte flow, offgas hydrogen concentration, offgas flow, cell temperature, anolyte cooling water flow, catholyte cooling water flow, cell applied voltage, and cell current. Conductivity and radioactivity monitoring for the cooling water were mentioned as a possible alarm condition. The staff also notes that cooling water inflow into the catholyte (e.g., from a leak) would change acid concentrations and potentially increase hydrogen generation rates.

In the revised CAR, an overtemperature event with the electrolyzer is identified in the category of a loss of confinement. The safety strategy for the facility worker and the environment utilizes prevention features. The PSSC is the process safety control subsystem. The safety function is to shut down process equipment before exceeding a temperature safety limit (i.e., the 70 °C (125 °F) design basis). The applicant stated that the temperature limit will be established by considering all material limits associated with the glovebox (NRC, 2002c). The applicant will perform final calculations and identify specific temperature setpoints in support of its ISA, based on the codes and standards identified in Section 11.6.7 of the revised CAR. This will assure that, subsequent to the shutdown of process equipment, normal convective cooling is sufficient (NRC, 2002c).

On page 11.6-16 of the revised CAR, setpoint design bases for the process safety control subsystem are identified as ANSI/ISA-67.04.01-2000. From clarifications provided during the August in-office review (NRC, 2002c), the applicant indicated that the setpoint analysis will consider electrolysis, potential exotherms from reactions, and natural cooling effects. The staff concludes that this approach provides assurance that the design basis temperature of 70 °C (125 °F) will not be exceeded, provides margin (i.e., vis-a-vis the boiling point of the solution, around 110 °C (230 °F)), and provides assurance that the control strategy will adequately prevent the overtemperature event. The staff has determined this to be acceptable for the CAR stage. Thus, Open Item AP-01 from the original DSER (NRC, 2002b) is now considered closed.

11.2.1.3.3 Hydrogen Production (KDB, KDD, KPA, KCA, KWG, and KWD)

The staff review noted that electrolytically generated hydrogen from overvoltage conditions and off-normal concentrations could produce hydrogen concentrations exceeding the lower flammability limit (LFL) if the scavenging airflow is based only upon radiolysis. In addition, overvoltage conditions could produce other undesirable effects, such as different reaction products and gases, flow oscillations, sparking, and greater heating.

In the revised CAR, the applicant identified the design basis for radiolytic hydrogen production as 25 percent of the LFL. The LFL is identified as 4 percent of hydrogen in air under normal conditions. A prevention strategy has been selected (see revised CAR p. 5.5-33), utilizing the PSSCs of the instrument air system (supplying scavenging air) and the offgas system (providing an exhaust path). The staff review notes that hydrogen is principally generated by radiolysis (on the anode side of the electrolyzer) and electrolysis (on the cathode side of the electrolyzer). The anode and cathode sides of the electrolyzer may be separated by membranes (i.e., the liquid phase) but will likely share a common header for venting to the offgas system.

Page 8-17 of the revised CAR discusses hydrogen produced by electrolysis. The design basis is stated as 25 percent of the LFL. A prevention strategy has been selected (see Section 5.5.2.4.6.13 of the revised CAR, p. 5.5-40), utilizing the process safety control subsystem as the PSSC. The safety function is to limit the generation of hydrogen from electrolysis by ensuring that the acid normality is sufficiently high and produces an offgas that does not exceed the design basis (25 percent of the LFL). Acid normality measurements would be instrumented, and normality would be maintained by the PSSC of chemical safety controls (NRC, 2003a). The applicant indicated that voltage control might also be used.

The applicant identified approaches for calculating LFLs for flammable gas mixtures and conditions removed from normal temperatures and pressures (NRC, 2003b; DCS, 2003f). The methodology for mixtures is based on a sum of fractions approach. Graphs were used to show the impact of temperature and pressure—pressure was not thought to be a concern for the MFFF because of the lack of high pressures within the buildings.

The staff notes that the general industrial practice is to use 25 percent of the LFL or less as the design basis for flammable gases in air (NRC, 1999; WSRC, 2000). A 25-percent limit is mentioned in Section 11.9.5.1 of the revised CAR for radiolytically generated hydrogen in and around process-type systems, using scavenging air. During off-normal or accident conditions, scavenging air is provided by the PSSC of the emergency scavenging air subsystem of the instrument air system. This emergency scavenging air is provided to those vessels where radiolytic hydrogen generation could exceed 4-percent hydrogen (the ambient LFL) in 7 days or less if venting and dilution were not maintained; the design basis for the scavenging air would still be 1-percent hydrogen (i.e., 25 percent of the ambient LFL).

As part of its review, the staff considered the following facts:

- The MOX standard review plan (SRP), NUREG-1718, mentions a 25-percent LFL limit for hydrogen in noninerted systems (NRC, 2000).

- The NRC used 25 percent of the LFL for hydrogen in a manner analogous for tank ullages in a radiochemical plant review (NRC, 1999). Consideration for this limit included

11-35

National Fire Protection Association (NFPA) code evaluation, radiolysis, chemical reactions, and uncertainties in ventilated process equipment in radiochemical facilities.

- The NRC inspects fuel cycle facilities based on not exceeding 25 percent of the LFL.

- The Department of Energy (DOE) uses 25 percent of the LFL for waste tanks and facilities at SRS and Hanford in a manner analogous to design basis.

- The SRS has hydrogen monitors installed in many ventilated tanks; these are set to alarm at 10 percent of the LFL (WSRC, 2000).

National Fire Protection Association (NFPA) 69, "Standard on Explosion Prevention Systems," issued 1997 (NFPA, 1997) and 801, "Standard for Fire Protection for Facilities Handling Radioactive Materials," issued 1998 (NFPA, 1998), emphasize 25 percent of LFL as an acceptable safety limit. The NFPA allows exceptions if they are justified (e.g., interlocks and well-defined systems and chemistry, multiple continuous monitoring); such exceptions can allow concentrations up to 60 percent of the LFL.

In the revised CAR, the applicant identified the design basis for hydrogen as NFPA 69 (NFPA, 1997). For hydrogen, the applicant identified the controls described below for groups of process equipment.

Group No. 5. Hydrogen Concentration Associated with Radiolysis

For inside equipment that has the potential for hydrogen concentration associated with radiolysis, the applicant proposed a design basis value of 25 percent of the LFL. This limit is in accordance with NFPA 69 (NFPA, 1997).

The safety function of the waste reception unit offgas treatment system is to provide an exhaust path for the removal of diluted hydrogen gas in process vessels. The instrument air system (scavenging air) has the safety function to provide sufficient scavenging air to dilute the hydrogen that is generated during radiolysis.

Group No. 6. Hydrogen Generated by Electrolysis within the Dissolution Process

For inside equipment that has the potential for hydrogen gas concentration associated with electrolysis within the dissolution unit, the applicant proposes a design basis value of 25 percent of the LFL. This limit is in accordance with NFPA 69 (NFPA, 1997).

The safety function of the process safety control system (PSCS) is to limit the generation of hydrogen by ensuring that the normality of nitric acid is sufficiently high to ensure that the offgas is not flammable.

The applicant's refined design basis of NFPA 69 for Groups 5 and 6 (maintain the combustible concentration at or below 25 percent of the LFL) is consistent with the previously cited NRC, DOE, and industry practice for similar applications, addresses combustible vapors present in these groups, and should provide adequate margin. The staff also interprets the design basis of NFPA 69 to apply to normal, off-normal, loss of power, and accident conditions. Consequently, the staff finds the 25 percent of LFL design basis acceptable for the CAR stage.

The control strategy for radiolysis is based on active dilution and active ventilation. This strategy has the ability to maintain concentrations of hydrogen below the design basis of 25 percent of the LFL and is acceptable to the staff for the construction authorization.

The strategy for electrolysis is based on acid normality control. The staff notes that acid normality can be monitored by density or other techniques that may be instrumented, with concentration adjusted by chemical additions. Additional administrative controls may be necessary; however, the applicant already has a PSSC entitled chemical safety controls for chemical makeup that could be used if needed. Specific details would be supplied by the applicant as part of any later license application. The staff conducted a review of the public literature and found curves relating hydrogen production to acid normality. Figure 11.2-1 gives two examples. Hydrogen can be produced over a wide range of acidities and cathodic materials; at catholyte concentrations above approximately 2 N HNO_3, measured hydrogen gas concentrations are below 1 percent (25 percent of the ambient LFL) in the evolved gases and thus would also be below 25 percent of the LFL in the ullage spaces above the electrolyzer. Consequently, the staff concludes that acid normality has the ability to be monitored and controlled in order to reduce combustible gases below 25 percent of the LFL, and this strategy is acceptable for construction authorization.

Therefore, Open Item AP-02 from the revised DSER (NRC, 2003c) has been adequately addressed and is now considered closed for the CAR phase.

The staff has reviewed the standards cited as design bases for combustible/hydrogen gas detectors (see Section 11.6.7 of the revised CAR). The staff notes that the proposed standards are used extensively in the industry in similar situations for combustible and hydrogen gas monitoring, as part of control, alarm, and emergency control systems. The staff finds the standards acceptable for the construction authorization stage.

11.2.1.3.4 Titanium Reactions (KDB and KDD)

The staff evaluation found that the applicant's proposed AP process uses oxidation-reduction chemistry based on the silver (I) to silver (II) couple. Silver (II) is corrosive, and special alloys are necessary for the electrolyzer equipment. The applicant stated that it intends to use titanium for the electrolyzer circuit and associated equipment that could be exposed to silver (II) ions (DCS, 2001). The applicant identified a negligible corrosion rate for titanium in the presence of silver (II) and nitric acid. The applicant intended to destroy silver (II) (i.e., by conversion into silver (I)) before the solutions contact other equipment in the process that is fabricated out of 300-series stainless steels. Destruction would be accomplished by the addition of peroxide, which reduces the silver (II) back to silver (I).

The staff finds that a higher alloy material, such as titanium (Bray, 1987, 1992; Bourges, 1985), is needed for adequate corrosion resistance in the presence of aggressive conditions that are likely to exist in this electrolyzer. However, titanium is a reactive metal, and the industry has developed guidelines for the safe use of such alloys, particularly for protection during wet/dry cycling and heating (DOE, 1994; NFPA, 2000; Mahnken, 2000; Poulson, 2000; PSC, 2001; TI, 2004). In addition, each electrolyzer operates with several hundred amps of current and multiple tens of volts, which could serve as an ignition source for the titanium in certain accident sequences. The staff notes that, under certain conditions, titanium fines or thin components exposed to hot sparks can ignite and burn, and incidents have been reported of unanticipated fires in titanium heat exchanger tube bundles, packing, and turbine blades (Mahnken, 2000). However, the staff also recognizes that bulk titanium components can burn only under extreme conditions.

The applicant described its safety approach for titanium events in the electrolyzer in its revised CAR (DCS, 2002e) and in supporting letters (DCS, 2003d, 2003g, 2004a; NRC, 2003d). The electrolyzer is a vertical cylinder with separate anode and cathode compartments. The compartments are concentric with the vertical axis. The cathode compartment is in the center while the anode compartment forms an annulus around it. The two compartments are separated from each other by a sintered silicon nitride (Si_3N_4) barrier (sometimes called a porous frit, or frit). Elastomeric materials are used to insulate and separate the cathode and anode structures, the anode and the ground, and the anode and the titanium shell.

The safety strategy is to prevent the titanium events during normal, abnormal, and shutdown and maintenance of the electrolyzer. During shutdown and maintenance, the applicant proposed to use administrative controls associated with the isolation of power to the electrolyzer when the electrolyzer is drained. For normal and abnormal operations, the applicant proposed a combination of active and passive engineered features to prevent the titanium events.

As noted previously, the anode and anodic compartment circuit connect to the positive terminal of the rectifier (direct current (dc) power) system. At the anode (a platinum electrode), the electricity oxidizes the silver (I) ions to silver (II) in the anolyte solutions, which subsequently react with and dissolve the plutonium dioxide powders. The cathode and cathodic compartment circuit connect to the negative terminal of the rectifier system. At the cathode (a tantalum electrode), reduction reactions occur, primarily forming nitrous acid from nitric acid.

The silicon nitride barrier between the anode and cathode compartments is nonconductive, has a high dielectric strength, and has sufficient porosity to allow ionic current transfer, but insufficient porosity to allow mixing of the anode and cathode solutions. Thus, the silicon nitride prevents the passage of material between the compartments while facilitating ionic electron transfer (the ions transfer the electric current, thus completing the circuit).

Teflon, other fluoropolymers, and polyethylene analogues have been mentioned as potential elastomeric materials. The applicant stated that insulation material used as guide sleeves between the anode and the titanium shell will be identified during the ISA process and will be capable of withstanding the environmental conditions of being submerged in the electrolyzer fluid. If appropriate, a maintenance/changeout frequency will be established, in accordance with Section 5.4.3 of the CAR, p. 5.4.10, regarding environmental design considerations and environmental qualification (EQ) requirements from the ISA.

The applicant identified the silicon nitride barrier and elastomeric materials as PSSCs—these represent passive engineered controls. In addition, the electrolyzer is geometrically safe from a criticality perspective, seismically designed (as per Section 11.12.3.3 of the revised CAR, which addresses B2 elements for seismic structural integrity), and can withstand turbulent flow without inducing vibration.

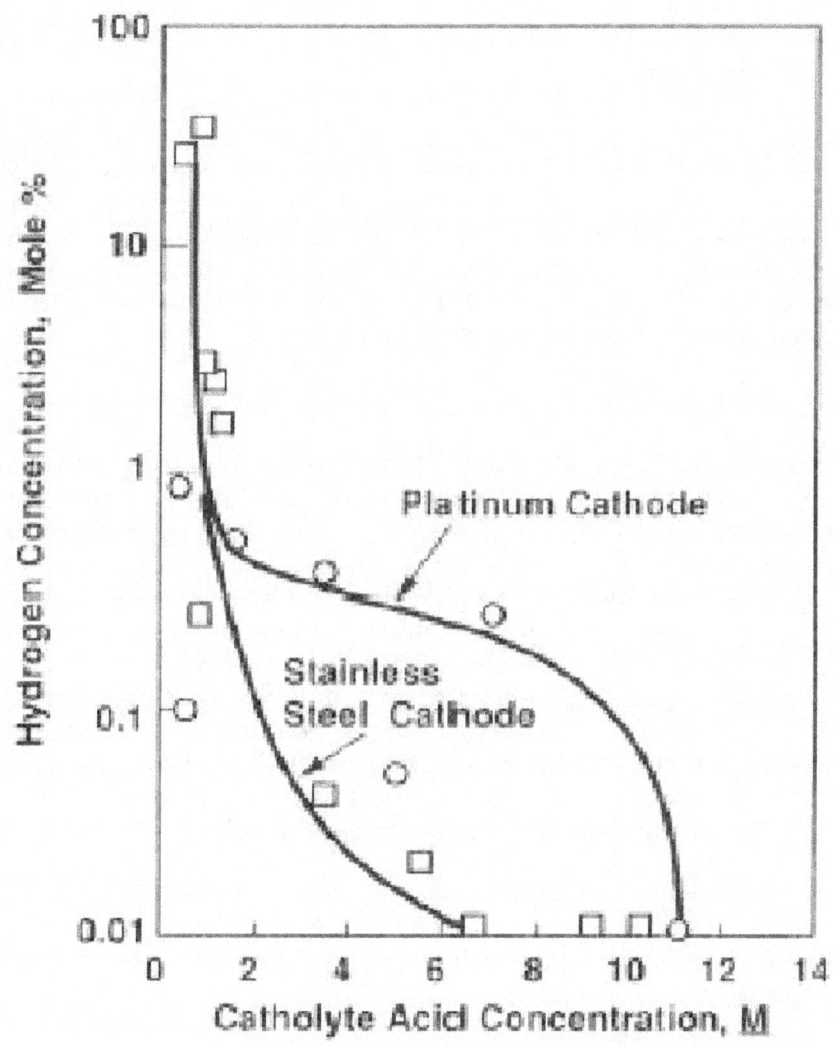

Figure 11.2-1 Acid Normality and Other Effects Upon Electrolytic Gas Generation

The applicant also noted that the nitric acid provides a liquid heat sink for cooling during operations, and that protection features exist to limit operations. There will be a current leakage detection system and a trip circuit on the rectifier, which represent the active engineered controls. The current leakage detection system is designed as a permissive signal and will stop the startup process (i.e., not allow the power to be turned on during startup) or cease operations (i.e., turn the power off to the electrolyzer) if the leakage current setpoint of 10 mA is exceeded. The trip circuit of the rectifier is another protective feature during operations. The rectifier normally supplies the electrolyzer with 400 A at 30 V. The trip setpoint is 420 A, and the rectifier is physically incapable of providing more than 900 A. The current leakage detection system and the trip circuit are part of the PSCS, an existing PSSC for active engineered controls.

The staff review finds that the overall safety strategy of prevention is appropriate given the postulated consequences.

Shutdown and Maintenance

The use of administrative controls that isolate power to electrolyzers during shutdown and maintenance is well established in the nuclear and chemical industries. In addition, there could be other requirements added to the procedures in the administrative control as a result of the ISA, such as hot-work limitations around titanium. However, these would be identified as part of any later DCS license application and would not change the fundamental strategy of using administrative controls. Thus, for the CAR stage, the PSSC of administrative control provides adequate assurances of safety and is acceptable.

Seismic Event During Operations

The electrolyzer structure is identified as the PSSC, with the safety functions of resisting seismic events (i.e., seismically designed), withstanding turbulent flow, not inducing vibrations, and maintaining geometry for criticality purposes. This represents a passive engineered control. The applicant subsequently clarified seismic design in Section 11.12.3.3 of the revised CAR. In addition, the applicant confirmed that the function of the seismic trip system would include isolation of power to the electrolyzers. The seismic trip system represents an active engineered control (part of the PSCS PSSC) that is in addition to the seismic design of the electrolyzers. The trip system isolates power to the electrolyzers, thus removing the potential energy and root cause for a titanium fire event. The staff concludes the electrolyzer structure and seismic trip system are independent controls and, combined with the low initiating frequency of the seismic event, have the ability to achieve the very high reliabilities necessary to render the titanium event highly unlikely. Thus, the staff finds this approach acceptable.

Non-seismic Event During Operations

The applicant identified passive and active engineered controls. The engineered controls consist of the following:

- The first passive PSSC is the sintered silicon nitride barrier, with the safety function of physically separating the cathode from the anode in the nitric acid solution by serving as a dielectric barrier.

- The second passive PSSC is the polytetrafluoroethylene (PTFE, sometimes called Teflon) insulator, which has the two safety functions of providing insulation/separation between the cathode and anode structures, and providing insulation/separation between the anode and the ground.

- The third passive PSSC consists of the guide sleeves, which have the safety function of insulating/separating the anode from the titanium shell. The actual material would be identified by the applicant during the ISA process and would be capable of withstanding the environmental conditions associated with being submerged in the electrolyzer fluid. As appropriate, the applicant indicated that a maintenance/changeout frequency will be established and documented as part of any later license application DCS may submit.

11-40

- The fourth PSSC (active engineered control) is the PSCS, with the safety functions of detecting current leakage (shut down if the leakage exceeds the design basis of 10 mA) and tripping the rectifier circuit if the design basis current of 420 A is exceeded.

The staff notes that the applicant has committed to the EQ of electrical and mechanical components, as outlined in Section 11.11 of this FSER. The performance of elastomeric and ceramic materials as highly reliable passive engineered controls in the environment of the electrolyzer will be demonstrated during the license application phase, or the applicant will identify other PSSCs. Inspection, maintenance, and changeout frequencies would also be identified by the applicant as part of any later license application DCS may submit.

The active engineered control has direct analogues with electrolyzers in the industry. The staff review found that control/leakage current and control/maximum current control combinations are frequently used and can be designed and fabricated to meet the 10 CFR 70.61 performance requirements. The applicant will provide design information that implements the design basis in any subsequent license application it may later submit. However, the leakage design basis of 10 mA is consistent with the detection of small currents in the mA range. Instrumentation is available for detecting smaller currents than this, and current trip instrumentation is available for almost any amperage rating. The staff also has found that trip currents in the 5–10 mA range are generally used for trip instrumentation for the protection of industrial workers and the prevention of sparks and component damage. The staff anticipates that the actual design at the license application stage will involve consideration of protective device characteristics, such as let-through currents and time/temperature effects in the titanium resulting from electrical faults and arcing. Thus, the staff concludes that the applicant's proposed design bases are consistent with available ranges of values for the protection of industrial workers, in that sparks and component damage would be prevented. The design bases are thus found to be acceptable.

There is an interrelationship between the passive and active controls. The first active safety function detects leakage current, which essentially results from the initiation of failure mechanisms in the passive controls. Failure of the active leak current detection may not prevent the complete failure of one or more of the passive controls, which in turn would need the trip circuit to function. The staff notes that safety control systems can be designed and constructed such that safety monitoring and safety trip systems can be fully independent. Actual independence of the controls will have to be demonstrated by the applicant as part of any later license application DCS may submit.

In conclusion, the staff finds that the passive engineered controls appear capable of rendering the titanium event unlikely, and the separate active engineered controls also appear capable of rendering the event at least unlikely. In addition, the safety controls have the ability to be independent from each other and not subject to common mode failure. Therefore, the combination of passive and engineered controls proposed by the applicant should be capable of rendering the event highly unlikely, and this is acceptable for the CAR stage. The staff further notes that the applicant may develop additional information and submit it as part of its later license application, demonstrating that an electrical-induced titanium fire is not credible or is highly unlikely. This closes Open Item AP-03 from the staff's revised DSER (NRC, 2003c).

11.2.1.3.5 Loss of Confinement of Process Solutions

The applicant discussed the control strategy for leaks and loss of confinement of process solutions in process cells and outside gloveboxes in Sections 5.5.2.1.6.2 and 5.5.2.1.6.11 of the revised CAR. In the revised CAR, the applicant identified a control strategy for loss of confinement events (leaks) of AP process vessels and pipes in process cells based on mitigation. This control strategy uses the process cell to protect the worker, and it uses the associated ventilation system to protect the site worker, the IOC, and the environment. The applicant intended to contain fluid leaks within the cell, and any airborne contamination would be treated with HEPA filtration before exhaust. For facility workers, the PSSC of process cell entry controls prevents the entry of personnel into process cells during normal operations and ensures that workers do not receive a dose in excess of limits while performing maintenance. The PSSC of the process cell has the safety function of containing leaks within the cell. The actual fluid leaks would not be prevented.

The process cell exhaust (ventilation) system is identified as the PSSC for protecting the site worker, the IOC, and the environment with the safety functions to provide filtration to limit the dispersion of radioactive material and to ensure that a negative pressure exists between the process cell areas and the C2 areas of the facility.

The staff review identified a potential event involving an acute chemical exposure to the facility, site workers, and the IOC from hazardous chemicals produced from licensed materials that leak from AP process vessels during such a loss of confinement event. Such a leak could occur because of erosion/corrosion of the vessels and piping. As discussed in Section 8.1.2.4.1 of this FSER, the applicant proposed controls for chlorine and nitrogen tetraoxide. Potential liquid phase leaks would consist of radioactive nitrate solutions which, once released from the vessels and pipes, would expose a large liquid surface area that allows a nitric acid and nitric oxide release into the cell's atmosphere. This material would not be removed by the HEPA filters on the exhaust system and would be released to the atmosphere.

The applicant indicated that the distance from the point of one of these in-cell releases to the worker is approximately equal to the 100-meter distance to the site worker; therefore consequence estimates of such releases for the site worker should bound any consequences for the facility worker (NRC, 2003a). The staff agrees with this qualitative distance analysis. This closes part of Open Item AP-13 from the original DSER (NRC, 2002b). However, for 379–757 L (100–200 gal) of radioactive nitrate solutions, IOC limits could be exceeded for several hundred meters. Some of the solutions might be at temperatures above ambient which could result in worker limits being exceeded for larger distances. Thus, the performance requirements of 10 CFR 70.61(b)(4) and 10 CFR 70.61(c)(4) might not be met. This potentially impacts the dissolution, dechlorination and dissolution, oxalic precipitation and oxidation, oxalic mother liquor, acid recovery, and liquid waste reception units. As discussed in Section 8.1.2.4.1 of this FSER, the applicant used lower airspeeds with a velocity-dependent evaporation model; some of the airspeeds are as low as 0.01 m/s (0.033 ft/s) (for comparison, a slight breeze is 1–2 m/s (3.3–6.6 ft/s)). As discussed in Section 8.1.2.4.1 of this FSER, the applicant has committed to a low chemical consequence limit and, if necessary, will identify planned features as IROFS to address the safety concern as part of any later license application it may submit. The staff concludes that this is an acceptable strategy, and Open Items CS-5b and AP-13 from the revised DSER (NRC, 2003c) are now closed.

The applicant identified a prevention strategy for process leaks in a C3 area outside of a glovebox. The PSSC is double-walled pipe for containing process fluids, and this protects the facility worker and the environment. The design basis is the fluid transport system (FTS), identified in Section 11.8.7 of the revised CAR. No PSSCs are anticipated for the site worker and the IOC because the estimated consequences are below the performance requirements of 10 CFR Part 70, "Domestic Licensing of Special Nuclear Material." However, any release would be mitigated by the C3 ventilation system.

The staff review finds that this approach relies upon practices and codes and standards typically used in the nuclear and chemical process industries to control similar hazards. By analogy, the approach should have the ability to address the potential concerns and is acceptable to the staff.

11.2.1.3.6 Oxalic Precipitation Concerns (KCA)

The applicant identified six categories of hazard events associated with the oxalic precipitation and oxidation unit. The types of events postulated in this unit include fire, explosion, loss of confinement, external exposure, load handling, and criticality. The safety strategy, including the PSSCs and design basis safety functions for controlling events within these categories, is discussed in Chapter 5 of this FSER.

The staff notes that the applicant's description mentions acidification of the residual mother liquors to avoid the precipitation and unanticipated accumulation of residual plutonium by the oxalate. This indicates a potential for a safety function (i.e., avoid plutonium precipitation and potentially related accident scenarios, such as erosion or plugging that could lead to loss of confinement). The staff notes that the design does not rely on concentration control to prevent a nuclear criticality in this unit and that the unit's calciner uses oxygen. The applicant has committed to standards for oxygen use and furnace applications (see Section 11.9.4 of the revised CAR). However, the calciner is likely to include components, such as bearings and seals, that have requirements to maintain their integrity. These components may be adversely affected and lose confinement integrity if operated at above-ambient temperatures in the presence of air or oxygen. The applicant identified nitrogen cooling of the calciner bearings as a means to protect them, presumably from the oxygen-rich environment, but has not identified this as a safety function. The issue of whether the nitrogen system is a PSSC because of its bearing cooling function has been discussed in Section 11.9 of this FSER.

Nitrogen cooling is identified as a means to extend bearing life; confinement is provided by the PSSC of the C4 confinement system.

As discussed in Section 8.1.2.4.1 of this FSER, the applicant proposed a control strategy for hazardous chemical releases from the potential loss of confinement of radioactive materials in this unit. Such a loss of confinement would result in an untreated flowpath (bypassing the offgas treatment system) for chemical releases from the nitric acid solutions, nitrate/oxalate mixture, and calciner gases. But any release of potentially hazardous chemicals from the calciner area would likely be bounded by potential releases from tanks in process cells because the source term in the process cell tanks would be larger. The staff finds that the applicant's proposed control strategy is acceptable.

The applicant identified a design basis for the plutonium dioxide powder produced by oxalic acid precipitation and oxidation. Without controls, the oxide powder may be substoichiometric or have entrained or absorbed solutions subject to radiolysis and thus be pyrophoric and present a hazard. The applicant proposed controls, which are discussed and evaluated below in Section 11.3.1.2.3.

11.2.1.3.7 Oxalic Mother Liquor Recovery

Prior experience with evaporators indicates the potential for the unintended accumulation of either solvent or plutonium, or both. The accumulation of material can result from changes in system chemistry (WVNS, 1999). Such unintended accumulation can pose the hazards of (1) inadvertent criticality, (2) erosion-corrosion from accumulated solids, and (3) the potential for red-oil events.

The applicant identified six categories of hazard events associated with the oxalic mother liquor recovery unit. The types of events postulated in this unit include fire, explosion, loss of confinement, external exposure, load handling, and criticality. The safety strategy, including the PSSCs and design basis safety functions for controlling events within these categories, is discussed in Chapter 5 of this FSER.

The staff notes that the applicant's description mentions acidification of the residual mother liquors in order to avoid the precipitation and unanticipated accumulation of residual plutonium by the oxalate. This indicates a potential safety function (i.e., avoid plutonium precipitation and potentially related accident scenarios, such as erosion or plugging that could lead to loss of confinement). The staff notes that the applicant is not relying on concentration control to prevent a nuclear criticality in this unit. A control strategy, PSSCs, and design bases have been identified for hazardous chemical releases from the potential loss of confinement of radioactive materials in this unit. At a minimum, this applies to the following:

- the distillate product stream
- the plutonium-containing stream returned to purification
- the evaporator itself and associated vessels

Section 8.1.2.4.1 of this FSER discusses the potential chemical releases from loss of confinement events. The staff finds that the control strategy and its related PSSCs and design bases constitute an acceptable approach.

11.2.1.3.8 Acid Recovery Unit

The acid recovery evaporators operate on the stream containing americium, uranium, and traces of plutonium. This is essentially a high alpha contaminated stream, and effective decontamination between the concentrates (bottoms products) and the distillate (overheads product) has safety implications. The NRC would anticipate separation requirements and/or specifications for these evaporators and their products. This is related to 10 CFR Part 20, "Standards for Protection Against Radiation," and will be reviewed as part of the review of any later license application DCS may submit.

11.2.1.3.9 Offgas Treatment Unit

The applicant identified the offgas treatment unit as a PSSC with the safety function of providing an exhaust path for the removal of gases in process vessels (Section 5.6 of the revised CAR).

Mechanical design bases are identified in Section 11.4.11.1.11 of the revised CAR. These include the following:

- two stages of HEPA filters

- spark arrestors and prefilters in each final filtration assembly

- each HEPA stage field tested to have an efficiency of 99.95 percent

- fire-rated dampers between designated fire areas

- in-place HEPA filter testing capability in accordance with ASME N510, "Nuclear Power Plant Air Cleaning Units and Components," issued 1980 (ASME, 1980), for the final filtration assemblies

- final filters and ductwork remain structurally intact during and after design basis earthquakes and withstand the effects of tornados

The staff requested information on the filtering line. The applicant provided additional information and identified the following design features of the offgas unit (DCS, 2001; RAI 142; Sections 11.3.2.11 and 11.3.2.13 of the revised CAR):

- Bubbling air scavenges tank ullage to maintain hydrogen concentrations at 1 percent or less.

- The system operates below the flash point of solvent vapors.

- Supplemental air is added to the system in order to further dilute any potential combustible concentration of gases and to maintain minimum volumetric throughput for the scrubbing and washing columns.

- The material of construction is stainless steel to resist the corrosive atmosphere.

- The HEPA filters are constructed of acid-resistant materials.

The offgas system handles vapors and gas mixtures that are potentially combustible in air streams, such as hydrogen, hydrazine, and TBP/dodecane (the solvent/diluent mixture—dodecane will likely be a petroleum distillation fraction and its constituents are more volatile than TBP). The applicant identified a setpoint/design basis/safety limit of 25 percent of the LFL in air for hydrogen (DCS, 2001; DCS, 2002f, RAI response 142). In RAI response 122 (DCS, 2001, 2002e), the applicant also identified a limit of 25 percent of the LFL for hydrogen in air from radiolysis in vessels containing plutonium. The staff notes that a limit of 25 percent of the LFL in air is routinely used by designers and operating facilities and is embodied in codes and standards. The applicant subsequently identified a design basis limit for hydrogen and other flammable gases and vapors, based on NFPA 69. This is discussed in Section 11.2.1.3.3 of this FSER. This should maintain potentially flammable gases and vapors at safe concentrations below their LFLs at all times, with adequate margin, and thus satisfactorily addresses the staff concerns. Consequently, Open Item AP-08 from the revised DSER is now considered closed (NRC, 2003c).

In response to RAI 142 (DCS, 2002f), the applicant did not identify temperatures below the flashpoint of solvent vapors. However, the applicant identified a safety strategy using a commitment to NFPA 69. This is discussed and found to be acceptable in Section 8.1.2.5 of this FSER. Consequently, Open Item AP-09 from the revised DSER is now considered closed (NRC, 2003c).

The process handles gases and vapors that are potentially reactive and toxic, such as nitrogen tetraoxide, nitric acid, nitric oxide, and hydrazine. The unplanned evolution of these gases—from and through radioactive solutions—via the offgas treatment unit could have potentially detrimental consequences that would likely exceed the performance requirements of 10 CFR Part 70 at considerable distances from the proposed facility. In addition, the applicant identified the removal of hazardous chemicals as a function of the offgas unit for protection of the HEPA filters (DCS, 2003a; NRC, 2003b). As discussed in Section 8.1.2.3.1 of this FSER, the applicant has agreed not to exceed low chemical consequence levels (see Tables 8-5 and 8-7). In addition, the offgas system includes components that have the ability to remove potentially toxic and reactive gases. In order to maintain the confinement barrier, the applicant implemented administrative controls to replace HEPA filters, such as those in the offgas system, after identified exposures to water or chemicals (DCS, 2003e). Therefore, the staff concludes that the applicant has provided strategies and design bases that have the ability to address the safety concerns regarding the offgas system, and this is acceptable for the CAR stage. Consequently, Open Item AP-10 from the revised DSER has been adequately addressed and is now considered closed (NRC, 2003c).

11.2.1.3.10 Corrosion Control (KDB and KDD)

Lower alloys can be inadvertently exposed to aggressive conditions. For example, stainless steel would likely experience uneven pitting corrosion that could lead to premature leaks and failures if it is routinely exposed to low concentrations of silver (II) ions. In a letter to the NRC, the applicant mentioned the use of stainless steel to resist corrosion in the offgas system and the use of acid resistant materials in the HEPA filters (DCS, 2001). The applicant proposed a generic corrosion control program as a PSSC. This appears to be based on general corrosion. The pitting corrosion that could occur from silver (II) ions might not be detected before failure by the proposed PSSC of a general corrosion control program; therefore, the potential exists for the corrosion leak to release plutonium compounds (i.e., a loss of confinement).

Corrosion-resistant materials would be needed to maintain confinement of radioactive and chemical species. The revised CAR identifies two administrative PSSCs that apply to corrosion. The first is chemical safety control. One of its safety functions is to ensure control of the chemical makeup of the reagents and to ensure segregation/separation of vessels/components from incompatible chemicals (i.e., planned corrosion exposures). The second PSSC involves the Material Maintenance and Surveillance Programs (Sections 5.5.2.1.6.2 and 5.6.2.4 of the revised CAR). The safety function of this PSSC is to detect and limit the damage resulting from corrosion and thus mitigate the effects to the facility worker and the environment. As discussed in Section 5.6.2.4 of the revised CAR, this PSSC can identify corrosion problems within the facility before the occurrence of catastrophic failures.

The administrative PSSCs are not required to prevent corrosion that could result in small leaks. Small is not defined but inferred to be a small percentage of inventory. However, the staff concludes that the design basis function of the corrosion function of the FTS PSSC should address instrumentation and/or monitoring of components that could be exposed to aggressive species in the offgas and other units and thus address small leaks. The FTS design bases are discussed in more detail in Section 11.8 of this FSER. Specific details would be provided in any later license application DCS may submit.

The staff concludes that this information on corrosion controls provided in the revised CAR follows established codes and practices in similar industries, has the ability to address the potential safety concerns, and, consequently, adequately addresses the staff's concerns about corrosion monitoring for preventing major failures. The staff finds this to be acceptable for the CAR stage. Therefore, Open Items AP-04 and AP-11 from the revised DSER have been adequately addressed and are now considered closed (NRC, 2003c).

11.2.1.3.11 Liquid Waste (KWD)

In Section 10.1.4 of the revised CAR, the applicant discussed waste minimization and waste management. Liquid and solid wastes produced at the proposed facility will be transferred to SRS for processing and disposal. The applicant has worked closely with SRS during the facility design phase and provided SRS with waste characterization information. The SRS has reviewed and evaluated the information in the context of the existing WAC. The applicant is committed to meeting the SRS WAC or providing a stream that qualifies for a WAC deviation and exemption. The WAC for the SRS waste solidification building has not been issued, but the applicant stated that the interface between it and SRS will ensure that the waste solidification building is designed to manage the facility high alpha waste stream and the depleted uranium stream. This was part of Open Item AP-05 in the original DSER. This portion of Open Item AP-05 (now designated as AP-05a) is considered closed.

The applicant stated that the alkaline waste stream will be acidified in a separate neutralization tank before being mixed in the high alpha waste tanks. Neutralization and acidification are performed to eliminate the potential for an explosion from azide formation that may form under alkaline conditions. In acidic media, the azides have a solubility limit greater than their concentration. Because the solubility limits of azides in alkaline media are lower, the alkaline media is acidified in order to increase the solubility limits. This ensures that the azides do not precipitate and create an explosion potential. Chapter 8 of the revised CAR and supplemental information provided by the applicant identify pH control as serving a safety function. Section 8.1.2.5 of this FSER provides further discussion of azide formation and an evaluation of the controls proposed by the applicant.

The applicant identified the high alpha activity and stripped uranium waste transfer lines as PSSCs (CAR Sections 5.5.2.3.6.5 and 10.5.2). These are double-walled stainless steel pipes that are seismically qualified and designed for leak detection and collection. The lines will be designed to accommodate mechanical and seismic loads. For load-handling events, the safety strategy relies on prevention. The PSSCs are the waste transfer lines. The safety function is to protect the lines from activities taking place outside the MFFF building. For external events (e.g., external fires, explosions, extreme winds, tornadoes, missiles, rain, and snow/ice loadings), the safety function is to prevent damage to the line. The design basis for both functions is ASME B31.3, "Process Piping," issued 1998 (ASME, 1998). This standard requires consideration of loads in the design of piping. The staff analysis notes that the code, the proposed approach with the waste transfer lines (i.e., double walled with leak detection), and the prevention strategy provide reasonable assurance that the design will not be damaged and release radionuclides outside of the MFFF building. The staff finds this to be acceptable. This item, identified as Open Item AP-06, is now closed.

The staff notes that an explicit inventory limit on waste is not specified in the revised CAR. Currently, the facility is designed to accommodate up to 90 days equivalent of most waste solutions (e.g., of the values in Table 11.2-1, because the storage of the LLW destined for the waste solidification building will likely be less than 90 days equivalent), although the applicant anticipated that there will be transfers of liquid wastes every 2 weeks. The applicant indicated that the facility will shut down before exceeding the liquid waste storage capacity. The staff interprets this to mean active waste generating operations would be curtailed at some setpoint before the tankage is completely full, until the potential backlog of waste at the facility is cleared. Actual setpoints would be defined by DCS as part of any license application it may later submit. The staff finds this approach acceptable for construction authorization.

[Text removed under 10 CFR 2.390]

Both are bounded by the safety assessment; therefore the staff finds this to be acceptable for the construction authorization. This item, identified as Open Item AP-05a, is now closed.

11.2.1.3.12 Sampling Systems

Table 11.3-34 in the revised CAR categorizes the sampling systems. Section 11.11 of the revised CAR provides information on the laboratory and indicates that a significant number of samples are required. The staff notes that laboratory personnel will most likely conduct the sampling. The applicant provided information on a PSSC entitled laboratory material controls that limits quantities of hazardous and radioactive materials in the laboratory. Section 5.6.2.7 of the revised CAR also states that procedures will be developed to establish limits on sample size and to ensure that laboratory operations are performed in accordance with safe laboratory operating practices. The staff finds that the outline of the sampling approaches appears to follow typical practices used in the chemical and nuclear industries.

In addition, DCS indicated that all sampling, with the exception of samples with very low levels of radioactivity, will be conducted within gloveboxes, and that there are no bag-in/bag-out operations because the sample containers would be pneumatically transferred from stations within the gloveboxes to the laboratory. The revised CAR identifies gloveboxes as PSSCs, with the safety function of maintaining confinement integrity for design basis impacts (Section 11.4.11.2). Laboratory material controls address the handling of the samples within the laboratory. Only the LLW samples would not be handled within gloveboxes; qualitatively, the staff concludes this does not present a challenge to the performance requirements of 10 CFR 70.61, "Performance Requirements." The staff finds that the applicant has identified PSSC and design basis information and finds that this issue has been adequately addressed for the construction authorization stage. This closes Open Item AP-12.

11.2.1.4 Design Basis of the PSSCs and Applicable Baseline Design Criteria

The plutonium isotopic composition design basis for AFS and PDCF/ARIES feeds is identified by the applicant as follows (see Section 11.3.7 of the revised CAR):

- ^{236}Pu < 1 ppb, at the origin of pit
- ^{238}Pu < 0.05%
- 90% < ^{239}Pu < 95%
- 5% < ^{240}Pu < 9%
- ^{241}Pu < 1% during lifetime of plant
- ^{242}Pu < 0.1%

The feed chemical impurities design bases were identified by the applicant and are listed in Tables 11.2-2 and 11.2-3 of this FSER. The radionuclide impurities design bases are listed in Table 11.2-4.

The americium content design basis for PDCF/ARIES feed material is identified by the applicant as follows:

$$\frac{^{241}Am}{Pu\ total\ +\ ^{241}Am} < 0.7\%\ during\ the\ lifetime\ of\ the\ plant$$

The feed plutonium dioxide powder to the AP process in the facility has a maximum density of less than 7 g/cc. Plutonium dioxide powder entering the facility can have a density up to 11.46 g/cc. Therefore, after receipt and storage, a density measurement is made. If necessary, a milling step is performed to ensure the density is below 7 g/cc.

The applicant stated that feed materials which have an impurity content that exceeds a value listed in Tables 11.2-2, 11.2-3, and 11.2-4 can, in some cases, be processed at the proposed facility. The applicant will evaluate these batches for safety before accepting them for processing at the facility.

The staff noted that these parameters and the values listed in Tables 11.2-2, 11.2-3, and 11.2-4 for the plutonium feed to the facility may affect the design and the safe operation of the facility. The applicant identified these as design bases. The PSSC is chemical safety controls, and the safety function is to "ensure control of the chemical makeup of the reagents and ensure segregation/separation of vessels/components from incompatible chemicals." According to the staff review, this is an administrative control based on the analysis of the feed to the facility. The staff also concludes that exceptional batches of plutonium dioxide that have impurity contents exceeding the design bases will be evaluated according to the facility change process in 10 CFR 70.72. The staff concludes that this approach provides assurance that the design basis plutonium feed parameters will not be exceeded and finds this to be acceptable for the construction authorization stage. Consequently, this part of Open Item AP-07 from the revised DSER has been adequately addressed and is now considered closed (NRC, 2003c). Section 11.2.1.3.1 of this FSER discusses closure of the other part of AP-07.

11-49

The applicant stated that the BDC specified in 10 CFR 70.64(a) are incorporated into the design and operation of the facility (Section 5.5.5.4 of the revised CAR). The applicant stated that information demonstrating compliance with these criteria is provided in the applicable chapters of the revised CAR. For chemical protection, 10 CFR 70.64(a)(5) states the following:

> Chemical protection. The design must provide for adequate protection against chemical risks produced from licensed material, facility conditions which affect the safety of licensed material, and hazardous chemicals produced from licensed material.

The applicant stated that reagents would be stored, and chemical mixtures would be prepared, in the reagent processing building and the reagent storage area of the AP area, and they are generally separated from each other and from radioactive materials. The applicant will avoid mixing incompatible materials by design, controls, and procedures. The AP and MP facilities are broken down into process functional units, which are made up of one or more subassemblies performing consistent and elementary tasks. The applicant stated that the breakdown into control functional units allows each entity to be operated relatively independently in the given operating mode (DCS, 2002e). The staff review notes that this separation and independence are consistent with accepted practices for safe operations.

The applicant will control process storage and operation conditions in order to prevent exothermic and potential autocatalytic reactions in the reagent processing building, as well as in the AP and MP areas. Autocatalytic and exothermic reactions of chemicals would be prevented through control of the process parameters (e.g., reactant concentration, temperature, catalyst concentration in solution, and pressure) that affect the reactions. Many of these controls have been identified as PSSCs with design bases. Codes and standards have been applied on a specific level, often as design bases (DCS, 2002e).

The applicant stated that there is reasonable assurance that the PSSCs will be sufficiently reliable and available based on the use of standard nuclear industry engineering practices. These practices are incorporated into the facility general design philosophy, design bases, system design, and commitments to applicable management measures. These practices ensure that applicable industry codes and standards are utilized, adequate safety margins are provided, engineering features are utilized to the extent practicable, the defense-in-depth philosophy is incorporated into the design, and PSSCs will be appropriately maintained (DCS, 2002e).

The staff review finds that the applicant has provided sufficient information to meet the requirements of 10 CFR 70.64(a)(5).

Related to chemical protection, the explosion protection BDC is stated as part of the fire protection BDC in 10 CFR 70.64(a)(3):

> Fire protection. The design must provide for adequate protection against fires and explosions.

The applicant stated that there is reasonable assurance that the PSSCs will be sufficiently reliable and available based on the use of standard nuclear industry engineering practices. These practices are incorporated into the facility general design philosophy, design bases, system design, and commitments to applicable management measures. These practices ensure that applicable industry codes and standards are utilized, adequate safety margins are provided, engineering features are utilized to the extent practicable, the defense-in-depth philosophy is incorporated into the design, and PSSCs will be appropriately maintained.

Table 11.2-2 Design Basis for Chemical Impurities in PDCF Feed Plutonium Dioxide

Chemical Component	Maximum Content (µg/g Pu)	Maximum Exceptional Content (µg/g Pu)	Chemical Component	Maximum Content (µg/g Pu)	Maximum Exceptional Content (µg/g Pu)
Ag	NA	10,000	Mg	500	10,000
Al	150	10,000	Mn	100	1,000
B	100	1,000	Mo	100	1,000
Be	100	2,500	N	400	400
Bi	100	1,000	Na	300	10,000
C	500	1,500	Nb	100	3,500
Ca	500	10,000	Ni	200	2,500
Cd	10	1,000	P	200	1,000
Cl	(+F < 250)	500	Pb	200	1,000
Co	100	10,000	S	250	1,000
Cr	100	500	Si	200	200
Cu	100	500	Sm	2	1,000
Dy	1	1,000	Sn	100	2,500
Eu	1	1,000	Ti	100	2,500
F	(+Cl < 250)	350	Th	100	100
Fe	500	2,500	V	300	2,500
Ga	12,000	12,500	W	200	2,500
Gd	3	250	Zn	100	1,000
In	20	1,000	Zr	50	1,000
K	150	10,000	Boron Equivalent	NA	
Li	400	10,000	Total Impurities	18,837	
NA = Not applicable or not available Maximum exceptional value means the maximum anticipated value for that element, with all others at the maximum value.					

The applicant stated that the BDC specified in 10 CFR 70.64(a) are incorporated into the design and operation of the facility (Section 5.5.5.4 of the revised CAR). The applicant further stated that information demonstrating compliance with these criteria is provided in the applicable chapters of the revised CAR. For chemical protection, 10 CFR 70.64(a)(5) states the following:

> Chemical protection. The design must provide for adequate protection against chemical risks produced from licensed material, facility conditions which affect the

Table 11.2-3 Design Basis for Chemical Impurities of AFS Plutonium Dioxide Feed Material

Chemical Component	Maximum Content for most (~75%) of Items (µg/g Pu)	Maximum Content exceeded only by 2% of Items (µg/g Pu)	Chemical Component	Maximum Content for most (~75%) of Items (µg/g Pu)	Maximum Content exceeded only by 2% of Items (µg/g Pu)
Ag	NA	10,000	I	NA	100
Al	4,000	15,000	In	20	2,500
Am	7,000 (100% ^{241}Am)	7,000 (100% ^{241}Am)	K	220,000	(Ca+K+Mg+Na) ≤40% Net weight
As	NA	100	La	NA	5,000
Au	NA	100	Li	5,000	10,000
B	100	1,000	Mg	70,000	(Ca+K+Mg+Na) ≤40% Net weight
Ba	5,000	10,000	Mn	1,000	2,000
Be	100	5,000	Mo	100	(Mo+Zr)<5,000
Bi	1,000	1,000	N	400	5,000
C	2,000	10,000	NO$_3$	NA	5,000
Ca	120,000	(Ca+K+Mg+Na) ≤40% Net weight	Na	130,000	(Ca+K+Mg+Na) ≤40% Net weight
Cd	1,000	1,000	Nb	100	3,500
Ce	NA	500	Ni	5,000	15,000
Cl	200,000	330,000	Np	500	1,000
Co	5,000	10,000	P	1,000	(P+S)≤10,000
Cr	3,000	8,000	Pb	200	5,000
Cu	500	3,000	Pd	NA	100
Dy	NA	NA	Pt	NA	100
Er	NA	500	Rb	100	5,000
Eu	NA	NA	S	330	(P+S)≤10,000
F	1,000	7,000	SO$_4$	1,000	(P+S)≤10,000
Fe	5,000	18,000	Sb	NA	100
Ga	12,000	15,000	Si	5,000	10,000
Gd	250	250	Sm	NA	NA
Ge	NA	100	Sn	1,000	10,000
Hf	50	1,000	Sr	5,000	10,000
Hg	NA	100	Ta	4,000	10,000

Table 11.2-3 (cont): Design Basis for Chemical Impurities of AFS Plutonium Dioxide Feed Material

Chemical Component	Maximum Content for most (~75%) of Items (µg/g Pu)	Maximum Content exceeded only by 2% of Items (µg/g Pu)	Chemical Component	Maximum Content for most (~75%) of Items (µg/g Pu)	Maximum Content exceeded only by 2% of Items (µg/g Pu)
Ti	100	3,000	V	300	1,000
Th	100	300	W	4,000	10,000
Tl	NA	100	Y	200	10,000
Enriched Uranium (EU)	EU≤30% Net weight	EU≤30% Net weight Annual max. value—50 kg (110 lbs) (^{235}U—93.2%)	Zn	1,000	10,000
Depleted Uranium, Natural Uranium	[TBD]	500,000	Zr	50	(Mo+Zr)≤5,000

safety of licensed material, and hazardous chemicals produced from licensed material.

The applicant stated that reagents would be stored, and chemical mixtures would be prepared, in the reagent processing building and the reagent storage area of the AP area, and they are generally separated from each other and from radioactive materials (DCS, 2002e). The applicant intends to avoid mixing incompatible materials by design, controls, and procedures. The AP and MP facilities are broken down into process functional units, which are made up of one or more subassemblies performing consistent and elementary tasks. The applicant stated that the breakdown into control functional units allows each entity to be operated relatively independently in the given operating mode (DCS, 2002e). The staff review notes that this separation and independence are consistent with accepted practices for safe operations.

The applicant will control process storage and operation conditions in order to prevent exothermic and potential autocatalytic reactions in the reagent processing building and in the AP and MP areas. Autocatalytic and exothermic reactions of chemicals would be prevented through control of the process parameters (e.g., reactant concentration, temperature, catalyst concentration in solution, and pressure) that affect the reactions. Many of these controls have been identified as PSSCs with design bases (DCS, 2002e). Codes and standards have been applied on a specific level, often as design bases.

The applicant stated that there is reasonable assurance that the PSSCs will be sufficiently reliable and available based on the use of standard nuclear industry engineering practices. These practices are incorporated into the facility general design philosophy, design bases, system design, and commitments to applicable management measures. These practices ensure that applicable industry codes and standards are utilized, adequate safety margins are provided, engineering features are utilized to the extent practicable, the defense-in-depth philosophy is incorporated into the design, and PSSCs will be appropriately maintained (DCS, 2002e).

Table 11.2-4 Radionuclide Impurities in the Feed Plutonium Dioxide

PDCF Type

Impurity	Isotope	Maximum Content µg/g Pu
Americium	^{241}Am—100%	7,000 µg/g Pu (Note 1)
Uranium (HEU)	^{235}U—93.2%	Standard value—5,000 µg/g Pu Maximum value—20,000 µg/g Pu for 10% of the delivered cans during 1 year Annual maximum value = 17 kg (37 lbs) (Note 2)
(1) At the plutonium design basis feed rate of 3.5 MTHM/yr, the americium annual quantity becomes 24.5 kg/yr (54 lb/yr). (2) The uranium standard maximum value corresponds to 17.5 kg/yr (39 lb/yr), while 10% at 20,000 and 90% at 5,000 (micrograms U/g Pu) correspond to 22.75 kg/yr (50 lb/yr).		

AFS Type

Impurity	Isotope	Maximum Content µg/g Pu
Americium	^{241}Am—100%	11,000 µg/g Pu
Enriched Uranium	^{235}U—93.2%	Maximum value—30% of can net weight Annual maximum value = 50 kg (110 lb)
Depleted Uranium and Natural Uranium	^{238}U	Maximum value—42% of can net weight (with 5% of enriched uranium)

The staff review finds that the applicant has provided sufficient information to meet the requirements of 10 CFR 70.64(a)(5). Using the SRP, the staff review notes that the previously identified open items in AP have been closed by additional safety strategies and information from the applicant, and the staff concludes that the applicant has satisfied this BDC.

Related to chemical protection, the explosion protection BDC is stated as part of the fire protection BDC in 10 CFR 70.64(a)(3):

Fire protection. The design must provide for adequate protection against fires and explosions.

The applicant states that there is reasonable assurance that the PSSCs will be sufficiently reliable and available based on the use of standard nuclear industry engineering practices. These practices are incorporated into the facility general design philosophy, design bases, system design, and commitments to applicable management measures. These practices ensure that applicable industry codes and standards are utilized, adequate safety margins are provided, engineering features are utilized to the extent practicable, the defense-in-depth philosophy is incorporated into the design, and PSSCs will be appropriately maintained (DCS, 2002e).

The staff review finds the applicant has now provided sufficient information to meet the requirements of 10 CFR 70.64(a)(3). Using the SRP, the staff review notes that the previously identified open items related to MP have been closed by additional safety strategies and information from the applicant, and the staff concludes that the applicant has now satisfied this BDC.

11.2.2 Evaluation Findings

In Section 11.3 and Chapters 5, 8, and 11 of the revised CAR, DCS provided design basis information for chemical process safety PSSCs identified for the proposed facility. Based on the staff's review of these revised CAR chapters and on supporting information provided by the applicant that is relevant to AP and chemical process safety, the staff finds that, for the reasons discussed above, DCS has now met the BDC set forth in 10 CFR 70.64(a)(3) for explosions, and 10 CFR 70.64(a)(5) for chemical safety. Further, the staff concludes, pursuant to 10 CFR 70.23(b), that the design bases of the PSSCs identified by the applicant will provide reasonable assurance of protection against natural phenomena and the consequences of potential accidents.

Several previously identified open items in the April 30, 2002, and the April 30, 2003, DSERs have now been closed in a satisfactory manner—AP-02, AP-03, AP-07, AP-08, AP-09, and AP-10.

11.2.3 References

(ASME, 1980) American Society of Mechanical Engineers. ANSI/ASME N50, "Testing of Nuclear Air-Cleaning Systems." New York, 1980.

(ASME, 1998) American Society of Mechanical Engineers. ASME B31.1, "Process Piping." New York, 1998.

(Bennedict, 1981) Bennedict, Pigfood, and Levi. *Nuclear Chemical Engineering.* McGraw-Hill, New York, 1981.

(Bourges, 1985) Bourges, J., C. Madic, G. Koehly, and M. Lecomte. "Dissolution of Plutonium Dioxide in Nitric Acid Medium by Electrogenerated Silver (II)," Y/TR-91/4, translated from the French *Journal of the Less Common Metals*, 122:303–311. 1986. Paper originally presented at Actinides 85, Aix-en-Provence, France, September 2–6, 1985 (CONF-8509147-28-Trans).

(Bray, 1987) Bray, L.A., J.L. Ryan, and E.J. Wheelwright. "Electrochemical Process for Dissolving Plutonium Dioxide and Leaching Plutonium from Scrap or Wastes." AIChE Symposium Series 254, Volume 83, *Electrochemical Engineering Applications.* R.E. White, R.F. Savinell, and A. Schneider eds. (CONF-861146-10), 1987.

(Bray, 1992) Bray, L.A., J.L. Ryan, E.J. Wheelwright, and G.H. Bryan. "Catalyzed Electrolytic Plutonium Oxide Dissolution: The Past 17 Years and Future Potential." Chapter 30 in *Transuranium Elements: A Half Century.* L.R. Morss and J. Fuger, eds. American Chemical Society, Washington DC, 1992.

(DCS, 2001) Hastings, P., Duke Cogema Stone & Webster. Letter to U.S. Nuclear Regulatory Commission, RE: Response to Request for Additional Information. August 31, 2001.

(DCS, 2002a) Hastings, P., Duke Cogema Stone & Webster. Letter to U.S. Nuclear Regulatory Commission, RE: Clarification of Responses to NRC Request for Additional Information. January 7, 2002.

(DCS, 2002b) Hastings, P., Duke Cogema Stone & Webster. Letter to U.S. Nuclear Regulatory Commission, RE: Clarification of Responses to NRC Request for Additional Information. March 8, 2002.

(DCS, 2002c) Hastings, P., Duke Cogema Stone & Webster. Letter to U.S. Nuclear Regulatory Commission, RE: Clarification of Responses to NRC Request for Additional Information. April 23, 2002.

(DCS, 2002d) Hastings, P., Duke Cogema Stone & Webster. Letter to U.S. Nuclear Regulatory Commission, RE: Evaluation of the Draft Safety Evaluation Report (DSER) on the Construction of a Mixed Oxide Fuel Fabrication Facility. July 9, 2002.

(DCS, 2002e) Ashe, K., Duke Cogema Stone & Webster. Letter to Docket Number 070-03098, U.S. Nuclear Regulatory Commission, RE: Mixed Oxide Fuel Fabrication Facility—Construction Authorization Request. October 31, 2002 (page changes through January 27, 2005).

(DCS, 2002f) Hastings, P., Duke Cogema Stone & Webster. Letter to U.S. Nuclear Regulatory Commission, RE: Requests for Additional Information, Clarifications, and Open Item Mapping into the Construction Authorization Request Revision. November 22, 2002.

(DCS, 2003a) Hastings, P., Duke Cogema Stone & Webster, Letter to U.S. Nuclear Regulatory Commission, RE: Responses to Financial Qualification, Fire Safety, Chemical Safety, Aqueous Processing, Material Processing and Ventilation Open Items/Additional NRC Questions on Construction Authorization Request (CAR) Revision. February 18, 2003.

(DCS, 2003b) Hastings, P., Duke Cogema Stone & Webster. Letter to U.S. Nuclear Regulatory Commission, RE: NRC to DCS Letter dated 13 February 2003, *February 2003 Monthly Open Item Status Report.* February 18, 2003.

(DCS, 2003c) Hastings, P., Duke Cogema Stone & Webster. Letter to U.S. Nuclear Regulatory Commission, RE: Construction Authorization Request Change Pages. April 10, 2003.

(DCS, 2003d) Hastings, P., Duke Cogema Stone & Webster. Letter to U.S. Nuclear Regulatory Commission, "Response to DSER Open Items AP-3 and CS-3." May 23, 2003.

(DCS, 2003e) Hastings, P., Duke Cogema Stone & Webster. Letter to U.S. Nuclear Regulatory Commission, "Response to DSER Open Items AP-10 and CS-2." May 30, 2003.

(DCS, 2003f) Hastings, P., Duke Cogema Stone & Webster. Letter to U.S. Nuclear Regulatory Commission, "Response to Request for Additional Information—Chemical Safety Open Items CS-09, AP-02, AP-08, and AP-09." September 29, 2003.

(DCS, 2003g) Hastings, P., Duke Cogema Stone & Webster. Letter to U.S. Nuclear Regulatory Commission, "Response to Request for Additional Information—DSER Open Items MP-01 (UO$_2$) and AP-03 (Titanium Fires)." October 10, 2003.

(DCS, 2004a) Hastings, P., Duke Cogema Stone & Webster. Letter to U.S. Nuclear Regulatory Commission, "Response to Request for Additional Information—DSER Open Item AP-03 (Titanium Fires)." March 12, 2004.

(DOE, 1994) U.S. Department of Energy. "Primer on Spontaneous Heating and Pyrophoricity," DOE-HDBK-1081-94. December 1994.

(Mahnken, 2000) Mahnken, G. "Watch Out for Titanium Tube Bundle Fires," *Chemical Engineering Progress*, Vol. 96, No. 4, pp 47–52. April 2000.

(NFPA, 1997) National Fire Protection Association. Standard 69, "Standard on Explosion Prevention Systems." 1997.

(NFPA, 1998) National Fire Protection Association. NFPA Standard 801, "Standard for Fire Protection for Facilities Handling Radioactive Materials," 1998.

(NFPA, 2000) National Fire Protection Association. NFPA Standard 481, "Standard for the Production, Handling, and Storage of Titanium." 2000.

(NFPA, 2001) National Fire Protection Association. Standard 2001, "Standard on Clean Agent Extinguishing Systems."

(NRC, 1999) Pierson, R.C., U.S. Nuclear Regulatory Commission. Memorandum to D. Clark Gibbs, U.S. Department of Energy, RE: Estimates of Hydrogen Generation From Wastes at the Proposed, TWRS-P Facility. April 21, 1999.

(NRC, 2000) U.S. Nuclear Regulatory Commission. NUREG-1718, "Standard Review Plan for the Review of an Application for a Mixed Oxide (MOX) Fuel Fabrication Facility." Washington, DC, August 2000.

(NRC, 2001) Persinko, A., U.S. Nuclear Regulatory Commission. Memorandum to E.J. Leeds, U.S. Nuclear Regulatory Commission, RE: 11/27-29/01 In-Office Review Summary of DCS Construction Authorization Request Supporting Documents for the MFFF. December 18, 2001.

(NRC, 2002a) Persinko, A., U.S. Nuclear Regulatory Commission. Memorandum to E.J. Leeds, U.S. Nuclear Regulatory Commission, RE: 2/13/02 Meeting Summary: MFFF Program Changes and Applicant Reorganization. February 27, 2002.

(NRC, 2002b) Pierson, R. U.S. Nuclear Regulatory Commission. Letter to R. Idhe, Duke Cogema Stone & Webster, "Draft Safety Evaluation Report on Construction of Proposed Mixed Oxide Fuel Fabrication Facility." April 30, 2002.

(NRC, 2002c) Persinko, A., U.S. Nuclear Regulatory Commission. Memorandum to M.N. Leach, U.S. Nuclear Regulatory Commission, RE: August 28–30, 2002 In-Office Review Summary of DCS Construction Authorization Request Supporting Documents for the MFFF. December 31, 2002.

(NRC, 2003a) Persinko, A., U.S. Nuclear Regulatory Commission. Memorandum to M.N. Leach, U.S. Nuclear Regulatory Commission, RE: December 10–12 Meeting Summary: Meeting with DCS to Discuss Mixed Oxide Fuel Fabrication Facility Revised Construction Authorization Report. January 31, 2003.

(NRC, 2003b) Persinko, A., U.S. Nuclear Regulatory Commission. Memorandum to M.N. Leach, U.S. Nuclear Regulatory Commission, RE: January 15–16 Meeting Summary: Meeting with DCS to Discuss Mixed Oxide Fuel Fabrication Facility Revised Construction Authorization Report. March 5, 2003.

(NRC, 2003c) Pierson, R., U.S. Nuclear Regulatory Commission. Letter to R. Idhe, Duke Cogema Stone & Webster, "Draft Safety Evaluation Report on Construction of Proposed Mixed Oxide Fuel Fabrication Facility, Revision 1." April 30, 2003.

(NRC, 2003d) Brown, D., U.S. Nuclear Regulatory Commission. Memorandum to J. Giitter, U.S. Nuclear Regulatory Commission, "October 16 & 21, 2003 Summary of Phone Call with the Applicant: Chemical Safety Open Items for the Mixed Oxide (MOX) Fuel Fabrication Facility." October 22, 2003.

(Poulson, 2000) Poulson, E. "Safety-Related Problems in the Titanium Industry in the Last 50 Years," *Journal of Metals*, Vol. 52, No. 5, pp 13–17. 2000.

(PSC, 2001) O'Connor, M.K., Process Safety Center. "A Fire in Titanium Structured Packing Involving Thermite Reactions." August 3, 2001. http://www.process-safety.tamu.edu/safety-alert/08_03_01.htm.

(TI, 2004) Titanium Industries, Inc. "Fire Prevention." November 16, 2004. http://www.titanium.com/titanium/tech_manual/tech15.cfm.

(WSRC, 2000) Hobbs, D.T. "Possible Explosive Compounds in the Savannah River Site Waste Tank Farm Facilities," WSRC-TR-91-444, Revision 3. February 15, 2000.

(WVNS, 1999) Baker, M.N. and H.M. Houston, West Valley Nuclear Services Company. "Liquid Waste Treatment System: Final Report," DOE/NE/44139-88. June 1999.

11.3 Mixed Oxide Process System Description

11.3.1 Conduct of Review

This section of the FSER contains the staff's review of MP safety described by the applicant in Section 11.2 of the revised CAR, with supporting process safety information from Chapters 5, 8, and 11 of the revised CAR (DCS, 2002d). The objective of this review is to determine whether the chemical process safety PSSCs and their design bases identified by the applicant provide reasonable assurance of protection against natural phenomena and the consequences of potential accidents. The staff evaluated the information provided by the applicant for chemical process safety by reviewing Chapter 8 of the revised CAR, other sections of the revised CAR, supplementary information provided by the applicant, and relevant documents available at the applicant's offices but not submitted by the applicant. The staff also reviewed technical literature as necessary to understand the process and safety requirements. The review of MP safety design bases and strategies was closely coordinated with the review of the radiation and chemical safety aspects of accident sequences described in the safety assessment of the design bases (see Chapter 5 of this FSER), the review of fire safety aspects (see Chapter 7 of this FSER), and the review of plant systems (see Chapter 11 of this FSER).

The staff reviewed how MP and chemistry information in the revised CAR addresses or relates to the following regulations:

- 10 CFR 70.23(b) states, as a prerequisite to construction approval, that the design bases of the PSSCs and the Quality Assurance Program be found to provide reasonable assurance of protection against natural phenomena and the consequences of potential accidents.

- 10 CFR 70.64 requires that BDC and defense-in-depth practices be incorporated into the design of new facilities or new processes at existing facilities. With respect to chemical protection, 10 CFR 70.64(a)(5) requires that the MFFF design provide for adequate protection against chemical risks produced from licensed material, facility conditions which affect the safety of licensed material, and hazardous chemicals produced from licensed material. Related to chemical protection, 10 CFR 70.64(a)(3) requires that the facility design provide for adequate protection against fires and explosions, such as those that could be initiated by or involve chemicals at the proposed facility.

The review for this CAR stage focused on the design basis of chemical process safety systems, their components, and other related information. For each chemical process safety system, the staff reviewed information provided by the applicant for the safety function, system description, and safety analysis. The review also encompassed proposed design basis considerations such as redundancy, independence, reliability, and quality. The staff used Chapter 8.0 of NUREG-1718 (NRC, 2000), as guidance in performing the review. As stated on page 8.0-2 of NUREG-1718, information contained in the application should be of sufficient quality and detail to allow for an independent review, assessment, and verification by NRC reviewers.

At NRC-licensed facilities, as stated in the "Memorandum of Understanding between the Nuclear Regulatory Commission and the Occupational Safety and Health Administration: Worker Protection at NRC-Licensed Facilities," (*Federal Register*, Vol. 53, No. 210, October 31, 1998, pp. 43950–43951), the NRC oversees chemical safety issues related to (1) radiation risk produced by radioactive materials, (2) chemical risk produced by radioactive materials, and (3) plant conditions that affect the safety and safe handling of radioactive materials. These chemical safety issues are important because they represent an increased radiation risk to workers. The NRC does not oversee facility conditions that result in an occupational risk but do not affect the safe use of licensed radioactive material.

The staff reviewed the revised CAR for the following areas applicable to process safety at the CAR stage and consistent with the level of design (NRC, 2000, p. 8.0-8):

- MP description
- hazardous chemicals and potential interactions affecting licensed materials
- MP chemical accident sequences
- MP chemical accident consequences
- MP safety controls

Additional documentation from the applicant and the literature was reviewed as necessary to understand the process and safety requirements. In addition, the revised CAR incorporates the BDC of 10 CFR 70.64(a) into the design and operations of the proposed facility (see Section 5.5.5.4 of the revised CAR, p. 5.5-67 (DCS, 2004)), and applicable sections of the revised CAR are intended to demonstrate compliance with these BDCs.

The staff utilized the guidance provided by Chapter 8.0 of NUREG-1718 for assistance in determining the compliance of the application with the regulation. The evaluation used the guidance in Section 8.4 of NUREG-1718 for determining acceptance with 10 CFR Part 70, consistent with the CAR stage and the level of the design. The evaluation is summarized in the sections that follow.

11.3.1.1 System Description of the MP Process

This section provides a description and overview of the MP, including design, operational, and process-flow information. This information is provided to support the hazard and accident analysis provided in Chapter 5 of this FSER, as well as to assist in understanding the overall design and function of the MP.

The MP area receives polished plutonium dioxide from the AP process, depleted uranium dioxide (i.e., uranium depleted in the ^{235}U isotope below the natural assay of 0.71 percent), and the required components for assembling light-water reactor MOX assemblies. The process mixes the plutonium and uranium dioxides to form MOX fuel pellets. The pellets are loaded into fuel rods, which are then assembled into MOX fuel assemblies for use in commercial reactors. The MP area is designed to process up to 70 metric tons heavy metal (uranium plus plutonium) annually. The safety functions of the PSSCs associated with the MP are discussed in Chapter 5 of the revised CAR.

The facility uses the advanced micronized master blend (A-MIMAS) process for the manufacture of MOX fuel assemblies. The A-MIMAS process uses a two-step, dry mixing procedure. In the first step, the plutonium dioxide powder is mixed with depleted uranium dioxide and recycled scrap powder to form a primary blend (master blend) with approximately 20 percent plutonium dioxide content of the total mass. This master blend is then micronized—reduced in particle size into a very fine powder. In the second step, the primary blend is forced through a sieve, poured into a jar, and mixed with more depleted uranium dioxide and scrap powder to obtain the final blend with the specified plutonium content (typically around 6 percent of the heavy metal content). The two-step mixing process is used to ensure a consistent product.

The MP consists of 43 process units or systems divided into six areas corresponding to the different segments of the process (see Figure 11.3-1).

Receiving Area

This area includes truck unloading, plutonium dioxide container handling, counting, and storage before and after transfer to the AP line. The function of the receiving area is to receive, unload, and store plutonium dioxide and uranium dioxide powder. The receiving area comprises the following units:

- uranium dioxide receiving and storage unit
- uranium dioxide drum emptying unit
- plutonium dioxide receiving unit
- plutonium dioxide 3013 storage unit
- plutonium dioxide buffer storage unit

Powder Area

This area has equipment for dosing MOX powder at the specified plutonium content in two steps for homogenizing and for pelletizing. The powder area receives uranium dioxide and plutonium dioxide powders and produces a mixture of specific plutonium content suitable for the production of MOX fuel pellets. The powder area is composed of the following units:

- plutonium dioxide can receiving and emptying unit
- primary dosing unit
- primary blend ball billing unit
- final dosing unit
- homogenization and pelletizing units
- scrap processing unit
- scrap ball milling unit
- powder auxiliary unit
- jar storage and handling unit

Pellet Process Area

In this area, MOX pellets are sintered, ground, and sorted. The function of the pellet process area is to receive, store, process, and handle fuel pellets. The pellet process area is composed of the following units:

- green pellet storage unit
- sintering units
- sintered pellet storage unit
- grinding units

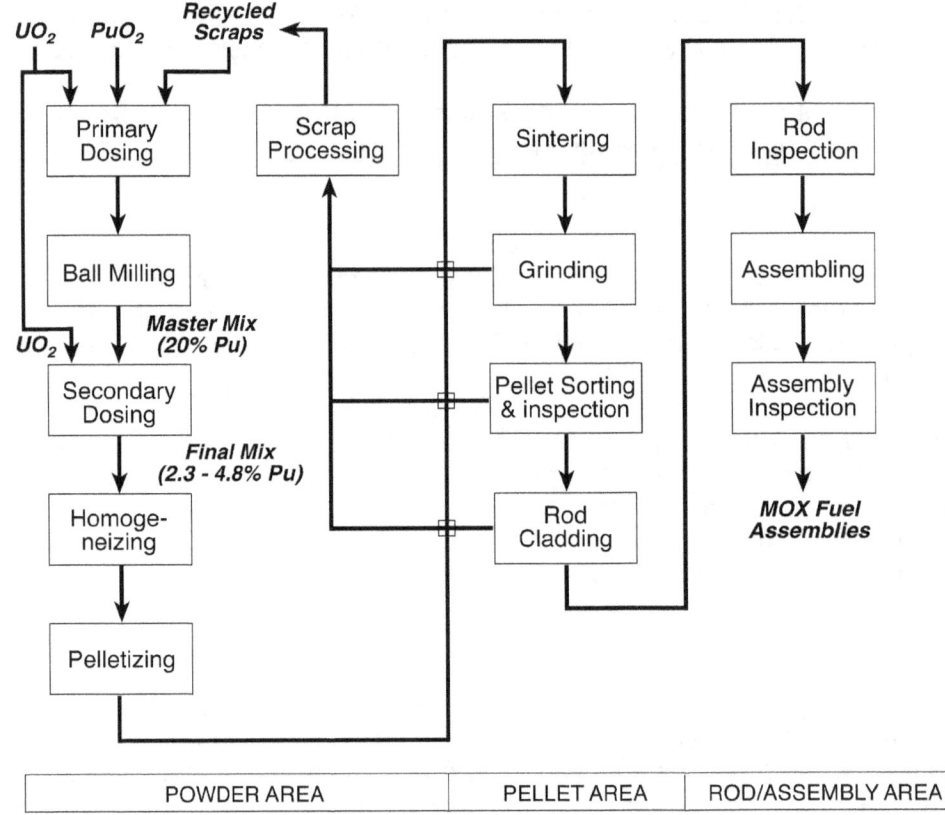

Figure 11.3-1 Overview of MP

- ground and sorted pellet storage unit
- pellet inspection and sorting units
- quality control and manual sorting units
- scrap box loading unit
- pellet repackaging unit
- scrap pellet storage unit
- pellet handing unit

Fuel Rod Process Area

In this area, pellets are loaded into rods, and the rods are inspected. The function of the fuel rod process area is to assemble, inspect, and store fuel rods. The fuel rod process area is composed of the following units:

- rod cladding and decontamination units
- rod tray loading unit
- rod storage unit
- helium leak test unit
- x-ray inspection units
- rod scanning unit

- rod inspection and sorting unit
- rod decladding unit

Assembly Area

In this area, rods are loaded into assemblies, and the assemblies are inspected and stored. The functions of the assembly area are to receive fuel rods and the required fuel assembly components and to assemble, inspect, and store completed MOX fuel assemblies. The assembly area is composed of the following units:

- assembly mockup loading unit
- assembly mounting unit
- assembly dry cleaning unit
- assembly dimensional inspection unit and assembly final inspection unit
- assembly handling and storage unit
- assembly packaging unit

Waste Area

In a separate waste area, solid radioactive waste generated during the MP is processed, stored, and packaged for shipment. The waste area consists of the following units:

- filter dismantling unit
- maintenance and mechanical dismantling unit
- waste storage unit
- waste nuclear counting unit

A detailed description of the main units is provided in Section 11.2 of the revised CAR.

11.3.1.2 Staff Review of MP Process Safety

11.3.1.2.1 Potential Uranium Dioxide Pyrophoricity and Burnback Concerns

The MP will blend the depleted uranium dioxide powder with the plutonium dioxide to form the matrix for the MOX fuel. Uranium dioxide powders are handled in conventional nuclear fuel fabrication facilities. The staff review noted a potential concern regarding the pyrophoric nature (sometimes referred to as burnback) of some fine uranium dioxide powders that can result in oxidation, damage to equipment (essentially a thermal oxidation and heating effect), and a potential release path resulting from the damage of confinement and filter systems (NRC, 1992). This is a known hazard, as such rapid oxidations have occurred in NRC-licensed facilities. Those events involved burnback reactions that started in process equipment and caused localized damage, and then spread through the ventilation system. After those events, relatively large quantities of uranium dioxide powders were found on the filters and equipment (including polycarbonate barriers and filters (prefilters and primary HEPA) filters)) was damaged. The hot uranium dioxide particles were carried through the ventilation system to the filters, where a combination of the hot particles and continued oxidation reactions damaged the HEPA filters. The health consequences were low because of rapid response by personnel and the relatively low radiological hazard of uranium (i.e., as compared to plutonium).

Burnback reactions evolve heat. The following depicts a typical reaction:

$$3 UO_2 + O_2 => U_3O_8$$

The reaction enthalpy is approximately -75 Kcal/mole U_3O_8 formed at 25 °C (77 °F) (i.e., exothermic—heat is released), using information from standard sources (e.g., Perry, 1997; Cordfunke, 1969). This equates to approximately 440 J/g of uranium. Usually, mixtures of oxides (e.g., U_3O_8 and UO_3) are formed and the actual heat release may vary slightly—the formation of UO_3 actually increases the heat released by about 50 percent. The heat released by the reactions increases slightly with increasing temperature. Table 11.3-1 summarizes the burnback reaction enthalpies.

Table 11.3-1 Standard Burnback Reaction Enthalpies

Temperature (°C)	UO_3 Reaction (Kcal/mol)	UO_3 Reaction (J/g U)	U_3O_8 Reaction (Kcal/mol)	U_3O_8 Reaction (J/g U)
25	-35.00	-615.59	-75.30	-441.46
100	-35.03	-616.12	-75.30	-441.45
150	-35.07	-616.78	-75.30	-441.47
200	-35.11	-617.54	-75.32	-441.57
250	-35.15	-618.31	-75.35	-441.75
300	-35.20	-619.08	-75.40	-442.03
350	-35.24	-619.81	-75.46	-442.40
400	-35.28	-620.50	-75.54	-442.87

Note: the negative sign indicates that the reaction is exothermic and releases heat.

The reaction initiation temperature and rate of heat release may also vary with the specific surface area of the powder. Finely divided uranium dioxide (surface area greater than 10 m²/g) can oxidize rapidly at room temperature. Coarser powders undergo oxidation at higher temperatures (e.g., 60 °C (140 °F), and several steps and intermediates may be involved. The burnback phenomena also requires optimal dispersion of a mixture of the particles in air. Uranium dioxide powders in a container or loaded on filters and exposed to air oxidize slowly (i.e., oxidant limited) unless this dispersion occurs. If the powder is too dispersed, no propagation/powder heating occurs, and the uranium dioxide oxidizes slowly (i.e., analogous to a fuel limited condition).

The applicant stated that the depleted uranium dioxide particle size will be approximately 100 microns before ball milling and greater than 2 microns after ball milling (DCS, 2003d). Therefore, from the time depleted uranium dioxide is received at the MFFF to when it is loaded into the homogenization unit, there is a substantially lower risk of depleted uranium dioxide burnback. However, after ball milling, the surface area of the depleted uranium dioxide is larger (closer to 10 m²/g), and the risk of burnback is higher.

The applicant discussed oxidation reactions of uranium dioxide in Section 8.5.1.6.1 of the revised CAR and identified this as a potential cause for a fire in a glovebox. The applicant also provided additional information on uranium dioxide powders in response to requests for additional information (RAIs) (DCS, 2001; DCS, 2002f) and identified temperature design bases for materials of construction in the facility. The temperature design bases have been reviewed by the staff and found to be acceptable (see Section 11.3.1.2.2 of this FSER).

Documentation for powder processing areas was reviewed during an in-office review, and no PSSCs or design bases were found to address the potential loss of confinement hazard (NRC, 2001). The applicant provided supplemental information on the subject (DCS, 2002b). The applicant stated that uranium dioxide is processed as a fine powder at low temperatures and within inert atmospheres; thus burnback does not occur during normal operations. The applicant further indicated that burnback could occur during off-normal conditions if the inert atmosphere has been replaced by air (the applicant has not currently identified a safety function for the inert atmosphere). The applicant stated that burnback has been taken into account in the thermal analysis of the facility during off-normal conditions and cites the RAI (DCS, 2001; DCS, 2002f) responses. No PSSCs or design bases were identified by the applicant to address loss of confinement concerns caused by burnback. In Section 8.5.1.6.1, the revised CAR does identify the following as additional protective features (DCS, 2004, pp. 8-26):

- Uranium dioxide is delivered to the facility site and stored in sealed, 114-L (30-gal) drums. Uranium dioxide is double bagged within the drums, under a nitrogen atmosphere.

- Uranium dioxide is maintained in a nitrogen atmosphere throughout the process.

- Fire detection and suppression systems are provided for gloveboxes (carbon dioxide injection) and process rooms (clean agent).

- Noncombustible or nonflammable materials will be used for process equipment construction and furnishing.

- Combustible materials will be controlled.

The staff concludes that a potential pyrophoric reaction or burnback of uranium dioxide cannot be dismissed, because it has occurred previously during fine uranium dioxide fuel powder processing and PSSCs had not been identified to address its potential effects upon confinement, such as the entrainment of the (potentially hot) powder into the ventilation system, deposition on filters, and damage to the filters by the hot powder particles and continuing oxidation reactions. This could potentially impact several units in the MP area that handle fine uranium dioxide powder by itself or blended with plutonium dioxide. Such a burnback event could result in damage to the final HEPA filters and loss of confinement, the release of large quantities of uranium oxides (a chemical toxicity concern), the release of plutonium powders from the commingled blend, and/or the initiation of other loss of confinement events such as fires.

The applicant provided additional information on the filter system (DCS, 2003a). The applicant again acknowledged that burnback reactions can occur but did not expect this to affect the final HEPA filters for the following reasons:

- During normal operations, uranium dioxide is not expected to be present on the final HEPA filters because the glovebox and very high depressurization exhaust system (VHD) intermediate HEPA filters prevent any significant quantities of uranium dioxide powder

11-65

from reaching the final HEPA filter housing and the first and second stage spark arrestor (previously identified as the high-strength roughing filter and prefilter) remove nearly all of the remaining particles before they reach the final HEPA filters (DCS, 2004).

- During a fire, uranium dioxide powder is one of many potential embers. Large embers are removed by the first and second stage spark arrestors. The remaining micron-size particles are either cooled by the time they reach the final HEPA filters or do not contain enough energy to degrade the performance of the final filters. The staff notes that the uranium dioxide powder is likely to start and remain as small particles.

- The energy associated with the burnback phenomena is small when compared to the energies involved in a facility area fire.

[Text removed under 10 CFR 2.390]

The applicant provided the following design information on the final filtration unit and housing design, which is identified as part of the filter system PSSC (DCS 2003a, Attachment 1; DCS 2003d; NRC 2003c; DCS, 2003b, Section 11.4.9):

- All of the ductwork is constructed of stainless steel.

- The first-stage roughing filter comprises stainless steel wire mesh in a stainless steel frame.

- The second-stage spark arrestor is stainless steel wire mesh interwoven with fiberglass. The filtration efficiency is 99 percent for particles greater than 1 micron in diameter.

- A two-stage HEPA filter completes the facility HEPA filter train.

The applicant expected to credit the following engineered features and administrative controls on the final filtration unit to ensure that the 10 CFR 70.61 performance requirements are met (DCS, 2003a, Attachment 1; DCS, 2003b, Sections 11.4.9 and 11.4.11.1.4):

- high-strength stainless steel mesh first-stage spark arrestor

- high-efficiency, high-strength second-stage spark arrestor (stainless steel/glass fiber prefilter)

- protected two-stage final HEPA filters with a structural integrity of greater than 10 inches of water (i.e., pressure drop)

- multiple redundant ventilation fan systems

- ventilation system design which ensures adequate airflow dilution

- ventilation system design which ensures a pressure drop of less than 10 inches of water across the HEPA filter elements

- preventive maintenance to ensure HEPA filter integrity

- low combustible loads

- fire areas protected by 2-hour-minimum rated fire barriers

The staff concludes that potential burnback events are not prevented and that fine uranium dioxide powders and uranium oxide embers from such events could travel through the ventilation system to the final HEPA filter housing.

The staff analyzed the potential impact from burnback events (NRC, 2004), summarized as follows:

- The burnback event can be initiated wherever uranium dioxide is stored or processed, provided that the uranium dioxide is finely divided, and sufficient energy (air temperature) is present to start the reaction. At the MFFF, the staff assumes that only the depleted uranium oxide which has been ball-milled is at risk.

- The sequence which results in damage to final HEPA filters from burnback is postulated to start in a process glovebox downstream of the ball milling unit. The depleted uranium dioxide could either be spilled or exposed to fire. Convective air currents would carry the unreacted, finely divided depleted uranium dioxide into the glovebox ventilation exhaust system. The unreacted, finely divided depleted uranium dioxide would travel to the final C4 HEPA filters plenum. The burnback event is postulated to occur after, or just before, the depleted uranium dioxide strikes the first stage of final C4 HEPA filters. The sudden release of reaction energy (heat) could damage the first stage of final C4 HEPA filters, resulting in a loss of pressure drop and loss of filter efficiency, across the first stage. Operations personnel would be notified of the loss by differential pressure-indicating switches. A second C4 train would be brought on line, and the affected train would be removed from service for maintenance.

- The consequences of both the spill-initiated and fire-initiated burnback events are low. However, the potential event results in damage to a PSSC and loss of its safety function.

The staff evaluated the hazard to final C3 and C4 HEPA filters that would be posed by a uranium burnback event occurring on or near those filters. The staff postulated that both a glovebox spill and a glovebox fire involving depleted uranium dioxide are credible means to disperse depleted uranium dioxide into the ventilation systems.

Multiple filters are present in the C3 and C4 ventilation systems; therefore the margins could be higher.

The consequence analysis demonstrates that a HEPA filter would survive uranium burnback after the maximum credible dispersal of depleted uranium dioxide powder in a glovebox spill or fire. The margin of safety is at least a factor of 16 against spills, and a factor of 5 against fires. This margin is more than sufficient, given the conservative assumptions in the staff's evaluation. Consequently, the staff finds the PSSC and design basis information associated with the ventilation system and final filtration units to be sufficient to address the potential burnback phenomena associated with the pyrophoric nature of some uranium dioxide powders. This closes the previously identified Open Item MP-1. The staff considers this to be related to the filter soot loading issue discussed in Chapter 7 of this FSER (former Open Item FS-1).

11.3.1.2.2 Potential Plutonium Dioxide Heating Effects

The staff has reviewed plutonium-handling areas for potential chemical safety concerns. The review noted concerns related to the potential heat generation by the plutonium dioxide—plutonium in the glovebox environment can easily reach equilibrium temperatures of 80 °C (176 °F) (DOE, 1998, Section 2.6.3.1).

Section 8.5.1.6.2 of the revised CAR and RAI responses (DCS, 2001; DCS, 2002f) provide supplementary information on plutonium thermal effects, as well as a summary of the design bases for decay heat and temperatures. The specific heat loads for plutonium were identified by the applicant as follows:

- unpolished plutonium—2.899 W/kg of unpolished plutonium dioxide powder

- polished plutonium—2.181 W/kg of polished plutonium dioxide powder

Using values from the literature (DOE, 1998), the staff review estimates heat loads of 2.5–3.5 W/kg of plutonium dioxide, depending on the isotopic ranges used. These values generally overlap the applicant's heat load estimates. Thus, the applicant's values are reasonable.

The applicant identified temperature design bases for the materials of construction in the facility, which are reproduced as Tables 11.3-2 and 11.3-3 (DCS, 2001; DCS, 2002f). The response identified the storage rooms, storage gloveboxes, and larger production units as having potentially large heat loads. The applicant's response mentions that the temperature design bases in Tables 11.3-2 and 11.3-3 will be met during normal operations, but might be exceeded during incidents where ventilation is not maintained for the plutonium dioxide storage area and the handling and storage tunnel. Consequently, in Section 5.5.2.1.6.9 of the revised CAR, the applicant identified the high depressurization exhaust system (HDE) (part of the C3 confinement system) as the PSSC with the safety function of providing exhaust so that temperatures in the 3013 cannister storage area are maintained within the design basis values. The staff notes that these design bases and approaches are consistent with accepted practice for steels (e.g.,ASME), concrete (e.g., ACI), and most plastics (e.g., Perry, 1997) and finds this approach to be acceptable. Polycarbonate is discussed separately in Section 11.4 of this FSER. The staff finds this approach to be acceptable for the construction authorization stage according to the acceptance criteria in Section 8.4.3.5 of the SRP.

Table 11.3-2 Applicant's Design Basis Temperature Criteria

Material	Situation	
	Normal Operating Temperature, °C (°F)	Hypothetical Maximum Operating Temperature, °C (°F)
Ordinary Concrete	60 (140)	100 (212)
Stainless Steel	425 (797)	425 (797)
BPP #9	80 (176)	100 (212)
BPP #10	100 (212)	100 (212)
NS41 Silicone Elastomer	180 (356)	180 (356)
Polycarbonate (Lexan)	35 (95) (thermal cycling) 50 (122)	70 (158)
	BPP = borated polyethylene plaster	

Table 11.3-3 Applicant's Additional Temperature Design Basis Criteria for Personnel Protection

Material	Normal Operating Temperature Limit, °C (°F)
Borate (colemanite) concrete	80 (176)
Kyowaglas—storage —operating	80 (176) 35 (95)
Fuel rods, pellets, and cladding	60 (140)

11.3.1.2.3 Potential Plutonium Dioxide Pyrophoricity and Burnback Concerns

The AP process produces plutonium (IV) oxalate, which generally requires a higher calcination temperature (about 400–500 °C (752–932 °F)) than either plutonium (III) or plutonium (VI) oxalates (about 200 °C (392 °F)). The staff found that plutonium dioxide powder from the calciner in the oxalic precipitation and oxidation unit does not have a design basis or specification, and the calcining operation does not have a design basis related to plutonium dioxide quality. The staff concerns fall into two areas:

(1)   The first concern involves oxidation reactions and powder dispersion because of the presence of substoichiometric oxides of plutonium dioxide (essentially PuO_{2-x}) that are ignited by the decay heat. Depending on conditions, plutonium can form varying oxides, some of which can be pyrophoric. In general, plutonium oxides with oxygen contents lower than the dioxide are potentially pyrophoric (DOE, 1998, Section 2.6.3.2).

(2)   The presence of water or other volatile species and impurities can increase pressure inside containers because of radiolysis and decay heat and ultimately cause overpressurization, resulting in an explosion and a loss of confinement.

The applicant discussed plutonium dioxide pyrophoricity in Section 8.5.1.6.2 of the revised CAR. This section states that plutonium dioxide supplied to the facility will meet the requirements of DOE-STD-3013-2000, "Stabilization, Packaging, and Storage of Plutonium-Bearing Materials," issued November 1999 (DOE, 1999), in order to ensure stability. For plutonium dioxide within the facility, a standard or specification is not identified. Supplemental information provided by the applicant (DCS, 2002b) states that plutonium dioxide is stable in air. The applicant stated that the formation of substoichiometric oxides is not a concern, as reduction of the plutonium (IV) oxalate does not occur. The staff notes that oxide formation depends on the process conditions, and, without a specification or basis, some substoichiometric plutonium dioxide from impurity reactions cannot be discounted. The powder will also be handled under a nitrogen atmosphere, but this was not identified as a PSSC.

The staff conducted a brief literature review and found that plutonium dioxide is often present as a substoichiometric oxide (i.e., PuO_{2-x}) and prone to absorb moisture unless it has been calcined and held at a temperature of about 900 °C (1652 °F) for 2 hours to stabilize (ceramicize) the material (DOE, 1998, Section 2.7). Unstabilized plutonium oxides may exhibit pyrophoric reactions in air because of their substoichiometry or the radiolysis of absorbed water, which could lead to a loss of confinement and release or initiate other events, such as fires.

The DOE experience with plutonium dioxide indicates that the pyrophoricity and stability (including entrained, attached, and absorbed compounds, such as water) of plutonium dioxide depend on the calcination conditions (e.g., time, temperature, and atmosphere), and that substoichiometric and unstable plutonium dioxide can initiate events (DOE, 1999). The staff review indicates that a lower calcination temperature may be involved at the facility, and standards and limits may be needed to control plutonium dioxide substoichiometry and instability. The staff did not find standards or limits described by the applicant that will assure that the formation of significant amounts of substoichiometric or unstable plutonium dioxide (from the calciner and in MP operations) is prevented.

The applicant identified nominal values of humidity (water) in the powders entering the MP area. For plutonium dioxide, 1 percent is used for normal situations and 3 percent is used for off-normal situations. These were not identified as design bases and PSSCs were not specified. The DOE 3013 standard (DOE, 1999) mentions a loss on ignition of 0.5 percent (i.e., weight loss from water vaporizing caused by heating of the material). Moderation (i.e., water content) is not controlled by the applicant in the calciner, homogenization, and MP buffer storage units. Water and volatile content have not been identified as design bases.

Furthermore, at the February 13, 2003, public meeting (NRC, 2003b), the applicant stated that a review was underway to determine if unstabilized plutonium dioxide would be received by the facility. The staff review of the calcining section of the AP process (see Section 11.2 of this FSER) did not identify any PSSCs or design bases for ensuring that stabilized plutonium dioxide powder would be produced. At the December 10–12, 2002, public meeting (NRC, 2003a), the applicant indicated the plutonium dioxide storage containers would be convenience cans, with gas vent filters and spring-loaded tops for pressure relief. These were not identified as PSSCs.

The applicant subsequently submitted additional information on the subject (NRC, 2001; DCS, 2003a). Leakage from the can is covered by the spill event discussed in Section 5.5.2.3.6.4 of the revised CAR on load-handling controls and C4 confinement. For pressure accumulation within the can, the lid is deterministically assumed to impact the glovebox. The material-handling control PSSC is used to mitigate this potential event. For this case, this PSSC may include the control of moisture content of the material, residence time of the canned material (e.g., in the range of months and years), and/or design pressure of the reusable can. The specific IROFS will be identified by DCS in its ISA summary and evaluated by the staff as part of any license application DCS may later submit.

In addition, the applicant indicated that overpressurization from the oxidation of plutonium (III) oxalate contained within the stored cans may be prevented through one of the following means:

- controls on plutonium oxalate furnace (calciner) parameters, such as residence time and minimum temperature to ensure complete oxidation and moisture content of plutonium oxalate entering the furnace

- experimental confirmation of the minimum water content accompanying plutonium (III) at the exit of the furnace to prevent any overpressurization because of the energy liberated during reoxidation (to plutonium (IV))

- measurement of the plutonium (III) content in the plutonium dioxide powder

The specific IROFS will be identified by DCS in its ISA Summary and evaluated by the staff as part of any license application DCS may later submit.

The staff review found that leakage or spillage from the can that does not challenge confinement is adequately addressed by the C4 confinement system, as discussed in Section 5.5.2.3.6.4 of the revised CAR. For the case with pressure accumulation, the material-handling control PSSC has the safety function to prevent impacts to the glovebox from (1) overpressurization from residual volatile species (and their radiolysis) and (2) overpressurization from the oxidation of plutonium (III). The description of the material handling control PSSC in Section 5.6.2.3 of the revised CAR did not mention any design bases. The staff notes that the applicant appeared to be considering both preventive and mitigative approaches for overpressurization events.

The applicant provided supplemental information on this subject (DCS, 2003b). This information identifies an additional safety function of the material handling controls PSSC to prevent potential overpressurization of the reusable plutonium dioxide cans resulting from radiolysis or oxidation of plutonium (III) oxalate, as well as its subsequent impact to the glovebox. The associated design basis is to ensure that the reusable can is designed to the maximum internal pressure calculated for these events, plus an additional 10 percent as the margin. The staff notes that this provides an approach with a defined margin and is consistent with the safety approach for FTS components and code requirements, as discussed in Section 11.8 of this FSER. The staff concludes that this provides an approach and range of design basis values consistent with normal industry practice and finds this acceptable for the construction authorization stage.

11.3.1.2.4 Sintering Furnace Concerns

The staff requested clarification and more information on the controls around the sintering furnaces, including the hydrogen detectors, as they appear to involve a complex mixture of hydrogen detectors, oxygen sensors, and pressure controls. In response (DCS, 2001; DCS, 2002f), the applicant provided a diagram that showed that part of the control range of hydrogen in argon is flammable in air and stated that the sensors would detect hydrogen and, at 25 percent of the LFL, would terminate hydrogen flow at the hydrogen/argon mixing station. In addition, fire detector(s) in the room would detect any fire and alarm but would not terminate hydrogen flow. The applicant explained that the sintering furnace would not be in a glovebox, and, as the room functioned as secondary confinement, the sintering room and the furnace would become the PSSCs for confinement (DCS, 2001; DCS, 2002f).

The sintering furnace has water-cooled walls and a moisture conditioning system for the furnace gases (the hydrogen/argon mixture). The in-office review of the preliminary hazard analysis and preliminary accident analysis did not find a potential steam explosion included (NRC, 2001). The applicant provided supplemental information on potential steam explosions (DCS, 2002b). The applicant stated that steam explosions have been identified during the facility safety analysis as a credible event. Ongoing safety analyses by the applicant have identified three types of scenarios that can lead to a steam explosion—(1) entry of water from the water cooling loop, (2) entry of water from the humidifying loop, and (3) steam generation within the water cooling systems. The applicant mentioned that a steam explosion involving a water-cooled furnace had previously occurred at Los Alamos National Laboratory, but this involved internal cooling coils, while the proposed furnaces would have external cooling coils. The applicant further stated that a cooling water leak will be demonstrated to be highly unlikely, specific IROFS features will be identified for the humidifying loop, and relief valves will render steam pressurization of the cooling water loop highly unlikely. Supporting information to demonstrate an even frequency of highly unlikely events is not included in the response.

The staff reviewed Sections 5.5.2.4.6.2 and 11.2.2.16 of the revised CAR for additional information. These sections identify a prevention strategy to address potential steam explosions. The PSSC is the PSCS with the safety function of isolating humidifier water flow on a high water level. Thus, the water supply to the humidifier would be terminated before the humidifier overflowing and potentially allowing liquid water to enter the sintering furnace via the gas supply side. This would be an active, engineered control. Such active engineered controls can be designed with multiple layers, subsystems, and components in order to achieve very high reliabilities for level control. The staff concludes that such an active engineered control could be designed and implemented to prevent water intrusion and a subsequent steam explosion; therefore, it finds it acceptable at the CAR stage. Sections 11.2.2.16 and 11.4.11.8 identify an additional function of the PSCS to shut down the sintering furnace (by electrical cutoff) upon a loss of cooling water flow, and to shut down zone heating if the related surface temperature is excessive (greater than 60 °C (140 °F))—this is identified as the design basis. There is also a backup cooling water supply, and the cooling water coils are on the outside of the furnace. The staff reviewed preliminary view and section drawings of the sintering furnace during an in-office review and concluded that (1) the coils are outside the sintering furnace shell, (2) the shell is thicker than the cooling water tube thickness, and (3) the coils are not confined within additional metal shells. The staff therefore concludes that the applicant can demonstrate that a steam explosion from a cooling water leak is either unlikely or highly unlikely and, if it were to occur, unlikely to penetrate the significantly thicker sintering furnace shell (i.e., if the event were to occur at all, it would be directed toward the unconfined areas which are in the opposite direction from the furnace itself). The staff finds that the applicant's approach for addressing steam explosions is acceptable for the CAR stage, according to Section 8.4.3.5 of the SRP.

The applicant did not perform any coverage or location/distance analyses for sensors and detectors. The applicant stated that it would verify that a hydrogen leak from the furnace would be detected and terminated by pressure detection. The applicant expected that, between the hydrogen monitors and pressure sensors, a hydrogen leak would be detected and the flow terminated (DCS, 2002a; NRC, 2001).

Section 11.6.7 of the revised CAR identifies the design bases for I&C PSSCs. This section states that combustible and hydrogen detectors will be selected in accordance with ISA-S12.13-Part I-1995, "Performance Requirements, Combustible Gas Detectors," issued 1995. Installation, operation, and maintenance of combustible gas detectors would be in accordance with ISA RP12.13-Part II-1987, "Installation, Operation, and Maintenance of Combustible Gas Detection Equipment," issued 1987. The staff has reviewed these standards for applicability. They include considerations of gas properties, sources, and detector locations, and they are a routine approach for the industry. A setpoint methodology is also included, based on ANSI/ISA-67.04.01-2000, "Setpoints for Nuclear Safety-Related Instrumentation," issued 2000 (ANSI, 2000). This is also a standard approach for the nuclear industry. The staff concludes that the design bases for hydrogen detection have been adequately described for the CAR stage.

The applicant indicated that hydrogen sensors in the room would detect any leaks and would terminate the flow of hydrogen flow to the furnace. In addition, pressure controls would detect any loss of pressure in the furnace caused by a leak and also terminate the hydrogen flow (NRC, 2001). The staff expressed concerns about the potential for small leaks to result in hydrogen burning that might go undetected and exacerbate radionuclide releases. Regarding the airlocks, the applicant stated that there would be interlocks to prevent both doors (inner and outer) from opening at the same time. In addition, oxygen sensors in the furnace would detect atmospheric intrusion and terminate the hydrogen flow; these have been identified as PSSCs.

The applicant provided supplemental information on PSSCs and design bases in the sintering furnace area that identified additional PSSCs and design bases (DCS, 2002c). In addition, Section 11.2.2.16 of the revised CAR states that, if hydrogen is detected in the furnace room, then the gas supply is automatically shifted to pure argon (i.e., the hydrogen flow is terminated). In Section 5.5.2.4.6.1, the revised CAR discusses a prevention strategy for hydrogen explosions. The PSSC is identified as the PSCS, with the safety function of preventing the formation of an explosive mixture of hydrogen within the facility associated with the use of the hydrogen-argon mixture. The revised CAR indicates that the applicant is performing detailed analyses of the hydrogen-argon system and associated furnace design and operations as part of the final design and license application. The revised CAR notes that potential specific controls for preventing hydrogen explosions around the sintering furnace (such as limiting the hydrogen content in the hydrogen-argon mixture, monitoring for oxygen within the furnace, monitoring for hydrogen outside of the furnace, and crediting dilution airflow associated with the HDE or VHD) have already been identified as PSSCs in other safety strategies; thus there would be little or no impact of the specific control selection upon design at the later stage. The staff finds the proposed approach to be an acceptable strategy for preventing explosions around the sintering furnace. The staff concludes that combinations of the specific controls already identified as PSSCs should be capable of performing the safety function of preventing explosive mixtures.

The staff notes that a specific approach based upon limiting the hydrogen content in the hydrogen-argon mixture may be the simplest and most effective approach.

In Section 8.5.1.1.1, the revised CAR states that the hydrogen design basis is 25 percent of the LFL of hydrogen in air. This value will not be exceeded during normal and off-normal conditions. Actual setpoints will be identified by DCS in its ISA summary and evaluated by the staff as part of any license application DCS may later submit.

The applicant provided a methodology for LFL determination, which is discussed in Section 11.2 of the FSER.

The staff review indicates that 25 percent of the LFL in a chemical flow system is normally considered a hydrogen limit analogous to a design basis (NRC, 1999). Other NRC-licensed fuel fabrication facilities typically use 25 percent of the LFL as an administrative limit. The SRS waste storage tank facilities alarm at 10 percent but use 25 percent of the LFL as their design basis (Hobbs, 2000). The staff concludes that a value of 25 percent of the LFL is usually used as a design basis or equivalent value in similar nuclear facility applications. The staff finds this to be acceptable for the CAR stage. This also closes Open Item AP-2 in Section 11.2 of the April 2003 DSER.

11.3.1.2.5 Design Basis of the PSSCs and Applicable Baseline Design Criteria

The design bases of MP associated with chemical processing have been discussed in Sections 11.3.1.2.1–11.3.1.2.4 of this FSER.

The applicant stated that the BDC specified in 10 CFR 70.64(a) are incorporated into the design and operation of the facility (Section 5.5.5.4 of the revised CAR). The applicant stated that information demonstrating compliance with these criteria is provided in the applicable chapters of the revised CAR. For chemical protection, 10 CFR 70.64(a)(5) states the following:

> Chemical protection. The design must provide for adequate protection against chemical risks produced from licensed material, facility conditions which affect the safety of licensed material, and hazardous chemicals produced from licensed material.

The applicant indicated that reagents are stored, and chemical mixtures are prepared, in the reagent processing building and the reagent storage area of the AP area, and they are generally separated from each other and from radioactive materials. The applicant will avoid mixing incompatible materials by design, controls, and procedures. The AP and MP facilities are broken down into process functional units, which are made up of one or more subassemblies performing consistent and elementary tasks. The applicant stated that the breakdown into control functional units allows each entity to operate relatively independently in the given mode. The staff review notes that this separation and independence are consistent with accepted practices for safe operations.

The applicant will control process storage and operation conditions in order to prevent exothermic and potential autocatalytic reactions in the reagent processing building and the AP and MP areas. Autocatalytic and exothermic reactions of chemicals would be prevented through the control of the process parameters (e.g., reactant concentration, temperature, catalyst concentration in solution, and pressure) that affect the reactions. Many of these controls have been identified as PSSCs with design bases. Codes and standards have been applied on a specific level, often as design bases.

The applicant stated that there is reasonable assurance that the PSSCs will be sufficiently reliable and available based on the use of standard nuclear industry engineering practices. These practices are incorporated into the facility general design philosophy, design bases, system design, and commitments to applicable management measures. These practices ensure that applicable industry codes and standards are utilized, adequate safety margins are provided, engineering features are utilized to the extent practicable, the defense-in-depth philosophy is incorporated into the design, and PSSCs will be appropriately maintained.

The staff review finds that the applicant has provided sufficient information to meet the requirements of 10 CFR 70.64(a)(5). Using the SRP, the staff review notes that the previously identified open items in MP have been closed by additional safety strategies and information from the applicant, and the staff concludes that the applicant has satisfied this BDC.

Related to chemical protection, the explosion protection BDC is stated as part of the fire protection BDC in 10 CFR 70.64(a)(3):

> Fire protection. The design must provide for adequate protection against fires and explosions.

The applicant stated that there is reasonable assurance that the PSSCs will be sufficiently reliable and available based on the use of standard nuclear industry engineering practices. These practices are incorporated into the facility general design philosophy, design bases, system design, and commitments to applicable management measures. These practices ensure that applicable industry codes and standards are utilized, adequate safety margins are provided, engineering features are utilized to the extent practicable, the defense-in-depth philosophy is incorporated into the design, and PSSCs will be appropriately maintained.

The staff review finds that the applicant has provided sufficient information to meet the requirements of 10 CFR 70.64(a)(3). Using NUREG-1781, the staff review notes that the previously identified open items regarding the MP process have been closed by additional safety strategies and information from the applicant, and the staff concludes the applicant has satisfied this BDC.

11.3.2 Evaluation Findings

In Section 11.3.7 of the revised CAR, DCS provided design basis information for the MP process that it identified as PSSCs for the facility. Based on the staff's review of the revised CAR and supporting information provided by the applicant relevant to the MP process, the staff finds that, for the reasons discussed above, DCS has met the BDC set forth in 10 CFR 70.64(a)(3) for explosions and 10 CFR 70.64(a)(5) for chemical safety. Further, the staff concludes, pursuant to 10 CFR 70.23(b), that the design bases of the PSSCs identified by the applicant will provide reasonable assurance of protection against natural phenomena and the consequences of potential accidents.

Several previously identified open items in the April 30, 2002, and April 30, 2003, DSERs have now been closed in a satisfactory manner—MP-1, MP-2, MP-3, and MP-4.

11.3.3 References

(ANSI, 2000) American National Standards Institute/Instrument Society of America. ANSI/ISA 67.04.01-2000, "Setpoints for Nuclear Safety-Related Instrumentation." Research Triangle Park, North Carolina, 2000.

(Cordfunke, 1969) Cordfunke, E.H.P. *The Chemistry of Uranium (Including its Applications in Nuclear Technology).* Elsevier, New York, 1969.

(DCS, 2001) Hastings, P., Duke Cogema Stone & Webster. Letter to U.S. Nuclear Regulatory Commission, RE: Response to Request for Additional Information. August 31, 2001.

(DCS, 2002a) Hastings, P., Duke Cogema Stone & Webster. Letter to U.S. Nuclear Regulatory Commission, RE: Clarification of Responses to NRC Request for Additional Information. January 7, 2002.

(DCS, 2002b) Hastings, P., Duke Cogema Stone & Webster. Letter to U.S. Nuclear Regulatory Commission, RE: Clarification of Responses to NRC Request for Additional Information, DCS-NRC-000083. February 11, 2002.

(DCS, 2002c) Hastings, P., Duke Cogema Stone & Webster. Letter to U.S. Nuclear Regulatory Commission, RE: Clarification of Responses to NRC Request for Additional Information. March 8, 2002.

(DCS, 2002d) Hastings, P., Duke Cogema Stone & Webster. Letter to U.S. Nuclear Regulatory Commission, RE: Clarification of Responses to NRC Request for Additional Information. April 23, 2002.

(DCS, 2002e) Ashe, K., Duke Cogema Stone & Webster. Letter to Docket Number 070-03098, U.S. Nuclear Regulatory Commission, RE: Mixed Oxide Fuel Fabrication Facility—Construction Authorization Request. October 31, 2002 (including page changes through February 9, 2005).

(DCS, 2002f) Hastings, P., Duke Cogema Stone & Webster. Letter to U.S. Nuclear Regulatory Commission, RE: Requests for Additional Information, Clarifications, and Open Item Mapping into the Construction Authorization Request Revision (DCS-NRC-000120). November 22, 2002.

(DCS, 2003a) Hastings, P., Duke Cogema Stone & Webster. Letter to U.S. Nuclear Regulatory Commission, RE: Mixed Oxide (MOX) Fuel Fabrication Facility Construction Authorization Request Change Pages. February 18, 2003.

(DCS, 2003b) Hastings, P., Duke Cogema Stone & Webster. Letter to U.S. Nuclear Regulatory Commission, RE: Mixed Oxide (MOX) Fuel Fabrication Facility Construction Authorization Request, DCS-NRC-000129, February 18, 2003 (DCS-NRC-000131). April 1, 2003.

(DCS, 2003c) Hastings, P., Duke Cogema Stone & Webster. Letter to U.S. Nuclear Regulatory Commission, RE: Response to Request for Additional Information—Chemical Safety Open Items CS-9, AP-02, AP-08, and AP-09, DCS-NRC-000157. September 29, 2003.

(DCS, 2003d) Hastings, P., Duke Cogema Stone & Webster. Letter to U.S. Nuclear Regulatory Commission, RE: Response to Request for Additional Information—DSER Open Items MP-01 (UO$_2$) and AP-03 (Titanium Fires), DCS-NRC-000161. October 10, 2003.

(DCS, 2004) Swiegart, R., Duke Cogema Stone & Webster. Letter to Docket Number 070-3098, U.S. Nuclear Regulatory Commission, Mixed Oxide Fuel Fabrication Facility, Response to Request for Additional Information. October 11, 2004.

(DOE, 1998) U.S. Department of Energy. DOE Std 1128-98, "Guide of Good Practices for Occupational Radiological Protection in Plutonium Facilities." Washington, DC, June 1, 1998.

(DOE, 1999) U.S. Department of Energy. DOE STD 3013-2000, "Stabilization, Packaging, and Storage of Plutonium-Bearing Materials." Washington, DC, November 1999.

(Hobbs, 2000) Hobbs, D.T. "Possible Explosive Compounds in the Savannah River Site Waste Tank Farm Facilities," WSRC-TR-91-444, Revision 3. February 15, 2000.

(NRC, 1992) U.S. Nuclear Regulatory Commission. Information Notice 92-14, "Uranium Oxide Fires at Fuel Cycle Facilities." Washington, DC, February 21, 1992.

(NRC, 1999) Pierson, R., U.S. Nuclear Regulatory Commission. Memorandum to D. Clark Gibbs, U.S. Department of Energy, RE: Estimates of Hydrogen Generation from Wastes at the Proposed TWRS-P Facility. April 21, 1999.

(NRC, 2000) U.S. Nuclear Regulatory Commission. NUREG-1718, "Standard Review Plan for the Review of an Application for a Mixed Oxide (MOX) Fuel Fabrication Facility." Washington, DC, August 2000.

(NRC, 2001) Persinko, A., U.S. Nuclear Regulatory Commission. Memorandum to E.J. Leeds, U.S. Nuclear Regulatory Commission, RE: November 27–29, 2001 In-Office Review Summary of DCS Construction Authorization Request Supporting Documents for the MFFF. December 18, 2001.

(NRC, 2003a) Persinko, A., U.S. Nuclear Regulatory Commission. Memorandum to M.N. Leach, U.S. Nuclear Regulatory Commission, RE: December 10–12 Meeting Summary: Meeting with DCS to Discuss Mixed Oxide Fuel Fabrication Facility Revised Construction Authorization Report. January 31, 2003.

(NRC, 2003b) Persinko, A., U.S. Nuclear Regulatory Commission. Memorandum to M.N. Leach, U.S. Nuclear Regulatory Commission, RE: February 6-7 Meeting Summary: Meeting with DCS to Discuss Mixed Oxide Fuel Fabrication Facility Revised Construction Authorization Report. March 5, 2003.

(NRC, 2003c) Persinko, A., U.S. Nuclear Regulatory Commission. Memorandum to K. Halvey Gibson, U.S. Nuclear Regulatory Commission, RE: July 29–August 1, 2003, Meeting Summary:

Meeting with Duke Cogema Stone & Webster to Discuss Open Items Related to Mixed Oxide Fuel Fabrication Facility. August 29, 2003.

(NRC, 2004) Giitter, J., J. Holonich, and R. Pierson, U.S. Nuclear Regulatory Commission. Memorandum to H. Peterson and M.N. Leach, U.S. Nuclear Regulatory Commission, RE: Safety Evaluation and Staff Positions on the Closure of Remaining Chemical Safety Open Items MP-1 Pertaining to the Potential for Uranium Burnback to Damage Final HEPA Filters. April 15, 2004.

(Perry, 1997) *Perry's Chemical Engineers' Handbook*, Seventh Edition. McGraw-Hill, New York, 1997.

11.4 Ventilation and Confinement Systems

11.4.1 Conduct of Review

This section of the FSER contains the staff's review of ventilation and confinement systems described by the applicant in Chapter 11 of the revised CAR (DCS, 2002). The objective of this review is to determine whether the ventilation and confinement PSSCs and their design bases identified by the applicant provide reasonable assurance of protection against natural phenomena and the consequences of potential accidents. The staff evaluated the information provided by the applicant for ventilation and confinement systems by reviewing Chapter 11 of the revised CAR, other sections of the revised CAR, and supplementary information provided by the applicant. The review of ventilation and confinement systems design bases and strategies was closely coordinated with the review of accident sequences described in the safety assessment of the design bases (see Chapter 5 of this FSER) and the review of other plant systems.

The staff reviewed how the information in the revised CAR addresses the following regulation:

• 10 CFR 70.23(b) states, as a prerequisite to construction approval, that the design bases of the PSSCs and the Quality Assurance Program be found to provide reasonable assurance of protection against natural phenomena and the consequences of potential accidents.

The review of the CAR focused on the design basis of ventilation and confinement systems, their components, and other related information. For ventilation and confinement systems, the staff reviewed information provided by the applicant for the safety function, system description, and safety analysis. The review also encompassed proposed design basis considerations such as redundancy, independence, reliability, and quality. The staff used Chapter 11.0 in NUREG-1718 (NRC, 2000) as guidance for performing the review.

11.4.1.1 System Description

11.4.1.1.1 Functions and Major Components

In the revised CAR, the applicant proposed a ventilation and confinement system to confine radioactive materials within process areas and gloveboxes and to ensure minimum dispersal of radioactive materials during routine operations and under accident conditions. The system was intended to meet the recommendations in RG 3.12, "General Design Guide for Ventilation

Systems of Plutonium Processing and Fuel Fabrication Plants," issued 1973 (NRC, 1973), and consists of the following:

- Ventilation zones are operated at pressure differentials designed such that air leakage occurs from areas of low radiation hazard into areas with the greater radiation hazard (NRC, 2000; Section 11.4.1.2 of the revised CAR) (see Figure 11.4-1).

- A system of final filter assemblies is consistent with ASME N509, "Nuclear Power Plant Air Cleaning Units and Components," issued 1980 (ASME, 1980a), consisting of HEPA filters, stainless steel/glass fiber prefilters, and stainless steel roughing filters intended to remove radioactive materials from process areas and occupied areas during routine operations and under accident conditions. The stainless steel/glass fiber prefilters and stainless steel roughing filters are designed to remove hot embers and a large percentage of the soot in order to protect the final HEPA filters from fire damage and excessive soot plugging (DCS, 2003c; NRC, 2000; Section 11.4.9 of the revised CAR).

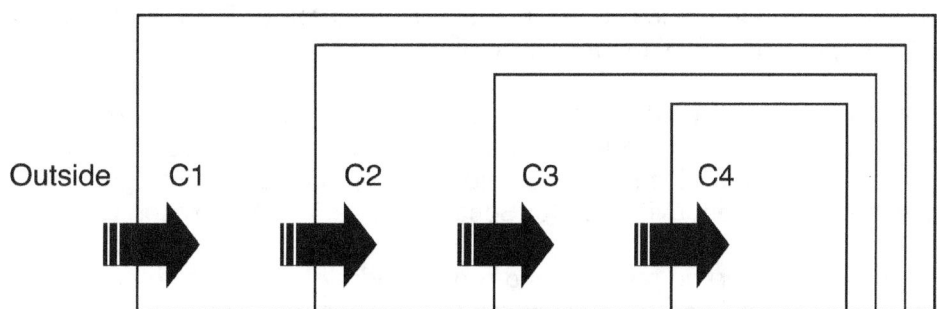

Figure 11.4-1 Ventilation Confinement Zones

- The design is in accordance with ASME N510 (ASME, 1980b) to allow testing and inservice surveillance in order to ensure operability and required functional performance (NRC, 2000; Section 11.4.11 of the revised CAR).

- Redundant PSSCs ensure that the performance requirements in 10 CFR 70.61 are met (NRC, 2000).

- The design provides sufficient capacity and capability under routine operations. Under accident conditions involving fires, the proposed design has sufficient capacity to maintain air temperatures at the HEPA filters less than 232 °C (450 °F) and to ensure that HEPA filters will remain operational. The applicant proposed to use air dilution of hot air from a fire area with air from other areas to protect the HEPA filters (see Section 7.1.5.5 of this FSER). The HEPA filters are capable of operating at temperatures up to 232 °C (450 °F) without severe degradation effects (NRC, 2000; Section 11.4.9 of the revised CAR).

- Under accident conditions, the applicant proposed to only take credit for the final filtration assemblies and not HEPA filters located on gloveboxes or at confinement zone boundaries (NRC, 2000; Section 11.4.9 of the revised CAR). The applicant proposed to use a release factor of 1×10^{-4} for the final filter assemblies in its accident safety analyses (Sections 5.5.3.2 and 11.4.9 of the revised CAR). The applicant based its proposal on having redundant HEPA filter banks that are protected by stainless steel roughing filters, and stainless steel/glass fiber prefilters in each redundant filter assembly with HEPA filters that have been tested to have an efficiency for the removal of 0.3 micron particles of at least 99.97 percent (DCS, 2003c). After installation, the HEPA filter banks will be leak tested in accordance with ASME N510 (ASME, 1980b) to ensure that leakage efficiency is at least 99.95 percent. The stainless steel roughing filters and stainless steel/glass fiber prefilters are noncombustible and have sufficient structural strength to collect greater than 90 percent of the soot generated in a design basis fire at the design flow rate to withstand the full differential pressure generated by the fans when completely clogged (DCS, 2003c). Under severe conditions, the applicant considered that each bank of HEPAs will remove at least 99 percent of the particulates and will provide an overall efficiency of 99.99 percent for the two combined banks.

- Monitoring instrumentation, alarms, and controls ensures that pressure differentials in confinement zones are maintained, alternative power supplies are actuated when needed, and the consequences of accidents are mitigated (NRC, 2000; Sections 11.4.2, 11.4.3, 11.4.4, and 11.4.7 of the revised CAR).

- The design provides for a safe air supply to the emergency control rooms, consisting of heaters, coolers, prefilters, HEPA filters, and acid gas/organic vapor cartridges to remove chemicals. This system will maintain a safe environment during emergencies for personnel and equipment (NRC, 2000; Section 11.4.2.6 of the revised CAR).

- The design provides for the removal and replacement of filters and other expected maintenance activities in order to minimize personnel exposures. The design allows for in-place testing of HEPAs in accordance with ASME N510 (ASME, 1980b) in order to ensure that HEPAs have been properly installed and are undamaged (NRC, 2000; Section 11.4.9.1 of the revised CAR).

- Gloveboxes consist of welded stainless steel enclosures with windows, alone and in interconnected groups, that act as a primary barrier to confine hazardous (radioactive, toxic, or flammable) materials and to provide structural support capable of protecting process equipment during a postulated seismic event. The MFFF personnel's access to equipment inside the gloveboxes is provided through access holes in the glovebox windows fitted with gloves that maintain the confinement boundary. Gloveboxes that contain powder or pellet forms are inerted with nitrogen gas to eliminate adverse effects of atmospheric oxygen on the process or fuel (NRC, 2000; Section 11.4.7 of the revised CAR).

- Glovebox window panels, viewing ports, or video cameras give visual access to the gloveboxes. Light fixtures provide illumination and are generally located outside the glovebox. The windows are clear rectangular panels that fit into gasketed frames that cover specifically designed openings in the glovebox shell. The windows proposed by the applicant are polycarbonate (Lexan®) that may have lead-impregnated polymer sheets or lead-glass panels to provide additional radiation protection. The gasket material is Neoprene, and the window panels range from small portholes to large, 1×1.5 m (3.3×4.9 ft) polycarbonate panels that are 10 mm (0.39 in.) thick. A panel of this size would weigh a maximum of 18 kg (40 lb), not including frames, gaskets, glove hardware, gloves, or panel-handling equipment. A panel of laminated safety glass of similar size would weigh approximately 36 kg (79 lb) (NRC, 2000; Section 11.4.7 of the revised CAR).

- Gloveboxes have pass-through connectors in glovebox shells that are used to bring processes and utilities (e.g., air, electricity, and water) inside the glovebox. These connectors are designed and tested to ensure that glovebox pressure integrity stays within the maximum leakage criteria. At mechanical interfaces in the glovebox shells, pass-throughs are designed to allow motion generated outside of the glovebox to be transferred to equipment inside of the glovebox while maintaining confinement boundary leak-tight criteria. Primary process equipment contains MOX product in various forms (i.e., powder, pellets, trays, and rods). These MOX forms are manufactured, transferred, stored, and maintained inside of gloveboxes (NRC, 2000; Section 11.4.7 of the revised CAR). Gloveboxes designated as PSSCs and relied on to maintain the confinement boundary include the following features (NRC, 2000; Section 11.4.11.2 of the revised CAR):

  — They use highly automated processes with hard-wired logic to avoid challenging the glovebox confinement.

  — They provide seismic support of process equipment and resist impacts and potential load drops outside of the gloveboxes.

  — They provide a stable base for barriers, stops, and guides that would prevent the impact from equipment and material moving inside of the gloveboxes.

- Gloveboxes are designed, fabricated, and welded in accordance with national codes and standards appropriate for their use (NRC, 2000; Section 11.4.7 of the revised CAR). The glovebox system uses continuously welded construction with the seams ground smooth, minimizes the holdup of powder, minimizes leakage paths, and facilitates the decontamination of equipment. Fabrication and welding codes include ANSI N690, "Specification for the Design, Fabrication, and Erection of Safety Related Steel Structures for Nuclear Facilities," issued 1994 (ANSI, 1994), and American Welding Society (AWS) D1.1, "Structural Welding Code," issued 1998 (AWS, 1998).

- The ability to safely shut down the primary process is facilitated by the seismic design for the glovebox and similar equipment, as well as structural support members (NRC, 2000; Section 11.4.7 of the revised CAR). Equipment geometry and alignment must be maintained in order to maintain confinement. The glovebox system is designed to prevent physical interaction with confinement boundary elements or PSSCs under worst-case loading conditions associated with normal, off-normal, accident, and design basis events in accordance with ANSI N690 (ANSI, 1994). The system will also be designed to meet the criteria provided in RG 1.100, Revision 2, "Seismic Qualification of Electric and Mechanical Equipment for Nuclear Power Plants," issued 1988 (NRC, 1988).

- Ductwork is designed, fabricated, and tested in accordance with ASME B31.3, "Process Piping," issued 1998 (ASME, 1998); ASME N509 (ASME, 1998); and U.S. Energy

11-81

Research and Development Administration (ERDA) 76-21, "Nuclear Air Cleaning Handbook," issued 1976 (ERDA, 1976).

11.4.1.1.2 Control Concepts

As described in Sections 11.4.2.2.4, 11.4.2.3.4, 11.4.2.4.4, 11.4.2.5.4, 11.4.2.6.4, 11.4.2.7.4, 11.4.3.4, 11.4.4.4, and 11.4.7.1.4 of the revised CAR, the ventilation and confinement system provides for I&C systems to monitor and control the ventilation and confinement PSSCs. These I&Cs include the following:

- Pressure I&Cs maintain proper negative pressures in each of the separate confinement zones. Negative pressures are established such that airflows from areas having a lower radiation hazard to areas having a higher radiation hazard. These controls include pressure monitoring instrumentation and associated interlocks with individual confinement systems to ensure that the proper negative pressures are maintained. Fan controls are also provided to activate redundant fans, if needed, to ensure proper negative pressures.

- Manual and automatic damper controls regulate air and gas flows within gloveboxes and confinement zones. These controls maintain proper confinement zone pressures and ensure proper airflows or isolation of gloveboxes in the event of glovebox breaks, in addition to the isolation of fire areas in the event of fires. Tornado dampers are provided to ensure that overpressures do not occur in the event of tornadoes or high winds.

- Controls are provided to transfer alternate power supplies. Switchgear controls ensure that standby, emergency, and uninterruptible power supplies (UPSs) are properly actuated if normal power supplies are lost.

- Instrumentation measures differential pressures across filter banks to alert operators if filters need to be replaced.

- Variable-speed controls for fan operation maintain proper pressure control in each confinement zone. Individual confinement systems are interlocked to ensure that airflows from zones of higher radiation hazard cannot flow to zones of lower radiation hazard.

- Air temperature instrumentation is provided for fire protection and to ensure that area temperature requirements are properly maintained for habitability and for meeting equipment requirements. Airflow instrumentation is provided to ensure proper system operation.

- Nitrogen and dry air supply flow control is provided to ensure that proper confinement zone negative pressures are maintained.

- Sintering furnace I&Cs are provided to ensure that proper reducing conditions are maintained in the sintering furnace and that potential explosive or flammable conditions are detected and prevented. In the event of leakage from a sintering furnace seal, the staff calculated that a worker exposed to plutonium releases from a failed furnace seal would receive a dose less than 0.1 Sv (10 rem). This dose is less than the limit for

11-82

intermediate consequence events in 10 CFR 70.61 for a facility worker (0.25 Sv (25 rem)).

11.4.1.2 System Interfaces

Individual confinement zone ventilation systems are interconnected to ensure that the proper negative pressures are maintained within confinement zones. Interlocks are provided to ensure that airflows from zones of higher radiation hazard cannot flow to zones of lower radiation hazard. The ventilation and confinement system also interfaces with the normal, standby, emergency, and uninterruptible power supplies so that systems can properly function in the event of power loss. Gloveboxes have functional and physical interfaces with ventilation systems; electrical systems; air, gas, chilled water, and demineralized water systems; chemical processing systems; and fire suppression systems (NRC, 2000; Sections 11.4.1, 11.4.1.2.3, 11.4.2.2.4, 11.4.2.2.5, 11.4.2.3.4, 11.4.2.3.5, 11.4.2.4.4, 11.4.2.4.5, 11.4.2.5.4, 11.4.2.5.5, 11.4.2.6.4, 11.4.2.6.5, 11.4.2.7.4, 11.4.2.7.5, 11.4.3.5, 11.4.4.4, 11.4.4.5, 11.4.7.1.4, and 11.4.7.1.5 of the revised CAR).

11.4.1.3 Design Bases of the PSSCs

Design bases for the ventilation and confinement system PSSCs are discussed in Section 11.4.11 of the revised CAR. Based on the applicant's safety assessment, PSSCs for the ventilation and confinement system are shown in Table 11.4-1.

The staff reviewed the design bases for the ventilation and confinement system PSSCs to ensure that there is reasonable assurance of protection against natural phenomena and the consequences of potential accidents. The design bases are consistent with RG 3.12 (NRC, 1973) and other industry air cleaning standards such as ASME N509, (ASME, 1980a), ASME N510 (ASME, 1980b), and ASME AG-1, "Code on Nuclear Air and Gas Treatment," issued 1991 (ASME, 1991). The staff also reviewed the design bases against the BDC in 10 CFR 70.64. The following discusses how the BDC are met.

The staff reviewed the revised CAR to ensure that the design bases would meet the quality standards and that management measures would be appropriately applied. Application of quality standards and management measures is discussed in Chapter 15 of this FSER.

For natural phenomena hazards, the staff reviewed the proposed design bases consistent with Chapter 5 of the revised CAR, which discusses the analyses performed to show that the facility is adequately designed against natural phenomena with consideration of the most severe documented historical events for the SRS. The ventilation and confinement system gloveboxes, ductwork, and filter assemblies are designed to withstand the design basis earthquake. In addition, tornado dampers ensure that the effects of tornadoes, hurricanes, and high winds applicable to the proposed site do not degrade ventilation and confinement system PSSCs.

The staff reviewed the design bases of the ventilation and containment system to ensure that it can withstand the effects of environmental conditions and the dynamic effects associated with normal operations, maintenance, testing, and postulated accidents. The ventilation and confinement system is designed to withstand fire and chemical effects. In-place testing and maintenance of HEPA filters are performed in accordance with ASME N509 (ASME, 1980a) and ASME N510 (ASME, 1980b). HEPA filters are designed to withstand pressure transients

11-83

Table 11.4-1 Confinement System Design Bases

PSSC	System Function	Controlling Parameters	Status
C4 Confinement System	Control releases of plutonium Remain operable during fires Maintain differential pressure between glovebox and C3 areas Maintain negative confinement airflow after small breaches	C4 zone pressure maintained at negative pressure with respect to C3 zone during normal operation and transients Redundant pressure sensors to maintain C4 pressures Designed to maintain exhaust safety function assuming single active component failure Final HEPA filter assembly release fraction—1×10^{-4} Final HEPA filter design temperature of 232 °C (450 °F) Two 100% capacity redundant assemblies of two HEPA filter banks before discharge Stainless steel roughing filters and stainless steel/glass fiber prefilters in each filtration assembly upstream of HEPA filters Four 100% capacity fans in C4 discharge system Manually activated fire-isolation valves between designated fire areas High-capacity flow to provide 0.635 m/s (125 ft/min) airflow velocity into a breach in the glovebox primary confinement In-place HEPA filter testing for final discharge filtration assemblies System design in accordance with RG 3.12, except heat removed upstream of the final HEPA filters during a fire is accomplished by airflow deletion, not water spray HEPA filter design; HEPA filter housing design, construction, and testing; and HEPA filter housing design and testing in accordance with ASME N509 HEPA filter design and testing; HEPA filter housing design and testing; ductwork and pipe flexible connections; and fan design, construction, and testing in accordance with ASME AG-1 Ductwork piping in accordance with ASME B31.3 Bubble-tight isolation valve construction and testing, HEPA filter housing testing, and HEPA filter testing in accordance with ERDA 76-21 Filter testing in accordance with ASME N510 with each HEPA stage having a leakage efficiency of 99.95% Piping, valves, and fittings associated with gloveboxes in accordance with ASME B31.3 Fan power from normal (non-PSSC), standby (non-PSSC), emergency (PSSC), and uninterruptible (PSSC) supplies Remains operational during and after facility fires, tornadoes, and design basis earthquakes	HEPA filter removal efficiency may be inadequate for severe accident conditions such as fires HEPA filter soot loading capacity may be inadequate during fires

Table 11.4-1 Confinement System Design Bases

PSSC	System Function	Controlling Parameters	Status
C3 Confinement System	Control releases of plutonium Remain operable during fires Ensure 3013 canister storage area temperature Provide cooling air to electrical rooms that support C3 and C4 exhaust system fans	C3 zone pressure maintained at negative pressure with respect to atmosphere during normal operation and transients HEPA filter release fraction—1×10^{-4} HEPA filter design temperature of 232 °C (450 °F) Designed to maintain exhaust safety function assuming single active component failure Two 100% capacity redundant assemblies of two HEPA filter banks before discharge Stainless steel roughing filters and stainless steel/glass fiber prefilters in each filtration assembly upstream of HEPA filters Two 100% capacity fans in C3 confinement system Automatic or manual fire-rated dampers between designated fire areas In-place HEPA filter testing for final discharge filtration assemblies System design in accordance with RG 3.12, removed upstream of the final HEPA filters during a fire, accomplished by airflow deletion, not water spray HEPA filter design; HEPA filter housing design, construction, and testing; and HEPA filter housing isolation dampers in accordance with ASME N509 HEPA filter design and testing; HEPA filter housing design and testing; ductwork and pipe flexible connections; and fan design, construction, and testing in accordance with ASME AG-1 Sheet metal ductwork design, construction, and testing; bubble-tight isolation damper construction and testing; HEPA filter housing testing; and HEPA filter testing in accordance with ERDA 76-21 Filter testing in accordance with ASME N510 with each HEPA stage having a leakage efficiency of 99.95% Tornado dampers Fan power from normal (non-PSSC), standby (non-PSSC), and emergency (PSSC) supplies Provides emergency cooling capability for selected areas Remains operational after facility fires, tornadoes, and design basis earthquakes	HEPA filter removal efficiency may be inadequate for severe accident conditions such as fires HEPA filter soot loading capacity may be inadequate during fires

Table 11.4-1 Confinement System Design Bases

PSSC	System Function	Controlling Parameters	Status
C2 Confinement Passive Barrier System	Limit the dispersion of radioactive material	Two HEPA filter banks before discharge Stainless steel roughing filters and stainless steel/glass fiber prefilters in each filtration assembly HEPA filter design temperature of 232 °C (450 °F) Automatic fire-rated dampers between designated fire areas In-place HEPA filter testing for final discharge filtration assemblies System design in accordance with RG 3.12, except heat removal is by airflow dilution HEPA filter design; HEPA filter housing design, construction, and testing; and HEPA filter housing isolation dampers in accordance with ASME N509 HEPA filter design and testing; HEPA filter housing design and testing; ductwork and pipe flexible connections; and fan design, construction, and testing in accordance with ASME AG-1 Sheet metal ductwork design, construction, and testing; bubble-tight isolation damper construction and testing; HEPA filter housing testing; and HEPA filter testing in accordance with ERDA 76-21 Filter testing in accordance with ASME N510 with each HEPA stage having a leakage efficiency of 99.95% Tornado dampers Fan power from normal (non-PSSC) and standby (non-PSSC) supplies Final filters and downstream ductwork remain structurally intact during and after tornadoes and design basis earthquakes	

11-86

Table 11.4-1 Confinement System Design Bases

PSSC	System Function	Controlling Parameters	Status
Process Cell Exhaust System	Effectively filter process cell exhaust Operate to ensure that a negative pressure exists between the process cell areas and C2 areas	HEPA filter release fraction—1×10^{-4} Two 100% capacity filtration units with two HEPA filter banks before discharge Stainless steel roughing filters and stainless steel/glass fiber prefilters in each final filtration assembly HEPA filter design temperature of 232 °C (450 °F) Two 100% capacity fans System design in accordance with RG 3.12, except heat removal is by airflow dilution Fan power from normal (non-PSSC), standby (non-PSSC), and emergency (PSSC) supplies Fire-rated dampers between designated fire areas Process cell pressure maintained subatmospheric during normal and transient operation Designed to maintain safety function assuming single component failure HEPA filter design; HEPA filter housing design, construction, and testing; and HEPA filter housing isolation dampers in accordance with ASME N509 HEPA filter design and testing; HEPA filter housing design and testing: ductwork and pipe flexible connections; and fan design, construction, and testing in accordance with ASME AG-1 Sheet metal ductwork design, construction, and testing; bubble-tight isolation damper construction and testing; HEPA filter housing testing; and HEPA filter testing in accordance with ERDA 76-21 Filter testing in accordance with ASME N510 with each HEPA stage having a leakage efficiency of 99.95% Remains operational after fires and tornadoes and design basis earthquakes	HEPA filter removal efficiency may be inadequate for severe accident conditions such as fires

Table 11.4-1 Confinement System Design Bases

PSSC	System Function	Controlling Parameters	Status
Emergency Control Room Air-Conditioning System	Ensure habitable conditions for operators Provide cooling to the emergency electrical rooms	Redundant systems each with an outside air intake with continuous monitoring for hazardous chemicals One 100% capacity filtration stages (using prefilter stage, two HEPA filter stages, and chemical filters) for each control room air supply and one 100% capacity booster fan for each intake One 100% capacity air handling unit per control room One 100% capacity pressurization fan and one 100% capacity booster fan One 100% capacity air conditioning unit for each emergency control room and for each emergency electrical and battery room Two 100% capacity exhaust fans for each emergency battery room HEPA filter design temperature of 232 °C (450 °F) Tornado dampers prevent overpressurization In-place HEPA filter testing for final discharge filtration assemblies HEPA filter design; HEPA filter housing design, construction and testing; and HEPA filter housing isolation dampers in accordance with ASME N509 HEPA filter design and testing; HEPA filter housing design, and testing; ductwork and pipe flexible connections; and fan design, construction, and testing in accordance with ASME AG-1 Sheet metal ductwork design, construction, and testing; bubble-tight isolation damper construction and testing; HEPA filter housing testing; and HEPA filter testing in accordance with ERDA 76-21 Filter testing in accordance with ASME N510 with each HEPA stage having a leakage efficiency of 99.95% Fan power from normal (non-PSSC), standby (non-PSSC), and emergency (PSSC) supplies Remains operational during and after facility fires and after tornadoes and design basis earthquakes	
Emergency Diesel Generator Ventilation System	Provide emergency generator ventilation	Two 50% capacity air conditioning units for each switchgear room One 100% capacity roof ventilator for engine room cooling during standby (engine fan cools room during engine operation) Fan power from normal (non-PSSC), standby (non-PSSC), and emergency (PSSC) supplies Remains operational after facility fires, tornadoes, and design basis earthquakes	

Table 11.4-1 Confinement System Design Bases

PSSC	System Function	Controlling Parameters	Status
Offgas Treatment System	Provide an exhaust path for the removal of gases in process vessels Provide an exhaust path for aqueous phase evaporative cooling in process vessels, thereby providing a mechanism for heat removal Provide venting of vessels/equipment that potentially contain TBP and its associated byproducts to prevent overpressurization in the case of excessive oxidation of TBP and/or its degradation products	HEPA filter release fraction—1×10^{-4} HEPA filter design temperature of 232 °C (450 °F) Two 100% capacity redundant assemblies of two HEPA filter banks before discharge with coolers and electric reheating coils on the filter inlets to reduce moisture carryover on the filters Two 100% capacity fans In-place HEPA filter testing for final discharge filtration assemblies in accordance with ASME N509 and N510 Piping and valves in accordance with ASME B31.3 Filter testing in accordance with ASME N510 with each HEPA stage having a leakage efficiency of 99.95% Fan power from normal (non-PSSC), standby (non-PSSC), and emergency (PSSC) supplies Remains operational during and after tornadoes and design basis earthquakes	
Gloveboxes and Glovebox Pressure Controls	Maintain glovebox pressure	Gloveboxes are designed to remain functional during and after a design basis earthquake Breach of confinement from load drops protected by automated processes	
Sintering Furnace Confinement Boundary	Control releases of plutonium	Seals designed for peak operating temperature of 316 °C (601 °F) No damage if furnace is shut down or if overheating or low cooling flow conditions occur Furnace shell and airlocks designed to withstand overpressures of 250 kPa (2.5 bar) Furnace shell leak tightness of 5×10^{-5} leakage at 15 kPa (2.2 psi) Moisture carryover controls to prevent steam generation Controls in place to minimize hydrogen hazards Redundant pressure controls to maintain furnace pressure Furnace designed to function during design basis earthquake	

Table 11.4-1 Confinement System Design Bases

PSSC	System Function	Controlling Parameters	Status
Tornado Dampers	Protect ventilation system from tornado effects	Resists design basis tornado effects Remains operational after facility fires and design basis earthquakes	
Canisters, casks, and containers	Control releases of plutonium	Designed to standards to prevent releases underapplicable handling and accident conditions	
Supply Air System	Supply emergency cooling air to canister storage area and electrical rooms Maintain confinement zone differential pressures	Provide supply air for emergency cooling HEPA filter stages for static confinement System design in accordance with RG 3.12 Sheet metal ductwork design, construction, and testing; bubble-tight isolation damper construction and testing; and HEPA filter testing in accordance with ERDA 76-21 Filter testing in accordance with ASME N510 with each HEPA stage having a leakage efficiency of 99.95%	

considering filter loadings and fan suction pressures. Filter replacement will be performed using bag-in, bag-out procedures to reduce the possibility of spreading contamination.

For potential accidents involving fires, the staff reviewed the proposed design bases consistent with Chapters 5 and 7 of the revised CAR, which describe the analyses and design features applicable to fire protection. Design features of the ventilation and confinement system for fire protection include filter assembly redundancy, use of dilution to mitigate the high-temperature effects of a fire, use of stainless steel roughing filters and stainless steel/glass fiber prefilters to prevent hot particles greater than 1 micron in size from contacting and starting fires on filters, and use of two redundant banks of HEPA filters in each filter assembly having a temperature rating of 232 °C (450 °F) and meeting the standards set by Underwriters Laboratories in UL 586, "High-Efficiency, Particulate, Air Filter Units," issued 1990 (DCS, 2003c; UL, 1990).

The applicant considered the effects of fire on the ventilation and confinement system. The applicant assumed that fires are restricted to single fire areas and will not spread (NRC, 2000; Section 11.4.8 of the revised CAR). The staff reviewed the Fire Protection Program and, as described in Section 7.1.5.5 of this FSER, concludes that during fires temperatures at the HEPA filters can be maintained less than 204 °C (400 °F) and that HEPA filters can be protected without severe temperature effects.

Under accident conditions, the applicant proposed to take credit only for the final filtration assemblies and not the HEPA filters located on gloveboxes or at confinement zone boundaries (NRC, 2000; Section 11.4.7.1.5 of the CAR). The applicant proposed to use a release factor of 1×10^{-4} for the final filter assemblies in its accident safety analyses (Sections 5.5.3.2 and 11.4.9 of the CAR). The applicant based its proposal on having redundant HEPA filter banks that are protected by stainless steel roughing filters and stainless steel/glass fiber prefilters in each redundant filter assembly with HEPA filters that have been tested to have an efficiency for the removal of 0.3 micron particles of at least 99.97 percent (DCS, 2003c). After installation, the HEPA filter banks will be leak tested in accordance with ASME N510 (ASME, 1980b) to ensure that leakage efficiency is at least 99.95 percent. The stainless steel roughing filters and stainless steel/glass fiber prefilters are noncombustible and have sufficient structural strength to collect greater than 90 percent of soot generated in a design basis fire at the design flow rate to withstand the full differential pressure generated by the fans when completely clogged (DCS, 2003c). Under severe conditions, the applicant considered that each bank of HEPAs will remove at least 99 percent of particulates and will provide an overall efficiency of 99.99 percent for the two combined banks. The staff reviewed this assumption in light of recommendations in NUREG/CR-6410, "Nuclear Fuel Cycle Facility Accident Analysis Handbook," issued 1998 (NRC, 1998), that a release factor for each filter assembly of no more than 1×10^{-2} should be used under severe conditions such as a fire. Because of the following factors, the staff considers that the proposed filtration system design has sufficient robustness to ensure that the HEPA filters will be protected during a fire (DCS, 2003a; DCS, 2003b):

- The filter assembly contains stainless steel roughing filters and stainless steel/glass fiber prefilters to minimize the possibility of hot embers contacting filters (DCS, 2003c). The roughing filters use a first-stage stainless steel wire mesh in stainless steel frames with a filtration efficiency of 60–70 percent for particles from 3–10 microns in size and 10–20 percent of 1–3 micron particles. The stainless steel/glass fiber prefilter uses a stainless steel and fiberglass mesh with an efficiency of 99 percent for particles greater than 2 microns. These filters are capable of preventing burning embers and ash from contacting the HEPA filters or starting fires on dust-loaded prefilters which could adversely affect the performance of the HEPA filters. Hot embers and ash particles that are smaller than 10 microns—the particle sizes not removed by the stainless steel roughing filters and stainless steel/glass fiber prefilters—would be unlikely to be carried through the ventilation system ductwork and remain hot enough to ignite collected dust on prefilters or HEPA filters (DCS, 2003c).

- HEPA filters have stainless steel frames in accordance with ASME AG-1 (ASME, 1991), have stainless steel face guards, meet UL 586 fire resistance standards (UL, 1990), and can operate at temperatures up to 232 °C (450 °F) and at differential pressures up to 2.49 kPa (10 kPa in water).

- The applicant's Fire Protection Program is designed to contain fires within a single fire area without spreading to adjacent areas. This program is based on using at least 2-hour fire barriers and combustible material controls (see Chapter 7 of this FSER). The applicant also analyzed uncertainties in the spread of fires by evaluating the temperatures and soot loadings if fires spread to the two worst-case fire areas. In these analyses, the applicant found that temperatures at the final filters of the C4 and C3 systems are less than 188 °C (370 °F), and the differential pressures across the HEPA filters would be less than 2.49 kPa (10 kPa in water).

Therefore, the staff considers that the filtration assemblies can perform their intended function under conditions of fire and that use of a removal efficiency of 99.99 percent is acceptable.

The applicant also performed a soot generation analysis to ensure that filter loadings during a fire would not adversely affect filter integrity. Using the Ballinger correlation (Ballinger, 1988) corrected for the proposed filter sizes, the applicant computed a maximum HEPA filter loading assuming the soot generation from the two fire areas producing the maximum amounts of soot. In these cases, greater than 90 percent of the soot generated during a fire is expected to be collected on the stainless steel roughing filters and the stainless steel/glass fiber prefilters, resulting in soot loadings at the HEPA filters of less than 350 g (12 oz) (DCS, 2003c). Under these loads, the differential pressures across the HEPA filters are less than 2.49 kPa (10 kPa in water) and would be sufficient to ensure that HEPA filters will not rupture from high particulate loading. The applicant also committed to performing tests to determine the specific soot loads applicable to the combustion of materials to be used at the MOX facility. Depending on the results, the number of filters could be increased to maintain adequate soot loads (DCS, 2003c).

The staff reviewed the design bases of the ventilation and confinement system to ensure that it provides adequate protection against chemical risks produced from licensed material, facility conditions that affect the safety of licensed material, and hazardous chemicals produced from licensed material. Gloveboxes are constructed of welded stainless steel to resist the corrosive effects of chemicals used in AP and MOX fuel fabrication processes. In addition, ductwork and filter assemblies upstream of the final filters are stainless steel, and filter materials will be designed to withstand the chemical effects resulting from normal operations. Accident analyses are further discussed in Chapter 5 of the revised CAR. Chemical safety is discussed in Chapter 8 of the revised CAR.

The staff reviewed the design bases of the ventilation and confinement system to ensure that it provides for emergency capability to control the release of licensed material during normal operations and under postulated accident conditions. Release of licensed material is controlled by the use of redundant HEPA filter banks in redundant filter assemblies. Individual HEPA filters are tested to ensure that individual HEPA filters are capable of removing at least 99.97 percent of 0.3 micron particles. Following installation, PSSC HEPA filters are in-place tested in accordance with ASME N510 (ASME, 1980b) to ensure that leakage around filter banks is less than 0.05 percent.

The staff reviewed the design bases of the ventilation and confinement system with respect to the electrical power supply. The C4 confinement system is supplied by normal, standby, emergency, and uninterruptible power supplies. The C3 exhaust system, process cell exhaust system, the emergency control room, and EDG systems are supplied by normal, standby, and emergency power supplies. The C2 confinement system and the supply air system are supplied by normal and standby power supplies. These diverse power supply systems will ensure the continued operation of ventilation and confinement system PSSCs. The staff's review of the electrical systems is provided in Section 11.5 of this FSER.

The staff reviewed the proposed design bases of ventilation and confinement system PSSCs to ensure that they provide for adequate inspection, testing, and maintenance to ensure availability and reliability to perform their function when needed. Redundant filter assemblies are provided so that single filter assemblies can be taken offline for maintenance, testing, and filter replacement. Dampers can be used to isolate individual filter assemblies and fans. The filter assembly design includes provisions for in-place testing of HEPA filters in accordance with ASME N510 (ASME, 1980b). Filter assemblies use bag-in/bag-out designs for filter replacement to minimize the possibility of spreading contamination.

The staff reviewed the proposed design bases of the ventilation and confinement system to ensure that it provides for criticality control and adherence to the double contingency principle. Based on experience from the MELOX site, the applicant assumed that up to 3 kg of plutonium dioxide could exist in the glovebox HEPA filter located in the pellet grinding glovebox, where material becomes airborne at a rate of 0.3 g/h (0.01 oz/h), assuming that the HEPA filters are replaced at 450-day intervals. This amount would be subcritical, as a quantity of 3 kg (6.6 lbs) of plutonium dioxide is substantially less than the minimum critical mass. The ANSI/ANS standard 8.1, "Nuclear Criticality Safety in Operations with Fissionable Materials Outside Reactors," issued 1988 (ANSI, 1988), contains single-parameter (i.e., always safe) subcritical limits for $^{239}PuO_2$ containing not more than 1.5 wt% water. At full density, the subcritical limit is 10.2 kg (22.5 lb); at half density, the subcritical limit is 27 kg (59.5 lb). This would bound the worst-case conditions that could be found in the HEPA filters, because the ANSI limits conservatively assume that all of the plutonium is ^{239}Pu (MOX plutonium will have at least 4 wt% ^{240}Pu), and the maximum density for unsintered plutonium dioxide powder (according to revised CAR Table 6-2) falls within the density range covered by the limits in ANSI/ANS-8.1.

The moisture level of powder is limited to 1 wt% water, according to Section 6.3.4.3.2.6 and Table 6-2 of the revised CAR. The MP gloveboxes are under an inert atmosphere, and this quantity of water (approximately 10 percent by volume) would be readily noticeable to operators. The amount of organic additives in the final blend is neutronically equivalent to 2 wt% water (discussed in Section 6.3.4.3.2.6 of the revised CAR). Although this slightly exceeds the bounding moderation level assumed in the ANSI/ANS-8.1 limits, this would not be sufficient moderation to exceed the subcritical mass limits. In addition, the plutonium dioxide would be spread over the surface of the filter medium and would not be accumulated in a spherical

11-93

geometry. Therefore, application of the subcritical limits in ANSI/ANS-8.1 is considered appropriate. The only remaining source of water would be the possible intrusion of condensation via the ventilation system. Because the ventilation system is designed as a dry system, condensation would not be present. Moreover, the ventilation system complies with the double contingency principle because the design incorporates multiple confinement zones before accumulation outside the ventilation system is possible. It would require multiple upset conditions to accumulate an unsafe mass of plutonium oxides outside the process gloveboxes through airborne migration. Therefore, the staff has reasonable assurance that the MFFF ventilation system will be criticality safe. The staff's review of nuclear criticality safety is provided in Chapter 6 of this FSER.

The staff reviewed the proposed design bases of the ventilation and confinement system to ensure that it provides for I&C systems to monitor and control the ventilation and confinement PSSCs. These I&Cs include (1) pressure I&Cs to maintain proper negative pressures in each of the separate confinement zones, (2) manual and automatic damper controls to regulate air and gas flows within gloveboxes and confinement zones, (3) controls for the transfer of alternate power supplies, (4) instrumentation to measure differential pressures across filter banks, (5) variable-speed controls for fan operation, (6) air temperature and airflow instrumentation, and (7) nitrogen and dry air supply controls. The staff's review of I&C systems is provided in Section 11.6 of this FSER.

11.4.2 Evaluation Findings

In Section 11.4 of the revised CAR, DCS provided design basis information for the Ventilation and Confinement Systems that it identified as PSSCs for the proposed MFFF. Based on the staff's review of the revised CAR and supporting information provided by the applicant relevant to the ventilation and confinement systems, the staff finds that it can conclude, pursuant to 10 CFR 70.23(b), that the design bases of the PSSCs identified by the applicant will provide reasonable assurance of protection against natural phenomena and the consequences of potential accidents.

11.4.3 References

(ANSI, 1988) American National Standards Institute/American Nuclear Society. Standard ANSI/ANS 8.1, "Nuclear Criticality Safety in Operations with Fissionable Materials Outside Reactors." New York, 1988.

(ANSI, 1994) American National Standards Institute. Standard ANSI N690, "Specification for the Design, Fabrication, and Erection of Safety Related Steel Structures for Nuclear Facilities." New York, 1994.

(ASME, 1980a) American Society of Mechanical Engineers. Standard ANSI/ASME N509, "Nuclear Power Plant Air Cleaning Units and Components." New York, 1980.

(ASME, 1980b) American Society of Mechanical Engineers. Standard ANSI/ASME N510, "Testing of Nuclear Air-Cleaning Systems." New York, 1980.

(ASME, 1991) American Society of Mechanical Engineers. Standard ASME AG-1, "Code on Nuclear Air and Gas Treatment." New York, 1991.

(ASME, 1998) American Society of Mechanical Engineers. Standard ASME B31.3, "Process Piping." New York, 1998.

(AWS, 1998) American Welding Society. Standard AWS D1.1, "Structural Welding Code." Miami, Florida, 1998

(Ballinger, 1988) Ballinger, M.Y., et al., "Aerosols Released in Accidents in Reprocessing Plants." *Nuclear Technology*, Vol. 81. May 1988.

(DCS, 2002) Ihde, R., Duke Cogema Stone & Webster. Letter to U.S. Nuclear Regulatory Commission, Mixed Oxide Fuel Fabrication Facility Construction Authorization Request Revision. October 31, 2002 (including page changes through February 9, 2005).

(DCS, 2003a) Hastings, P., Duke Cogema Stone & Webster. Letter to U.S. Nuclear Regulatory Commission, RE: NRC to DCS letter dated 13 February 2003, "February 2003 Monthly Open Item Status Report." February 18, 2003.

(DCS, 2003b) Hastings, P., Duke Cogema Stone & Webster. Letter to U.S. Nuclear Regulatory Commission, RE: Construction Authorization Request Change Pages. April 10, 2003.

(DCS, 2003c) Hastings, P., Duke Cogema Stone & Webster. Letter to U.S. Nuclear Regulatory Commission, RE: Evaluation of Draft Safety Evaluation Report (DSER) on Construction of Proposed Mixed Oxide Fuel Fabrication Facility, Revision 1. May 30, 2003.

(ERDA, 1976) U.S. Energy Research and Development Administration. ERDA 76-21, "Nuclear Air Cleaning Handbook." Oak Ridge, Tennessee, 1976.

(NRC, 1973) U.S. Nuclear Regulatory Commission. Regulatory Guide 3.12, "General Design Guide for Ventilation Systems of Plutonium Processing and Fuel Fabrication Plants." Washington, DC, 1973.

(NRC, 1988) U.S. Nuclear Regulatory Commission. Regulatory Guide 1.100, Revision 2, "Seismic Qualification of Electric and Mechanical Equipment for Nuclear Power Plants." Washington, DC, 1988.

(NRC, 1998) U.S. Nuclear Regulatory Commission. NUREG/CR-6410, "Nuclear Fuel Cycle Facility Accident Analysis Handbook." Washington, DC, 1998.

(NRC, 2000) U.S. Nuclear Regulatory Commission. NUREG-1718, "Standard Review Plan for the Review of an Application for a Mixed Oxide (MOX) Fuel Fabrication Facility." Washington, DC, 2000.

(UL, 1990) Underwriters Laboratories. Standard UL 586, "High-Efficiency, Particulate, Air Filter Units." Northbrook, Illinois, 1990.

(UL, 1995) Underwriters Laboratories. Standard UL 900, "Air Filter Units." Northbrook, Illinois, 1995.

11.5 Electrical Systems

11.5.1 Conduct of Review

This section of the FSER contains the staff's review of electrical systems described by the applicant in Chapter 11 of the revised CAR (DCS, 2002b). The objective of this review is to determine whether the electrical PSSCs and their design bases identified by the applicant provide reasonable assurance of protection against natural phenomena and the consequences of potential accidents. The staff evaluated the information provided by the applicant for electrical systems by reviewing Chapter 11 of the revised CAR, other sections of the revised CAR, supplementary information provided by the applicant, and relevant documents available at the applicant's offices but not submitted by the applicant. The review of electrical systems design bases and strategies was closely coordinated with the review of the I&C and electrical aspects of accident sequences described in the safety assessment of the design bases (see Chapter 5 of this FSER), as well as the review of other plant systems.

The staff reviewed how the information in the revised CAR addresses the following regulations:

- 10 CFR 70.23(b) states, as a prerequisite to construction approval, that the design bases of the PSSCs and the Quality Assurance Program must be found to provide reasonable assurance of protection against natural phenomena and the consequences of potential accidents.

- 10 CFR 70.64 requires that BDC and defense-in-depth practices be incorporated into the design of new facilities. With respect to electrical systems, 10 CFR 70.64(a)(7) requires that the MFFF design "provide for continued operation of essential utility services," which includes electrical power.

The review for this construction approval focused on the design bases of electrical systems, their components, and other related information. For each electrical system, the staff reviewed information provided by the applicant for the safety function, system description, and safety analysis. The review also encompassed proposed design basis considerations such as redundancy, independence, reliability, and quality. The staff used Chapter 11.0 in NUREG-1718 (NRC, 2000) as guidance for performing the review.

11.5.1.1 System Description

The facility electrical systems include the normal power, standby alternating current (ac) power, and emergency power systems. The facility communications system, although not an electrical system, is briefly described herein.

11.5.1.1.1 Function

The electrical systems (as described in Section 11.5, "Electrical Systems," of the revised CAR) are to provide reliable electrical power for normal operation and safe shutdown and monitoring when normal electrical power is lost. Specifically, electrical systems provide support to and are essential for the operation of PSSCs such as the HDE, emergency control room air-conditioning system, C4 confinement system (including very high depressurization), emergency control system, EDG fuel oil system (EDGFOS), and emergency generator ventilation system.

11.5.1.1.2 Major Components

11.5.1.1.2.1 Normal Power System. The normal power system distributes both ac and dc power throughout the facility but is not a PSSC. Control of the breakers for the normal power system is provided from the utility control system or locally.

Normal AC Power System

The normal ac power system consists of two separate 13.8-kV feeds from the F-area substation. Each feeder with a separate 13.8-kv/4.16-kV transformer and an associated distribution switchgear bus is capable of supplying 100 percent of the facility's loads. Each 4.16-kv switchgear bus supplies power to five 4.16-kV/480-v load center transformers, one emergency bus, and various 4.16-kV motors. Small loads are supplied power from 480-v to 120/240-v transformers and associated distribution panels throughout the facility.

Normally, the two 4.16-kV switchgear buses are isolated from each other. If an offsite feed is lost and the other feeder is available, an automatic delayed transfer connects the two switchgear buses together to supply power from the one remaining offsite feeder. Additional crossties utilizing series breakers are provided between the corresponding buses of most 480-v load centers and are operated manually to facilitate maintenance.

Normal DC Power System

The normal dc system supplies power for the trip/close coils of low- and medium-voltage circuit breakers in the normal power distribution system and other dc loads. The ungrounded system consists of two 125-vdc lead-acid batteries, each with an associated charger and distribution panel. Each battery has sufficient capacity to supply its loads for 1 hour, and each charger can supply its loads plus recharge its associated battery to fully charged within 24 hours. Annunciators for low battery voltage, low charging current, and battery charger failure along with battery voltage indication are provided in the utilities control rooms.

11.5.1.1.2.2 Standby AC Power System. The standby ac power system supplies power to loads that are required for safe shutdown of the process on the loss of normal power but is not a PSSC. The system also provides for a quick resumption of production to minimize the economic impact to the facility. Upon the loss of an offsite feeder and failure of the automatic bus transfer, loads are shed for the deenergized 4.16-kV switchgear bus and power is supplied to standby loads by the standby ac power system. The system consists of two separate diesel generators with associated auxiliary equipment and two 120/208-v UPSs.

Standby Diesel Generators

Both generators are connected, when required, to the deenergized 4.16-kV switchgear bus via the 4.16-kV paralleling bus and are sized to carry all the facility life-safety loads together (critical loads (non-IROFS) and loads on the associated emergency bus), plus other loads that are important to facility production. The standby diesel generators are automatically started in about 10 seconds, and loads are sequenced on the bus based on standby power requirements. The diesel generators can be stopped and/or started locally or from the utility and remote utility control rooms. Synchronizing capability is provided for full-load testing. Each diesel is provided with its own fuel oil day tank which stores fuel for 8 hours of operation, up to 2500 L (660 gal).

An additional fuel oil supply tank is provided with enough fuel for both standby diesel generators to operate for 24 hours at full load. Starting capability is provided by redundant batteries sized for five cranking cycles, each with a charger capable of recharging its associated battery within 12 hours following a duty-cycle discharge.

Uninterruptible Power Supplies

The two 120/208-vac UPSs provide power to the programmable logic controllers (PLCs) used for process control and instrumentation loads. Each UPS feeds an associated essential bus by a battery/charger through parallel static inverters. When the static inverters are unavailable, the corresponding essential bus is fed from a 480-v normal bus via a regulating transformer. In addition, a maintenance bypass is provided.

11.5.1.1.2.3 <u>Emergency Power System</u>. The emergency power system distributes both ac and dc power to PSSCs and is itself also a PSSC. Control of the breakers for the emergency power system is provided in the emergency control room or locally.

11.5.1.1.2.3.1 <u>Emergency AC Power System</u>. The emergency ac power system supplies power to redundant PSSCs when an offsite power source is lost, the bus transfer fails, and the standby diesel generators do not start. Under those conditions or a degraded voltage condition, the 4.16-kv emergency bus is isolated from the normal 4.16-kV bus, loads are stripped from the bus as required, the associated EDG is started, and emergency loads are automatically sequenced onto the emergency bus. The system consists of two redundant and independent EDGs (including their support systems) with their associated 4.16-kV emergency bus, a 480-v emergency system, 480-v UPSs, and a 120-v vital ac power system. The system's equipment is qualified for the design basis seismic event and for the expected normal, off-normal, and design basis accident environmental conditions.

Emergency Diesel Generators

Each EDG is connected, when required, to its associated separate and independent 4.16-kV emergency bus to provide emergency power to large PSSC loads, such as the HDE fans, and to lower voltage levels for other PSSC loads, such as the UPSs for the VHD fans and the vital 120-vac power system. Each diesel generator is automatically started on the loss of normal and standby power or degraded voltage to its associated emergency bus and accepts loads in approximately 10 seconds. The generators can be stopped and/or started locally or from their associated emergency control room. Synchronizing capability is provided for full-load testing. Each diesel is sized to start and accelerate its connected loads and is provided with its own fuel oil day tank which stores fuel for 8 hours of operation, up to 2500 L (660 gal). A separate fuel oil supply tank is provided for each diesel generator with enough fuel for 7 days. The EDGs are the same type as the standby diesel generators. Starting capability is provided by redundant batteries sized for five cranking cycles, each with a charger capable of recharging its associated battery within 12 hours following a duty-cycle discharge. An EDG can be taken out of service for maintenance while the standby diesel generators and the redundant emergency ac power system (including the redundant EDG) remain available.

Emergency 480-v System

The 480-v emergency system supplies power to small 480-v loads and to emergency motor control centers (MCCs). The system consists of two redundant 4.16-kv/480-v transformers, each connected to a separate and independent 480-v emergency bus for power distribution. Redundant buses and MCCs permit maintenance of one train's equipment with the opposite train in service.

Emergency 480-v UPS

A dedicated 480-v UPS (two per each train of 4.16-kv emergency ac power) provides uninterruptible power to one of the four VHD glovebox extraction fans. Each UPS consists of a rectifier and an inverter with a battery bank capable of carrying a VHD fan for 1 hour. Power is assumed to be restored to the UPS from either the associated emergency or the standby diesel generators in 1 hour or less.

Vital 120 VAC Power System

The 120-v vital ac power system provides I&C power to PSSC loads. The system consists of two independent and redundant UPSs, each supplying an associated vital bus by a battery/charger (part of the emergency dc power system) through a static inverter. When the static inverter is unavailable, the corresponding vital bus is fed from an associated 480-v emergency bus via a regulating transformer. A maintenance bypass is also provided. Each UPS battery has a 1-hour capacity and is located in a separate emergency battery room. The remaining 120-v vital ac power system components for each UPS are located in separate rooms.

11.5.1.1.2.3.2 <u>Emergency DC Power System</u>. The emergency dc system supplies power for the trip/close coils of low- and medium-voltage circuit breakers in the emergency power distribution system, for some emergency lighting, and for other PSSC dc loads. The ungrounded system consists of two redundant and independent 125-vdc lead-acid batteries, each with an associated charger and distribution panel. Each battery has sufficient capacity to supply its loads for 1 hour, and each charger can supply its loads plus recharge its associated battery to fully charged from a minimum charged state within 24 hours. Physical separation and electrical isolation prevent a fault in one train from affecting the opposite train. Annunciators for low battery voltage, low charging current, and battery charger failure along with battery voltage indication are provided in the utility control rooms.

11.5.1.1.2.4 <u>Communication Systems</u>. Communication systems provide voice and data communications for normal operations, emergency response, and security and are supplied power from the essential 120/208-vac buses. The applicant has stated that the communications systems are not PSSCs because communication systems are not required to perform safety functions during design basis events.

Normal Communication System

Part of the normal communication system consists of the telephone voice systems, including the public switched telephone system and dedicated nonpublic branch exchange (PBX) telephone lines, that are installed throughout the facility site. A public address system, consisting of central amplifiers and electronics, remote microphones, and building speakers, will be installed to provide intrasite and F-Area public address announcements. A wireless trunk system will be used for communication between operations and maintenance personnel, and a fiber optic data highway will be provided throughout the facility.

Emergency Communication System

The emergency communication system consists of dedicated, non-PBX telephone lines between the following:

- facility operations support center and SRS area emergency coordinators

- facility fire alarm system and SRS operations center

- facility operations support center and SRS safety system/alarm public announcement system

- facility operations support center and SRS operations center, emergency operations center, and technical support center.

In addition, each safe haven location is provided with telephone, intercom, and public address systems.

11.5.1.2 System Interfaces

The normal, standby, and emergency power systems provide electrical power to equipment throughout the facility and thus interface with most facility systems and components, especially PSSCs such as the HDE fans and systems listed in Section 11.5.1.1.1 of this FSER. Environmental conditions for proper operation of the electrical systems are maintained by the HVAC systems, and protection against fires is provided by the fire protection systems. The electrical systems and their components are controlled by various manual and automatic controls in the process, utility, and emergency control systems. Fuel oil is provided to the diesel generators in the electrical systems via interfaces with the EDGFOS and standby diesel generator fuel oil system.

11.5.1.3 Design Bases of the PSSCs and Applicable Baseline Design Criteria

For design basis requirements, the electrical systems designated as PSSCs are to be available and provide reliable electrical power for normal operation and safe shutdown and monitoring during credible external events. The electrical systems designated as PSSCs should remain functional when subjected to severe natural phenomena and environmental and dynamic effects, consistent with the BDC in 10 CFR 70.64(a)(2) and 10 CFR 70.64(a)(4), respectively, and be adequately protected from fires in accordance with 10 CFR 70.64(a)(3). Additionally, the electrical systems must support the continued operation of essential utility services, consistent with the BDC in 10 CFR 70.64(a)(7).

The electrical systems should also be designed with I&Cs for monitoring and controlling each bus and major component, consistent with the BDC in 10 CFR 70.64(a)(10). Pursuant to 10 CFR 70.64(b), the electrical systems should be designed using defense-in-depth practices, with a preference for engineered controls over administrative controls.

To ensure that the design basis requirements and BDC are met, the applicant has committed to specific industry standards and staff guidance as discussed in the following sections.

11.5.1.3.1 Emergency AC Power System

The emergency ac power system (as a PSSC) will be designed with redundancy, independence, physical separation, no single-failure vulnerability, and failsafe performance with sufficient capacity, capability, periodic testing and maintenance, and adequate protective relaying to perform its safety function and maintain qualification for natural phenomena and environmental and dynamic effects (see Section 11.11 of this FSER). It will also be designed using guidance from the following sources (to the extent described in the revised CAR and the applicant's letters dated August 31, 2001, December 5, 2001, and November 22, 2002 (DCS, 2001b, 2001c, 2002c)):

- for overall system design:

  — Institute of Electrical and Electronics Engineers (IEEE), IEEE Std 308-1991, "IEEE Standard Criteria for Class 1E Power Systems for Nuclear Generating Stations," (IEEE, 1992b)

  — RG 1.32, Revision 2, "Criteria for Safety-Related Electric Power Systems for Nuclear Power Plants," issued 1977 (NRC, 1977)

- for equipment seismic qualification (SQ):

  — IEEE Std 344-1987, "IEEE Recommended Practices for Seismic Qualification of Class 1E Equipment for Nuclear Generating Stations," issued 1987 (IEEE, 1987)

  — RG 1.100, 1988 (NRC, 1988)

- for periodic testing:

  — IEEE Std 338-1987, "IEEE Standard Criteria for Periodic Testing of Nuclear Power Generating Station Class 1E Power and Protection Systems," issued 1987 (IEEE, 1987)

  — RG 1.118, Revision 3, "Periodic Testing of Electric Power and Protection Systems," issued 1994 (NRC, 1994)

- for single failure:

  — IEEE Std 379-1994, "IEEE Standard Application of the Single Failure Criterion to Nuclear Power Generation Station Safety Systems," issued December 16, 1994 (IEEE, 1994)

- for qualification of cables installed in open trays:

- — IEEE Std 383-1974, "IEEE Standard for Type Test of Class 1E Electric Cables, Field Splices and Connections for Nuclear Power Generating Stations," issued April 15, 1974 (IEEE, 1974)

- • for electrical independence and separation:

  - — IEEE Std 384-1992, "Standard Criteria for Independence of Class 1E Equipment and Circuits," issued December 1, 1992 (IEEE, 1992a)

  - — RG 1.75, Revision 2, "Physical Independence of Electric Systems," issued September 1978 (NRC, 1978)

- • for equipment protection:

  - — IEEE Std 741-1997, "IEEE Standard Criteria for the Protection of Class 1E Power Systems and Equipment in Nuclear Power Generating Stations," issued September 25, 1997 (IEEE, 1997b)

- • for design and installation of batteries:

  - — IEEE Std 484-1996, "IEEE Recommended Practice for Installation Design and Installation of Vented Lead-Acid Batteries for Stationary Applications," issued July 17, 1996 (IEEE, 1996b)

The EDGs will be designed using guidance from the following sources (to the extent described in the revised CAR and the applicant's letters dated August 31, 2001, December 5, 2001, and November 22, 2002 (DCS, 2001b, 2001c, 2002c)):

- • for fuel oil quality:

  - — ANSI/ASTM D975-94, "Standard Specification for Diesel Fuel Oils," issued October 10, 1995 (ANSI, 1995)

- • for overall design and qualification:

  - — IEEE Std 387-1995, "IEEE Standard Criteria for Diesel Generator Units Applied as Standby Power Supplies for Nuclear Power Generating Stations," issued April 12, 1996 (IEEE, 1996a)

  - — RG 1.9, Revision 3, "Selection, Design, Qualification, and Testing of Emergency Diesel Generator Units Used as Class 1E Onsite Electric Power Systems at Nuclear Power Plants," issued July 1993 (NRC, 1993)

The emergency 480-v UPS and 120-v vital ac power system will be designed using guidance from the following sources (to the extent described in the revised CAR and the applicant's letter dated August 31, 2001, and November 22, 2002 (DCS, 2001b, 2002c)):

- • For overall design:

— IEEE Std 944-1986, "IEEE Recommended Practice for the Application and Testing of Uninterruptible Power Supplies for Power Generating Stations," issued June 30, 1986 (IEEE, 1986)

11.5.1.3.2 Emergency DC Power System

The emergency dc power system (as a PSSC) will be designed with redundancy, independence, physical separation, no single-failure vulnerability, and failsafe performance with sufficient capacity; capability; periodic testing and maintenance; adequate protective relaying to perform its safety function; and qualification for natural phenomena, environmental, and dynamic effects (see Section 11.11 of this FSER). It will also be designed using guidance from the following (to the extent described in the revised CAR and the applicant's letters dated August 31, 2001, December 5, 2001, March 8, 2002, and November 22, 2002 (DCS, 2001b, 2001c, 2002a, 2002c)):

– for overall system design:

— IEEE Std 308-1991 (IEEE, 1992b)

— IEEE Std 946-1992, "IEEE Recommended Practice for the Design of Safety-Related DC Auxiliary Power Systems for Nuclear Power Generating Stations," issued February 25, 1993 (IEEE, 1993)

— RG 1.32, Revision 2 (NRC, 1977)

• for equipment SQ:

— IEEE Std 344-1987 (IEEE, 1987)

— RG 1.100, Revision 2 (NRC, 1988)

• for periodic testing:

— IEEE Std 338-1987 (IEEE, 1988)

— IEEE Std 450-1995, "IEEE Recommended Practice for Maintenance, Testing, and Replacement of Vented Lead-Acid Batteries for Stationary Applications," issued May 31, 1995 (IEEE, 1995a)

— RG 1.118, Revision 3 (NRC, 1994)

• for single failure:

— IEEE Std 379-1994 (IEEE, 1994)

• for electrical independence and separation:

— IEEE Std 384-1992 (IEEE, 1992a)

—	RG 1.75, Revision 2 (NRC, 1978)

- for design and installation of batteries:

	—	IEEE Std 484-1996 (IEEE, 1996b)

	—	IEEE Std 485-1997, "IEEE Recommended Practice for Sizing Lead-Acid Batteries for Stationary Applications," issued September 3, 1997 (IEEE, 1997a)

	—	NFPA 111, "Standard for Stored Electrical Energy Emergency and Standby Power Systems," issued 1996 (NRC, 1996)

- for qualification of cables:

	—	IEEE Std 383-1974 (IEEE, 1974)

In Section 5.5.5 of the revised CAR, the applicant described its general design philosophy used in formulating the preliminary design of the facility. Pursuant to 10 CFR 70.64(b), in order to ensure that, to the extent practicable, engineered controls are relied upon over administrative controls, DCS has established a hierarchy of controls. In further adherence to 10 CFR 70.64(b), the applicant stated that it has incorporated defense-in-depth practices at new facilities in its preliminary facility design. Defense-in-depth is defined in the 10 CFR 70.64 footnote. In Section 5.5.5.2 of the revised CAR, DCS stated that it has incorporated defense-in-depth practices through use of the single-failure criterion. Under this criterion, PSSCs are required to be capable of carrying out their functions in the event that any single active component fails, whether such failure occurs within the applicable system or in an associated system that supports the component's operation.

Accordingly, the facility electrical systems should be designed using defense-in-depth practices employing the single-failure criterion, including redundancy, independence, separation, and fail-safe for electrical systems identified as PSSCs through the use of the industry standards listed above. For example, the emergency ac power system, which provides power if offsite power is lost, is designed with redundant and independent EDGs that are physically separated from each other. This is discussed in Sections 11.5.1.3.1 and 11.5.1.3.2 of this FSER.

11.5.2 Evaluation Findings

In Section 11.5.7 of the revised CAR, DCS provided design basis information for electrical systems that it identified as PSSCs for the facility. Based on the staff's review of the revised CAR and the supporting information provided by the applicant, as well as the applicant's commitments to the industry standards and guidance discussed in the sections above for electrical systems, the staff finds that DCS has met the BDC set forth in 10 CFR 70.64(a)(7). The staff concludes, pursuant to 10 CFR 70.23(b), that the design bases of the PSSCs evaluated in this FSER section will provide reasonable assurance of protection against natural phenomena and the consequences of potential accidents. The staff further concludes, from its review of the revised CAR and supporting information submitted by DCS, that the preliminary design of the facility electrical systems meet the defense-in-depth provisions and the preference to engineered controls over administrative controls stated in 10 CFR 70.64(b).

11.5.3 References

(ANSI, 1995) American National Standards Institute/American Society for Testing Materials. D975-94, "Standard Specification for Diesel Fuel Oils." Philadelphia, Pennsylvania. October 10, 1995.

(DCS, 2001a) Hastings, P., Duke Cogema Stone & Webster. Letter to U.S. Nuclear Regulatory Commission, RE: MOX Fuel Fabrication Facility Site Geotechnical Report (DCS01–WRS–DS–NTE–G–0005–C). August 10, 2001.

(DCS, 2001b) Hastings, P., Duke Cogema Stone & Webster. Letter to U.S. Nuclear Regulatory Commission, RE: Response to Request for Additional Information (DCS-NRC-000059). August 31, 2001.

(DCS, 2001c) Hastings, P., Duke Cogema Stone & Webster. Letter to U.S. Nuclear Regulatory Commission, RE: Clarification of Responses to NRC Request for Additional Information (DCS–NRC–000074). December 5, 2001.

(DCS, 2002a) Hastings, P., Duke Cogema Stone & Webster. Letter to U.S. Nuclear Regulatory Commission, RE: Clarification of Responses to NRC Request for Additional Information (DCS–NRC–000085). March 8, 2002.

(DCS, 2002b) Ihde, R, Duke Cogema Stone & Webster. Letter to U.S. Nuclear Regulatory Commission, RE: Mixed Oxide Fuel Fabrication Facility Construction Authorization Request Revision (DCS-NRC-000114). October 31, 2002.

(DCS, 2002c) Hastings, P., Duke Cogema Stone& Webster. Letter to U.S. Nuclear Regulatory Commission, RE: Requests for Additional Information, Clarifications and Open Item Mapping into the Construction Authorization Request Revision (DCS-NRC-000120). November 22, 2002.

(IEEE, 1974) Institute of Electrical and Electronics Engineers. Std 383-1974, "IEEE Standard for Type Test of Class 1E Electric Cables, Field Splices and Connections for Nuclear Power Generating Stations." New York, April 15, 1974.

(IEEE, 1986) Institute of Electrical and Electronics Engineers. Std 944-1986, "IEEE Recommended Practice for the Application and Testing of Uninterruptible Power Supplies for Power Generating Stations." New York, June 30, 1986.

(IEEE, 1987) Institute of Electrical and Electronics Engineers. Std 344-1987, "IEEE Recommended Practices for Seismic Qualification of Class 1E Equipment for Nuclear Generating Stations." New York, August 3, 1987.

(IEEE, 1988) Institute of Electrical and Electronics Engineers. Std 338-1987, "IEEE Standard Criteria for Periodic Testing of Nuclear Power Generating Station Class 1E Power and Protection Systems." New York, August 22, 1988.

(IEEE, 1992a) Institute of Electrical and Electronics Engineers. Std 384-1992, "Standard Criteria for Independence of Class 1E Equipment and Circuits." New York,, December 1, 1992.

(IEEE, 1992b) Institute of Electrical and Electronics Engineers. Std 308-1991, "IEEE Standard Criteria for Class 1E Power Systems for Nuclear Generating Stations." New York, August 26, 1992.

(IEEE, 1993) Institute of Electrical and Electronics Engineers. Std 946-1992, "IEEE Recommended Practice for the Design of Safety-Related DC Auxiliary Power Systems for Nuclear Power Generating Stations." New York, February 25, 1993.

(IEEE, 1994) Institute of Electrical and Electronics Engineers. Std 379-1994, "IEEE Standard Application of the Single Failure Criterion to Nuclear Power Generation Station Safety Systems." New York, December 16, 1994.

(IEEE, 1995a) Institute of Electrical and Electronics Engineers. Std 450-1995, "IEEE Recommended Practice for Maintenance, Testing, and Replacement of Vented Lead-Acid Batteries for Stationary Applications." New York, May 31, 1995.

(IEEE, 1995b) Institute of Electrical and Electronics Engineers. Std 765-1995, "IEEE Standard for Preferred Power Supply (PPS) for Nuclear Power Generating Stations." New York, July 11, 1995.

(IEEE, 1996a) Institute of Electrical and Electronics Engineers. Std 387-1995, "IEEE Standard Criteria for Diesel Generator Units Applied as Standby Power Supplies for Nuclear Power Generating Stations." New York, April 12, 1996.

(IEEE, 1996b) Institute of Electrical and Electronics Engineers. Std 484-1996, "IEEE Recommended Practice for Installation Design and Installation of Vented Lead-Acid Batteries for Stationary Applications." New York, July 17, 1996.

(IEEE, 1997b) Institute of Electrical and Electronics Engineers. Std 485-1997, IEEE Recommended Practice for Sizing Lead-Acid Batteries for Stationary Applications." New York, September 3, 1997.

(IEEE, 1997a) Institute of Electrical and Electronics Engineers. Std 741-1997, "IEEE Standard Criteria for the Protection of Class 1E Power Systems and Equipment in Nuclear Power Generating Stations." New York, September 25, 1997.

(NFPA, 1996) National Fire Protection Association Inc. NFPA 111, "Standard for Stored Electrical Energy and Standby Power Systems." Quincy, Massachusetts, 1996.

(NRC, 1977) U.S. Nuclear Regulatory Commission. Regulatory Guide 1.32, Revision 2, "Criteria for Safety-Related Electric Power Systems for Nuclear Power Plants." Washington, DC, February 1977.

(NRC, 1978) U.S. Nuclear Regulatory Commission. Regulatory Guide 1.75, Revision 2, "Physical Independence of Electric Systems." Washington, DC, September 1978.

(NRC, 1988) U.S. Nuclear Regulatory Commission. Regulatory Guide 1.100, Revision 2, "Seismic Qualification of Electric and Mechanical Equipment for Nuclear Power Plants." Washington, DC, June 1988.

(NRC, 1993) U.S. Nuclear Regulatory Commission. Regulatory Guide 1.9, Revision 3, "Selection, Design, Qualification, and Testing of Emergency Diesel Generator Units Used as Class 1E Onsite Electric Power Systems at Nuclear Power Plants." Washington, DC, July 1993.

(NRC, 1994) U.S. Nuclear Regulatory Commission. Regulatory Guide 1.118, Revision 3, "Periodic Testing of Electric Power and Protection Systems." Washington, DC, September 1994.

(NRC, 2000) U.S. Nuclear Regulatory Commission. NUREG-1718, "Standard Review Plan for the Review of an Application for a Mixed Oxide (MOX) Fuel Fabrication Facility." Washington, DC, 2000.

11.6 Instrumentation and Control Systems

11.6.1 Conduct of Review

This chapter of the FSER contains the staff's review of I&C systems described by the applicant in Chapter 11 of the revised CAR (DCS, 2002b). The objective of this review is to determine whether the I&C PSSCs and their design bases identified by the applicant provide reasonable assurance of protection against natural phenomena and the consequences of potential accidents. The staff evaluated the information provided by DCS for I&C systems by reviewing Chapter 11 of the revised CAR, other sections of the revised CAR, supplementary information provided by the applicant, and relevant documents available at the applicant's offices but not submitted by the applicant. The review of I&C system design bases and strategies was closely coordinated with the review of the electrical and I&C aspects of accident sequences described in the safety assessment of the design bases (see Chapter 5 of this FSER) and the review of other plant systems.

The staff reviewed how the I&C information in the revised CAR addresses the following regulations:

- 10 CFR 70.23(b) states, as a prerequisite to construction approval, that the design bases of the PSSCs and the Quality Assurance Program be found to provide reasonable assurance of protection against natural phenomena and the consequences of potential accidents.

- 10 CFR 70.64 requires that BDC and defense-in-depth practices be incorporated into the design of new facilities. With respect to I&C, 10 CFR 70.64(a)(10) requires that the MFFF design "provide for inclusion of instrument and control systems to monitor and control the behavior of items relied on for safety."

The review for this construction approval focused on the design basis of I&C systems, their components, and other related information. For each I&C system, the staff reviewed information provided by the applicant for the safety function, system description, and safety analysis. The review also encompassed proposed designbasis considerations such as redundancy, independence, reliability, and quality. The staff used Chapter 11 in NUREG-1718, "Standard Review Plan for the Review of an Application for a Mixed Oxide (MOX) Fuel Fabrication Facility," issued 2000, as guidance in performing the review.

11.6.1.1 System Description

The facility I&C systems include the MP and AP process control, utility control, and emergency control systems.

11.6.1.1.1 Function

The I&C systems (as described in Section 11.6, "Instrumentation and Control Systems," of the revised CAR) are to monitor and control the manufacturing process systems, the plant utility systems, and the plant safety and emergency systems. Specifically, I&C systems are to monitor and control plant parameters during normal and transient conditions to ensure the quality of the produced product and ensure that limits (including safety limits) are not exceeded. They also provide control signals to plant equipment to prevent the occurrences of faulted conditions and mitigate the consequences of faulted conditions should they occur.

11.6.1.1.2 Major Components

11.6.1.1.2.1 MP and AP Process Control Systems. The MP and AP process control systems use an automated, distributed processing control system strategy with the manufacturing process translated into control algorithms for each process step. The systems include the normal, protective, and safety control subsystems which ensure the final product conforms to specifications and reduce plant waste and risk. Specifically, the normal control subsystem controls the manufacturing process, the protective control subsystem protects personnel (industrial safety) and equipment, and the safety control subsystem ensures that safety limits will not be exceeded and that undesirable operational conditions are prevented or mitigated.

The process control systems are built around PLCs which, in turn, are built around software-controlled microprocessors. There are two independent safety channels in the safety control subsystem, both of which are separate and independent from the other two control subsystems.

All channels in the safety control subsystem in the MP or AP process control system will be designated PSSCs. The design basis for the safety control subsystem is described in FSER Section 11.6.1.3.1 for a PSSC.

11.6.1.1.2.2 Utility Control Systems. The utility control systems also use a PLC/ microprocessor-based automated, distributed processing control strategy and include separate normal, protective, and auxiliary control subsystems which ensure that the plant support (utility) systems operate within specifications. Specifically, the normal control subsystem controls operation of the support systems, the protective control subsystem protects personnel and equipment, and the auxiliary control subsystem is a backup to the normal control system and prevents challenges to the emergency control system. The single channels of the normal and auxiliary control subsystems are separate and independent from each other and the protective control subsystem. The auxiliary control subsystems that are part of the utility control systems are not credited by DCS towards meeting the performance requirements of 10 CFR Part 70, "Domestic Licensing of Special Nuclear Material," and have not been identified as PSSCs.

11.6.1.1.2.3 Emergency Control System. The emergency control system is a PSSC and consists of manual and some automatic controls for power, ventilation, and required safety functions. Specifically, the emergency control system ensures particular plant support systems operate when needed to mitigate the consequences of hazardous occurrences. The system is a hard-wired system with no software control and is designed to continue to operate during and following specific design basis events. It is divided into two separate and independent trains, each with a separate control room.

11.6.1.1.2.4 <u>Data Communications Networks</u>. The data communications networks employ Ethernet technology and provide for communications among various systems and components such as normal controllers, workstations, the manufacturing management and information system (MMIS), the manufacturing status system, and the computer-aided diagnosis system. For the MP and AP process control systems, Fieldbus technology or hard-wired methods are used to link normal controllers to motor control centers and sensor/instruments. The X-terminal network (XTN) links display terminals, workstations, the MMIS, manufacturing status system, and the computer-aided diagnosis system. The local industrial network (LIN) links workstations, controllers, the MMIS, manufacturing status system, and the computer-aided diagnosis system. The immediate control network (ICN) links the workstations and normal controllers. Separate connections are used to link each device to each network.

For the utility control systems, Fieldbus technology or hard-wired methods are used to link normal controllers to MCCs and sensor/instruments. Dual redundant networks link the workstations and normal controllers. Separate connections are used to link each device to each network.

11.6.1.1.2.5 <u>Control Rooms</u>. The facility includes various control rooms with workstations (industrial personal computers) to monitor and control the operation of the AP and MP functional units and utility system. Access to the control system data communication network is provided in the control rooms, and emergency stop switches typically provide manual control capability. Human factors engineering principles are used in the design of the control rooms (see FSER Chapter 12).

The major control rooms are discussed in the following sections:

•	AP and MP control rooms

	The AP and MP control rooms provide central locations for monitoring, supervising, and controlling the manufacturing and processing operations. In these control rooms, operators access production control information to verify that process automation is performing satisfactorily or to receive notification of a problem. Video displays of the actual functional unit conditions are provided. If an AP or MP manufacturing process functional unit is not operating, manning of the associated control room, for the particular manufacturing process steps, is not required.

	In addition to providing for the control of AP systems, the AP control room is always manned and provides for the control of the normal and safety utilities systems, the fire detection systems, and the health physics systems.

- utilities control room

  In addition to the controls for the utilities systems located in the AP control room, the utilities control room provides support system monitoring and controlling capability when the controls in the AP control room for the utilities systems are unavailable. Support systems monitored and controlled from these two redundant locations include the HVAC systems, the electrical distribution system, and various other process support systems such as the hot and chilled water system.

- emergency control rooms

  Two separate emergency control rooms (Train A and Train B) each with an independent ventilation system are located in the shipping and receiving building. Solid state or traditional electromechanical control devices in these control rooms are hard wired to PSSCs in the emergency control system. These controls have priority over the normal, auxiliary, and protective controls.

11.6.1.1.2.6 Sensors. The facility functional units are provided with sensors and instruments (parts of the process control systems) to monitor and measure operational conditions and parameters such as temperature, pressure, mass, component identification bar codes, machine tool and valve positions, etc. Some of the information is provided as inputs for the automatic control of actuators (fans, pumps, heaters, conveyers, etc.) and other devices which perform steps in the manufacturing process.

The utility control systems are also provided with sensors and instruments to monitor the performance of support systems such as the electrical power and HVAC systems. These sensors also provide input signals for the automatic controllers for the various utility control systems.

11.6.1.1.2.7 Controllers. The AP process, MP, and the utility control systems have controllers which receive inputs from sensors and instruments. Specific controllers are discussed below:

- normal controllers

  The PLCs provide for the control of normal operations of the functional and utility units. These normal controllers, located in rooms close to the process they control, contain communication devices and are connected to the control system networks so that data flow between the controllers and the distributed input/output, the normal controllers of other functional and utility units, the workstations, the MMIS, and the computer-aided diagnosis system. This data communication between controllers and systems allows coordination of activities. These controllers are not PSSCs.

- protective controllers

  The protective controllers are dedicated controllers that protect operations personnel from injury and functional unit equipment from damage resulting from inappropriate operation such as overspeed, overtorque, and overtravel. These controllers are autonomous and are not connected to the control system networks. They may be PLCs or traditional electromechanical relays and are physically located in or near MCCs.

11-110

Protective sensors are hard-wired to the controllers, with output control signals hardwired to control circuits in MCCs or power panels. Protective controllers have priority over control signals from the normal controllers and provide performance data to the normal controllers which, in turn, provide data to workstations via the ICN. Operators cannot directly access the protective controllers and cannot routinely intervene in their operation. These controllers are not PSSCs.

- safety controllers

Safety controllers in the process control of the functional units serve specific and limited functions for preventing and mitigating certain accidents. These controllers are autonomous and are not connected to the control system network. Performance data from the safety controllers are provided to the normal controllers which, in turn, provide data to workstations via the ICN. Any safety controller in the MP or AP process control system may be designated a PSSC as a result of the facility safety analysis. The design basis for that controller will then be in accordance with FSER Section 11.6.1.3.1.

The safety controllers for the functional units may be PLCs or traditional relays. They are physically located in rooms separate from the normal controllers and are hard-wired to the sensors and actuator control circuits in MCC or power distribution panels. The control signals from the safety controllers have priority over signals from the normal controllers. If a safety controller does not detect appropriate functional unit conditions, it will block commands issued by the normal controllers. Operators cannot directly access the safety controllers and cannot routinely intervene in their operation with the exception of the scrap jar isotopic concentration data which are manually loaded into the safety controllers.

Upon receipt of a fire condition signal from the facility fire detection system, fire safety controllers direct the normal controllers to immediately close the fire doors. Following a delay, the fire safety controller will close the fire doors if the normal controllers fail to take action. The fire barrier is a PSSC.

- auxiliary controllers

Auxiliary controllers in selected utility systems provide backup monitoring and control capability if the normal control systems are unavailable. They are also installed in separate locations from the normal controllers and are hard-wired to dedicated workstations located in the AP control room. Auxiliary controllers in the utility control system have not been identified as PSSCs.

- emergency controllers

The facility emergency controllers in the emergency control system are solid-state controllers or traditional relay logic circuits with control switches located on emergency control panels in the emergency control rooms. Inputs to emergency controllers from the sensors and outputs from the controllers to MCCs are hard wired. The emergency controllers provide manual control for selected systems. These controllers are PSSCs.

11.6.1.1.2.8 <u>MMIS</u>. The MMIS is a realtime system that tracks the product inventory and maintains/updates the manufacturing status system database files as the product moves through the manufacturing process. It tracks quantities of product that are allowed in any area at a given time and authorizes the movement of product into or out of a functional unit. The MMIS also is a server for the XTN.

11.6.1.1.2.9 <u>Manufacturing Status System</u>. The manufacturing status system is identical to the MMIS and maintains a copy of the MMIS database. The system is used to sort, analyze, and report on data collected by the MMIS.

11.6.1.1.2.10 <u>Computer-Aided Diagnosis System</u>. The computer-aided diagnosis system monitors the operation of the normal controllers for the function units and the ICN. The system is independent from the controllers and has its own software. It provides a primary diagnostic capability when there is a problem with the production process and works with the software used in the PLCs to determine the state of the PLC when the problem occurred.

11.6.1.1.2.11 <u>Process Computers</u>. Process computers are dedicated microprocessor-based computers that control and handle data for the more complex process measurement systems such as the laser optical micrometer. These computers operate the measurement systems, provide data signal conditioning, and send data to the PLCs controlling the associated functional unit.

11.6.1.1.2.12 <u>Seismic Monitoring System</u>. The seismic monitoring system provides data for the evaluation of the confinement structure and other PSSCs and automatically shuts down the MP and AP process during a high seismic event.

11.6.1.2 System Interfaces

The facility I&C systems are primarily electronic systems that interface with mechanical, electrical, and process systems. The I&C systems provide signals to control electromechanical and pneumatic actuators and data signals for information display, processing, and storage. The control systems and their associated components, such as electromechanical actuators, are supplied electrical power by the facility electrical systems. The pneumatic actuators are provided air by either the instrument air or plant air systems. The various control rooms provide the primary human-system interface for all the facility systems. Environmental conditions for proper operation of the I&C systems are maintained by the HVAC systems, and protection against fires is provided by the fire protection systems.

11.6.1.3 Design Bases of the PSSCs and Baseline Design Criteria

For design basis requirements, the I&C systems are to be available and reliable for normal operation and safe shutdown and monitoring (see FSER Section 11.6.1.1.1 for a discussion of the function performed by the I&C systems). With respect to I&C, 10 CFR 70.64(a)(10) requires that the facility design "provide for inclusion of instrument and control systems to monitor and control behavior of items relied on for safety." The I&C systems designated as PSSCs are designed to remain functional when subjected to severe natural phenomena and environmental and dynamic effects, consistent with the baseline design criteria in 10 CFR 70.64(a)(2) and 10 CFR 70.64(a)(4), respectively, and be adequately protected from fires per

10 CFR 70.64(a)(3). Additionally, the I&C systems must support continued operation of essential utility services, consistent with the baseline design criterion in 10 CFR 70.64(a)(7).

Pursuant to 10 CFR 70.64(b), the I&C systems are to be designed using defense-in-depth practices, with a preference for engineered controls over administrative controls.

To ensure that the design basis requirements and the baseline design criteria are met, DCS has committed to specific industry standards and staff guidance as discussed in the following section.

11.6.1.3.1 I&C Systems (PSSCs)

The I&C systems identified by DCS as PSSCs are the emergency control system and the AP and MP safety control subsystems (DCS, 2002b). These PSSCs will be designed with provisions for periodic testing, redundancy, independence, no single-failure vulnerability, fail-safe failure mode, proper instrument spans/setpoints/control ranges, status monitoring, and qualification for natural phenomena and environmental and dynamic effects (see FSER Section 11.11). Additionally, these systems will be designed using guidance (to the extent described in the revised CAR and DCS letters dated August 31, 2001, December 5, 2001, January 7, 2002, November 22, 2002, and February 11, 2003 (DCS, 2001a, 2001b, 2002a, 2002c, 2003) from the documents listed below.

- for overall system design (including seismic monitoring and trip system):

  — IEEE Std 603-1998, "IEEE Standard Criteria for Safety Systems for Nuclear Power Generating Stations," issued 1998 (IEEE, 1998f)

- for single failure:

  — IEEE Std 379-1994 (IEEE, 1994a)

- for software programmable electronic systems:

  — Electric Power Research Institute (EPRI) TR-106439, "Guideline on Evaluation and Acceptance of Commercial Grade Digital Equipment for Nuclear Safety Applications," issued October 1996 (EPRI, 1996)

  — International Electrotechnical Commission (IEC) 61131-3 (1993-03), "Programmable Controllers -Part 3: Programming Languages," issued 1993 (IEC, 1993)

  — IEEE Std 7-4.3.2-1993, "IEEE Standard Criteria for Digital Computers in Safety Systems of Nuclear Power Generating Stations," issued 1993 (IEEE, 1993)

  — IEEE Std 730-1998, "Software Quality Assurance Plans," issued 1998 (IEEE, 1998e)

  — IEEE Std 828-1998, "IEEE Standard for Software Configuration Management Plans," issued 1998 (IEEE, 1998d)

- IEEE Std 830-1998,"IEEE Standard Recommended Practice for Software Requirements Specifications," issued 1998 (IEEE, 1998c)

- IEEE Std 1012-1998, "IEEE Standard for the Software Verification and Validation," issued 1998 (IEEE, 1998b)

- IEEE Std 1028-1997, "IEEE Standard for Software Reviews," issued 1997 (IEEE, 1997a)

- IEEE Guide 1042-1987, "Software Configuration Management," issued 1987 (IEEE, 1987c)

- IEEE Std 1074-1997, "IEEE Standard for Developing Software Life Cycle Processes," issued 1997 (IEEE, 1997b)

- IEEE Std 1228-1994, "IEEE Standard for Software Safety Plans," issued 1994 (IEEE, 1994b)

- NUREG/CR-6090, "The Programmable Logic Controller and Its Application in Nuclear Power Plants," issued February 1999 (NRC, 1999a)

- NUREG/CR-6463, "Review Guidelines on Software Languages for Use in Nuclear Power Plant Safety Systems," issued June 1996 (NRC, 1996)

- RG 1.168, "Verification, Validation, Reviews, and Audits for Digital Computer Software Used in Safety Systems of Nuclear Power Plants," issued September 1997 (NRC, 1997a)

- RG 1.169, Configuration Management Plans for Digital Computer Software Used in Safety Systems of Nuclear Power Plants," issued September 1997 (NRC, 1997b)

- RG 1.172, "Software Requirements Specifications for Digital Computer Software Used in Safety Systems of Nuclear Power Plants," issued September 1997 (NRC, 1997c)

- RG 1.173, "Developing Software Life Cycle Processes for Digital Computer Software Used in Safety Systems of Nuclear Power Plants," September 1997 (NRC, 1997d)

- Safety Evaluation by the Office of Nuclear Reactor Regulation, "EPRI Topical Report TR-106439," issued May 1997 (NRC, 1997e)

- for electrical independence and separation:

  - IEEE Std 384-1992 (IEEE, 1992)

— NUREG-0800, Standard Review Plan, Branch Technical Position HICB-11, "Guidance on the Application and Qualification of Isolation Devices," issued June 1997 (NRC, 1997f)

— RG 1.75, Revision 2 (NRC, 1978)

- for equipment SQ:

  — IEEE Std 344-1987 (IEEE, 1987b)

  — RG 1.100, Revision 2 (NRC, 1988)

- for setpoints:

  — ANSI/ISA-67.04.01-2000 (ANSI, 2000a)

  — RG 1.105, Revision 3, "Setpoints for Safety-Related Instrumentation," issued December 1999 (NRC, 1999b).

- for human-system interface:

  — IEEE Std 1023-1988, "IEEE Guide for the Application of Human Factors Engineering to Systems, Equipment, and Facilities of Nuclear Power Generating Stations," issued 1988 (NRC, 1988)

  — NUREG-0700, Revision 2, "Human System Design Review Guidelines," issued May 2002 (NRC, 2002)

- for the seismic monitoring and trip system (recording function):

  — RG 3.17-1974, "Earthquake Instrumentation for Fuel Reprocessing Plants," issued 1974 (NRC, 1974)

- for periodic testing:

  — IEEE Std 338-1987 (IEEE, 1987a)

  — NUREG-0800, Standard Review Plan, Branch Technical Position HICB-17, "Guidance on Self-Test and Surveillance Test Provisions," issued June 1997 (NRC, 1997g)

  — RG 1.118, Revision 3 (NRC, 1995)

- for reduction of electromagnetic and radio frequency interference:

  — IEEE Std 518-1982, "IEEE Guide for the Installation of Electrical Equipment to Minimize Electrical Noise Inputs to Controllers from External Sources," issued 1982 (IEEE, 1982)

— IEEE Std 1050-1996, "Guide for Instrumentation and Control Equipment Grounding in Generating Stations," issued 1996 (IEEE, 1996)

— RG 1.180, "Guidelines for Evaluating Electromagnetic and Radio-Frequency Interference in Safety-Related Instrumentation and Control Systems," issued January 2000 (NRC, 2000)

- for design of data communications networks:

— ANSI/IEEE 802.3 Standards Series, "IEEE Standards for Local Area Networks: Carrier Sense Multiple Access with Collision Detection (CSMA/CD) Access Method and Physical Layer Specifications," issued 2000 (ANSI, 2000b)

- for design of combustible gas detectors:

— ISA-S12.13-Part 1-1995 (ISA, 1995)

— ISA RP12.13-Part II-1987 (ISA, 1987)

In Section 5.5.5 of the revised CAR, the applicant described its general design philosophy used in formulating the preliminary design of the facility. Pursuant to 10 CFR 70.64(b), in order to ensure that engineered controls are relied upon over administrative controls, to the extent practicable, DCS has established a hierarchy of controls. In further adherence to 10 CFR 70.64(b), DCS stated that it has incorporated defense-in-depth practices in its preliminary facility design. Defense in depth is defined in the 10 CFR 70.64 footnote. In Section 5.5.5.2 of the revised CAR, DCS stated that it has incorporated defense-in-depth practices through use of the single-failure criterion. Under this criterion, PSSCs are required to be capable of carrying out their functions in the event that any single active component fails, whether such failure occurs within the applicable system, or in an associated system that supports the component's operation.

Accordingly, the facility I&C systems are to be designed using defense-in-depth practices using the single-failure criterion including redundancy, independence, separation, and fail safe for the I&C systems identified as PSSCs through use of the industry standards listed above. As an example, the emergency control system is designed with redundant, separate, and independent trains with fail safe failure modes (see FSER Section 11.6.1.1.2.3).

11.6.2 Evaluation Findings

In Section 11.6.7 of the revised CAR, DCS provided design basis information for I&C systems that it identified as PSSCs for the facility. Based on the staff's review of the revised CAR and supporting information provided by the applicant and the applicant's commitments to the industry standards and guidance discussed in the sections above for I&C systems, the staff finds that DCS has met the BDC set forth in 10 CFR 70.64(a)(10). The staff concludes, pursuant to 10 CFR 70.23(b), that the design bases of the PSSCs evaluated in this FSER section will provide reasonable assurance of protection against natural phenomena and the consequences of potential accidents. The staff further concludes, from its review of the revised CAR and supporting information submitted by DCS, that the preliminary design of the facility I&C systems meet the defense-in-depth provisions and the preference for engineered controls over administrative controls as stated in 10 CFR 70.64(b).

11.6.3 References

(ANSI, 2000a) American National Standards Institute/Instrument Society of America. 67.04.01-2000, "Setpoints for Nuclear Safety-Related Instrumentation." Research Triangle Park, North Carolina, 2000.

(ANSI, 2000b) American National Standards Institute/Institute of Electrical and Electronics Engineers. 802.3 Standards Series, "IEEE Standards for Local Area Networks: Carrier Sense Multiple Access with Collision Detection (CSMA/CD) Access Method and Physical Layer Specifications." New York, 2000.

(DCS, 2001a) Hastings, P., Duke Cogema Stone & Webster. Letter to U.S. Nuclear Regulatory Commission, RE: Response to Request for Additional Information (DCS-NRC-000059). August 31, 2001.

(DCS, 2001b) Hastings, P., Duke Cogema Stone & Webster. Letter to U.S. Nuclear Regulatory Commission, RE: Clarification of Responses to NRC Request for Additional Information (DCS–NRC–000074). December 5, 2001.

(DCS, 2002a) Hastings, P., Duke Cogema Stone & Webster. Letter to U.S. Nuclear Regulatory Commission, RE: Clarification of Responses to NRC Request for Additional Information. January 7, 2002.

(DCS, 2002b) Ihde, R., Duke Cogema Stone & Webster. Letter to U.S. Nuclear Regulatory Commission, RE: Mixed Oxide Fuel Fabrication Facility Construction Authorization Request Revision (DCS-NRC-000114). October 31, 2002.

(DCS, 2002c) Hastings, P., Duke Cogema Stone & Webster. Letter to U.S. Nuclear Regulatory Commission, RE: Requests for Additional Information, Clarifications, and Open Item Mapping into the Construction Authorization Request Revision (DCS-NRC-000120). November 22, 2002.

(DCS, 2003) Hastings, P., Duke Cogema Stone & Webster. Letter to U.S. Nuclear Regulatory Commission, RE: Responses to Site Description, Instrumentation, and Criticality Open Items/Additional NRC Questions on Construction Authorization Request (CAR) Revision. February 11, 2003.

(EPRI, 1996) Electric Power Research Institute. TR-106439, "Guideline on Evaluation and Acceptance of Commercial Grade Digital Equipment for Nuclear Safety Applications." Palo Alto, California, October 1996.

(IEC, 1993) International Electrotechnical Commission. 61131-3 (1993-03), "Programmable Controllers—Part 3: Programming Languages." Geneva, Switzerland, 1993.

(IEEE, 1982) Institute of Electrical and Electronics Engineers. Std 518-1982, "IEEE Guide for the Installation of Electrical Equipment to Minimize Electrical Noise Inputs to Controllers from External Sources." New York, 1982.

(IEEE, 1987a) Institute of Electrical and Electronics Engineers. Std 338-1987, "IEEE Standard Criteria for Periodic Testing of Nuclear Power Generating Station Class 1E Power and Protection Systems." New York, 1987.

(IEEE, 1987b) Institute of Electrical and Electronics Engineers. Std 344-1987, "IEEE Recommended Practices for Seismic Qualification of Class 1E Equipment for Nuclear Generating Stations." New York, 1987.

(IEEE, 1987c) Institute of Electrical and Electronics Engineers. Guide 1042-1987, "Software Configuration Management." New York, 1987.

(IEEE, 1988) Institute of Electrical and Electronics Engineers. Std 1023-1988. "IEEE Guide for the Application of Human Factors Engineering to Systems, Equipment, and Facilities of Nuclear Power Generating Stations." New York, 1998.

(IEEE, 1992) Institute of Electrical and Electronics Engineers. Std 384-1992, "Standard Criteria for Independence of Class 1E Equipment and Circuits." New York, 1992.

(IEEE, 1993) Institute of Electrical and Electronics Engineers. Std 7-4.3.2-1993, "IEEE Standard Criteria for Digital Computers in Safety Systems of Nuclear Power Generating Stations." New York, 1993.

(IEEE, 1994a) Institute of Electrical and Electronics Engineers. Std 379-1994, "IEEE Standard Application of the Single Failure Criterion to Nuclear Power Generation Station Safety Systems." New York, 1994.

(IEEE, 1994b) Institute of Electrical and Electronics Engineers. Std 1228-1994, "IEEE Standard for Software Safety Plans." New York, 1994.

(IEEE, 1996) Institute of Electrical and Electronics Engineers. Std 1050-1996, "Guide for Instrumentation and Control Equipment Grounding in Generating Stations." New York, 1996.

(IEEE, 1997a) Institute of Electrical and Electronics Engineers. Std 1028-1997, "IEEE Standard for Software Reviews." New York, 1997.

(IEEE, 1997b) Institute of Electrical and Electronics Engineers. Std 1074-1997, "IEEE Standard for Developing Software Life Cycle Processes." New York, 1997.

(IEEE, 1998b) Institute of Electrical and Electronics Engineers. Std 1012-1998, "IEEE Standard for the Software Verification and Validation." New York, 1998.

(IEEE, 1998c) Institute of Electrical and Electronics Engineers. Std 830-1998, "IEEE Standard Recommended Practice for Software Requirements Specifications." New York, 1998.

(IEEE, 1998d) Institute of Electrical and Electronics Engineers. Std 828-1998, "IEEE Standard for Software Configuration Management Plans." New York, 1998.

(IEEE, 1998e) Institute of Electrical and Electronics Engineers. Std 730-1998, "Software Quality Assurance Plans." New York, 1998.

(IEEE, 1998f) Institute of Electrical and Electronics Engineers. Std 603-1998, "IEEE Standard Criteria for Safety Systems for Nuclear Power Generating Stations." New York, 1998.

(ISA, 1987) Instrument Society of America. ISA RP12.13-Part II-1987, "Installation, Operation, and Maintenance of Combustible Gas Detection Instruments." 1987.

(ISA, 1995) Instrument Society of America. ISA-S12.13-Part 1-1995, "Performance Requirements, Combustible Gas Detectors." 1995.

(NRC, 1974) U.S. Nuclear Regulatory Commission. Regulatory Guide 3.17-1974, "Earthquake Instrumentation for Fuel Reprocessing Plants." Washington, DC, 1974.

(NRC, 1978) U.S. Nuclear Regulatory Commission. Regulatory Guide 1.75, Rev. 2, "Physical Independence of Electric Systems." Washington, DC, September 1978.

(NRC, 1988) U.S. Nuclear Regulatory Commission (NRC). Regulatory Guide 1.100, Rev. 2, "Seismic Qualification of Electric and Mechanical Equipment for Nuclear Power Plants." Washington, DC, June 1988.

(NRC, 1995) U.S. Nuclear Regulatory Commission. Regulatory Guide 1.118, Rev. 3, "Periodic Testing of Electric Power and Protection Systems." Washington, DC, April 1995.

(NRC, 1996) U.S. Nuclear Regulatory Commission. NUREG/CR-6463, "Review Guidelines on Software Languages for Use in Nuclear Power Plant Safety Systems." Washington, DC, June 1996.

(NRC, 1997a) U.S. Nuclear Regulatory Commission. Regulatory Guide 1.168, "Verification, Validation, Reviews, and Audits for Digital Computer Software Used in Safety Systems of Nuclear Power Plants." Washington, DC, September 1997.

(NRC, 1997b) U.S. Nuclear Regulatory Commission. Regulatory Guide 1.169, "Configuration Management Plans for Digital Computer Software Used in Safety Systems of Nuclear Power Plants." Washington, DC, September 1997.

(NRC, 1997c) U.S. Nuclear Regulatory Commission. Regulatory Guide 1.172, "Software Requirements Specifications for Digital Computer Software Used in Safety Systems of Nuclear Power Plants." Washington, DC, September 1997.

(NRC, 1997d) U.S. Nuclear Regulatory Commission. Regulatory Guide 1.173, "Developing Software Life Cycle Processes for Digital Computer Software Used in Safety Systems of Nuclear Power Plants." Washington, DC, September 1997.

(NRC, 1997e) U.S. Nuclear Regulatory Commission. Safety Evaluation by the Office of Nuclear Reactor Regulation, "EPRI Topical Report TR-106439." Washington, DC, May 1997.

(NRC, 1997f) U.S. Nuclear Regulatory Commission. NUREG-0800, Standard Review Plan, Branch Technical Position HICB-11, "Guidance on the Application and Qualification of Isolation Devices." Washington, DC, 1997.

(NRC, 1997g) U.S. Nuclear Regulatory Commission. NUREG-0800, Standard Review Plan, Branch Technical Position HICB-17, "Guidance on Self-Test and Surveillance Test Provisions." Washington, DC, 1997.

(NRC, 1999a) U.S. Nuclear Regulatory Commission. NUREG/CR-6090, "The Programmable Logic Controller and Its Application in Nuclear Power Plants." Washington, DC, February 1999.

(NRC, 1999b) U.S. Nuclear Regulatory Commission. Regulatory Guide 1.105, Revision 3, "Setpoints for Safety-Related Instrumentation." Washington, DC, December 1999.

(NRC, 2000) U.S. Nuclear Regulatory Commission. Regulatory Guide 1.180, "Guidelines for Evaluating Electromagnetic and Radio-Frequency Interference in Safety-Related Instrumentation and Control Systems." Washington, DC, January 2000.

(NRC, 2001) U.S. Nuclear Regulatory Commission. NUREG/CR-6597, "Results and Insights on the Impact of Smoke on Digital Instrumentation and Control." Washington, DC, January 2001.

(NRC, 2002) U.S. Nuclear Regulatory Commission. NUREG-0700, Rev. 2, "Human System Design Review Guidelines." Washington, DC, May 2002.

11.7 <u>Material Transport Systems</u>

11.7.1 Conduct of Review

This chapter of the FSER contains the staff's review of the material transport systems described by the applicant in Chapter 11.0 of the revised CAR (DCS, 2002a). The objective of this review is to determine whether the material transport systems' PSSCs and their design bases identified by the applicant provide reasonable assurance of protection against natural phenomena and the consequences of potential accidents. The staff evaluated the information provided by DCS for material transport systems by reviewing Chapter 11.0 of the revised CAR, other sections of the revised CAR, supplementary information provided by the applicant, and relevant documents available at the applicant's offices but not submitted by the applicant. The review of material transport systems' design bases and strategies was closely coordinated with the review of accident sequences described in the safety assessment of the design bases (see Chapter 5 of this FSER) and the review of other plant systems.

The staff reviewed how the information in the revised CAR addresses the following regulation:

- 10 CFR 70.23(b) states, as a prerequisite to construction approval, that the design bases of the PSSCs and the Quality Assurance Program must be found to provide reasonable assurance of protection against natural phenomena and the consequences of potential accidents.

The review of the revised CAR focused on the design bases of material transport systems, their components, and other related information. For material transport systems, the staff reviewed information provided by the applicant for the safety function, system description, and safety analysis. The review also encompassed proposed design basis considerations such as redundancy, independence, reliability, and quality. The staff used Chapter 11.0 in NUREG-1718 (NRC, 2000) as guidance in performing the review.

11.7.1.1 System Description

Revised CAR Section 11.7 describes the functional requirements and design bases for equipment designed to transfer MOX fuel production material that is in a dry, solid form. Examples of such forms are plutonium dioxide and uranium oxide powders, master blends of MOX powder, production batches of MOX powder, green pellets, sintered pellets, and fuel rods/assemblies. The equipment described in this section is located inside the MP area and shipping and receiving area. Because of the nature of the equipment (i.e., it is inside and attached to, or supported by, gloveboxes), certain parts of this review relate to gloveboxes. The staff's review of the confinement system and gloveboxes can be found in Section 11.4 of this FSER. Descriptions of the MP and AP process are provided in revised CAR Sections 11.2 and 11.3, respectively, and are discussed in the corresponding FSER sections. A description of the process control system is given in FSER Section 11.6.

Different material-handling equipment is used to transport material in the facility, depending on the form of the product, the container used to carry it, and the configuration of the process equipment that receives the container. Fuel material in the MP is in one of five forms—powder, pellets, rods, assemblies, or waste. The material-handling equipment is used in (1) the receipt and opening of plutonium dioxide and uranium dioxide containers in the powder area, (2) pellet processing, (3) fuel rod and fuel assembly fabrication, (4) fuel assembly inspection and storage, (5) fresh fuel cask loading, and (6) loading the fresh fuel casks onto over-the-road trailers.

Material-handling equipment that can be used during these processes includes, but is not limited to, scales, pallet trucks, fork lifts, drum-tilting devices, storage frames, handling monorails, pouring stations, feeding lines and control valves, hoppers, monorail cranes, standard and vibrating conveyors, turntables, traveling cranes, bridge cranes, gloveboxes, storage arrays, pneumatic transfer stations, airlocks, hoppers, impactors, funnels, can opening/closing devices, support frames, elevators, clamping devices, grippers, inspection stands, cleaning stations, jars, molybdenum boats, sintering furnaces, three-dimensional stackers, trolleys, winches, transfer tunnels, grinders, tray stackers, tilting tables, and air pallets. Various containers are also included in the list of material-handling equipment, including DOE Standard 3013 (DOE, 2000) containers and transport casks, transfer containers, waste containers, and MOX fresh fuel casks.

DOE Standard 3013 containers provide primary and secondary confinement for plutonium received at the facility. This standard applies to plutonium-bearing metals and oxides containing at least 30 wt% plutonium and uranium, as well as to plutonium-oxide materials with significant chloride contamination. The 3013 container is made up of an outer can, inner can, and convenience can. The outer and inner cans make up the primary and secondary containment, respectively.

Waste containers will be used to hold and ship MOX transuranic wastes. The waste transfer containers hold waste drums that provide the primary and secondary confinement for the waste. The drums are bag-lined before loading and sealed with a gasketed cover. The drums are also provided with filters to prevent pressurization and prevent the release of wastes from the drum.

MOX fresh fuel casks will contain multiple fresh MOX fuel assemblies for shipping. The fresh fuel casks will be qualified to meet the requirements of 10 CFR Part 71, "Packaging and

Transportation of Radioactive Material." The NRC safety review of the fresh fuel casks for transportation is being performed separately from the facility review.

11.7.1.1.1 Function

The functions that the material transport systems are designed to perform include the following:

- transferring MOX fuel material and components from one point in the process to another, in accordance with process throughput, positioning tolerance, mechanism reliability, and radiological shielding requirements

- maintaining structural integrity and control of process containers to ensure that the confinement boundary is not breached

- maintaining structural integrity and control of process containers to ensure that criticality control functions are performed

- working with fire barriers, as required, to transfer material across process atmosphere or fire barrier boundaries

- transferring tooling and equipment spare parts during maintenance operations from point to point within the glovebox system

The material-handling equipment operation during those processes begins when the plutonium powder arrives in a DOE safe, secure trailer at the shipping and receiving area at the facility. The 3013 containers, on a shipping pallet, are transferred from the truck to the shipping bay laydown area by forklift. The pallet is unpacked on a turntable and transported by roller conveyor to the 3013 storage area. Likewise, for the depleted uranium dioxide receiving and storage unit, uranium dioxide is delivered to the secured warehouse building in palletized drums. From there, the drums are sent to the MOX process area and are staged in a buffer area near the uranium dioxide drum-emptying room. The 3013 containers are transferred to the transfer cask opening area to remove the overpack. A hoist lifts the 3013 package (of approximately 9.1 kg (20 lb)) onto a small roller conveyor. From there, the plutonium dioxide/3013 storage crane transfers the 3013 package to plutonium dioxide/3013 storage racks. When removed from storage, the package rides by conveyor to the decanning unit, which is fully enclosed in a glovebox. Inside this glovebox, the 3013 can is moved both horizontally and vertically, and the outer can is removed. The inner can is transferred by pneumatic transfer tube to Level 4 of the AP building where the inner can is opened. Following this operation, the convenience can is opened. The opened convenience can is rotated and emptied into a homogenizer located in a glovebox immediately below the can opening glovebox. From there, the homogenized plutonium is transferred by pneumatic lift to the electrolyzer on Level 3 of the AP building. The electrolyzer marks the beginning of the AP chemical processing of the plutonium, which is discussed in detail in Sections 8 and 11.2 of this FSER.

Following AP chemical processing, the "wet" MOX material is returned to the MP. The MOX material will remain within gloveboxes from this point until it emerges in a completed, sealed fuel rod. The MP is described and evaluated in Section 11.3 of this FSER.

Table 11.7-1 PSSCs for Loss of Confinement/Dispersal of Nuclear Materials Events

Loss of Confinement/ Dispersal of Nuclear Material Events Related to Material-Handling Equipment	Identified PSSC	For the protection of the...			
		Facility Worker	Site Worker *	IOC*	Environ- ment
Corrosion	Material Maintenance & Surveillance Program[†]	✔	-	-	✔
Small breaches in glovebox boundary or backflow from utility lines	C4 confinement system	✔	-	-	✔
Leaks of AP process vessels or pipes within process cells	Process cell	✔	-	-	-
	Process cell entry controls	✔	-	-	-
	Process cell exhaust system	-	✔	✔	✔
Rod-handling operations	Material-handling equipment**	✔	-	-	-
	Material-handling controls**	✔	-	-	-
	Facility worker action	✔	-	-	-
Breaches in containers outside gloveboxes caused by handling operations in C2 areas	3013 canister**	✔	✔	✔	✔
	Transfer container**	✔	✔	✔	✔
	Material-handling controls**	✔	✔	✔	✔
Breaches in containers outside gloveboxes caused by handling operations in C3 areas	3013 canister**	✔	-	-	-
	Transfer container**	✔	-	-	-
	Facility worker controls	✔	-	-	-
	Material-handling controls**	✔	-	-	-
	C3 confinement system	-	✔	✔	✔

* There may be confinement systems or barriers not listed in this table that provide defense-in-depth protection for the site worker, IOC, or the environment for which no credit is technically being taken by DCS.

** These items are material transport system PSSCs evaluated in this section of the FSER. The remaining items are discussed in other sections, as appropriate, of this document.

[†] Material Maintenance and Surveillance Programs have been identified by the applicant as a PSSC (see Section 5.6.2 of the MFFF revised CAR, Revision 1).

11.7.1.1.2 Major Components

Major components include transfer containers, process equipment, confinement systems, MOX fuel transport casks, and waste containers. The applicant stated in revised CAR Section 5.0 that the material-handling system may have a plutonium dispersal hazard if the static barrier of the primary confinement system is damaged because of a loss of confinement/dispersal of nuclear material event or a load-handling event. The applicant has identified the material-handling PSSCs, shown in Tables 11.7-1 and 11.7-2, which reduce the risk of the postulated events to ensure the protection of facility workers, site workers, individuals outside the MFFF controlled area (IOC), and the environment.

The applicant stated in Chapter 5 of the revised CAR that the material-handling system may have a load-handling hazard from the presence of lifting or hoisting equipment used during normal operations or maintenance activities in the facility. A load-handling event could occur when either a lifted load containing radioactive materials is dropped, or the load or lifting equipment impacts equipment containing radioactive material. Heavy load drops and other load-handling events, as specifically defined in NUREG-1718 (NRC, 2000) and NUREG-0612, are discussed in the revised CAR Section 11.10 and are evaluated in Section 11.10 of this FSER.

11.7.1.1.3 Control Concept

The MP and AP process control systems use a distributed processing control system strategy, with the manufacturing process translated into control algorithms for each process step. The systems include normal, protective, and safety control subsystems that ensure the final product conforms to manufacturing specifications and minimize plant waste and risk. The normal control subsystem controls the manufacturing process, and the protective control subsystem maintains industrial safety (protects personnel) and protects equipment. The safety control subsystem ensures that safety limits will not be exceeded and that undesirable operational conditions are prevented or mitigated.

11.7.1.2 Design Bases of PSSCs

This section describes the PSSC design bases contained in the revised CAR.

Material-Handling Equipment and Controls

The applicant states in Section 11.7.7 of the revised CAR that material-handling equipment and support structural members will be designed to prevent physical interaction with confinement boundary elements or PSSCs under worst-case loading associated with normal, upset, and design basis events. To achieve this design objective, the applicant intends to apply design principles such as redundant brakes, with fail-safe design, on lifting equipment; structural oversizing of mechanical drive equipment; overspeed detection; mechanical stops; overtorque detection; electrical interlocks; component sizing; magnetic grippers; glovebox hoods; and shielding. The applicant also states that, as an additional safety function, material- handling controls will be designed to prevent the potential overpressurization of reusable plutonium oxide cans resulting from radiolysis or oxidation of plutonium (III) oxalate, thereby preventing potential glovebox impacts from such overpressurizations. Material-handling controls are also discussed in FSER Chapter 5.

Table 11.7-2 PSSCs for Load-Handling Events

Load-Handling Events Related to Material-Handling Equipment	Identified PSSC	For the protection of the...			
		Facility Worker	Site Worker*	IOC*	Environment
AP Process Cells	Process cell	✔			
	Process cell entry controls	✔			
	Process cell exhaust system		✔	✔	✔
AP/MP C3 Glovebox Areas	Material handling controls**	✔			✔
	Material handling equipment**	✔			✔
	Glovebox	✔			✔
	Facility worker controls	✔			
	C3 confinement system		✔	✔	
C1 and/or C2 Areas/3013 Canister	3013 canister**	✔	✔	✔	✔
	Material handling controls**	✔	✔	✔	✔
C1 and/or C2 Areas/3013 Transport Cask	3013 transport cask**	✔	✔	✔	✔
	Material handling controls**	✔	✔	✔	✔
C1 and/or C2 Areas/Fuel Rod	Facility worker action	✔			
C1 and/or C2 Areas/MOX Fuel Transport Cask	MOX fuel transport cask**	✔			✔
	Material handling controls**	✔			✔
C1 and/or C2 Areas/Waste Container	Facility worker action	✔			
C1 and/or C2 Areas/Transfer Container	Transfer container**	✔	✔	✔	✔
	Material handling controls**	✔	✔	✔	✔
C1 and/or C2 Areas/Final C4 HEPA Filter	Material handling controls**	✔	✔	✔	✔
C4 Confinement/Spill inside Glovebox	C4 confinement system	✔	✔	✔	✔
C4 Confinement/Outside of MFFF Building	Waste transfer line	✔	✔	✔	✔
C4 Confinement/Facilitywide	MFFF building structure	✔	✔	✔	✔
	Material handling controls**	✔	✔	✔	✔

* There may be confinement systems or barriers that provide defense in depth protection for the site worker, IOC, or the environment for which no credit is technically being taken by DCS.

** These items are material transport system PSSCs evaluated in this section of the FSER. The remaining items are discussed in other sections, as appropriate, of this document.

The design bases of material-handling PSSCs are qualified by the following codes and standards:

- ANSI/AISC N690-1994 (AISC, 1994) for design of components required to maintain structural integrity

- ASME B30.2-1996, "Top Running Bridge, Single or Multiple Girder, Top Running Trolley Hoist Overhead and Gantry Cranes," issued 1996 (ASME, 1996), for design of overhead cranes

- ASME B30.16-1998, "Overhead Hoists," issued 1998 (ASME, 1998), for design of hoisting equipment

- ASME NOG-1-1998, "Rules for Construction of Overhead and Gantry Cranes (Top Running Bridge, Multiple Girder)," issued 1998 (ASME 1998b), for seismic design

- AWS D1.1-1998 (AWS, 1998a) for structural integrity and fabrication

- AWS D1.6-1999, "Structural Welding Code—Stainless Steel," issued 1999 (AWS, 1999), for structural integrity and fabrication

- AWS D14.1-1997, "Specification for Welding Industrial and Mill Cranes and Other Material Handling Equipment," issued 1997 (AWS, 1997), for structural integrity and fabrication

- Crane Manufacturers Association of America, CMAA-70-1994, "Specifications for Top Running Bridge and Gantry Type Multiple Girder Electric Overhead Traveling Cranes," issued 1994 (CMAA, 1994a), for design of bridge cranes

- CMAA-74-1994, "Specification for Top Running and Under Running Single Girder Electric Overhead Traveling Cranes Utilizing Under Running Trolley Hoist," issued 1994 (CMAA, 1994b), for design of bridge cranes

- NUREG-0554, "Single Failure Proof Cranes at Nuclear Power Plants," issued May 1979 (NRC, 1979), for design of bridge cranes

3013 Canister and Transport Cask

The 3013 canister consists of individually sealed outer and inner containers which isolate the stored materials from the environment. The 3013 canister is designed to meet the requirements of DOE-STD-3013-2000, "Stabilization, Packaging, and Storage of Plutonium-Bearing Materials" (DOE, 2000). The outer can is designed and qualified for a 9-m (30-ft) drop onto a flat, unyielding surface while remaining leak-tight, and the inner can is designed to remain leak-tight after a drop of 1.3 m (4 ft) onto a flat unyielding surface. The outer and inner cans are designed to withstand pressures of 4927 kPa (699 psig) and 790 kPa (100 psig) and are hydrostatically tested before use at a pressure 1.5 times that of the design pressure. Both containers must be fabricated from ductile, corrosion-resistant materials, such as 300 series stainless steel or better. Closure welding of the stainless steel must be done in such a way as to minimize the sensitization of stainless steel to stress-corrosion cracking. Heat generation limits the mass of plutonium contained in the containers to less than or equal to 19 watts (1.1 Btu/min). Both of the containers are designed to hold the material for a maximum of 50 years.

The 3013 transport cask is a PSSC designed and qualified to protect the 3013 canister against transportation accidents. It will be certified to meet the applicable requirements contained in 10 CFR 71.73, "Hypothetical Accident Conditions."

Transfer Containers

Transfer containers are designated as PSSCs that will be used to hold and ship MOX transuranic wastes. The waste transfer containers are designed, constructed, and qualified to meet the requirements of the U.S. Department of Transportation Specification 7A of 49 CFR 178.350, "Specification 7a; General Packaging, Type A."

MOX Fuel Transport Casks

Fresh fuel assemblies will be transported in MOX fuel transport casks. The casks may be stacked in storage frames in the truck shipping bay. To accommodate the load, the frames will be designed for the full weight of the shipping package including seismic effects. The fresh fuel casks will be qualified to meet the requirements of 10 CFR Part 71. The casks will not be used in the facility before being licensed by the NRC, and, if approved, they will be handled in the proposed facility according to the certification.

Seismic Design

Material-handling equipment designated as PSSCs is designed and qualified according to national codes and standards, enabling the equipment to perform its safety function during normal operations, upset conditions, and design basis events. The ability to safely shut down the primary process is facilitated by the seismic design for the material-handling equipment and structural support members. Equipment geometry and alignment must be maintained in order to have an orderly shutdown of the system. The system is designed to prevent physical interaction with confinement boundary elements or PSSCs under worst-case loading associated with normal, off-normal, accident, and design basis events. The system will be designed to meet the criteria provided in RG 1.100, Revision 2 (NRC, 1988); IEEE Standard 344-1987 (IEEE, 1987); and ANSI/AISC N690-1994 (AISC, 1994). By adhering to IEEE Standard 344-1987, DCS is satisfying the NRC's regulations pertaining to SQ of mechanical equipment. Thus, thermal loads, attached piping loads, and live loads, such as fluid sloshing, are considered for equipment operability. In addition, applied loads are required to meet or exceed accelerations corresponding to their installed locations.

The seismic monitoring system is designed to satisfy the criteria provided in RG 3.17-1974 (NRC, 1974). The design basis of the seismic monitoring system is that it provides sufficient data to evaluate the response of the confinement structure and other PSSCs to a seismic event and initiate a shutdown of process systems in the event of a high seismic event. The seismic system will meet the requirements of IEEE 603-1998 (IEEE, 1998).

The staff has reviewed the DCS commitment to these codes and standards and finds that they provide an acceptable basis for the seismic design of the MFFF material transport system.

Evaluation of Capacity

The staff evaluated the information provided by DCS in its revised CAR regarding the capacity of the proposed material-handling equipment. The material-handling equipment is designed such that, in the event of accident or off-normal conditions, the equipment is designed to de-power or return to a shutdown condition. The throughput, or capacity, of the equipment peaks during normal operations or maintenance. The revised CAR describes the design of the equipment as being sized to handle the required throughput of shipping packages, containers, canisters, drums, casks, cans, powder, pellets, scrap, rods, and assemblies necessary for normal and off-normal operating conditions and maintenance. The revised CAR also discusses that the active systems, such as motors, power transmission systems and pass-throughs, carriers, actuators, end effectors, structural supports, sensors, and control systems will be based on the material throughput requirements for each process unit.

Capacities of the material-handling equipment vary based on the operational throughput needed for the equipment, the design and qualification of the equipment, and other specific design criteria. The material-handling equipment designated as PSSCs must also retain its loads under all credible accidents and design basis natural phenomena events. Therefore, the capacity of these PSSCs may be greater than non-PSSCs. To accomplish these design requirements, equipment will be designed to (1) maintain clearance between equipment and the confinement boundary under all conditions, (2) include physical stops to prevent uncontrolled motion of payloads from breaching containment in the event of over-travel or seismic conditions, (3) ensure that actuating mechanisms, such as grippers, are designed to retain the payload under all conditions including loss-of-power and credible seismic events, and (4) maintain appropriate margins of safety in hoisting equipment. Capacity of equipment is not directly discussed by design codes. However, the staff accepts that, if PSSC equipment is built to the design codes referenced in the revised CAR, it will be designed to handle all loads, events, and configurations while maintaining its safety function.

Material-handling equipment intended to suspend loads from flexible cables is designed using codes for cranes, monorails, and hoists, which include appropriate minimum factors of safety. The capacities of hoisting equipment identified as PSSCs are further de-rated according to safety factors from NUREG-0554 applicable to single-failure-proof cranes. Process equipment used during maintenance is further de-rated to 65 percent of capacity according to project-specific design criteria. Hoists will be designed in accordance with ASME B30.16-1998 (ASME, 1998a). Material-handling equipment that runs on fixed tracks is designed using structural design codes. Equipment classified as PSSCs is qualified in accordance with load combinations and acceptance criteria provided in the ANSI/AISC N690-1994 structural design code (AISC, 1994).

The staff has reviewed the description of the capacity of the material-handling equipment. Based on the description of the system provided in the revised CAR and the use of industry codes and standards, the staff finds the design to be acceptable to handle the expected volume of radioactive material during normal operating and accident conditions.

Evaluation of Redundancy and Diversity

The revised CAR describes the passive design of the equipment to handle shipping packages, containers, canisters, drums, casks, cans, powder, pellets, scrap, rods, and assemblies necessary for normal and off-normal operating conditions and maintenance. The staff reviewed the system design basis including the provision of active systems, such as motors, power transmission systems and pass-throughs, carriers, actuators, end effectors, structural supports, sensors, and controls. Material-handling equipment includes devices that suspend loads from flexible cables and material-handling equipment that runs on fixed tracks. Some of this equipment is designed to work external to a glovebox, and some is designed to work internal to a glovebox environment.

Redundancy and diversity in the design are accomplished by various factors of safety, types of equipment, and the layering of active and passive controls that protect the confinement boundary. The facility design for material-handling equipment inside gloveboxes includes redundant brakes, with fail-safe design, on lifting equipment; structural oversizing of mechanical drive equipment; overspeed detection; mechanical stops; overtorque detection; electrical interlocks; and component sizing based on worst-case loading combinations. Various containers and casks are also discussed in the facility design. Each of these containers or casks is designed for different applications and is certified under Federal regulations before use.

The staff has reviewed the applicant's description of the material-handling equipment regarding system redundancy and diversity. On the basis of standard industry practices, the staff finds the design to be acceptable to prevent the release of radioactive materials to the environment. The staff notes that casks and canisters are likewise acceptable if they are used within their certification basis or if appropriate compensatory measures are taken with consideration to the hazard to the facility and site worker, the IOC, and the environment.

Evaluation of Safe Shutdown

The staff evaluated the information provided by DCS in the revised CAR regarding the ability to safely shut down the proposed material-handling equipment during normal, accident, and maintenance conditions. The material-handling equipment is designed such that in the event of an accident or off-normal condition, the equipment is designed to de-power in a fail-safe condition. Emergency power is not provided to the material-handling equipment. For example, during a loss of power, hoist brakes passively activate, and end effectors, such as magnetic grippers, passively fail to a closed or retain load condition. In this way, the design ensures that in any accident or off-normal condition, all system loads are maintained and the confinement boundaries are not challenged by dropped or unrestrained loads. The staff has reviewed the applicant's description of the material-handling equipment's ability to safely shut down and finds this approach to be acceptable.

Evaluation of Welded Construction

In NUREG-1718 (NRC, 2000), the regulatory acceptance criteria for the material transport system specifically mentions that tank and piping systems should be of welded construction to the fullest extent possible. For the purposes of this review, this guidance is applied to gloveboxes (a material- handling PSSC). Continuously welded construction means the seams are ground smooth, which facilitates cleaning and minimizes holdup of powder, pellets, dust, or debris. The specification of welded construction also minimizes leakage paths and facilitates decontamination of gloveboxes. The applicant's proposed design for gloveboxes is evaluated in FSER Section 11.4. Based on the use of industry codes for the design and construction of welded material-handling equipment, the staff finds this design basis to be acceptable.

Evaluation of Passive Features/Remote Operation

As discussed previously, the material-handling equipment is designed such that, in the event of an accident or off-normal condition, the equipment will de-power in a fail-safe condition. Emergency power is not provided to the material-handling equipment. For example, during a loss of power, hoist brakes passively activate, and end effectors, such as magnetic grippers, passively fail to a closed or retain load condition. In this way, the design ensures that in any accident or off-normal condition, all system loads are maintained and the confinement boundaries are not challenged by dropped or unrestrained loads. The material- handling equipment is designed with engineered features to prevent active failures from impacting the glovebox walls. Based on the applicant's commitment in CAR Section 11.7.7 to these passive engineered features, the staff finds that the design provides an adequate level of protection against active failures.

For most operations, the process control system is designed to control the material-handling equipment during normal process conditions. In the event of an off-normal or accident condition, additional control system elements are capable of overriding the normal process controllers to mitigate the potential hazardous condition. This equipment and its functions are described in further detail in Section 11.6 of this FSER. During maintenance, process equipment and controllers are deenergized, and equipment may be selectively energized under manual control of facility personnel engaged in the maintenance activities. Based on the control system design's being for remote operation of process equipment, the staff finds this design for remote operation to be acceptable.

Evaluation of Radiation Safety

The staff review and evaluation of the radiation safety program is discussed in detail in Section 9 of this FSER. In general, the design basis for the material-handling equipment hardware for radiation safety entails the use of design configurations to minimize powder/dust or debris; mounting of stainless steel casings on structural supports to prevent powder/dust retention; easy visibility and accessibility of parts for cleaning; use of sealed bearings or leak-free coupling mechanisms; use of appropriate surface quality or coatings for equipment in contact with powder; limiting lubricant use to the extent practical; continuous and smoothly ground internal welds; re-entrant corners of large relative radius; and sealed powder-handling channels. Based on commitments to industry standards for surface finish and the general equipment design for minimization of powder holdup in equipment (e.g., Section 11.7.6 of the CAR), the staff finds these provisions to be acceptable.

Evaluation of Corrosion Resistance

The material-handling equipment proposed by the applicant is to be made primarily of stainless steel with the appropriate surface finishes to resist corrosion. The corrosion of carbon steel parts that cannot be painted will be prevented by a glovebox environment of nitrogen or dry air. For other areas, components may be coated or painted for corrosion resistance and ease of decontamination. Outside gloveboxes, painting systems will be used for materials located in C3b rooms to facilitate decontamination. In addition, the material condition of the equipment will be monitored by the Material Maintenance and Surveillance Programs. On the basis of industry codes and standards that specify system design accounting for corrosion as a standard industrial practice, the staff finds the design to be acceptable.

Evaluation of Personnel Protection

Section 11.4.7.2 of NUREG-1718 (NRC, 2000) states that the need for hoods, gloveboxes, and shielding for personnel protection should be evaluated. These systems are generally required for processing operations involving more than gram quantities of plutonium or general operations involving 50 micrograms or more of plutonium in respirable form. In Section 11.8.2 of the revised CAR, the applicant states that personnel protection for these materials is provided by process cells and welded equipment confinement. Process cells contain equipment that handles radioactive materials in chemical solutions, and that equipment is of a fully welded construction and does not require routine maintenance. Equipment containing radioactive materials in the powder process (MP) is contained in gloveboxes in process rooms that provide equivalent confinement to fully welded equipment in process cells. In Section 11.4.7.1.3 of the revised CAR, the applicant documented its analysis of the accident scenarios for fire and impact events with gloveboxes. Based on the applicant's analyses and commitments described above, the staff finds this design for personnel protection to be acceptable.

11.7.2 Evaluation Findings

In Chapter 11.7 of the revised CAR, DCS provided design basis information for the material transport systems that it identified as PSSCs for the proposed facility. Based on the staff's review of the revised CAR and supporting information provided by the applicant relevant to the material transport systems, the staff concludes, pursuant to 10 CFR 70.23(b), that the design bases of the PSSCs evaluated in this FSER section will provide reasonable assurance of protection against natural phenomena and the consequences of potential accidents.

11.7.3 References

(AISC, 1989) American National Standards Institute/American Institute of Steel Construction. "Manual of Steel Construction, Allowable Stress Design," 9th Edition. 1989.

(AISC, 1994) American National Standards Institute/American Institute of Steel Construction (ANSI/AISC). N690, "Specification for the Design, Fabrication, and Erection of Steel Safety-Related Structures for Nuclear Facilities." 1994.

(ASME, 1995) American Society of Mechanical Engineers. B46.1, "Classification and Designation of Surface Qualities." 1995.

(ASME, 1996) American Society of Mechanical Engineers. B30.2, "Top Running Bridge, Single or Multiple Girder, Top Running Trolley Hoist Overhead and Gantry Cranes." 1996.

(ASME, 1998a) American Society of Mechanical Engineers. B30.16, "Overhead Hoists." 1998.

(ASME, 1998b) American Society of Mechanical Engineers. NOG-1, "Rules for Construction of Overhead and Gantry Cranes (Top Running Bridge, Multiple Girder)." 1998.

(AWS, 1997) American Welding Society. D14.1, "Specification for Welding Industrial and Mill Cranes and Other Material Handling Equipment." 1997.

(AWS, 1998a) American Welding Society. D1.1, "Structural Welding Code—Steel." 1998.

(AWS, 1998b) American Welding Society. D1.3, "Structural Welding Code—Sheet Steel." 1998.

(AWS, 1998c) American Welding Society. D9.1, "Sheet Metal Welding Code." 1998.

(AWS, 1999) American Welding Society. D1.6, "Structural Welding Code—Stainless Steel." 1999.

(CMAA, 1994a) Crane Manufacturers Association of America. 70, "Specifications for Top Running Bridge and Gantry Type Multiple Girder Electric Overhead Traveling Cranes." 1994.

(CMAA, 1994b) Crane Manufacturers Association of America. 74, "Specification for Top Running and Under Running Single Girder Electric Overhead Traveling Cranes Utilizing Under Running Trolley Hoist." 1994.

(DCS, 2001) Hastings, P., Duke Cogema Stone & Webster. Letter to U.S. Nuclear Regulatory Commission RE: Clarification of Responses to NRC Request for Additional Information (DCS-NRC-000074). December 5, 2001.

(DCS, 2002a) Ihde, R., Duke Cogema Stone & Webster. Letter to Document Control Desk, U.S. Nuclear Regulatory Commission, RE: Mixed Oxide Fuel Fabrication Facility—Construction Authorization Request. October 31, 2002 (including page changes through February 9, 2005).

(DCS, 2002b) Hastings, P., Duke Cogema Stone & Webster. Letter to Document Control Desk, U.S. Nuclear Regulatory Commission, RE: Clarification of Responses to NRC Request for Additional Information. February 11, 2002.

(DOE, 2000) U.S. Department of Energy. Standard-3013-2000, "Stabilization, Packaging, and Storage of Plutonium-Bearing Materials." Washington, DC, 2000.

(IEEE, 1987) Institute of Electrical and Electronics Engineers. Standard 344, "IEEE Recommended Practice for Seismic Qualification of Class 1E Equipment for Nuclear Power Generating Stations." 1987.

(IEEE, 1998) Institute of Electrical and Electronics Engineers. Standard 603, "IEEE Standard Criteria for Safety Systems for Nuclear Power Generating Station." 1998.

(NRC, 1974) U.S. Nuclear Regulatory Commission. Regulatory Guide 3.17, "Earthquake Instrumentation for Fuel Reprocessing Plants." Washington, DC, 1974.

(NRC, 1979) U.S. Nuclear Regulatory Commission. NUREG-0554, "Single Failure Proof Cranes at Nuclear Power Plants." Washington, DC, May 1979.

(NRC, 1988) U.S. Nuclear Regulatory Commission. Regulatory Guide 1.100, Rev. 2, "Seismic Qualification of Electric and Mechanical Equipment for Nuclear Power Plants." Washington, DC, 1988.

(NRC, 2000) U.S. Nuclear Regulatory Commission. NUREG-1718, "Standard Review Plan for the Review of an Application for a Mixed Oxide (MOX) Fuel Fabrication Facility." Washington, DC, August 2000.

11.8 Fluid Transport Systems

11.8.1 Conduct of Review

This chapter of the FSER contains the staff's review of the fluid transport systems described by the applicant in Chapter 11.0 of the revised CAR (DCS, 2002a). The objective of this review is to determine whether the fluid transport PSSCs and their design bases identified by the applicant provide reasonable assurance of protection against natural phenomena and the consequences of potential accidents. The staff evaluated the information provided by the applicant for fluid transport systems by reviewing Chapter 11.0 of the revised CAR, other sections of the revised CAR, supplementary information provided by the applicant, and relevant documents available at the applicant's offices but not submitted by the applicant. Additional documentation from the literature was reviewed as necessary to understand the process and safety requirements. The review of fluid transport systems design bases and strategies was closely coordinated with the review of other chapters in this FSER (fire protection in Chapter 7, chemical safety in Chapter 8, and safety assessment of the design bases in Chapter 5).

The staff reviewed how the information in the revised CAR addresses the following regulation:

*   10 CFR 70.23(b) requires that the design bases of the PSSCs and the Quality Assurance Program must provide reasonable assurance of protection against natural phenomena and the consequences of potential accidents before construction of the PSSCs is approved.

The review for this construction approval focused on the design bases of fluid transport systems, their components, and other related information. For fluid transport systems, the staff reviewed and evaluated information provided by the applicant for the safety function, system description, and safety analysis. The review also encompassed proposed design basis considerations such as redundancy, independence, reliability, and quality. The staff used Chapter 11.0 in NUREG-1718 (NRC, 2000), particularly Section 11.4.7, and industry codes and standards as guidance in performing the review.

The MFFF fluid transport systems include systems that handle process and utility fluids. Other fluid-containing support systems are discussed in Section 11.9 of this FSER.

The NRC staff reviewed the revised CAR submitted by DCS for the following areas applicable to the fluid transport systems at the construction approval stage and consistent with the level of design:

- system description
- system function
- major components
- control concepts
- system interfaces
- design bases

Section 11.4.7 of NUREG-1718 was the primary guidance used for this review. Regarding the proposed facility fluid transport systems, specific design considerations given in the revised CAR are to demonstrate the following:

- Capacity is adequate to handle the expected volume of radioactive material during normal operating and accident conditions.

- Redundancy or diversity of components prevents the release of radioactive materials to the environment or contributes to the safe operation of the fluid transport systems.

- The fluid transport system can be safely shut down during normal and accident conditions. Provisions for emergency power are included for critical process components.

- Tank and piping systems are of welded construction to the fullest extent possible.

- Tank and piping systems are designed to take advantage of gravity flow to reduce the potential for contamination associated with pumping and pressurization.

- Criticality will not occur under normal and credible accident conditions.

- All system components expected to be in contact with strong acids or caustics are corrosion resistant.

- Piping is designed to minimize entrapment and buildup of solids in the system.

- The systems are evaluated to determine the need for hoods, gloveboxes, and shielding for personnel protection. Generally, wet processing operations involving gram quantities of plutonium and any operations involving 50 micrograms of respirable plutonium are conducted in a glovebox.

- Surface finishes of materials in the work areas have satisfactory decontamination characteristics for their particular application.

- Fluid transport systems maintain functionality when subjected to tornadoes, tornado missiles, earthquakes, floods, and any other natural phenomena deemed to be credible as further established in the ISA to be performed by the applicant.

In the FSER discussions that follow, the system descriptions are provided as well as function, major components, control concepts, and system interfaces. These discussions include but are not limited to PSSCs, to provide an understanding of the system. The design bases of PSSCs are discussed in Section 11.8.1.3 of this FSER.

11.8.1.1 System Description

The fluid transport system contains the hardware portion of the AP wet process that contains the dissolution unit, the purification cycle, solvent recovery, oxalic precipitation, and precipitation and drying for the various stages of plutonium polishing. All wastes generated are stored and sent by pipeline to DOE SRS for disposition.

In general, the fluid transport system consists of components that contain radioactive process fluids, radioactive waste products, and utility fluids. The PSSCs in the process equipment include vessels, tanks, pulsed columns, heat exchangers, pumps, piping, and valves. Other process equipment includes electrolyzers, airlifts, drip pots, sampling lines and pots, spargers, gravity feeds, dosing wheels, magnetic stirrers, extraction screws, and stationary and rotary filters.

Fluid transport components are designed for the most severe credible service conditions. Fluids containing radioactive materials are designed to always be within at least two levels of confinement according to the general principles described in revised CAR Section 11.4 and reviewed in the FSER. Fluid-bearing components within process cells that are not easily accessed for routine inspection will be designed with corrosion allowances and provided with drip trays sized to hold the contents of the largest vessel in a criticality-safe condition. Sump pumps on the drip trays are monitored for activity signaling a leak. The system is designed to accommodate flushing and high-pressure decontamination to remove sediment buildup and any blockages and to maintain ALARA principles, before system maintenance.

The fluid transport system for the AP process is diverse in the control concepts for the process control systems that govern the fluid transport system. It is composed of the normal, protective, and safety control subsystems. The normal control subsystem controls the facility manufacturing and processing operations. The protective control subsystem provides protection for equipment and personnel. The safety control subsystem is designed to ensure that safety limits will not be exceeded and that undesired operational conditions or events will not occur or will be properly mitigated. For more information on the AP process, refer to Section 11.2 of this FSER. Section 11.6 of the FSER provides more details on the design and operation of the I&C system.

Fluid transfer systems containing hazardous fluids are contained within trenches, rooms, or double-walled piping systems or are accessible for inspection and are of a fully welded construction. All piping components designated as PSSCs are designed to withstand the design basis earthquake loads. Isolation valves may be of the butterfly, gate, plug, or ball type. The valves are specified for service after consideration of the chemical characteristics of the fluid, piping material of construction, and operating conditions.

Section 11.4.7 of NUREG-1718 (NRC, 2000), used for the review of the fluid transport systems, specifically mentions that tank and piping systems should be of welded construction to the fullest extent possible. For process equipment, radiological fluids are maintained within at least two levels of confinement. Components containing fluids that are located in process cells are specified with corrosion allowances, and welded joints are radiographed, as appropriate. Components that are not of fully welded construction are installed in a glovebox. Piping components carrying radiological fluids between two confinements are either of fully welded double-wall construction or are installed in gloveboxes. Welding requirements are contained in the ASME Boiler and Pressure Vessel (B&PV) Code, Section VIII, 1996 Edition through the 1998 Addenda (ASME, 1998), ASME B31.3 (ASME, 2001a), and their referenced AWS codes. Welding work will be performed according to the fluid transport system (FTS) category (see FSER Table 11.8-1) and quality assurance plan quality level of the system being welded.

Materials used for the construction of this equipment are specified in accordance with ASME and ASTM specifications. ASME materials are used in the fabrication of equipment and piping components built to the requirements of ASME B&PV Code, Section VIII, 1996 Edition through the 1998 Addenda, and ASME B31.3. In general, design of equipment to these standards means that the components are designed for the most severe service conditions. Included in the severe service conditions are pressure, temperature, stress, material compatibility, and corrosion.

11.8.1.1.1 Function

The AP process can be segmented into the following four operational areas:

(1)   plutonium purification process—includes the decanning unit, dissolution unit, purification cycle, oxalic precipitation and oxidation unit, homogenization unit, and canning unit

(2)   recovery processes—includes the solvent recovery cycle, oxalic mother liquor recovery unit, and acid recovery unit

(3)   waste storage—includes the liquid waste reception unit

(4)   offgas treatment—includes the offgas treatment unit

This section concentrates on the MOX equipment located in the plutonium purification process. The plutonium purification process separates impurities from the fissile material. This process is a radiochemical process where the fissile material commingles with inorganic and organic solutions at various concentrations. The plutonium purification process is divided into six discrete steps:

(1)   The decanning unit is a mechanical operating unit where can opening and powder transfer operations are automatically performed, primarily in gloveboxes.

(2)   Plutonium dioxide from the decanning unit is electrochemically dissolved with silver (Ag^{2+}) in nitric acid. The plutonium valence of the resulting plutonium nitrate solution is altered by hydrogen peroxide from (VI) to (IV). The plutonium nitrate solution is transferred to the purification cycle feed tank.

(3)   The plutonium nitrate solution from the dissolution unit is extracted into the organic phase from impurities (e.g., gallium, americium) in the flux. The organic stream is scrubbed with dilute nitric acid, its valence is further reduced to Pu (III), and it is stripped back into the aqueous solution. The plutonium nitrate is oxidized to the tetravalent state by nitrogen oxide fumes.

(4)   In the oxalic precipitation and oxidation unit, plutonium is precipitated with excess oxalic acid in vortex precipitators.

(5)   In the homogenization unit, the plutonium oxalate precipitate and the mother liquors flow from the precipitator and are channeled to a flat filter where they are filtered, washed, and vacuum-dewatered. The oxalate is dried and calcined.

(6)   The canning unit is a mechanical operating unit where the calcined plutonium dioxide is sampled and packaged for use in the MP.

These steps are described and evaluated in detail in Chapter 8 and Section 11.3 of this FSER.

11.8.1.1.2 Major Components

The major components of the fluid transport systems are part of the primary process and are located in the AP area of the facility. In addition to piping and valves, the major components of the fluid transport systems include the following:

- Welded process equipment includes vessels, tanks, process columns, and heat exchangers. In general, fully welded process equipment is located in process cells. Storage tanks vary in design at different stages of the primary process. Storage tanks include annular tanks, stab tanks, and conventional tanks. These tanks are fabricated using fully welded construction. Other welded process equipment includes various small tanks used in the AP process, such as separating pots, leak detection pots, barometric seal pots, pulse column pots, drip pots, condensate pots, and demisters. The AP process columns are also of a fully welded construction. Examples of these columns are pulsed, rectification, packed or scrubbing columns, and tray columns for process distillation. Various AP process heat exchangers used in radiological service are also of a fully welded construction. These heat exchangers may be evaporators, condensers, and jacketed heaters/coolers designed to transfer process heat.

- Partially welded process equipment and prime movers include filters, mixing tanks, and precipitators. Partially welded equipment classified as FTS Category 1 is housed inside gloveboxes (see FSER Section 15.1 for a more detailed discussion). Other process prime movers include pumps, low-pressure airlifts, ejectors, and siphons. Pump types include centrifugal and positive displacement dosing pumps.

11.8.1.1.3 Control Concept

The AP process control systems are designed to ensure that the product of the manufacturing process will conform to the product specifications while simultaneously minimizing risk to the facility worker, site worker, and IOC. They are composed of the normal, protective, and safety control subsystems. The normal control subsystem controls the facility's normal manufacturing and processing operations. The protective control subsystem provides protection for personnel and equipment. The safety control subsystem is designed to ensure that safety limits will not be exceeded and that undesired operational conditions or events will not occur or will be mitigated. Section 11.6 of the revised CAR discusses the facility instrumentation and control systems in more detail.

In general, each unit is controlled by one or several PLCs associated with a monitoring workstation located in the AP control room. All units are operated in an automatic mode. The operator may also intercede via a manual mode in which the interlocks are active in case of trouble in the automatic mode or for maintenance operations. The MMIS collects the information coming from all process units to control the position and the exchange of special nuclear material (SNM), as well as the traceability and the quality of the products.

Process storage and operation conditions are controlled to prevent exothermic and potential autocatalytic reactions in the AP area. Autocatalytic and exothermic reactions of chemicals, precipitation, and criticality are prevented through control of the process parameters (e.g., reactant concentration, temperature, catalyst concentration in solution, and pressure) that affect the reactions.

11.8.1.2 System Interfaces

The individual systems that interface with the fluid transport systems include the mechanical utility systems, bulk gas system, and reagent systems as described in the revised CAR Section 11.9, "Fluid Systems."

The mechanical utility systems that interface with the fluid transport systems include the process chilled water system, the demineralized water system, the process hot water system, the process steam and condensate systems, and the instrument air system. The bulk gas systems that interface with the fluid transport systems include the nitrogen system, the argon/helium system, the helium system, and the oxygen system. The reagent systems that interface with the fluid transport system include the nitric acid system, the silver nitrate system, the TBP system, the hydroxylamine system, the sodium hydroxide system, the oxalic acid system, the diluent system, the sodium carbonate system, the hydrogen peroxide system, the hydrazine system, the manganese nitrate system, the decontamination system, and the nitrogen oxide system.

11.8.1.3 Design Bases of PSSCs

This section describes the design basis commitments made by DCS for the PSSCs of the fluid transport system. The fluid transport system PSSCs identified by the applicant include backflow prevention features, material maintenance and surveillance programs, seismic isolation valves, waste transfer line, and double-walled pipe.

The applicant established an FTS categorization to describe the combination of component and material codes, SCs, and quality levels. These categories are summarized in Table 11.8-1. Design, engineering, construction, and operation of the components classified in each category must conform to applicable industry codes and standards. Fluid transport system PSSCs are categorized as either FTS 1 or 2, while non-PSSCs are either FTS 3 or 4. Tables 11.8-2 through 11.8-5 list the design basis codes and standards that will apply to each of the FTS categories. Design basis parameters for the fluid transport system are presented in Table 11.8-6.

By adhering to industry codes and standards, the design bases of the fluid transport system PSSCs conform to the regulatory acceptance criteria established in Section 11.4.7.2 of NUREG-1718 (NRC, 2000). These criteria are summarized in Section 11.8.1 of this FSER.

The DCS safety assessment, in Section 5.5 of the revised CAR, lists two failure modes related to fluid transport equipment that could lead to a loss of confinement/dispersal of nuclear material events. They are (1) corrosion occurring from within or from the outside of process equipment and (2) leaks of AP process vessels or pipes within process cells. The material maintenance and surveillance programs and the qualifications of the fluid transfer systems to limit corrosion have been identified as PSSCs that reduce the radiological risk to facility workers from corrosion. The safety function of the material maintenance and surveillance programs is to detect and limit the damage resulting from corrosion (principally associated with failures from corrosion occurring outside of process equipment). The safety function of the fluid transfer systems is to limit corrosion through the use of materials compatible with the environment and system fluids (principally associated with failures from corrosion occurring inside process equipment). No PSSCs are required to protect the IOC and site worker from this type of accident because of the low unmitigated consequences. The revised CAR lists the design corrosion allowances for equipment not accessible to inspection, and therefore not available for the material maintenance and surveillance program, which can provide additional protection for the facility worker against the consequences of corrosion.

The process cell has been identified to reduce the risk to facility workers from leaks of AP process vessels or pipes within process cells. The safety function of the process cell is to contain leaks within process cells. The process cell entry controls have also been identified as PSSCs. The safety function of the process cell entry controls is to prevent the entry of personnel to process cells during normal operations and to maintain worker radiation doses within limits while performing maintenance in process cells. No PSSCs are required to protect the IOC and site worker from this type of accident because of the low unmitigated consequences. However, the C3 confinement system provides defense-in-depth protection to the IOC and site worker.

Evaluation of Capacity

The staff evaluated the information provided by DCS in its revised CAR regarding the capacity of the proposed fluid transport systems. The staff notes that fluid transport systems with safety functions are identified as PSSCs. The PSSCs in the fluid transport systems are designed to have adequate capacity to handle the volume of radioactive material during normal operation and design basis accident conditions. The components' design bases are defined from thermal and hydraulic calculations and will consider the physical and chemical properties of the process fluid. The design flow and volumetric capacity design criteria were provided and discussed in Table 11.8-6. The nominal capacity of the process is 24 kg/day (52.9 lb/day). Solvents are recovered from the process as they are generated. Process heat is removed by intermediate cooling loops to the process chilled water system. Process heat is provided by the process steam system through the process hot water system. Both the process heating and cooling systems are sized to provide reliable and sufficient heating and cooling to the facility process systems. Drip trays in process cells are designed to contain radiological fluids, sized to contain the contents from the largest vessel in the cell, and shaped to maintain the leaked fluid in a criticality-safe geometry.

Table 11.8-1 Categorization of the Fluid Transport System

FTS Category	Description of Components in the Category
Category 1	Includes components that are PSSCs that contain process fluids with significant quantities of plutonium or americium. See Table 11.8-2 for applicable codes and standards. The application of the specific criteria for the material, fabrication, examination, testing, and installation were derived from applicable codes and standards and augmented by operating experience at the French La Hague facility.
Category 2	Includes components that are PSSCs that contain process fluids with trace quantities of plutonium or americium or nonradiological fluids. See Table 11.8-3 for applicable codes and standards. Positive material identification, inspection, and test requirements are used in engineering and procurement specifications for Category 2 components.
Category 3	Includes components that are non-PSSCs that may contain process fluids with trace quantities of plutonium or americium or nonradiological fluids that play a significant role for plant production reliability. See Table 11.8-4 for applicable codes and standards.
Category 4	Includes components as well as facility services that maintain production reliability. See Table 11.8-5 for applicable codes and standards.

Table 11.8-2 Design Basis Codes and Standards for Category 1 Fluid Transport System Components

FTS Category 1	Design Basis Codes and Standards
Process Vessels	ASME B&PV Code, Section VIII, 1996 Edition through the 1998 Addenda (ASME, 1998), Division 1 or 2, for lethal service with enhanced positive material identification and test and inspection requirements
Pumps	ASME B73.1, "Specifications for Horizontal End Suction Centrifugal Pumps for Chemical Process," issued 2001 (ASME, 2001b), and ASME B73.2, "Specifications for Vertical In-line Centrifugal Pumps for Chemical Process," issued 2001 (ASME, 2001c), enhanced design specification of ASME materials with enhanced positive material identification and test and inspection requirements Specialty pumps per manufacturer's standards (e.g., submerged rotor seal-less pumps) API Standard 610, "Centrifugal Pumps for Petroleum, Heavy Duty Chemical and Gas Industry Services," issued 1995 (API, 1995b)
Piping	ASME B31.3 (ASME, 2001a) Category M, with enhanced positive material identification, and test and inspection requirements
Valves	ASME B31.3 (ASME, 2001a) Category M
Other Criteria	SC-1 and Quality Level-1 for all PSSCs

Table 11.8-3 Design Basis Codes and Standards for Category 2
Fluid Transport System Components

FTS Category 2	Design basis Codes and Standards
Process Vessels	ASME B&PV Code, Section VIII, 1996 Edition through the 1998 Addenda (ASME, 1998), Division 1 or 2 Compressed Gas Association (CGA) S-1.1, "Pressure Relief Device Standards—Part 1—Cylinders for Compressed Gases," issued 2001 (CGA, 2001a), and CGA S-1.3, "Pressure Relief Device Standards—Part 3—Stationary Storage Containers for Compressed Gases," issued 2001 (CGA, 2001b)
Pumps	ASME B73.1 (ASME, 2001b) and ASME B73.2 (ASME, 2001c) Specialty pumps per manufacturer's standards (e.g., submerged rotor seal-less pumps) API Standard 610 (API, 1995b) NFPA 20, "Standard for the Installation of Stationary Pumps for Fire Protection," issued 2003 (NFPA, 2003)
Piping	ASME B31.3 (ASME, 2001a) with enhanced test and inspection requirements
Valves	ASME B31.3 (ASME, 2001a)
Other Criteria	SC-1 and Quality Level-1 for all PSSCs

Table 11.8-4 Design Basis Codes and Standards for Category 3
Fluid Transport System Components

FTS Category 3	Design basis Codes and Standards
Process Vessels	ASME B&PV Code, Section VIII, 1996 Edition through the 1998 Addenda (ASME, 1998), Division 1 or 2 CGA S-1.1 - (CGA, 2001a) and CGA S-1.3 (CGA, 2001b) UL-142, "Standard for Safety for Steel Aboveground Tanks for Flammable and Combustible Liquids"
Pumps	ASME B73.1 (ASME, 2001b) and ASME B73.2 (ASME, 2001c) and Hydraulic Institute standards Specialty pumps per manufacturer's standards (e.g., submerged rotor seal-less pumps) NFPA 20 (NFPA, 2003)
Piping	ASME B31.3 (ASME, 2001a)
Valves	ASME B31.3 (ASME, 2001a)
Other Criteria	CS or SC-2, as applicable and Quality Level 2, 3, or 4, as applicable

Table 11.8-5 Design Basis Codes and Standards for Category 4 Fluid Transport System Components

FTS Category 4	Design Basis Codes and Standards
Process Vessels	ASME B&PV Code, Section VIII, 1996 Edition through the 1998 Addenda (ASME, 1998), Division 1 or 2 CGA S-1.1 (CGA, 2001a) and CGA S-1.3 (CGA, 2001b) UL-142
Pumps	ASME B73.1 (ASME, 2001b) and ASME B73.2 (ASME, 2001c) and Hydraulic Institute standards Specialty pumps per manufacturer's standards (e.g., submerged rotor seal-less pumps)
Piping	ASME B31.3 (ASME, 2001a)
Valves	ASME B31.3 (ASME, 2001a)
Other Criteria	CS or SC-2, as applicable and Quality Level 2, 3, or 4, as applicable

Normal electrical power is removed from process prime movers during an earthquake event. Those electrical loads requiring power for safe shutdown are supplied by emergency uninterrupted power. (See Section 11.5 of this FSER for additional discussion and the staff evaluation of the facility electrical system.) Based on the commitment to design-bases codes and standards (e.g., Tables 11.8-1 and 11.8-2 of the CAR), the staff has determined that fluid transport systems will provide adequate capacity to handle the expected volume of radioactive material during normal operating and accident conditions.

Evaluation of Redundancy and Diversity

Section 11.4.7.2 of NUREG-1718 (NRC, 2000) states that the applicant is to describe the redundancy and diversity of components required to prevent the release of radioactive materials to the environment or needed for safe operation of the fluid transport system. The fluid transport system is designed with multiple layers of confinement and is supplemented by administrative programs designed to monitor the integrity of these systems. Radiological fluids are maintained within at least two layers of confinement. Piping systems are double walled with leak detection systems if they are not located in process cells or gloveboxes. Drip trays with sump monitors are designed to detect leakage.

[Text removed under 10 CFR 2.390]

Various types of methods are used to transfer fluids, such as pumps, gravity, steam jet lifts, siphons, and airlifts. Hydraulic seals as well as check valves and isolation valves of various types are used to isolate systems. Redundancy and diversity in the design are the result of the various safety factors and types of equipment provided in the design and the layering of active and passive controls that protect the fluid transport system.

Table 11.8-6 Design Bases for the Fluid Transport System

FTS Component	Design Pressure	Design Temperature	Design Flow & Volumetric Capacity
Storage tanks	The highest value of the following: (1) max. pressure—normal operations + 10% (2) max. pressure—transient condition	The highest value of the following: (1) max. temp.—normal operations (2) max. temp.—transient condition	Maximum flow in normal operating conditions + 20% Volumetric capacity for vessels determined on the basis of process safety and operation reliability requirements
Process columns	Same as above	Same as above	Same as above
Heat exchangers	Same as above	Same as above	Same as above
Pumps	Same as above	Same as above	Same as above
Prime movers such as air lifts, ejectors, jets, siphons	Same as above	Same as above	Process hydraulic calculation
Piping and valves	The highest value of the following: (1) max. pressure—normal operations + 10% (2) max. pressure in normal operating condition +0.9 bar (13 psig) (3) max. pressure—transient condition	The highest value of the following: (1) max. temp.—normal operations + 15 °C (27 °F) (2) max. temp.—transient condition	Pipe sizing per line velocity and pressure drop requirements

The fluid transport system is diverse in the control concepts for the process control systems that govern the fluid transport systems. The AP process control systems are designed to ensure that the product of the manufacturing process will conform to the product specifications with minimal waste and risk. They are composed of the normal, protective, and safety control subsystems. The normal control subsystem controls the facility normal manufacturing and processing operations. The protective control subsystem provides protection for personnel and equipment. The safety control subsystem is designed to ensure that safety limits will not be exceeded and that undesired operational conditions or events will not occur or will be mitigated. The staff has reviewed the DCS description of the fluid transport equipment and finds these systems to be diverse. On the basis of standard industry practices, the staff finds the design to be acceptable.

Evaluation of Safe Shutdown

The staff evaluated the information provided by DCS in its revised CAR regarding the ability to safely shut down the proposed fluid transport systems during normal and accident conditions and maintenance. The AP process control system, the electrical power system, and basic system design criteria primarily control the ability of the proposed systems to shut down safely. The AP process instrumentation and control systems are designed to ensure that the product of the manufacturing process will conform to the product specifications with minimal waste and risk. They are composed of the normal, protective, and safety control subsystems. The normal control subsystem controls the facility's normal manufacturing and processing operations. The protective control subsystem provides protection for personnel and equipment. The safety control subsystem is designed to ensure that safety limits will not be exceeded and that undesired operational conditions or events will not occur or will be mitigated. The PSSCs for the I&C systems are the safety subsystems of the MP and AP process control systems and utility control system and the hard-wired emergency control system. The I&C systems monitor and control plant parameters during normal and transient conditions to ensure that limits are not exceeded and to ensure the required quality of the product. They also provide signals to control equipment to prevent and mitigate faulted conditions. All emergency control equipment is qualified for design basis seismic events and normal, off-normal, and design basis accident environmental conditions. (See Section 11.6 of this FSER for additional detail on I&C system safety.)

Normal electrical power is removed from process prime movers during an earthquake event. Those electrical loads requiring power for safe shutdown are supplied by emergency uninterrupted power. (See the discussion of electrical systems safety in Section 11.5 of this FSER for more detail.)

Basic system design criteria are applied to provide the first line of protection against events or hazards posed by process fluids. Radiological fluids are transferred using gravity flow, airlifts, air jets, steam jets, and siphons when practicable. Separator or "knockout pots" are needed in lines in which fluid transfer is made by air or vacuum lift. The separated fluid is allowed to flow by gravity into the desired component while the airflow vents at the top of the pot. This design prevents backflow siphoning. Steam jet lift transfer system piping is terminated in the receiving vessel vent space to provide an air gap that prevents backflow siphoning. Siphons are used to initiate gravity transfer of fluids in applications where flow rate is not critical. The siphon transfers liquid from the higher upstream tank to the lower downstream tank. The elevation difference between tanks prevents backflow.

Hydraulic seals are used to prevent backflow of process fluid to auxiliary systems during reagent addition. The liquid seal or plug is maintained by the piping configuration. The seal is implemented by a "U" bend in piping or by hydraulic seal pots. The hydraulic seal design ensures that the seal remains filled with liquid at all times, that the seal withstands internal pressure differences between connected vessels, and that siphon action does not occur.

Check valves are used only in the process fluid pressure boundary. The check valve design is based on effective pressure drop, type of seating material, pressure and flow reversal response time, mounting requirements, and reliability and maintainability. Redundant isolation valves that are PSSCs are used to automatically isolate utility and reagent fluids in the process area when earthquake conditions are detected. These PSSC isolation valves close in the event of valve or actuator failure. Isolation valve selection is based on process hydraulics, control system characteristics, mounting requirements, and other valve specifications. These valves may be of the butterfly, gate, plug, or ball type. The valves will be specified for service after consideration of the chemical characteristics of the fluid, piping material of construction, and operating conditions. The valves will be designed and constructed according to good engineering practices and in accordance with applicable codes, such as the following:

- ASME B16.10, "Face-to-Face and End-to-End Dimensions of Valves," issued 2000 (ASME, 2000)

- API 598, "Valve Inspection and Testing," issued October 1996 (API, 1996)

- API 600, "Bolted Bonnet Steel Gate Valves for Petroleum and Natural Gas Industries—Modified National Adoption of ISO 10434:1998," issued October 2001 (API, 2001a)

- API 602, "Compact Steel Gate Valves—Flanged, Threaded, Welding, and Extended Body Ends," issued October 1998 (API, 1998)

- API 603, "Corrosion-Resistant, Bolted Bonnet Gate Valves—Flanged and Butt-welding Ends," issued April 2001 (API, 2001b)

- API 608, "Metal Ball Valves—Flanged, Threaded and Butt-welding Ends," issued September 1995 (API, 1995a)

- API 609, "Butterfly Valves: Double Flanged, Lug- and Water-Type," issued May 1997 (API, 1997)

The valves and their supports will also be designed to withstand and remain operable during the design basis earthquake, as they are intended to prevent uncontrolled flooding of the fuel fabrication building as a result of a seismic event. The safety function of the isolation valves is to maintain safe isolation between controlled areas and uncontrolled areas that may contain radioactive materials. Based on the proposed design and use of industry practices, the staff finds that the redundancy and diversity of the fluid transport systems adequately prevent the release of radioactive materials to the environment.

Evaluation of Welded Construction

The regulatory acceptance criteria used for the review of this system specifically mentions that tank and piping systems be of welded construction to the fullest extent possible. For process equipment, radiological fluids are maintained within at least two levels of confinement. Components containing fluids that are located in process cells are specified with corrosion allowances, and welded joints are radiographed, as appropriate. Fluid transfer systems containing hazardous fluids are contained within trenches, rooms, or double-walled piping systems or are accessible for inspection and are of a fully welded construction. Components that are not of fully welded construction are installed in a glovebox. Piping components carrying radiological fluids between two confinements are either of fully welded double-wall construction or are installed in gloveboxes. Welding requirements are contained in the ASME B&PV Code, Section VIII, 1996 Edition through the 1998 Addenda (ASME, 1998), ASME B31.3 (ASME, 2001a), and their referenced AWS codes. Welding work will be performed according to the FTS category and quality assurance plan quality level of the system being welded. Requirements for the qualification of welders and the welding process are specified in both the ASME B&PV Code, Section VIII, Division 1 and 2, 1996 Edition through the 1998 Addenda, and the ASME B31.3 codes. Other general design and technical specifications that will be used for welding are ASME Sections II, V, and IX; ASTM codes; and ANSI/AWS D-10.4 and B-3.0.

Material used for the construction of this equipment is specified in accordance with ASME and ASTM material specifications. ASME materials are used in the fabrication of equipment and piping components built to the requirements of ASME B&PV Code, Section VIII, 1996 Edition through the 1998 Addenda, and ASME B31.3. ASTM materials are also used for the fabrication of other components. In general, design of equipment to these standards means that the components are designed for the most severe service conditions. Included in the severe service conditions are pressure, temperature, stress, material compatibility, and corrosion. The staff has reviewed the DCS design basis for welding and finds it acceptable, based on the information submitted referencing appropriate codes and standards for the design and construction of the fluid transport system.

Evaluation of Passive Features That Address Cross Contamination

The staff evaluated the information provided by DCS in its revised CAR regarding the passive features designed to prevent contamination of the fluid transport systems. The facility design basis for the fluid transport systems contains a significant number of passive features that help to prevent cross contamination.

Fluid transport components are hydraulically designed for the most severe service conditions. Radiological fluids are maintained within at least two levels of confinement. Fluid transfer systems containing hazardous fluids are contained within trenches, rooms, or double-walled piping systems or are accessible for inspection and are of a fully welded construction. All piping components designated as PSSCs are designed to withstand the design basis earthquake loads.

Separator or knockout pots are specified for piping in which fluid transfer is made by air or vacuum lift. The separated fluid is allowed to flow by gravity into the desired component, while the airflow vents at the top of the pot. This design prevents backflow siphoning. Steam jet lift transfer system piping is terminated in the receiving vessel vent space to provide an air gap that prevents backflow siphoning. Siphons are used to initiate gravity transfer of fluids in applications where flow rate is not critical. The siphon transfers liquid from the higher upstream tank to the lower downstream tank. The elevation difference between tanks prevents backflow. Knockout pots, steam jet lifts, and elevation differences between tanks are passive features that help to prevent cross contamination.

Hydraulic seals are used to prevent backflow of process fluid to auxiliary systems during reagent addition. The seal is implemented by a "U" bend in piping or by hydraulic seal pots. The hydraulic seal design ensures that the seal remains filled with liquid at all times, that the seal withstands internal pressure differences between connected vessels, and that siphon action does not occur. Hydraulic seals are additional passive features that help to prevent cross contamination.

Check valves are used only in the process fluid pressure boundary. Redundant isolation valves, which are PSSCs, are used to automatically isolate utility and reagent fluids in the process area when earthquake conditions are detected. The isolation valves close in the event of valve or actuator failure. Check valves in the process fluid pressure boundary and isolation valves that fail to the safe position are examples of passive safety features that help to prevent cross contamination. DCS has committed to the design basis previously discussed. The staff has reviewed the facility design basis for cross contamination and, based on nuclear industry experience, concludes that the design is acceptable.

Evaluation of Radiation Safety

The staff review and evaluation of the Radiation Safety Program are discussed in detail in Chapter 9 of this FSER. In general, the design basis for the fluid transport systems hardware for radiation safety is as follows:

- Process equipment is designed to handle fissile material in accordance with radiation safety principles.

- Stainless steel and other material designed to be compatible with process fluids will be used to prevent corrosion.

- Parts are easily visible and accessible for cleaning, and the material specified has appropriate surface quality or coatings, if needed.

- Process cell drip trays will be designed to contain the contents of the largest vessel in the process cell in a criticality-safe configuration.

- Decontamination will be done by flushing and high-pressure washing to remove sediment buildup or blockages, should they occur.

- The fluid transport systems layout is intended to minimize the potential for entrapment and buildup of radioactive materials.

Based on the DCS commitment to design the system to minimize the entrapment and buildup of radioactive materials (CAR Section 11.8.3.4), the staff finds this design to be acceptable.

Evaluation of Corrosion Resistance

The facility fluid transport systems construction materials are selected based on compatibility with the physical and chemical characteristics of the process fluids. In general, stainless steel of type 304L or 316L and alloys of titanium and zirconium are used for FTS Category 1 components. Components of FTS Category 2 and 3 that handle acidic or alkaline fluids are generally constructed from 304L or 316L stainless steel. Material used for the construction of this equipment is specified in accordance with ASME and ASTM material specifications. ASME materials are used in the fabrication of equipment and piping components built to the requirements of ASME B&PV Code, Section VIII, 1996 Edition through the 1998 Addenda (ASME, 1998), and the ASME B31.3 (ASME, 2001a) code. ASTM materials are also used for the fabrication of other components. In general, design of equipment to these standards means that the components are designed for the most severe service conditions. Included in the severe service conditions are pressure, temperature, stress, and corrosion. Based on the DCS commitment to design the system to be resistant to corrosion, and on the commitment to use material maintenance and surveillance programs to detect and limit the damage from corrosion (as discussed in revised CAR Section 5.5.2.1.6.2), the staff finds this design to be acceptable.

Evaluation of Personnel Protection

NUREG-1718 (NRC, 2000) requires the evaluation of the need for hoods, gloveboxes, and shielding for personnel protection. These items are generally required for wet processing operations involving more than gram quantities of plutonium or general operations involving 50 micrograms or more of plutonium in respirable form. Process cells contain equipment that handles radioactive materials in chemical solutions; this equipment is of a fully welded construction and does not require routine maintenance. Equipment containing radioactive materials in the powder process (MP) is contained in gloveboxes in process rooms that provide equivalent confinement to fully welded equipment in process cells. The staff has reviewed the DCS list of equipment and agrees that the equipment involving more than gram quantities of plutonium or general operations involving 50 micrograms or more of plutonium in respirable form are properly contained in either gloveboxes or fully welded process equipment. Therefore, based on the facility design following the guidelines of NUREG-1718, the staff finds this design basis to be acceptable.

Evaluation of Functionality During Severe Natural Phenomena

Fluid transport systems designated as PSSCs are designed and qualified according to national codes and standards enabling them to perform their safety function during normal operations, upset conditions, and design basis events. These codes and standards are delineated in Tables 11.8-1, 11.8-2, and 11.8-3 of this FSER. The seismic monitoring system is designed to satisfy the criteria provided in RG 3.17-1974 (NRC, 1974). The design basis of the seismic monitoring system is that it provides sufficient data to evaluate the response of the confinement structure and other PSSCs to a seismic event and initiate a shutdown of process systems in the event of a high seismic event. The seismic system will meet the requirements of IEEE Standard 603, "IEEE Standard Criteria for Safety Systems for Nuclear Power Generating Stations," issued 1998 (IEEE, 1998).

The SQ requirements are applied using a graded classification program that considers the relative importance of the safety function and the structural behavior of the PSSC. (See the discussion of quality assurance classes and grading in Section 15.1 of this FSER for details on implementation.) This classification system is defined in RG 3.14, "Seismic Design Classification for Plutonium Processing and Fuel Fabrication Plants," issued 1973 (NRC, 1973). The SQ of MFFF mechanical equipment will meet the guidance of RG 1.100, Revision 2 (NRC, 1998).

The SC-I classification is applicable to facility SSCs and the supporting SSCs that are required to withstand the effects of an earthquake and remain functional to the extent that they will prevent the uncontrolled release of radioactive materials. The SC-I classification applies to all PSSCs that must perform a safety function during and/or after the design basis earthquake to comply with the facility safety assessment as discussed in Chapter 5 of this FSER. The SC-II designation applies to the portions of systems whose continued function is not required but whose failure could reduce the functioning of any plant feature of an SC-I SSC. Items that are neither SC-I or SC-II are not classified with respect to SC. Components that form an interface between SC-I and non-SC-I components are to be classified as SC-I. The quality assurance (QA) plan will apply to these components as previously discussed. Based on the applicant's commitments to design and qualify fluid transport system PSSCs to perform their safety functions as discussed in this section, the staff finds this design basis to be acceptable.

11.8.2 Evaluation Findings

In Section 11.8 of the revised CAR, DCS provided design basis information for the fluid transport systems that it identified as PSSCs for the proposed facility. Based on the staff's review of the revised CAR and supporting information provided by the applicant relevant to the fluid transport systems, the staff concludes, pursuant to 10 CFR 70.23(b), that the design bases of the PSSCs evaluated in this FSER section will provide reasonable assurance of protection against natural phenomena and the consequences of potential accidents.

11.8.3 References

(AISC, 1994) American National Standards Institute/American Institute of Steel Construction. N690, "Specification for the Design, Fabrication, and Erection of Safety Related Steel Structures for Nuclear Facilities." 1994.

(API, 1995a) American National Standards Institute/American Petroleum Institute. Standard 608, "Metal Ball Valves—Flanged, Threaded and Butt-welding Ends." September 1995.

(API, 1995b) American National Standards Institute/American Petroleum Institute. Standard 610, "Centrifugal Pumps for Petroleum, Heavy Duty Chemical and Gas Industry Services." August 1995.

(API, 1996) American National Standards Institute/American Petroleum Institute. Standard 598, "Valve Inspection and Testing." October 1996.

(API, 1997) American National Standards Institute/American Petroleum Institute. API Standard 609, "Butterfly Valves: Double Flanged, Lug- and Wafer-Type." May 1997.

(API, 1998) American National Standards Institute/American Petroleum Institute. Standard 602, "Compact Steel Gate Valves—Flanged, Threaded, Welding, and Extended Body Ends." October 1998.

(API, 2001a) American National Standards Institute/American Petroleum Institute. Standard 600, "Bolted Bonnet Steel Gate Valves for Petroleum and Natural Gas Industries— Modified National Adoption of ISO 10434:1998." October 2001.

(API, 2001b) American National Standards Institute/American Petroleum Institute. Standard 603, "Corrosion-Resistant, Bolted Bonnet Gate Valves—Flanged and Butt-welding Ends." April 2001.

(ASME, 1998) American Society of Mechanical Engineers Boiler and Pressure Vessel Code (B&PV Code). Section VIII, "Design and Fabrication of Pressure Vessels." 1996 Edition through the 1998 Addenda.

(ASME, 2000) American Society of Mechanical Engineers. ASME B16.10, "Face-to-Face and End-to-End Dimensions of Valves." 2000.

(ASME, 2001a) American Society of Mechanical Engineers. ASME B31.3, "Process Piping." 2001.

(ASME, 2001b) American Society of Mechanical Engineers. ASME B73.1, "Specifications for Horizontal End Suction Centrifugal Pumps for Chemical Process." 2001.

(ASME, 2001c) American Society of Mechanical Engineers. ASME B73.2, "Specifications for Vertical In-line Centrifugal Pumps for Chemical Process." 2001.

(CGA, 2001a) Compressed Gas Association. Standard S-1.1, "Pressure Relief Device Standards, Part 1—Cylinders for Compressed Gases." 2001.

(CGA, 2001b) Compressed Gas Association. Standard S-1.3, "Pressure Relief Device Standards, Part 3—Stationary Storage Containers for Compressed Gases." 2001.

(DCS, 2002a) Ihde, R., Duke Cogema Stone & Webster. Letter to Document Control Desk, U.S. Nuclear Regulatory Commission, RE: Mixed Oxide Fuel Fabrication Facility—Construction Authorization Request. October 31, 2002.

(DCS, 2002b) Hastings, P., Duke Cogema Stone & Webster. Letter to Document Control Desk, U.S. Nuclear Regulatory Commission, RE: Clarification of Responses to NRC Request for Additional Information. February 11, 2002.

(IEEE, 1987) Institute of Electrical and Electronics Engineers. Standard 344, "IEEE Recommended Practice for Seismic Qualification of Class 1E Equipment for Nuclear Power Generating Stations." 1987.

(IEEE, 1998) Institute of Electrical and Electronics Engineers. Standard 603, "IEEE Standard Criteria for Safety Systems for Nuclear Power Generating Stations." 1998.

(NFPA, 2003) National Fire Protection Association. Standard 20, "Standard for the Installation of Stationary Pumps for Fire Protection." 2003.

(NRC, 1973) U.S. Nuclear Regulatory Commission. Regulatory Guide 3.14, "Seismic Design Classification for Plutonium Processing and Fuel Fabrication Plants." Washington, DC, 1973.

(NRC, 1974) U.S. Nuclear Regulatory Commission. Regulatory Guide 3.17, "Earthquake Instrumentation for Fuel Reprocessing Plants." Washington, DC, 1974.

(NRC, 1998) U.S. Nuclear Regulatory Commission. Regulatory Guide 1.100, Rev. 2, "Seismic Qualification of Electric and Mechanical Equipment for Nuclear Power Plants." Washington, DC, 1988.

(NRC, 2000) U.S. Nuclear Regulatory Commission. NUREG-1718, "Standard Review Plan for the Review of an Application for a Mixed Oxide (MOX) Fuel Fabrication Facility." Washington, DC, August 2000.

11.9 Fluid Systems (Bulk Materials, Reagents, and Gases)

11.9.1 Conduct of Review

This chapter of the FSER contains the staff's review of the fluid systems described by the applicant in Chapter 11.0 of the revised CAR (DCS, 2002a). The objective of this review is to determine whether the fluid systems' PSSCs and their design bases identified by the applicant provide reasonable assurance of protection against natural phenomena and the consequences of potential accidents. The staff evaluated the information provided by the applicant for fluid systems by reviewing Chapter 11.0 of the revised CAR, other sections of the revised CAR, supplementary information provided by the applicant, and relevant documents available at the applicant's offices but not submitted by the applicant. The review of fluid systems' design bases and strategies was closely coordinated with the review of other chapters in this FSER (fire protection in Chapter 7, chemical safety in Chapter 8, and safety assessment of the design bases in Chapter 5).

The staff reviewed how the information in the revised CAR addresses the following regulation:

- 10 CFR 70.23(b) requires that the design bases of the PSSCs and the Quality Assurance Program provide reasonable assurance of protection against natural phenomena and the consequences of potential accidents before construction of the PSSCs is approved.

The review for this construction approval focused on the design bases of fluid systems, their components, and other related information. For fluid systems, the staff reviewed and evaluated the information provided by the applicant for the safety function, system description, and safety analysis. The review also encompassed proposed design basis considerations such as redundancy, independence, reliability, and quality. The staff used Chapter 11.0 in NUREG-1718 (NRC, 2000) as guidance in performing the review.

The discussions that follow provide the system descriptions, as well as function, major components, control concepts, and system interfaces. These discussions include but are not limited to PSSCs to provide an understanding of the system. Design bases of the PSSCs are provided in Section 11.9.1.2 of this FSER.

The NRC staff reviewed the revised CAR submitted by DCS for the following areas applicable to the fluid systems at the construction approval stage and consistent with the level of design:

- system description
- system function
- major components
- control concepts
- system interfaces
- design bases

Regarding the MFFF fluid systems, specific design considerations given in the CAR are to demonstrate adherence to regulations, and the acceptance criteria from Section 11.4.4.2 of NUREG-1718, set forth below:

- transfer of heat loads to an appropriate heat sink under normal, off-normal, and accident conditions

- adequate water supply under normal, off-normal, and accident conditions

- adequate component redundancy; the capability to isolate components, systems, or piping for maintaining system safety function under varying system configuration; and the capability of integrated system control

- the applicant's commitment to supporting management measures (including tests and other verification methods) that ensure the structural integrity and system leak-tightness (including the prevention of cross contamination (radioactive and chemical)), the operability and adequate performance of active system components, and the capability of the system to perform required functions during normal and accident situations

- capability for withstanding environmental hazards resulting from pipeline breaks and dynamic effects associated with flow instability and attendant loads and measures to prevent such dynamic conditions from occurring

- capacity and capability for detecting leaks and cross contamination (radioactive and chemical), for inservice component inspection and system maintenance, and for operational functional testing of the system and its components

- system capability to maintain functionality when subject to tornadoes, tornado missiles, earthquakes, floods, and any other appropriate severe natural phenomena

11.9.1.1 Description (Bulk Material, Reagent, and Gas Systems)

The facility fluid systems consist of bulk material, reagent, and gas systems designed to support the production function of the facility. DCS has identified the seismic isolation valves, EDGFOS, and the emergency scavenging air system as PSSC fluid systems. Non-PSSC systems may contain PSSC components, such as seismic isolation valves. Because these non-PSSC systems contain and could interact with PSSCs, they are summarized but not evaluated in this section. Only PSSC systems are evaluated. The fluid systems are divided into three generalized categories—mechanical utility systems, reagent systems, and bulk gas systems.

Mechanical Utility Systems

The mechanical utility systems consist of the HVAC chilled water system, process chilled water system, demineralized water system, process hot water system, process steam and condensate systems, plant water system, the EDGFOS, standby diesel generator fuel oil system, service air system, instrument air system (which includes the emergency scavenging air subsystem), breathing air system, and the radiation monitoring vacuum system. Besides those systems mentioned at the beginning of this section, most of these systems are non-PSSCs but contain components, such as seismic isolation valves, that are PSSCs.

The HVAC chilled water system consists of an external loop to supply chilled water to the HVAC supply air system fan coils and an internal cooling loop to provide cooling water to individual area fan coils. The primary ventilation cooling coils are located upstream of the building intake

HEPA filters, thereby placing them outside of the building radiological boundary. Placing them in this location allows the condensate formed to be discharged directly. The process chilled water system consists of an external cooling loop that provides cooling water to multiple intermediate heat exchangers. In-leakage to the system will be monitored for contamination by radiation detectors located on a continuous bypass flowpath in the common chilled water return line of each internal cooling loop. The demineralized water system receives, stores, and transfers pressurized and gravity-fed demineralized water to process equipment and utility systems. The process hot water system is a closed-loop circulating system that supplies hot water to process equipment in the solvent recovery and purification cycles and the oxalic precipitation and oxidation units. The process steam system transfers and regulates the primary steam supplied by SRS and the acid recovery unit secondary steam to the AP area. The process condensate system collects and transfers primary condensate and acid recovery unit secondary condensate. Radiation detection is included in the design for the secondary steam/condensate closed loop systems. The plant water system consists of a supply header that supplies industrial-grade water from SRS to utility, MP, and AP users. The standby diesel generator fuel oil system supplies fuel oil to the standby generators that are used to provide an onsite power source for major electrical loads. The service air system supplies pressurized air to the facility service air headers for maintenance and utility use and makeup to the instrument air system. The breathing air system is an independent air supply system that is used for personnel breathing in an emergency. The radiation monitoring vacuum system consists of multiple, parallel vacuum pumps, a common header, and associated piping connected to parallel continuous air monitors and samplers designed to evaluate airborne radioactivity in the facility.

The HVAC chilled water system, process chilled water, demineralized water, process steam and condensate systems, and plant water system contain seismic isolation valves, designated as PSSCs, that close during a seismic event to prevent flooding in the facility.

The EDGFOS is a PSSC that supports the EDGs. This system receives, samples, stores, and supplies fuel oil to the EDGs. The EDGFOS consists of two independent fuel oil subsystems, each capable of fuel receipt, long-term storage, and transferring an adequate supply of fuel oil to its associated EDG operating at 100-percent load continuously for 7 days. The tanks will be protected in a below-grade concrete vault adjacent to the EDG building. Each supply line has its own storage tank, transfer pump, day tank, purification system, strainers, filters, instrumentation and control elements, and piping. The supply lines will be installed in a protective enclosure located adjacent to the EDG. The EDGFOS will supply No. 2 fuel oil to the day tanks that are sized to store 2500 L (660 gal) each. The EDG exhaust system is a PSSC and is made up of piping and silencers supplied in accordance with the manufacturer's general design and supplied as part of the vendor-supplied skid-mounted generator set. The function of the EDGFOS is to receive, store, and transfer fuel oil to the EDGs used to provide emergency power to PSSCs in the facility. The function of the EDG exhaust system is to remove fumes and dampen noise produced by the EDGs. The major components of the EDGFOS are the fuel oil storage tank, the fuel oil day tanks, the fuel oil transfer pumps, isolation valves, and the EDG exhaust system. The EDGFOS employs fuel oil day tank level control and various level instrumentation and alarms. The EDGFOS interfaces with the EDGs and the emergency electrical power supply.

The instrument air system is a non-PSSC and is designed to supply instrument air to the instruments, air-operated pneumatic valves and HVAC dampers, scavenging air for plutonium vessel level measurement and equipment atmosphere scavenging, cooling air for gloveboxes and pelletizing press bellows, and nitrogen system backup. Instrument air is a dried and filtered air specifically designed for instrumentation, valves, and emergency scavenging air; the emergency scavenging air subsystem is a PSSC. Instrument air is further dried for use as bubbling air for scavenging AP gloveboxes and miscellaneous equipment. A portion of the instrument air will be further dried to get super-dry process air for various instruments and for air being supplied directly to the AP process stream and in glovebox applications. The instrument air system will be designed to meet or exceed the standards of ANSI/ISA S7.0.01-1996, "Quality Standard for Instrument Air."

The facility seismic isolation valves, which are PSSCs, automatically isolate during a seismic event. This system interfaces with the service air system, very high depressurization exhaust system, I&C system, and seismic detectors. The applicant has considered related industry experience in developing the design basis of non-safety-related air systems that support safety-related equipment at the facility.

Reagent Systems

The reagent systems consist of the nitric acid system, the silver nitrate system, TBP system, HAN system, sodium hydroxide system, oxalic acid system, diluent system, sodium carbonate system, hydrogen peroxide system, hydrazine system, manganese nitrate system, decontamination system, nitrogen oxide system, aluminum nitrate system, zirconium nitrate system, and uranyl nitrate system. In general, these systems are not PSSCs but contain components, such as isolation valves, that are PSSCs.

The nitric acid system provides 13.6 N nitric acid to the AP process and for the preparation of hydrazine, oxalic acid, manganese nitrate, and silver nitrate reagents. The acid is stored in tote tanks in the reagent processing building. The system includes isolation valves, designated PSSCs, that automatically isolate during a seismic event.

The silver nitrate system provides silver nitrate to the electrolyzers in the dissolution unit. Silver nitrate, demineralized water, and 13.6 N nitric acid are mixed in a preparation tank in the reagent processing area and pumped to a distribution tank in the AP area.

The TBP system provides TBP for solvent extraction in the purification cycle of the AP process and solvent washing in the solvent recovery cycle. The system includes isolation valves, designated PSSCs, that automatically isolate during a seismic event.

The HAN system supports the stripping of plutonium in the AP or purification process. The system includes isolation valves, designated PSSCs, that automatically isolate during a seismic event. The HAN system interfaces with the normal power system, seismic detectors, and the purification cycle.

The sodium hydroxide system makes the sodium hydroxide soda for washing in the solvent recovery cycle. The sodium hydroxide system interfaces with the solvent recovery cycle and the normal power supply.

The oxalic acid system provides oxalic acid for converting plutonium nitrate to plutonium oxalate in the oxalic precipitation and oxidation unit. The oxalic acid system design includes isolation valves, designated PSSCs, that automatically isolate during a seismic event. The oxalic acid system interfaces with the normal power system, seismic detectors, nitric acid system, demineralized water system, and oxalic precipitation and oxidation unit.

The diluent system supplies diluent (branched dodecane) for washing in the purification and solvent recovery cycles and diluent for preparation of the 30-percent TBP solvent solution. This system interfaces with the purification cycle, solvent recovery cycle, TBP system, and normal power system. The diluent system design includes isolation valves, designated PSSCs, that automatically isolate during a seismic event.

The sodium carbonate system is used in the solvent recovery cycle to wash and adjust the pH in the liquid waste reception unit in the AP process. Isolation valves, designated PSSCs, automatically isolate during a seismic event. The sodium carbonate system interfaces with the normal power system, seismic detectors, liquid waste reception unit, demineralized water system, and solvent recovery cycle.

The hydrogen peroxide system is designed to provide hydrogen peroxide for valence adjustment of the dissolution solution in the dissolution unit. Isolation valves, designated PSSCs, automatically isolate during a seismic event. The hydrogen peroxide system interfaces with the normal power system, seismic detectors, and the silver recovery unit.

The hydrazine system supplies hydrazine hydrate to be used in the preparation of hydrazine nitrate. The hydrazine nitrate is mixed with HAN and used in the purification cycle of the AP process. Tanks are vented to a hydrogen peroxide scrubber system before release to the atmosphere. All components of this system are located in the reagent processing building and do not require seismic isolation from the material or AP buildings.

The manganese nitrate system supplies manganese nitrate to the oxalic mother liquor recovery unit. The manganese nitrate is used as a catalyst. The manganese nitrate system interfaces with the normal power supply, oxalic mother liquor recovery unit, and the nitric acid system.

The decontamination system supplies decontamination solution to the AP process. The decontamination system interfaces with the nitric acid system, the demineralized water system, the oxalic mother liquor recovery unit, the normal power supply, purification cycle, solvent recovery cycle, homogenization unit, acid recovery unit, silver recovery unit, liquid waste reception unit, dissolution unit, and offgas treatment unit.

The nitrogen oxide system supplies nitrous fumes (nitrogen dioxide and dinitrogen tetroxide) to the AP process to assist in the oxidation of plutonium nitrate in the purification cycle. Nitrogen oxide detectors will be installed in rooms where fumes could be present. Isolation valves, designated PSSCs, automatically isolate during a seismic event. The system interfaces with the purification cycle, normal power supply, instrument air system, and seismic detectors.

The aluminum nitrate system supplies aluminum nitrate to the laboratory where it is mixed with 1.5 N nitric acid. The mixture is added to the aluminum nitrate buffer tank and transferred by dosing pump to a purification cycle seal pot. The aluminum nitrate system interfaces with the normal power supply, nitric acid system, and purification cycle.

The zirconium nitrate system supplies zirconium nitrate to avoid fluoride corrosion of titanium vessels in the purification cycle and acid recovery unit. The zirconium nitrate penetrations into the MFFF are provided with double isolation valves to automatically isolate during a seismic event.

The uranyl nitrate system supplies depleted uranyl nitrate to the plutonium feed stream and uranium waste stream for reducing the isotopic composition of uranium. This system interfaces with the nitric acid system, demineralized water system, dechlorination dissolution and dissolution units, purification unit, normal power supply, and decontamination system.

Bulk Gas Systems

The bulk gas systems consist of the nitrogen system, the argon/hydrogen system, helium system, oxygen system, and methane/argon (P10) system. In general, these systems are not PSSCs but contain components, such as seismic isolation valves, that are PSSCs.

The nitrogen system is a non-PSSC system suppled by an onsite nitrogen production system. Liquid nitrogen is stored in a tank in the MP area to supply germanium to the operating and reserve detectors. Nitrogen gas is also used for MP area glovebox inerting to eliminate the adverse effects of atmospheric oxygen on the process or fuel, and also for scavenging the sintering furnace airlock. The sintering furnace airlock transfers molybdenum boats with sintered pellets from an argon/hydrogen atmosphere, and the nitrogen flow in the airlock is intended to eliminate the argon/hydrogen atmosphere before returning to the MP gloveboxes. The graphite rear bearing of the calcination furnace is cooled and the atmosphere inerted with nitrogen to lengthen the life of the bearing. DCS stated that this bearing provides a containment function. Nitrogen gas is also used to scavenge the hydrazine and hydroxylamine tanks and is designed to eliminate the concern for the flammability of hydrazine vapors in these tanks. A liquid nitrogen storage tank and ambient air vaporizer, designed to supply at least 2 days supply of nitrogen, provide backup for the onsite nitrogen production system. The nitrogen system is designed to operate at ambient temperature, at a pressure of 300–790 kPa (3–7.9 bar) and a flow rate of 495–561 m^3/h (17,500–19,800 ft^3/h). The estimated annual usage of nitrogen is 103,900 m^3 (3.67 million ft^3). The nitrogen system must supply gas to the gloveboxes to meet the required oxygen impurity level inside the glovebox of less than 2000 ppm. Two nitrogen buffer tanks are included in the design with 34.2 m^3 (1209 ft^3) and 0.31 m^3 (11 ft^3) capacities. The backup nitrogen storage tank is 34 m^3 (9000 gal) capacity. In CAR Section 11.9.2.1.1, DCS stated that none of the functions of the nitrogen system are credited in the facility safety analyses, and the system is therefore not designated as a PSSC.

The argon/hydrogen system is a non-PSSC system that provides argon/hydrogen mixture to an operating electric sintering furnace, to furnace airlocks, and to the laboratory. The system will be designed to NFPA 50A, "Standard for Gaseous Hydrogen Systems at Consumer Sites." The argon/hydrogen system is designed to mix the gases to make a 95:5 mixture (a proprietary range is given) used in scavenging the operating electric sintering furnace and airlocks in the MP area. It is also used in the laboratory. An emergency system, made of argon/hydrogen in compressed cylinders at 4.3 to 8 percent hydrogen, serves as backup to the normal supply system. This emergency supply of pre-mixed argon and hydrogen is stored in a tube trailer with a capacity of 1500 m^3 (53,000 ft^3) for a 24-hour backup supply. Pure argon provides secondary backup and pure nitrogen provides tertiary backup to the argon/hydrogen system. The argon is supplied from two argon vaporizing packages sourced from two 11 m^3 (3000 gal) bulk liquid argon storage tanks, for a 2-week supply. Hydrogen gas is supplied from a tube trailer holding 1200 m^3 (43,000 ft^3) located in the facility gas storage area. The gas is supplied at 20,000 kPa

(200 bar) and pressure-reduced in two stations to normal operating pressure before being mixed with argon. The mixture of the two gases is controlled by PSSC proportioning control valves and a PSSC static mixer. The hydrogen concentration in the buffer tank is continuously monitored by redundant detectors, identified as PSSCs, to ensure that the lower explosion limit is not exceeded. The argon/hydrogen mixture is supplied to the operating furnace at 200 kPa (2 bar). The argon/hydrogen system is designed to operate at ambient temperature, at a pressure of 20,000 kPa (200 bar) and a flow rate of 68 m³/h (2400 ft³/h). The estimated annual usage of argon in the sintering furnaces and laboratory is 462,000 m³ (16.3 million ft³), and the estimated annual usage of hydrogen in the sintering furnace is 11,400 m³ (403,000 ft³). The major components of the argon/hydrogen system are the liquid argon bulk storage system, the hydrogen tube trailer, the backup argon/hydrogen cylinders, inline mixing stations, and the buffer tanks, pressure transmitters, pressure control valves, alarms, piping, and isolation valves. The argon/hydrogen system is controlled by argon and hydrogen levels after mixing, automatically switching from normal supply to backup supply on low pressure. Gas flow is maintained by flow control valves. Pressure control valves regulate gas pressure to the sintering furnaces and laboratory, redundant hydrogen monitors, and the facility seismic isolation valves, designated as PSSCs, that actuate in the event of a seismic event. The argon/hydrogen system interfaces with the sintering furnaces, normal power supply, standby power supply, laboratory, and facility seismic detectors.

The helium system is supplied from tube trailers located in the gas storage area. The safety functions of this system are to isolate during a seismic event and prevent backflow into the system. The major components of the helium system are the tube trailer, pressure regulator, pressure switches, alarms, relief valves, isolation valves, piping, and switchover circuit. Penetrations into the facility have seismic isolation valves, designated as PSSCs, that automatically isolate the system during a seismic event. The helium system interfaces with the rod cladding and decontamination unit, rod welding scavenging, laboratory, and seismic detectors.

The oxygen system provides oxygen from cylinders, stored in the north end of the loading dock at the reagent processing building, to the calcination furnace in the homogenization unit in the AP area and to the laboratory in the MP area. The major components of this system are the cylinders, pressure regulators, pressure switches, alarms, three-way valve, relief valves, isolation valves, piping, and instrumentation. The safety functions of this system are to isolate during a seismic event and prevent backflow into the system. The oxygen system interfaces with the calcination furnace, the MP area laboratory, and seismic detectors that actuate the seismic isolation valves, designated as PSSCs, during seismic events.

The methane/argon (P10) system is composed of 10 percent methane and 90 percent argon. DCS states that the system is supplied by one tube trailer holding a 6-week supply (1300 m³ (45,000 ft³) of argon-methane gas). The gas is found in the gas storage area and the laboratory.

11.9.1.2 Design Bases of the PSSCs

This section lists the referenced design basis codes and standards for proposed facility fluid systems identified as PSSCs.

Design Bases for PSSCs

Vessels/components are segregated/separated from incompatible chemicals. The chemical makeup of the reagent introduced into cells or AP reagent rooms is controlled to prevent explosions caused by chemical reactions. All piping, chemicals, and equipment will be clearly marked to prevent errors. Pressure vessels will be located away from PSSCs or otherwise protected so that a failure of any vessel will have no impact on the ability of the PSSC to perform its safety function.

Design Basis for Seismic Isolation Valves

Piping penetrations in the facility process building walls that could pose a fire, explosion, confinement, or flooding risk, or that could leak radioactive materials into the environment, will be isolated by redundant, automatic, seismically qualified isolation valves. The redundant isolation valves are designed to receive isolation signals from redundant seismic monitors. An isolation will occur when a signal is sent to the pneumatic operator of the isolation valve to vent. Valve position indicators will be included in the design of the process and utility control room. The seismic monitoring system is designed to satisfy the criteria provided in RG 3.17-1974 (NRC, 1974). The design basis of the seismic monitoring system is that it provides sufficient data to evaluate the response of the confinement structure and other PSSCs to a seismic event and initiate a shutdown of process systems in the event of a high seismic event. The seismic system will meet the requirements of IEEE Standard 603 (IEEE, 1998). The SQ of these valves will be done in accordance with IEEE Standard 344 (IEEE, 1987). (See discussion of the acceptability of this standard in Section 11.5 of this FSER and NRC RG 1.100, Revision 2 (NRC, 1988).) These valves have been identified as PSSCs and will be handled according to the facility Quality Assurance Program quality level appropriate to each application of an isolation valve.

The seismic isolation valves can be of the butterfly, gate, plug, or ball type. The design is determined based on the chemical characteristics of the fluid, the construction material for associated piping, and operating conditions. The valve design will use good engineering practices and will be in accordance with applicable code practices. The valves will be supported in such a way as to remain operable during design basis events.

Instrument air quality is very important to maintaining the reliability of pneumatic isolation valves. DCS has committed to design, operate, and maintain the instrument air system to the standard given in ANSI/ISA S7.0.01-1996 or better.

Evaluation of Seismic Isolation Valves

During this evaluation, the staff reviewed applicable Federal regulations, national codes and standards, the MOX SRP, other industry and NRC staff guidance, and available operational history. The staff's consideration and use of these documents are discussed in the following evaluation.

As discussed previously, piping penetrations in the facility process building walls that could pose a fire, explosion, confinement, or flooding risk, or that could leak radioactive materials into the environment, will be isolated by redundant, automatic, seismically qualified isolation valves. These valves have been designated as PSSCs. They will receive signals to isolate from redundant seismic monitors. The valves will be ordered, built, delivered, installed, and used in accordance with the facility Quality Assurance Program quality level appropriate to each application.

The staff evaluation of the design basis for the seismic isolation valves began with a review of the DCS proposed design basis against national codes and standards and industry practices. The staff also applied the guidance of the MOX SRP to the review.

The isolation of fluid systems from the MFFF building is achieved by redundant seismic isolation valves that are designed to fail to a closed position. An isolation will occur when a signal is sent to the pneumatic operator of the isolation valve to vent. The venting of the air from the operator removes the power from the system, resulting in the operator moving to the closed position. Valve position indicators will be included in the design of the process and utility control room. The SQ of these valves will be done in accordance with IEEE Standard 344 (IEEE, 1987). (See discussion of the acceptability of this standard in Section 11.5 of this SER and NRC RG 1.100, Revision 2 (NRC, 1988). DCS has committed, in its mechanical equipment SQ program (CAR Section 11.12), to consider attached piping loads, thermal loads, and live loads such as tank sloshing in its design. The applicant has committed to using standards for the SQ of isolation valves which are commensurate with standards applicable to commercial nuclear power stations. The staff finds these commitments acceptable.

The isolation valves will be designed to passively return to a closed, or isolated, position in the event of a failure of the valve actuator or the air supply, or a loss of power. Based on the varying sizes, designs, materials of construction, methods of operation, and fail-safe design, the staff finds that the seismic isolation valves have the redundancy and diversity of components required for isolation and to prevent release of radioactive material to the environment.

DCS has committed to design fluid systems, including the seismic isolation valves and other components (CAR Section 11.8.7), with consideration to the process fluids contained therein and will specify materials with the appropriate corrosion resistance. The applicant will also consider proper surface finishes and decontamination characteristics for their particular application in a system. Based on these considerations and their alignment with the requirements of the ASME B&PV Code, Section VIII (ASME, 1995), for materials and design methodology, the staff finds this design basis to be acceptable.

The seismic isolation valves will be designed and qualified to maintain functionality when subjected to severe natural phenomena such as tornadoes, tornado missiles, earthquakes, floods, and any other appropriate phenomena as established in the ISA. The seismic isolation valves will actuate on a seismic event that is one-third of the design basis earthquake ground motion, that is one-third of the safe-shutdown earthquake ground motion. This design basis meets the requirements for nuclear power generation facilities given in Section IV(a)(2)(i)(A) of Appendix S, "Earthquake Engineering Criteria for Nuclear Power Plants," to 10 CFR Part 50, "Domestic Licensing of Production and Utilization Facilities." The staff finds this design basis to be acceptable, since the facility design for actuation of the seismic isolation valves is based on the ground motion found to be acceptable at nuclear power generation facilities.

The seismic monitoring system is designed to satisfy the criteria provided in RG 3.17-1974. The design basis of the seismic monitoring system is that it provides sufficient data to evaluate the response of the confinement structure and other PSSCs to a seismic event and initiate a shutdown of process systems in the event of a high seismic event. Because the seismic system will meet the requirements of IEEE Standard 603 (IEEE, 1998), the staff finds this design basis to be acceptable.

Instrument air quality is important to maintaining the reliability of pneumatic isolation valves. DCS has committed to design, operate, and maintain the instrument air system to the standard given in ANSI/ISA S7.0.01-1996 (as described in CAR Section 11.9.1.10). In general, the staff agrees with the use of the instrument air system as the motive force for the seismic isolation valves. On the basis of the applicant's commitment to ANSI/ISA S7.0.01-1996, and the applicant's consideration of relevant industry experience with nonsafety-related air systems (discussed in Section 11.9.1.1 of this FSER), the staff finds this design basis to be acceptable.

<u>Design Basis of the Emergency Diesel Generator Fuel Oil System</u>

Both the main emergency fuel oil supply vault and the EDG building are designed to withstand hazardous natural phenomena such as earthquakes, floods, and tornadoes. The EDG piping and equipment are seismically qualified to prevent failures from causing a loss of emergency electrical power supply. The SQ will be performed in accordance with IEEE Standard 344-1987 (IEEE, 1987) (see discussion of the acceptability of this standard in Section 11.5 of this FSER, and NRC RG 1.100, Revision 2 (NRC, 1988).) Each EDGFOS has a bulk tank that is sized for 7 days of use (plus extra operational margin for testing). The storage capacity was sized for an EDG running at 100- percent power for 7 days plus some reserve capacity. The EDG design, with redundant transfer pumps, fuel oil lines, associated piping, and exhaust equipment, is intended to prevent a single failure from causing a loss of the emergency electrical power supply. This equipment has been identified as a PSSC and will be handled according to the facility Quality Assurance Program quality level appropriate to the importance of each piece of equipment. The equipment will comply with the requirements of IEEE Standard 308-1991 (IEEE, 1991); ANS 59-51-1997, "American Nuclear Society Fuel Oil Systems for Safety Related Emergency Diesel Generators"; and ASTM D75-94, "Standard Specification for Diesel Fuel Oils," issued 1974 (ASTM, 1974). The EDG exhaust system function is to collect and remove exhaust efficiently and silently. The engine silencer will be capable of attenuating noise from a running EDG to acceptable Occupational Safety and Health Administration (OSHA) levels for worker hearing protection in an industrial environment. The EDG exhaust system will be included in the generator set from the vendor as a packaged unit and will comply with NFPA-37, "Standard for the Installation and Use of Stationary Combustion Engines and Gas Turbines," issued 1998 (NFPA, 1998a), and NFPA 110, "Standard for Emergency and Standby Power Systems," issued 1999 (NFPA, 1999).

Fuel filtration or conditioning is proposed at five places in the EDGFOS, including (1) on receipt of new fuel from the remote filling station, (2) at the transfer pump suction, (3) in a bypass flow purification system, (4) between the day tank and the engine, and (5) at the engine itself. The filter on the remote filling station for the main tank will be a multi-element, differential pressure instrumented, inlet filter rated at 2 μm (0.08 mils). The basis for this filter is to prevent coagulated fuel and particulates from entering the storage tank assuming a low-pressure input flow of 5.0–6.3 L/s (80–100 gpm). The transfer system will use a rotary screw positive displacement pump to filter and circulate fuel oil from a main storage tank to the associated day tank. The transfer system filters will be differential pressure instrumented, dual basket switchable stainless steel suction strainer with automatic transfer and mesh filters rated to remove 100 percent of 1.8 mils (45 μm) particles and 98 percent of particles at 0.7 mils (17 μm).

The flow rate for the transfer pump is approximately 1.6 L/s (25 gpm). The bypass flow purification system will redirect approximately 0.9–1.3 L/s (15–20 gpm) of fuel through the purification filters and dewatering devices. The bypass flow purification system is a vendor package containing filters and water-removal equipment that is connected between the transfer system supply and return lines. The transfer pump will provide the flow through the system. The transfer pump will be operated on a periodic basis to maintain fuel oil condition. The purification package will condition the full bypass flow of fuel oil while the transfer pump is operating.

The immediate use day tank includes an engine supply line with a differential pressure instrumented, switchable stainless steel mesh dual-element strainer with automatic transfer valve. The mesh filters will be rated to remove 100 percent of particles at 0.7 mils (18 μm) and 98 percent of particles at 0.2 mils (5 μm). These filters are sized to prevent the clogging of the engine-mounted fuel filters, that are also sized at 0.2 mils (5 μm). The engine fuel pump provides motive force for this fuel flow. The typical flow in the fuel oil system between the engine and the day tank is approximately 21 L/min (330 gph), with the EDG using 4.5 L/min (72 gph) at 100-percent load and 16.3 L/min (258 gph) returning to the day tank.

The entire system of filters and screens is designed to turn over, filter, and purify the entire volume of the main storage and day tanks every 12 hours. The filtration system is designed to remove abrasive particulates and water that could damage the system pumps and the EDG.

Evaluation of Emergency Diesel Generator Fuel Oil System

To complete this evaluation, the staff reviewed national codes and standards, such as IEEE Standard 308 (IEEE, 1991), NFPA-37 (NFPA, 1998a), and NFPA 110 (NFPA, 1999). The staff reviewed the seismic design of the EDGs. The staff also reviewed the design basis and provisions for maintaining a clean, reliable source of fuel oil for the EDGs by comparing the proposed design to standard industry practices and historical diesel generator failure information published by the NRC.

As discussed in the system description, the EDGFOS is to be designed to supply adequate capacity to handle the expected volume of material during normal, off-normal, and accident conditions. Each EDGFOS bulk tank capacity is based on an EDG running at 100-percent load for 7 days (plus extra operational margin for testing). This design meets the requirements of IEEE Standard 308. The flow rate for the transfer pump is approximately 1.6 L/s (25 gpm). The bypass flow purification system will redirect approximately 0.9–1.3 L/s (15–20 gpm) of fuel through the purification filters and dewatering devices. The immediate-use day tank includes an engine supply line with a differential pressure instrumented, switchable stainless steel mesh dual-element strainer with automatic transfer valve. The typical flow in the fuel oil system between the engine and the day tank is approximately 21 L/min (330 gph), with the EDG using 4.5 L/min (72 gph) at 100-percent load and 16.3 L/min (258 gph) returning to the day tank. The engine silencer will be capable of attenuating noise from a running EDG to acceptable OSHA levels for worker hearing protection in an industrial environment. The EDG exhaust system will be included in the generator set from the vendor as a packaged unit. Based on this review, the staff believes that the designs of the EDGFOS and the EDG exhaust system are acceptable.

The EDGFOS and EDG exhaust systems do not interface with systems that carry radioactive material; therefore, the consideration of the redundancy and diversity of components required to prevent release of radioactive material to the environment is not required. The evaluation of the adequacy of the EDG system for its intended safety purpose is found in Section 11.5 of this FSER.

The seismic design of the EDGFOS and the EDG exhaust systems assures that these systems will be able to safely shut down during normal operations and accident conditions. Based on the description of the system and the structures that will contain the system, the EDGFOS is adequately designed to maintain functionality when subjected to severe natural phenomena such as tornadoes, tornado missiles, earthquakes, floods, and any other appropriate phenomena as established in the ISA. The EDG engine fuel oil pump is a PSSC that will support the EDGs starting from rest and accepting a load within 10 seconds at a temperature of -7 °C (20 °F). Based on the above, the staff finds that the design provisions for the EDGFOS and the EDG exhaust system are acceptable to support continued operation of the EDGs.

The staff has reviewed the design and configuration of the EDGFOS system tanks and piping. Because of the nature and location of the system, the staff finds that it cannot be designed to take advantage of gravity flow in order to perform its primary function. However, the redundancy and diversity of the system and the seismic design, applied to the design as standard industry practices, will assure a positive flow of fuel oil into, and waste gas removal from, the EDGs. The EDGFOS and EDG exhaust systems are designed to minimize the likelihood that nuclear material can enter these systems. Neither of these systems interfaces with systems that contain nuclear material; they are, in fact, completely external to the facility buildings. This provides reasonable assurance that contamination or criticality will not occur in these systems. DCS has committed to design these systems and components with consideration to the process fluids contained and to specify materials with the appropriate corrosion resistance where necessary. The staff finds the design of the system tanks, piping, and pumping systems to be acceptable.

The staff has evaluated the EDG exhaust system to determine the provisions for personnel protection. The engine silencer will be capable of attenuating noise from a running EDG to acceptable OSHA levels for worker hearing protection in an industrial environment. The EDG exhaust system will be included in the generator set from the vendor as a packaged unit and will comply with NFPA-37 (NFPA, 1998a) and NFPA 110 (NFPA, 1999). On the basis of the system design, industry codes and standards, and standard industry practices, the staff finds the EDG exhaust system to be acceptable.

The electrical capacity of the EDGs has been reviewed by the staff, and Section 11.5 of this FSER presents this evaluation.

Design Basis for Emergency Scavenging Air Subsystem

The emergency scavenging air subsystem of the instrument air system is designed to provide scavenging air for dilution of hydrogen that may be produced by radiolysis. The equipment most likely to produce hydrogen by radiolysis is primarily those tanks containing a high plutonium concentration solution. The emergency pressurized air supply tanks and piping are seismically qualified in accordance with IEEE Standard 344-1987 (IEEE, 1987) (see the discussion of the acceptability of this standard in Section 11.5 of this SER) and NRC RG 1.100, Revision 2 (NRC, 1988), and will be manually operated following a design basis event. The emergency scavenging air subsystem, a PSSC, is physically separated from the instrument air system and is automatically activated following the loss of the normal instrument air supply by opening one of the two parallel, fail-open, air-operated valves. The system is intended to be fitted to vessels and equipment in the AP areas that will produce hydrogen by radiolysis and that could reach the LFL of hydrogen in air (4 percent) in 7 days. The emergency scavenging air system is designed to provide a 7-day supply of emergency scavenging air, based on consideration of a reasonable time to repair or restore the normal bubbling air supply following failure of the instrument air system. The 1-hour supply for the receiver/buffer tanks is sized based on the estimated time to facilitate a normal process shutdown following loss of instrument air.

Evaluation of Emergency Scavenging Air Subsystem

In order to complete this evaluation, the staff reviewed national codes and standards, such as IEEE Standard 308 (IEEE, 1991), NFPA 801 (NFPA, 1998b), RG 3.12 (NRC, 1973b), RG 1.100, Revision 2 (NRC, 1988), and IEEE Standard 344-1987 (IEEE, 1987).

The emergency scavenging air subsystem is a PSSC designed to provide emergency scavenging air for dilution of hydrogen that may be produced by radiolysis. The equipment susceptible to this condition is primarily that containing a high plutonium concentration solution. Ordinarily, extra-dry instrument air is used by the bubbling air system to scavenge the hydrogen produced in equipment. The bubbling air system is designed to prevent the system from exceeding 25 percent of the LFL (or 1 percent) for hydrogen in the tanks. In the event of a failure of any of these systems, the emergency scavenging air subsystem is designed to compensate for the loss of the bubbling air system. The emergency scavenging air subsystem is automatically activated following a loss of instrument air. The emergency scavenging air system is designed to supply emergency scavenging air to all process vessels that could reach a hydrogen concentration of 4 percent within 7 days. Each bank of the emergency scavenging air system pressurized air supply main and backup banks of compressed air cylinders and piping are designed for 100-percent capacity for 7 days. The two banks are connected by an air pressure controller and transfer valves that switch from the primary bank to the secondary bank on low system pressure. The emergency scavenging air subsystem is not physically connected to the normal scavenging air supply.

The NRC staff reviewed the requirements for other safety-related backup systems, including national codes and standards such as IEEE Standard 308. In that standard, the required storage of fuel for the standby power supplies (such as EDGs) is either 7 days or the time required to replenish the fuel and test it. DCS has committed to a 7-day fuel supply for the EDGs, as described in CAR Section 11.5.7. Therefore, based on the IEEE standard and the DCS commitment to providing a 7-day fuel supply for the EDGs, the staff finds that a 7-day supply of emergency scavenging air is acceptable. The design basis for the maximum hydrogen level allowed by the emergency scavenging air subsystem is calculated to prevent the hydrogen concentration from exceeding 1 percent by volume. This level is equivalent to 25 percent of the LFL for hydrogen in air and, as discussed in Chapter 7 of this FSER, is generally an accepted limit. The applicant has committed to NFPA-801, which specifies that hydrogen concentrations be kept at or below 25 percent of the LFL in air. Therefore, the staff finds the emergency scavenging air subsystem design basis for the maximum allowed hydrogen concentration in vessels susceptible to hydrogen generation to be acceptable based on the DCS commitment to NFPA-801.

The emergency scavenging air subsystem is designed to provide two trains each capable of supplying 100 percent of the scavenging air requirements for 7 days. Based on the provision of two 100-percent capacity trains of emergency scavenging air, and the design of the system air pressure control to switch supply banks on low system pressure, the staff finds that the subsystem has the adequate redundancy and diversity of components required to prevent release of radioactive material to the environment and that its design is adequate for safe operation of the system.

The emergency scavenging air subsystem is independent of the instrument air system and is automatically actuated following loss of the scavenging air function in certain tanks susceptible to the buildup of flammable levels of hydrogen resulting from radiolysis. The subsystem does not rely on emergency power and is designed to operate during accident conditions. The staff concludes that, because of the passive nature of the subsystem, it can be safely shut down during any normal or accident conditions.

The emergency scavenging air subsystem is isolated from process fluids by the air spaces at the top of the tanks it is designed to serve. In accident situations, the system is pressurized above the pressure in the process equipment, so the flow of material is expected to be into the tank. The staff concludes that, based on the positive pressure of the subsystem, and the location of the supply nozzles relative to process fluids, the subsystem is adequately designed to minimize the likelihood of contamination and holdup of hazardous nuclear materials.

The design of the emergency scavenging air subsystem falls into FTS Category 2, discussed in FSER Section 11.8. Category 2 is for PSSC components that may contain trace quantities of plutonium or americium or nonradiological fluids that play a significant role in plant production reliability. This means that the system is designed to ASME B&PV and B31.3 codes standards. This includes consideration of piping material for corrosion protection and piping layout. Enhanced positive material identification, inspection, and test requirements are used in the procurement specifications for FTS Category 2 components. The staff concludes that these design criteria adequately address equipment design to resist corrosion, piping design, and layout.

The design basis of the emergency scavenging air subsystem includes layout and design of the system to minimize the potential for entrapment and buildup of radioactive materials. Therefore, the staff concludes that the equipment is made of materials with the proper surface finishes and decontamination characteristics for their particular application.

The SRP guidance indicates that equipment is to be adequately designed to maintain functionality when subjected to severe natural phenomena such as tornadoes, tornado missiles, earthquakes, floods, and any other appropriate phenomena as established in the ISA. The emergency scavenging air system is seismically designed and qualified in accordance with NRC RG 1.100, Revision 2 (NRC, 1988), and IEEE Standard 344-1987 (IEEE, 1987). Because the design conforms to that RG and industry standard, the staff concludes that the applicant adequately addressed the features that apply to this PSSC.

11.9.2 Evaluation Findings

In revised CAR Section 11.9, DCS provided design basis information for the fluid systems that it identified as PSSCs for the proposed facility. Based on the staff's review of the revised CAR and supporting information provided by the applicant relevant to the fluid systems, the staff concludes, pursuant to 10 CFR 70.23(b), that the design bases of the PSSCs evaluated in this

FSER section will provide reasonable assurance of protection against natural phenomena and the consequences of potential accidents.

11.9.3 References

(ASME, 1995) American Society of Mechanical Engineers. Boiler and Pressure Vessel Code. Section VIII, "Rules for Construction of Division 1 Pressure Vessels." 1995 Edition through the 1996 Addenda.

(ASME, 2000) American Society of Mechanical Engineers. ASME B16.10, "Face-to-Face and End-to-End Dimensions of Valves." 2000.

(ASME, 2001) American Society of Mechanical Engineers. ASME B31.3, "Process Piping." 2001.

(ASTM, 1974) American Society of Testing and Materials. ASTM D75, "Standard Specification for Diesel Fuel Oils." 1974.

(DCS, 2002a) Ihde, R., Duke Cogema Stone & Webster. Letter to Document Control Desk, U.S. Nuclear Regulatory Commission, RE: Mixed Oxide Fuel Fabrication Facility—Revised Construction Authorization Request. October 31, 2002 (including page changes through February 9, 2005).

(DCS, 2002b) Hastings, P., Duke Cogema Stone & Webster. Letter to Document Control Desk, U.S. Nuclear Regulatory Commission, RE Mixed Oxide Fuel Fabrication Facility —Evaluation of the Draft Safety Evaluation Report (DSER) on Construction of a Mixed Oxide Fuel Fabrication Facility. July 9, 2002.

(IEEE, 1987) Institute of Electrical and Electronics Engineers. IEEE Standard 344, "IEEE Recommended Practice for Seismic Qualification of Class IE Equipment for Nuclear Power Generating Stations." 1987.

(IEEE, 1991) Institute of Electrical and Electronics Engineers. IEEE Standard 308, "IEEE Standard Criteria for Class IE Power Systems for Nuclear Power Generating Stations." 1991.

(IEEE, 1998) Institute of Electrical and Electronics Engineers. IEEE Standard 603, "IEEE Standard Criteria for Safety Systems for Nuclear Power Generating Stations." 1998.

(NFPA, 1998a) National Fire Protection Association. NFPA 37, "Standard for the Installation and Use of Stationary Combustion Engines and Gas Turbines." 1998.

(NFPA, 1998b) National Fire Protection Association. NFPA 801, "Standard for Fire Protection for Facilities Handling Radioactive Materials." 1998.

(NFPA, 1999) National Fire Protection Association. NFPA 110, "Standard for Emergency and Standby Power Systems." 1999.

(NRC, 1973a) U.S. Nuclear Regulatory Commission. Regulatory Guide 3.14, "Seismic Design Classification for Plutonium Processing and Fuel Fabrication Plants." Washington, DC, 1973.

(NRC, 1973b) U.S. Nuclear Regulatory Commission. Regulatory Guide 3.12, "General Design Guide for Ventilation Systems of Plutonium Processing and Fuel Fabrication Plants." Washington, DC, 1973.

(NRC, 1974) U.S. Nuclear Regulatory Commission. Regulatory Guide 3.17, "Earthquake Instrumentation for Fuel Reprocessing Plants." Washington, DC, 1974.

(NRC, 1988) U.S. Nuclear Regulatory Commission. Regulatory Guide 1.100, Rev. 2, "Seismic Qualification of Electric and Mechanical Equipment for Nuclear Power Plants." Washington, DC, 1988.

(NRC, 2000) U.S. Nuclear Regulatory Commission. NUREG-1718, "Standard Review Plan for the Review of an Application for a Mixed Oxide (MOX) Fuel Fabrication Facility." Washington, DC, August 2000.

(NRC, 2002) Persinko, A., U.S. Nuclear Regulatory Commission. Letter to P. Hastings, Duke COGEMA Stone & Webster, RE: Response to DCS Letter Dated July 9, 2002 (DCS-NRC-0010000). October 30, 2002.

11.10 Heavy Lift Cranes

11.10.1 Conduct of Review

This chapter of the FSER contains the staff's review of the heavy lift cranes described by the applicant in the revised CAR (DCS, 2002). The objective of this review is to determine whether heavy lift crane PSSCs and their design bases identified by the applicant provide reasonable assurance of protection against natural phenomena and the consequences of potential accidents. The staff evaluated the information provided by the applicant for heavy lift cranes by reviewing Chapter 11.0 and other sections of the revised CAR, supplementary information provided by the applicant, and relevant documents available at the applicant's offices but not submitted by the applicant. The review of the design bases and strategies of the heavy lift cranes was closely coordinated with the review of other chapters in the FSER (fire protection in Chapter 7, chemical safety in Chapter 8, and safety assessment of the design bases in Chapter 5) and the review of other plant systems.

The review for this construction approval focused on the design basis of heavy lift cranes, their components, and other related information. For heavy lift cranes, the staff reviewed information provided by the applicant on the safety function, system description, and safety analysis. The review also encompassed proposed design basis considerations such as redundancy, independence, reliability, and quality. The staff used Section 11.4.8 in NUREG-1718 (NRC, 2000) as guidance in performing the review of heavy lift cranes.

The FSER discussions that follow provide the system descriptions, as well as functions, major components, control concepts, and system interfaces.

11.10.1.1 System Description

The applicant has not identified cranes, heavy lift or otherwise, as PSSCs in the revised CAR for the MFFF. However, material-handling equipment and controls have been identified as PSSCs.

DCS will identify IROFS in its ISA summary to be submitted as part of any later license application it may submit.

For the facility, heavy lift cranes are designated as those cranes designed to lift greater than 816 kg (1800 lb), as defined in NUREG-0612, "Control of Heavy Loads at Nuclear Power Plants." This load definition limits, to a very few, the number of cranes or lifting devices designated as heavy lift cranes. Other cranes or lifting devices may be identified as the design matures. The currently identified applications of heavy lift cranes in the facility are bridge cranes in the fresh fuel cask shipping truck bay, the assembly loading area, and the fresh fuel cask handling area; a bridge crane stacker for waste drum handling; a bridge crane for handling empty plutonium dioxide shipping package pallets; and a maintenance hoist in the EDG building.

The facility general design philosophy is stated as preventing lifts of heavy loads above PSSCs. Heavy lift cranes are not designated as PSSCs because equipment being lifted would be qualified for drops from the maximum lift height of the cranes. For example, the qualification of the MOX fresh fuel package is to maintain confinement integrity of the package for a drop from a height of 9.1 m (30 ft) onto an unyielding surface. DCS states that the maximum lift height for any piece of equipment in the facility is 4.9 m (16 ft). However, it is possible for cranes to impact PSSCs without a load drop. In this case, the PSSCs identified in the revised CAR are the material-handling controls discussed in Section 11.7 of this FSER. The term "material-handling controls" includes controls on material-handling equipment and administrative controls. Examples of these controls include safe travel paths, procedures and training to limit crane operations or to properly prepare loads or nearby equipment before crane use, and a radiation protection program to ensure workers are protected during maintenance activities. Specific material-handling controls will be identified in the ISA summary.

11.10.1.1.1 Function

The function of the facility heavy lift cranes is to lift and move critical loads in a manner consistent with the loads' qualification, to limit inadvertent movement, and to retain loads under normal, accident, and design basis natural phenomena conditions, such as earthquake and loss of power events. Critical loads are loads of a type that, if dropped and the contents released, could result in unacceptable radiological dose consequences to the worker, the IOC, or the environment. Other noncritical loads may also be moved by this equipment at any time. The heavy lift cranes' functions include controlling movement of the cranes and loads during operation and maintenance so they will not impact other PSSCs.

11.10.1.1.2 Major Components

The major components of the heavy lift cranes are the crane structures, drive motors for positioning and hoisting, the operator's cab or local operating station, and the lights and control systems for the cranes. Crane equipment consists of slings, lifting frames, and other below-the-hook lifting devices. Cranes may also be equipped with auxiliary hoists to perform routine lifting of smaller loads. The capacity of each crane is such that it should meet or exceed the lifted load; the lifted load includes the equipment to be moved plus the weight of the slings and other lifting devices plus applicable design margins specified in the crane design codes and standards.

11.10.1.1.3 Control Concepts

Most of the facility's heavy lift cranes will be operated locally by crane operators who are in visual contact with the crane and the load. The crane control stations are local, and the controls are conventional and required in the industry codes and standards, such as ASME B30.2 (ASME, 1996b). The stacker crane in the waste drum stacker area will be designed to operate automatically, with provisions for manual control as necessary. Similarly, the crane in the assembly loading area will function in automatic mode for some of its operations.

11.10.1.2 System Interfaces

The heavy lift cranes interface with the building structure; the rails on which the cranes run; the girders that span the work area; the electric drives for the bridge, trolley, and hoist; local operating stations; and the lights and control systems for the cranes. The crane may be controlled by radio communication with the operator. In these cases, the interactions between the crane controls and the building security, process monitoring, and control systems will be considered.

11.10.1.3 Design Bases of PSSCs

Heavy lift cranes or hoists at the MFFF have not been identified as PSSCs in Section 11.10 of the revised CAR. There are only two PSSCs over which heavy loads may be lifted—(1) the building floor, which has been evaluated for dynamic loads, including drop loads, and (2) the emergency generator during maintenance operations. Any nonheavy loads over PSSCs will be handled in accordance with material-handling controls, which are PSSCs. The applicant stated in Section 11.10.7 of the revised CAR that the drop of any load handled by a heavy lift crane would not result in consequences exceeding those presented in the 10 CFR 70.61 performance requirements.

Section 5.0 of the revised CAR discusses load-handling events. A load-handling hazard is postulated to occur from the presence of lifting or hoisting equipment used in normal or maintenance activities. Examples of load-handling events include a dropped lifted load, a lifted load impacting other equipment during the lift, or the load-handling equipment impacting other nearby items. An event of this type could damage handled loads, thereby dispersing nuclear material or chemicals, damaging nearby equipment resulting in a loss of confinement or loss of subcritical conditions, or damaging IROFS. Such load-handling events are evaluated in Chapter 5 of this FSER.

11.10.2 Evaluation Findings

In Section 11.10 of the revised CAR, the applicant did not designate heavy lift cranes as PSSCs. Based on the staff's review of the revised CAR and supporting information provided by the applicant relevant to the heavy lift cranes, the staff concurs that heavy lift cranes need not be designated as PSSCs.

11.10.3 References

(ANSI, 1993) American National Standards Institute. ANSI N14.6, "Radioactive Materials—Special Lifting Devices for Shipping Containers Weighing 10,000 lbs or More." 1993.

(ASME, 1996a) American Society of Mechanical Engineers. ASME B30.9, "Slings." 1996.

(ASME, 1996b) American Society of Mechanical Engineers. B30.2, "Top Running Bridge, Single or Multiple Girder, Top Running Trolley Hoist Overhead and Gantry Cranes." 1996.

(ASME, 1998) American Society of Mechanical Engineers. ASME B30.16, "Overhead Hoists (Underhung)." 1998.

(CMAA, 1994) Crane Manufacturers Association of America. 70, "Specifications for Top Running Bridge and Gantry Type Multiple Girder Electric Overhead Traveling Cranes." 1994.

(DCS, 2002) Ihde, R., Duke Cogema Stone & Webster. Letter to Document Control Desk, U.S. Nuclear Regulatory Commission, RE: Mixed Oxide Fuel Fabrication Facility—Construction Authorization Request. October 31, 2002 (including page changes through February 9, 2005).

(DOE, 2000) U.S. Department of Energy(DOE). DOE-STD-3013, "Stabilization, Packaging, and Storage of Plutonium-Bearing Materials." September 2000.

(NRC, 1980) U.S. Nuclear Regulatory Commission. NUREG-0612, "Control of Heavy Loads at Nuclear Power Plants." Washington, DC, July 1980.

(NRC, 2000) U.S. Nuclear Regulatory Commission. NUREG-1718, "Standard Review Plan for the Review of an Application for a Mixed Oxide (MOX) Fuel Fabrication Facility." Washington, DC, August 2000.

11.11 Environmental Qualification

11.11.1 Conduct of Review

This chapter of the FSER contains the staff's review of EQ described by the applicant in Chapter 11.0 of the revised CAR (DCS, 2002). The objective of this review is to determine whether the EQ of electrical and mechanical equipment designated as PSSCs and the EQ design bases identified by the applicant provide reasonable assurance of protection against natural phenomena and the consequences of potential accidents. The staff evaluated the information provided by the applicant for EQ of electrical and mechanical equipment by reviewing Chapter 11 of the revised CAR, other sections of the revised CAR, supplementary information provided by the applicant, and relevant documents available at the applicant's offices but not submitted by the applicant. The review of the EQ of electrical and mechanical equipment design bases and strategies was closely coordinated with the review of the electrical, I&C, and mechanical aspects of accident sequences described in the safety assessment of the design bases (see Chapter 5 of this FSER) and the review of other plant systems.

The staff reviewed how the information in the revised CAR addresses the following regulations:

- 10 CFR 70.23(b) and the Quality Assurance Program state, as a prerequisite to construction approval, that the design bases of the PSSCs must be found to provide reasonable assurance of protection against natural phenomena and the consequences of potential accidents.

- 10 CFR 70.64 requires that BDC and defense-in-depth practices be incorporated into the design of new facilities. With respect to the EQ of electrical and mechanical equipment, 10 CFR 70.64(a)(4) requires that the MFFF design "must provide for adequate protection from environmental conditions and dynamic effects associated with normal operations, maintenance, testing, and postulated accidents that could lead to loss of safety functions."

The review for this construction approval focused on the design basis of the electrical and mechanical equipment that are PSSCs specifically related to the EQ requirement. The staff used Chapter 11.0 in NUREG-1718 as guidance in performing the review.

11.11.1.2 System Description

The electrical and mechanical equipment evaluated in this section are PSSCs and are major components of systems (also PSSCs) such as the confinement systems (described in FSER Section 11.4), the electrical systems (described in FSER Section 11.5), the I&C systems (described in FSER Section 11.6), the material transport system (described in FSER Section 11.7), the fluid transport systems (described in FSER Section 11.8), the fluid systems (described in FSER Section 11.9), and the heavy lift cranes (described in FSER Section 11.10).

11.11.1.3 Design Basis of the PSSCs and Applicable Baseline Design Criteria

In addition to the design basis requirements for the electrical and mechanical equipment designated as PSSCs, which encompass the PSSCs within the systems listed in FSER Section 11.11.2, the electrical and mechanical equipment that are designated PSSCs are to remain functional when subjected to the environmental and dynamic effects referenced in the 10 CFR 70.64(a)(4) BDC. For EQ purposes, this electrical and mechanical equipment will be designed as described in the revised CAR and the staff's memorandum dated January 31, 2003 (NRC, 2003), based on IEEE Standard 323-1983, "IEEE Standard for Qualifying Class 1E Equipment for Nuclear Power Generating Stations" (IEEE, 1983).

DCS stated that electrical and mechanical PSSCs will be qualified to anticipated environments, including qualification to environments if the event causing the harsh environment is not prevented (NRC, 2003). Pursuant to this commitment, DCS will be required to demonstrate that each piece of electrical and mechanical equipment identified as a PSSC will perform its safety functions under the environmental and dynamic service conditions in which it will be required to function and for the length of time its function is required. In addition, non-PSSCs will be required to be able to withstand environmental stresses caused by environmental and dynamic service conditions under which their failure could prevent satisfactory accomplishment of safety functions by PSSCs. The staff finds that these commitments satisfy the requirements of 10 CFR 70.64(a)(4) and notes that this approach is consistent with how EQ requirements are applied at nuclear power plants (NRC, 1984).

11.11.2 Evaluation Findings

Based on the staff's review of the revised CAR, the supporting information provided by the applicant, and the applicant's commitments to the industry guidance referenced above, the staff finds, pursuant to 10 CFR 70.64(a)(4), that for EQ, the design basis of the proposed facility provides for adequate protection against environmental condition hazards. The staff concludes, pursuant to 10 CFR 70.23(b), that the design basis of the PSSCs relevant to EQ of electrical and mechanical equipment will provide reasonable assurance of protection against natural phenomena and the consequences of potential accidents.

11.11.3 References

(DCS, 2002) Ihde, R., Duke Cogema Stone & Webster. Letter to U.S. Nuclear Regulatory Commission, Mixed Oxide Fuel Fabrication Facility Construction Authorization Request Revision. October 31, 2002 (including page changes through February 9, 2005).

(IEEE, 1983) Institute of Electrical and Electronics Engineers. Std. 323-1983, "IEEE Standard for Qualifying Class 1E Equipment for Nuclear Power Generating Stations." New York, September 30, 1983.

(NRC, 1984) U.S. Nuclear Regulatory Commission. Regulatory Guide 1.89, Revision 1, "Environmental Qualification of Certain Electric Equipment Important to Safety for Nuclear Power Plants." Washington, DC, June 1984.

(NRC, 2000) U.S. Nuclear Regulatory Commission. NUREG-1718, "Standard Review Plan for the Review of an Application for a Mixed Oxide (MOX) Fuel Fabrication Facility." Washington, DC, August 2000.

(NRC, 2003) Persinko, Andrew, U.S. Nuclear Regulatory Commission. Memorandum to Melvyn N. Leach, U.S. Nuclear Regulatory Commission, RE: December 10–12, 2002, Meeting Summary: Meeting with Duke Cogema Stone & Webster to Discuss Mixed Oxide Fuel Fabrication Facility Revised Construction Authorization Request. January 31, 2003.

11.12 Seismic Qualification of Equipment, Systems, and Components

11.12.1 Conduct of Review

This chapter of the FSER contains the staff's review of SQ described by the applicant in Section 11.12 of the revised CAR (DCS, 2002a). The objective of this review is to determine whether the SQ of electrical and mechanical equipment and their SQ design bases identified by the applicant provide reasonable assurance of protection against natural phenomena and the consequences of potential accidents. Seismic issues related to the ground motion are evaluated in FSER Section 1.3.1.5. The staff evaluated the information provided by the applicant for SQ of electrical and mechanical equipment by reviewing Chapter 11 of the revised CAR, other sections of the revised CAR, supplementary information provided by the applicant, and relevant documents available at the applicant's offices but not submitted by the applicant. The review of the SQ of electrical and mechanical equipment design bases and strategies was closely coordinated with the review of the electrical, I&C, and mechanical aspects of accident sequences described in the safety assessment of the design bases (see Chapter 5 of this FSER) and the review of other plant systems.

The staff reviewed how the information in the revised CAR addresses the following regulations:

- 10 CFR 70.23(b) states, as a prerequisite to construction approval, that the design bases of the PSSCs must be found to provide reasonable assurance of protection against natural phenomena and the consequences of potential accidents.

- 10 CFR 70.64 requires that BDC and defense-in-depth practices be incorporated into the design of new facilities. With respect to natural phenomena hazards, 10 CFR 70.64(a)(2) requires that the MFFF design provide for adequate protection against such hazards, with consideration of the most severe documented historical events for the MFFF site.

The review for this construction approval focused on the design basis of the electrical and mechanical equipment related to the SQ requirement. The staff used Chapter 11.0 in NUREG-1718 as guidance in performing the review.

11.12.1.1 System Description

The electrical and mechanical equipment evaluated in this section are PSSCs and are major components of systems (also PSSCs) such as the confinement systems (described in FSER Section 11.4), the electrical systems (described in FSER Section 11.5), the I&C systems (described in FSER Section 11.6), the material transport system (described in FSER Section 11.7), the fluid transport systems (described in FSER Section 11.8), the fluid systems (described in FSER Section 11.9), and the heavy lift cranes (described in FSER Section 11.10).

11.12.1.2 Design Basis of the PSSCs and Applicable Baseline Design Criteria

The applicant's methodology for seismic classification of SSCs consists of SC and seismic performance requirements. The SC classification contains SC-I which applies to all PSSCs that must perform safety functions during and/or after the design basis earthquake to comply with the MFFF safety assessment as described in Chapter 5 of the revised CAR; and SC-II which applies to systems which are not required after a design basis earthquake but whose failure could adversely impact the ability of an SC-I system to perform its safety function. The seismic performance requirements segregate SC-I and SC-II by the safety functions they must perform during or after a seismic event such as remaining active during or after a seismic event, maintaining a pressure boundary or structural integrity, and not failing in a way that compromises a PSSC (SC-II).

The analysis requirements provided in the revised CAR consist of seismic design parameters that were developed with the determination of the design basis earthquake and procedures for the use of seismic inertial response techniques. For the seismic parameters, in-structure response spectra will be generated in accordance with one of the methods cited in Section 3.4 of ASCE 4-98. The applicant's qualification methodology allows the inertial response of PSSCs to be determined using dynamic analysis or equivalent static analysis for elements that can be represented by a single-degree-of-freedom or a simple multiple-degree-of-freedom model.

The SQ requirements as specified in the revised CAR allow either analysis or shake-table testing for most seismic performance categories. Shake table testing is required for components where analysis alone is insufficient to ensure operability after a seismic event (e.g., electrical components).

In the CAR (DCS, 2002a) and by letter dated August 31, 2001 (DCS, 2002b), DCS stated that electrical and mechanical equipment will be seismically qualified using IEEE Standard 344-1987 (IEEE 1987) and NRC RG 1.100, Revision 2 (NRC 1988).

DCS stated (DCS, 2002b) that mechanical equipment will be qualified to IEEE 344-1987, including NRC additions to the 1987 IEEE standard stated in RG 1.100. Therefore, mechanical equipment qualification will consider attached piping loads, thermal loads, and live loads such as fluid sloshing. In addition, applied loads will meet or exceed accelerations corresponding to their installed locations.

The staff concludes that these commitments are consistent with the application of SQ requirements at nuclear power plants. The staff further concludes that the facility design provides for adequate protection against seismic hazards with consideration of the most severe documented historical events for the facility site and satisfies the requirements of 10 CFR 70.64(a)(2).

11.12.2 Evaluation Findings

Based on the staff's review of the revised CAR, the supporting information provided by the applicant, and the applicant's commitments to the guidance referenced above, the staff finds, pursuant to 10 CFR 70.64(a)(2), that for SQ of equipment, the design basis of the proposed facility provides for adequate protection against seismic hazards, and that the applicant has adequately considered the most severe documented historical seismic events for the MFFF site. The staff concludes, pursuant to 10 CFR 70.23(b), that the design basis of the PSSCs relevant to the SQ of electrical and mechanical equipment will provide reasonable assurance of protection against natural phenomena and the consequences of potential accidents.

11.12.3 References

(DCS, 2002a) Ihde, R.H., Duke Cogema Stone & Webster. Letter to U.S. Nuclear Regulatory Commission, RE: Mixed Oxide Fuel Fabrication Facility Construction Authorization Request Revision. October 31, 2002 (including page changes through February 9, 2005).

(DCS, 2002b) Hastings, P.S., Duke Cogema Stone & Webster. Letter to U.S. Nuclear Regulatory Commission, RE: Construction Authorization Request: Clarification of Responses to NRC Request for Additional Information. February 11, 2002.

(IEEE, 1987) Institute of Electrical and Electronics Engineers. Std 344-1987, "IEEE Recommended Practices for Seismic Qualification of Class1E Equipment for Nuclear Generating Stations." 1987.

(NRC, 1988) U.S. Nuclear Regulatory Commission. Regulatory Guide 1.100, Rev. 2, "Seismic Qualification of Electric and Mechanical Equipment for Nuclear Plants." Washington, DC, August 2000.

(NRC, 2000) U.S. Nuclear Regulatory Commission. NUREG-1718, "Standard Review Plan for the Review of an Application for a Mixed Oxide (MOX) Fuel Fabrication Facility." Washington, DC, August 2000.

This page intentionally left blank

12. HUMAN FACTORS ENGINEERING FOR PERSONNEL ACTIVITIES

12.1 Conduct of Review

This chapter of the final safety evaluation report (FSER) contains the staff's review of the human factors engineering (HFE) plans, processes, and analyses performed by Duke Cogema Stone & Webster (DCS or the applicant) in Chapter 12 of the revised construction authorization request (CAR), dated October 31, 2002 (DCS, 2002c). The objectives of this review are to (1) establish that HFE is being applied to personnel activities identified as a principal structure, system, component (PSSC) (PSSCs and items relied on for safety (IROFS) include activities of personnel that are relied on to prevent potential accidents that could exceed the performance requirements in Title 10, Section 70.61, "Performance Requirements," of the *Code of Federal Regulations* (10 CFR 70.61)), consistent with the findings of the revised CAR, and (2) determine whether PSSCs and their design bases identified by DCS provide reasonable assurance of protection against natural phenomena and the consequences of potential accidents. The staff evaluated the information provided by the applicant for HFE by reviewing Chapter 12 of the revised CAR, other sections of the revised CAR, and supplementary information provided by the applicant in its letter dated November 22, 2002 (DCS, 2002d). The review of HFE was closely coordinated with the review of the instrumentation and control (I&C) and electrical aspects of accident sequences described in the safety assessment of the design bases (see Chapter 5 of this FSER) and the review of other plant systems.

The staff reviewed how the information in the revised CAR addresses the following regulation:

- 10 CFR 70.23(b) states, as a prerequisite to construction approval, that the design bases of the PSSCs and the Quality Assurance Program must be found to provide reasonable assurance of protection against natural phenomena and the consequences of potential accidents.

The scope of the HFE review included (1) a description of the safety-significant personnel actions, the associated human system interfaces, and the consequences of incorrectly performing or omitting actions for each personnel activity, (2) the applicant's plans for the HFE design review, (3) review of operating experience at existing facilities that are similar to the proposed mixed oxide (MOX) fuel fabrication facility (MFFF or the facility), (4) function and task analysis, (5) human-system interface (HSI) design, inventory, and characterization, (6) staffing, (7) procedure development, (8) training program development, and (9) verification and validation.

The staff used applicable portions of Chapter 12.0 in NUREG-1718, "Standard Review Plan for the Review of an Application for a Mixed Oxide (MOX) Fuel Fabrication Facility," issued August 2000, as guidance in performing the review.

12.1.1 Identification of Personnel Actions

The applicant discussed the nature of personnel actions at the proposed facility in Section 12.1 of the revised CAR. Control of the operations of the facility relies to a great extent on automated systems to ensure production quality and facility safety. In general, the operations staff is expected to perform the following types of tasks:

- initiate batch or continuous operations

- monitor the progress of the operations

- perform or initiate performance of quality control checks at preprogrammed hold points in the process

- monitor and confirm the status of confinement systems, fluid systems, and other facility systems

- respond to or recover from off-normal conditions

In the revised CAR, DCS discussed the human factors/human performance activities associated with maintenance of automated systems which would be used in the facility and did not identify any safety-significant human-system maintenance interfaces. The applicant stated that the integrated safety analysis (ISA) process will identify the sensors, instruments, and actuators that are classified as IROFS. The appropriate HSI requirements and the human performance requirements will be established as part of the DCS application for an operating license under 10 CFR Part 70, "Domestic Licensing of Special Nuclear Material." Activities associated with the maintenance or operation of the instruments, sensors, and actuators which are later classified as IROFS will be evaluated for human factors attributes using the criteria of Institute of Electrical and Electronics Engineers (IEEE) Standard 1023, "IEEE Guidelines for the Application of Human Factors Engineering to Systems and Equipment, and Facilities of Nuclear Power Generating Facilities," issued 1988 (IEEE, 1988), with the recognition that there are conditions, systems, operating requirements, and consequences unique to a nuclear power plant and not found in a fuel fabrication facility. The applicant also committed to use and review the criteria of the Design Review Checklist in NUREG/CR-6636, "Maintainability of Digital Systems: Technical Basis and Human Factors Guidelines," issued March 2000 (NRC, 2000d), for applicability to the digital controls of the PSSCs as part of its application for a license to possess and use licensed material. The applicant also stated that NUREG-0700 and all of the NUREG/CR references in Chapter 12.0 of NUREG-1718 would be used as guidance as part of its application for a possession and use license for human performance activities associated with maintenance of the facility automated systems. The staff finds this information acceptable because it provides the clarification requested regarding the human performance activities associated with maintenance at the facility.

In the revised CAR, the applicant described the criteria and basis used for determining that the protective control subsystem does not constitute a significant HSI and defined what "significant" means. The applicant stated that the protective control subsystem is designed to satisfy industrial safety requirements and is not a PSSC and provided additional information describing the design of the protective control subsystem's HSI. In the revised CAR, DCS more explicitly defined what is meant by "significant human-system interface" for the protective control subsystem, evaluated the potential for personnel errors of commission that might result in overriding or defeating safety systems, and provided cross-references to appropriate parts of Chapter 11 of the revised CAR. The staff finds this information acceptable because it provides the clarification requested regarding the protective control system, its human system interface, and potential personnel errors of commission.

The applicant stated in its revised CAR Section 12.1 that the facility would have a high level of automation with operators mainly monitoring the operation of systems and exercising supervisory control only when necessary. In a June 21, 2001, letter to the applicant, the U.S. Nuclear Regulatory Commission (NRC) staff asked the applicant to describe how its staff are alerted to undesirable conditions at control stations that are not normally staffed, and what criteria are used to decide when appropriate operations staff need to be at these remote locations for appropriate and timely response (NRC, 2001). By letter dated August 31, 2001 (DCS, 2001), DCS provided supplemental information stating that the performance of systems in automated areas would be constantly monitored by automated supervisory systems. One of the attributes of the functional units which would be monitored by the supervisory systems is the state of an automated activity. If the activity is not concluded in an anticipated state or within an expected time, or if a continuous process is not within allowed limits, an alarm would be generated in the control room for that functional unit.

The design of the facility establishes several different control rooms, and controls of the various functional units of the facility are grouped together into these control rooms. If a functional unit is in operation, the control room associated with that functional unit would be occupied. If none of the functional units assigned to a particular control room is operating, that control room would probably not be occupied. For example, control outputs for the fissile material mass accounting system would not be needed if there are no movements into, out of, or within a glovebox; similarly, the mass measurement system and mass limit alarms would not be meaningful in this situation. Signals for functions appropriate only to an operational unit would be transmitted to the control room that is assigned to that function.

[Text removed under 10 CFR 2.390]

Staffing evaluations will be completed as part of the DCS application for a 10 CFR Part 70 operating license and will be derived from the staffing requirements that exist in the La Hague and MELOX facilities. The applicant will provide this information in the license application for the facility. The staff finds this information acceptable because it provides the clarification requested regarding alerting staff to undesirable conditions at control stations not normally staffed, which control rooms would be staffed, and the development of staffing requirements derived from La Hague and MELOX. By letter dated November 22, 2002, the applicant stated that it was not necessary to incorporate this information in the revised CAR (DCS, 2002d).

In revised CAR Section 12.1, the applicant stated that, in general, omission of an operator action would not result in adverse conditions, and that errors in operator actions would generally be bounded by the other deterministic design basis-assumptions. The applicant clarified what is meant by "in general" and described by example the other deterministic design basis assumptions. The applicant stated that no scenario has been identified where omission of an operator action would result in adverse conditions, and errors in operator actions have been anticipated in the system design while considering other deterministic design basis accident assumptions and scenarios. The applicant also more explicitly defined what is meant by "other deterministic design basis accident assumptions and scenarios" and considered and evaluated the potential for personnel errors of commission that might result in overriding or defeating safety systems. The staff finds this information acceptable because it provides the clarification requested regarding other deterministic design basis assumptions and potential for personnel errors of commission.

12.1.2 Human Factors Engineering Design Planning

The applicant discussed HFE design planning in Section 12.2 of the revised CAR. The design of HFE includes the identification of HFE programmatic goals and scope and a description of the plans for HFE review, including HFE team makeup and processes for conducting HFE reviews. The HFE principles are applied to the facility design based on the guidelines of IEEE-1023 (IEEE, 1988). In a request for additional information dated June 21, 2001 (NRC, 2001), the NRC staff asked the applicant to verify a commitment made in an April 25, 2001, meeting with the NRC staff, to use NUREG-0711 to further guide its human factors design basis development work during construction and evaluate a subsequent revision to IEEE-1023. In letters dated August 31, 2001, and January 7, 2002 (DCS, 2001; DCS, 2002a; RAI 228), the applicant provided supplemental information which stated that NUREG-0711, Revision 2, "Human Factors Engineering Program Review Model," issued May 2002 (NRC, 2002b), would continue to be reviewed for HFE criteria that may be applicable to the design of the HSI for the control systems of PSSCs in the facility. DCS also stated that it would evaluate any future revision to IEEE-1023 for applicability to the design of the facility. The NRC staff finds this response acceptable because it provides the clarification requested on the use of IEEE-1023 and NUREG-0711. By letter dated November 22, 2002, the applicant stated that it was not necessary to incorporate this information in the revised CAR (DCS, 2002d).

In the revised CAR, DCS identified and described the meaning of "facility baseline design" and cross-referenced other appropriate chapters of the revised CAR. The applicant stated that facility baseline design is synonymous with the technical basis defined in the configuration management policies in Section 15.2.1 of the revised CAR. The staff finds this information acceptable because it defines the meaning of facility baseline design.

In the revised CAR, the applicant identified and described the aspects of the facility design that reduce the risk of errors or challenges to PSSCs and how these aspects are evaluated. DCS stated that the facility is designed to maximize the use of automation, thus minimizing human operations and interactions with the facility SSCs. By reducing these interactions, the probability of a human-caused error being introduced is reduced. The applicant also stated that it would consider both human errors of omission and commission in its evaluation of the probability of human error and describe these results as part of the license application. The staff finds this information acceptable because it provides both a process and rationale for maximizing the use of automation to reduce the probability of human errors of omission or commission.

In revised CAR Section 12.1.2, DCS also described how the HFE team will conduct its activities and where the team resides within the organization, with organizational roles and responsibilities clearly defined. The applicant discussed the activities of each of three phases of the HFE process—preliminary design, final design, and construction and startup. In the preliminary design phase, the facility control system architecture, control philosophies, and HSIs were developed with emphasis on the proven control methods from MELOX and La Hague. The original design and ongoing evolution of these facilities incorporated various degrees of human factors methods and reflect several years of safe operation. To supplement their use as a reference design, operational experience is incorporated into the facility design through a combination of lessons-learned evaluations (focusing on operability and maintainability issues and involving current operations and maintenance personnel) and review of the facility design on an ongoing basis by experienced operations staff. The applicant described, by example, how operating experience of the La Hague and MELOX facilities is incorporated in the facility design process, and provided examples of lessons-learned evaluations that show how the facility, as a proposed next generation facility, effectively incorporates this operating experience. DCS provided a presentation with examples of significant MELOX and La Hague operating events involving human performance to the NRC staff in a meeting at the NRC on October 11, 2001. The applicant submitted supplemental information to document its October 11, 2001, meeting presentation and provided additional examples of significant MELOX and La Hague operating events involving human performance. The staff finds this information acceptable because it provides the clarification requested regarding specific examples of operating experience at MELOX and La Hague which are being incorporated in facility design.

In the revised CAR, DCS also stated that, as part of the license application, criteria for HFE will be identified in facility design basis documents and will be applied throughout the final design for aspects of operation and maintenance of the facility. The task analysis, which will be completed as part of the application for a 10 CFR Part 70 operating license, will reflect the personnel activities relied on for safety as identified as part of the development of the ISA. During the detailed design of the HSIs, inventory and characterization of the interfaces will be performed. Evaluation of the characteristics of the HSIs will use the review criteria of NUREG-0700, Revision 2, "Human-System Interface Design Review Guidelines," issued May 2002 (NRC 2002a), as the basis.

12.1.3 Operating Experience

The applicant discussed operating experience in Section 12.3 of the revised CAR, as well as in Section 12.2 of the revised CAR (see Section 12.1.2 of this FSER above). In Section 12.3 of the revised CAR, the applicant stated that as a result of selection of existing facilities with successful operating histories as a reference design for the facility, and the ongoing involvement of operations and engineering personnel from these facilities in the development of facility design, no additional formal operating experience review is anticipated. The applicant clarified what is meant by "no additional formal operating experience review is anticipated" for the facility based on the operational experience at the La Hague and MELOX facilities previously incorporated in the facility design. Lessons learned from operating experience will be a continuing activity throughout construction, detailed design, and operation. The applicant also stated that there would be ongoing involvement of operations and engineering personnel from the MELOX and La Hague facilities in the development and design of the facility, thus providing a capability for evaluating and including results of operating experience as appropriate for the facility. The staff finds this an acceptable way to incorporate ongoing operational experience into the facility design.

12.1.4 Function and Task Analysis

The applicant discussed function and task analysis in Section 12.4 of the revised CAR, as well as in Sections 12.2.3.1 and 12.2.3.2. Operational tasks are well established for the MELOX and La Hague facilities for the purposes of preliminary design of the facility. The facility is an automated facility, and the tasks assigned to humans involve primarily initiating, verifying, and monitoring system status. The task analysis will be completed as part of the license application and will reflect the personnel activities relied on for safety identified as part of the ISA.

12.1.5 Human-System Interface Design, Inventory, and Characterization

The applicant discussed HSI design, inventory, and characterization in Section 12.5 of the revised CAR, as well as in Section 12.2.3. The HSI design, inventory, and characterization for the facility are initially based on the MELOX and La Hague designs. As part of the license application, detailed design of the HSI, inventory, and characterization of the interfaces will be completed.

12.1.6 Other Considerations

The applicant discussed staffing, procedure development, and training in Section 12.6 of the revised CAR and stated that these issues will be addressed in the HFE plan to be developed as part of the license application. The HFE verification and validation is discussed in Sections 12.2.3, 12.2.3.2, and 12.2.3.3 of the revised CAR. As part of the license application, HFE verification and validation activities will be conducted to support construction and startup. The HSI design will be verified in accordance with the configuration management and design control processes discussed in Chapter 15 of the revised CAR. A final personnel activities review will be performed during startup testing. This review will be an integrated system validation of personnel activities relied on for safety including, but not limited to, HSIs, procedure development, training development, staffing, and maintenance tasks. The human performance activities identified in the functional allocations and task analysis will be updated in the license application to reflect the results of the ISA.

12.1.7 Design Bases of the PSSCs

In Chapter 5 of the revised CAR, DCS identified administrative controls and HSIs as PSSCs, to be implemented by appropriate procedures, training, and management measures. These PSSCs are the human factors PSSCs for the facility. The applicant has stated that the facility is being designed to maximize the use of automation, thus minimizing human operations and interactions with the facility SSCs. By reducing these interactions, the probability of a human-caused error being introduced is reduced. The applicant has also committed to using guidance as appropriate from the following standards and NUREGs as additional design bases for reducing human error:

- IEEE Std. 1023 (IEEE, 1988)

- NUREG-0700, Revision 2 (NRC, 2002a)

- NUREG-0711, Revision 2 (NRC, 2002b)

12.2 Evaluation Findings

In Chapter 12 of the revised CAR, the applicant described the general design philosophy (hierarchy of controls) and defense-in-depth practices (double contingency protection for criticality events; single-failure criterion including redundancy, independence, separation, and fail safe for PSSCs; plus other noncredited PSSCs) applied during the preliminary design of the facility. Based on that information and the discussion provided in the sections above for HFE for personnel activities, the staff concludes that the applicant's HFE plans, processes, and analyses provide reasonable assurance that the design bases of the relevant PSSCs identified by DCS will protect against natural phenomena and the consequences of potential accidents, and are thus adequate to approve the revised CAR, pursuant to 10 CFR 70.23(b). The applicant will be required to submit more detailed evaluation of HFE as part of any later license application it may submit.

12.3 <u>References</u>

(DCS, 2001) Hastings, P., Duke Cogema Stone & Webster. Letter to U.S. Nuclear Regulatory Commission, RE: Response to Request for Additional Information—Construction Authorization Request. August 31, 2001.

(DCS, 2002a) Hastings, P., Duke Cogema Stone & Webster. Letter to U.S. Nuclear Regulatory Commission, RE: Clarification of Responses to NRC Request for Additional Information. January 7, 2002.

(DCS, 2002b) Hastings, P., Duke Cogema Stone & Webster. Letter to U.S. Nuclear Regulatory Commission, RE: Clarification of Responses to NRC Request for Additional Information. February 11, 2002.

(DCS, 2002c) Ihde, R., Duke Cogema Stone & Webster. Letter to U.S. Nuclear Regulatory Commission, Mixed Oxide Fuel Fabrication Facility Construction Authorization Request Revision. October 31, 2002 (including page changes through February 9, 2005).

(DCS, 2002d) Hastings, P., Duke Cogema Stone & Webster. Letter to U.S. Nuclear Regulatory Commission, RE: Response to Request for Additional Information, Clarification, and Open Item Mapping into the Construction Authorization Request Revision. November 22, 2002.

(DOD, 1989) U.S. Department of Defense. MIL-STD-1472D, "Human Engineering Design Criteria for Military Systems, Equipment and Facilities." Washington, DC, March 1989.

(IEEE, 1988) Institute of Electrical and Electronics Engineers. Std. 1023, "IEEE Guidelines for the Application of Human Factors Engineering to Systems and Equipment and Facilities of Nuclear Power Generating Facilities." 1988.

(NRC, 1999) U.S. Nuclear Regulatory Commission. "Domestic Licensing of Special Nuclear Material (10 CFR Part 70)." *Federal Register.* Vol. 64, No. 146. pp. 41338–41357. July 30, 1999.

(NRC, 2000a) U.S. Nuclear Regulatory Commission. NUREG/CR-6633, "Advanced Information Systems Design: Technical Basis and Human Factors Review Guidance." Washington, DC, March 2000.

(NRC, 2000b) U.S. Nuclear Regulatory Commission. NUREG/CR-6634, "Computer-Based Procedure Systems: Technical Basis and Human Factors Review Guidance." Washington, DC, March 2000.

(NRC, 2000c) U.S. Nuclear Regulatory Commission. NUREG/CR-6635, "Soft Controls: Technical Basis and Human Factors Review Guidance." Washington, DC, March 2000.

(NRC, 2000d) U.S. Nuclear Regulatory Commission. NUREG/CR-6636, "Maintainability of Digital Systems: Technical Basis and Human Factors Review Guidance." Washington, DC, March 2000.

(NRC, 2000e) U.S. Nuclear Regulatory Commission. NUREG/CR-6637, "Human Systems Interface and Plant Modification Process: Technical Basis and Human Factors Review Guidance." Washington, DC, March 2000.

(NRC, 2001) Giitter, J., U.S. Nuclear Regulatory Commission. Letter to P. Hastings, Duke Cogema Stone & Webster, RE: MOX Fuel Fabrication Facility Construction Authorization—Request for Additional Information. June 21, 2001.

(NRC, 2002a) U.S. Nuclear Regulatory Commission. NUREG-0700, Rev. 2, "Human-System Interface Design Review Guidelines." Washington, DC, May 2002.

(NRC, 2002b) U.S. Nuclear Regulatory Commission. NUREG-0711, Rev. 2, "Human Factors Engineering Program Review Model." Washington, DC, May 2002.

13. SAFEGUARDS

13.1 Physical Protection

13.1.1 Conduct of Review

This chapter of the final safety evaluation report (FSER) reviews the physical protection commitments made by Duke Cogema Stone & Webster (DCS or the applicant) in Chapter 13 of the revised construction authorization request (CAR) (DCS, 2002b) and supplemental information provided by the applicant. The staff used applicable portions of NUREG-1718, "Standard Review Plan for the Review of an Application for a Mixed Oxide (MOX) Fuel Fabrication Facility," issued August 2000, as guidance in performing the review (NRC, 2000). The staff also reviewed the applicant's letter of March 8, 2002 (DCS, 2002a), which responded to a request for information. The objective of the review is to determine whether the applicant has committed to developing and implementing a physical protection plan that meets the requirements in Title 10, Section 73.20, "General Performance Objectives and Requirements," of the *Code of Federal Regulations* (10 CFR 73.20); 10 CFR 73.45, "Performance Capabilities for Fixed Site Physical Protection Systems"; and 10 CFR 73.46, "Fixed Site Physical Protection Systems, Subsystems, Components, and Procedures," and Appendices B, C, G, and H to 10 CFR 73.46.

13.1.2 Evaluation Findings

The applicant committed to submit, as part of any later license application, (1) a physical security plan that meets the requirements of 10 CFR 73.20, 73.45, and 73.46 and Appendix B, "General Criteria for Security Personnel", Appendix C, "Licensee Safeguards Contingency Plans," Appendix G, "Reportable Safeguards Events," and Appendix H, "Weapons Qualification Criteria," to 10 CFR Part 73, "Physical Protection of Plants and Materials," and (2) a design that meets the design basis threat (NRC, 2003) (note that the attachments to this letter contain classified information and are not publicly available).

The staff concludes that the applicant has provided adequate commitments to submit, as part of its license application, a physical security plan that will meet the applicable physical protection requirements in 10 CFR Part 73.

The U.S. Nuclear Regulatory Commission (NRC) is conducting a comprehensive review of safeguards programs and design basis threats as a result of the September 11, 2001, events. When this review is completed, a determination will be made with respect to the effect on the mixed oxide fuel fabrication facility design.

13.1.3 References

(DCS, 2002a) Hastings, P., Duke Cogema Stone & Webster. Letter to U.S. Nuclear Regulatory Commission, RE: Clarification of Responses to NRC Request for Additional Information. March 8, 2002.

(DCS, 2002b) Ihde, R., Duke Cogema Stone & Webster. Letter to U.S. Nuclear Regulatory Commission, RE: Mixed Oxide Fuel Fabrication Facility Construction Authorization Request Revision. October 31, 2002 (including page changes through February 9, 2005).

(NRC, 2000) U.S. Nuclear Regulatory Commission. NUREG-1718, "Standard Review Plan for the Review of an Application for a Mixed Oxide (MOX) Fuel Fabrication Facility." Washington, DC, January 2000.

(NRC, 2003) Pierson, R., U.S. Nuclear Regulatory Commission. Letter to P. Hastings, Duke Cogema Stone & Webster, RE: Design Basis Threat. July 28, 2003.

13.2 Material Control and Accounting

13.2.1 Conduct of Review

The staff reviewed the applicant's treatment of material control and accounting (MC&A) issues in the revised CAR (DCS, 2002) for the MFFF using the guidance in Section 13.2 of NUREG-1718 (NRC, 2000). The purpose of this review was to establish whether the applicant's design basis information for systems relevant to MC&A, and related commitments, will lead to an acceptable Fundamental Nuclear Material Control (FNMC) Plan that will meet the requirements of 10 CFR Part 74, "Material Control and Accounting of Special Nuclear Material."

The material control system must be designed to protect against, detect, and respond to the loss or diversion of nuclear materials. The material accounting system must be designed to determine the quantities of nuclear materials in a licensee's possession, maintain knowledge of such materials, verify the presence of such materials, and detect the loss or diversion of such materials. The staff reviewed and evaluated information provided by the applicant that is related to the design bases for the proposed MC&A program. The following physical aspects of the MC&A design were reviewed:

- process monitoring
- item monitoring
- alarm resolution
- quality assurance and accounting
- international safeguards

13.2.1.1 Process Monitoring

The applicant identified a detailed segmentation of the main processes into process control subunits which are capable of monitoring the status of licensed material in process. The applicant also integrated MC&A features into the proposed facility's manufacturing management information system (MMIS). The staff evaluated the applicant's process monitoring program and finds that this program consists of adequate design features in its abrupt loss detection capability, including process subdivision and measurement points, material control tests, location categorization, material substitution, trend analysis, material exemptions, and research and development operations.

13.2.1.2 Item Monitoring

The applicant committed to an item monitoring program, which establishes item identification and the basis for verifying the presence and integrity of licensed nuclear materials. The applicant also applied MC&A aspects of this program to the facility design and the highly automated remote-controlled process and manufacturing facility features, including its MMIS. The staff has reviewed the elements of this item monitoring program and finds that the applicant's program is capable of providing timely plant-wide detection of the loss of items and real-time status of nuclear materials, including a system of item identification, item classification, tamper-safing procedures, material accessibility, item accounting and control procedures, item measurements, sample items, and item verification tests.

13.2.1.3 Alarm Resolution

The applicant identified features of an alarm resolution system which is capable of resolving the nature and cause of any MC&A alarm within an approved time period, notifying the NRC of any unresolved alarms, and establishing and maintaining the ability to respond to any suspected thefts of licensed nuclear material. The staff has reviewed the elements of this alarm resolution program and finds that the applicant's program commitments demonstrate it will have the ability to respond to and resolve MC&A alarms of potential loss of nuclear materials, including alarm resolution procedures, decision rules, response time, item discrepancies, alarm reporting responsibilities, and any suspected thefts.

13.2.1.4 Quality Assurance and Accounting

The applicant identified feasible approaches and methods with respect to the 11 elements of its MC&A Quality Assurance Program, which includes management structure, personnel training and qualification, measurement systems, measurement control, physical inventory, records system and maintenance, shipments and receipts, scrap material control, human errors, independent assessment, and material custodial responsibilities. The staff has reviewed the elements of these quality assurance and accounting programs and finds that these program commitments are appropriate and acceptable.

13.2.1.5 International Safeguards

The applicant outlined design features for potential future interaction involving the International Atomic Energy Agency and related international safeguards agreements and regulatory requirements. The staff finds that these design features are adequate.

13.2.2 Evaluation Findings

The staff concludes that the applicant provided adequate commitments to submit, as part of its application for a license to possess and use licensed materials, an FNMC Plan that will meet the 10 CFR Part 74 requirements. The staff finds that, at the construction authorization approval stage, MC&A issues have been adequately addressed.

13.2.3 References

(DCS, 2002) Ihde, R., Duke Cogema Stone & Webster. Letter to U.S. Nuclear Regulatory Commission, RE: Mixed Oxide Fuel Fabrication Facility Construction Authorization Request Revision. October 31, 2002 (including page changes through February 9, 2005).

(NRC, 1995) U.S. Nuclear Regulatory Commission. NUREG-1280, "Standard Format and Content Acceptance Criteria for the Material Control and Accounting (MC&A) Reform Amendment, Rev. 1." Washington, DC, April 1995.

(NRC, 2000) U.S. Nuclear Regulatory Commission. NUREG-1718, "Standard Review Plan for the Review of an Application for a Mixed Oxide (MOX) Fuel Fabrication Facility." Washington, DC, January 2000.

14. EMERGENCY MANAGEMENT

14.1 Conduct of Review

This chapter of the final safety evaluation report (FSER) reviews the emergency management and related information provided by Duke Cogema Stone & Webster (DCS or the applicant) during a meeting with the staff on January 4, 2001 (NRC, 2001), further related information presented by the applicant in Chapters 5, 11, and 14 of the revised construction authorization request (CAR) (DCS, 2002), and information in the revised CAR Table 5.6-1 (principal structures, systems, and components (PSSCs) related to emergency conditions). The staff used the applicable portions of Chapter 14.0 in NUREG-1718, "Standard Review Plan for the Review of an Application for a Mixed Oxide (MOX) Fuel Fabrication Facility," issued 2000 (NRC, 2000), as guidance in performing the review. The staff reviewed how the information in the revised CAR, and the other information referenced above, addresses the following regulations:

- 10 CFR 70.23(b) states, as a prerequisite to construction approval, that the design bases of the PSSCs and the Quality Assurance Program must be found to provide reasonable assurance of protection against natural phenomena and the consequences of potential accidents.

- 10 CFR 70.64, "Requirements for New Facilities or New Processes at Existing Facilities," requires that baseline design criteria (BDC) and defense-in-depth practices be incorporated into the design of new facilities. With respect to emergency capability, 10 CFR 70.64(a)(6) requires that the design of the proposed facility provide for emergency capability to maintain control of (1) licensed material and hazardous chemicals produced from licensed material, (2) evacuation of onsite personnel, and (3) onsite emergency facilities and services that facilitate the use of available offsite services.

As discussed further below, the staff evaluated the applicant's safety assessment provided in Chapter 5 of the revised CAR (including the PSSCs related to emergency conditions referenced in revised CAR Table 5.6-1) to ensure that the applicant considered all appropriate accident initiators and to ensure that the BDC for emergency capability were met. The staff found that the applicant considered hazards from internal, external, and natural phenomena. Those accident initiators that were not used as parts of the design basis were adequately explained. While DCS did not specifically cite 10 CFR 70.64(a)(6) in the revised CAR, it did address how it planned to meet all of the BDC in the slides it presented at a meeting on January 4, 2001 (NRC, 2001). In the presentation, the applicant indicated that it would meet the BDC for emergency capability through emergency planning design criteria and the site work task agreement (WTA), including integration with the U.S. Department of Energy (DOE) Savannah River Site (SRS) Emergency Plan; the Emergency Onsite Mixed Oxide Fuel Fabrication Facility Evacuation Plan; utilization of existing onsite facilities and services at SRS, coordinated through WTA; and safe havens provided to personnel while controlling potential losses of licensed material.

The applicant's assessment of potential loss of control of radioactive material is discussed in Section 5.5.2.1, "Loss of Confinement/Dispersal of Nuclear Material Events," of the revised CAR. In Section 5.5.2.10.6.2, "Events Involving Hazardous Chemicals and Radioactive Material," the applicant concluded that the chemical events would be bounded by the radioactive release events and no additional PSSCs in this area would be required. The staff finds that the applicant has adequately considered all of the event categories in its analysis, for emergency planning purposes.

In revised CAR Section 5.5.2.1.6, the applicant identified the following 10 potential loss-of-confinement events:

(1)   over-temperature

(2)   corrosion

(3)   small breaches in glovebox confinement boundary or backflow from a glovebox through utility lines

(4)   leaks of aqueous polishing process vessels or pipes within process cells

(5)   canister handling operations

(6)   rod handling operations

(7)   breaches in containers outside gloveboxes caused by handling operations

(8)   over/underpressurization of glovebox

(9)   excessive temperature caused by decay heat from radioactive materials

(10) glovebox dynamic exhaust failure

The applicant identified the following 12 PSSCs either to prevent these hypothetical events from occurring, or to mitigate the potential consequences:

(1)   the safety instrumentation and controls system
(2)   C3 confinement system
(3)   material maintenance and surveillance programs
(4)   the C4 confinement system
(5)   the process cell
(6)   process cell entry controls
(7)   3013 canister outer can opening device
(8)   M 3013 canister
(9)   material-handling equipment and material-handling controls
(10) glove box pressure controls
(11) the high-depressurization exhaust system
(12) training and procedures

The staff's evaluation of the applicant's methodology concerning loss of confinement is presented in Chapter 5 of this revised FSER. The staff notes that 10 CFR 20.1801, "Security of Stored Material," and 10 CFR 20.1802, "Control of Material Not in Storage," have specific requirements for control of licensed material, and the applicant will need to address these requirements as part of any later license application it may submit.

[Text removed under 10 CFR 2.390]

[Text removed under 10 CFR 2.390]

14.2 Evaluation Findings

In Section 14 of the revised CAR, DCS provided design basis information pertaining to emergency management measures for the proposed facility. Based on the staff's review of the revised CAR and supporting information provided by the applicant relevant to emergency management, the staff finds that DCS, at the construction authorization stage, has met the BDC for emergency capability set forth in 10 CFR 70.64(a)(6). Accordingly, the staff concludes, pursuant to 10 CFR Section 70.23(b), that the design bases of the PSSCs pertaining to emergency management provide reasonable assurance of protection against natural phenomena and the consequences of potential accidents.

14.3 References

(DCS, 2002b) Ihde, R., Duke Cogema Stone & Webster. Letter to U.S. Nuclear Regulatory Commission, RE: Mixed Oxide Fuel Fabrication Facility Construction Authorization Request Revision. October 31, 2002 (including page changes through February 9, 2005).

(NRC, 1998) U.S. Nuclear Regulatory Commission. NUREG/CR-6410,"Nuclear Fuel Cycle Facility Accident Analysis Handbook." Washington, DC, 1998.

(NRC, 2000) U.S. Nuclear Regulatory Commission. NUREG-1718, "Standard Review Plan for the Review of an Application for a Mixed Oxide (MOX) Fuel Fabrication Facility." Washington, DC, January 2000.

(NRC, 2001) Persinko, A., U.S. Nuclear Regulatory Commission. Memorandum to E. Leeds, U.S. Nuclear Regulatory Commission, RE: January 4–5, 2001, Summary of Meeting with Duke Cogema Stone & Webster to Discuss Design Basis for the Mixed Oxide Fuel Fabrication Facility. January 24, 2001.

15. MANAGEMENT MEASURES

This chapter of the final safety evaluation report (FSER) contains the staff's review of management measures described by the Duke Cogema Stone & Webster (DCS or the applicant) in Chapter 15 of the revised construction authorization request (CAR) (DCS, 2002b). Management measures are defined in Title 10, Section 70.4, "Definitions," of the *Code of Federal Regulations* (10 CFR 70.4) as functions, performed by a licensee, generally on a continuing basis, that are applied to items relied on for safety (IROFS), to provide reasonable assurance that the items are available and reliable to perform their functions, when needed. As further stated in the 10 CFR 70.4 definition, management measures include configuration management (CM), maintenance, training and qualifications, procedures, audits and assessments, incident investigations, records management, and other quality assurance (QA) elements. Management measures are made applicable to the revised CAR review by 10 CFR 70.64(a)(1). For purposes of establishing quality standards, 10 CFR 70.64(a)(1) requires that the design of the proposed mixed oxide (MOX) fuel fabrication facility (MFFF, or the facility) "must be developed and implemented in accordance with management measures, to provide adequate assurance that IROFS will be available and reliable to perform their function when needed."

In the MOX Project Quality Assurance Plan (MPQAP), Revision 2, the applicant had previously addressed the QA program description requirements (which are referenced in FSER Section 15.1). The staff issued a safety evaluation report (SER) approving the MPQAP, including additional Duke Cogema, Stone & Webster (DCS) clarification and commitments regarding two QA issues, on October 1, 2001 (NRC, 2001, Section 15.1). Subsequently, DCS revised the MPQAP to address the additional clarifications and commitments and submitted Revision 3 of the MPQAP (DCS, 2002a, Section 15.1). The staff verified that the MPQAP, Revision 3, adequately incorporated all of the DCS clarifications and commitments noted in the SER and approved the MPQAP, Revision 3, for construction activities (NRC, 2003, Section 15.1). The staff concluded, pursuant to 10 CFR 70.23(b), that the MPQAP, as applied to all structures, systems, and components (SSCs) of the proposed facility, will provide reasonable assurance of protection against natural phenomena and the consequences of potential accidents. The scope of this 10 CFR 70.23(b) finding was described in the conclusion of the MPQAP SER as pertaining to the construction of the SSCs for the proposed facility, including all related design procurement and fabrication activities. As further stated in the conclusion to the MPQAP SER, this finding did not pertain to any startup testing or operation of the proposed MFFF. Accordingly, since the staff in its October 1, 2001, SER, approved the applicant's general QA program as described above, this FSER focuses on the additional QA elements and other management measures described in revised CAR Chapter 15 and evaluates these management measures pursuant to 10 CFR 70.64(a)(1). The objective of this FSER review is to determine whether the proposed management measures, together with the previously approved MPQAP, establish a QA program which provides reasonable assurance of protection against natural phenomena and the consequences of potential accidents, as required for revised CAR approval by 10 CFR 70.23(b). For the reasons stated in the following FSER sections, the staff concludes, pursuant to 10 CFR 70.23(b), that the proposed management measures set forth in the revised CAR, together with the previously approved MPQAP, establish a QA program which provides reasonable assurance of protection against natural phenomena and the consequences of potential accidents.

The staff evaluated the information provided by the applicant for management measures by reviewing Chapter 15 of the revised CAR, other sections of the revised CAR, supplementary information provided by the applicant, and relevant documents available at the applicant's offices but not submitted by the applicant. The review of management measures was closely coordinated with the review of accident sequences described in the safety assessment of the design bases (see Chapter 5 of this FSER). Note that management measures will be applied to IROFS, and that IROFS need not be specified until the applicant submits its integrated safety analysis (ISA) summary (see 10 CFR 70.65(b)(6)). Thus, the staff will perform a more detailed evaluation of management measures as part of the review of any later license application DCS may submit.

The review of the applicant's description of management measures is addressed in the following sections in the order that the applicant presented them in the revised CAR, beginning with QA. The staff used Chapter 15.0 in NUREG-1718, "Standard Review Plan for the Review of an Application for a Mixed Oxide (MOX) Fuel Fabrication Facility," issued 2000, as guidance in performing the review.

15.1 Quality Assurance

15.1.1 Conduct of Review

The staff reviewed the QA descriptions and commitments of the applicant's revised CAR (DCS, 2002b) for the MFFF in accordance with NUREG-1718. The purpose of this review is to establish that the applicant has a plan to integrate its implementation of management measures into its approved MPQAP, which is applicable to the design, fabrication, and construction of all SSCs, including those designated as principal SSCs (PSSCs) (NRC, 2001). The following regulations apply specifically to QA:

- 10 CFR Part 21, "Reporting of Defects and Noncompliance," describes regulatory requirements for identifying, controlling, and reporting defects in a facility, activity, or basic component supplied to a facility licensed under the Atomic Energy Act which could create a substantial safety hazard.

- 10 CFR 70.64(a)(1) states that, for quality standards, the facility design must be developed and implemented in accordance with management measures, to provide adequate assurance that IROFS will be available and reliable to perform their function when needed.

- 10 CFR 70.64(a)(8) states that the design of IROFS must provide for inspection and testing to ensure that IROFS will be available and reliable to perform their functions when needed.

- 10 CFR 70.23(b) states, as a prerequisite to revised CAR approval, that the QA program must be found to provide reasonable assurance of protection against natural phenomena and the consequences of potential accidents. This section further states that the criteria in Appendix B, "Quality Assurance Criteria for Nuclear Power Plants and Fuel Reprocessing Plants," (referred to hereafter as Appendix B) to 10 CFR Part 50, "Domestic Licensing of Production and Utilization Facilities," will be used in determining whether the

QA program provides reasonable assurance of protection against natural phenomena and the consequences of potential accidents.

In the revised CAR Section 15.1, the applicant stated that the application of management measures will ensure that the PSSCs are available and reliable to perform their intended design functions.

The applicant submitted MPQAP, Revision 2, as the required description of a QA program that meets the general QA requirements of Appendix B. The applicant has committed to compliance with the provisions of Parts I and II of American Society of Mechanical Engineers (ASME) NQA-1-1994, "Quality Assurance Program Requirements for Nuclear Facilities," as revised by the ASME NQA-1a-1995 Addenda, issued 1994 (ASME, 1994), and U.S. Nuclear Regulatory Commission (NRC) RG 1.28, Revision 3, "Quality Assurance Program Requirements (Design and Construction), issued 1985 (NRC, 1985)." These ASME NQA-1 and G provisions are hereafter referred to as NQA-1. The staff reviewed, compared, and evaluated the MPQAP against the Appendix B requirements and the NQA-1 provisions, as well as NUREG-1718 guidance. The QA areas reviewed by the staff include the applicant's descriptions in the MPQAP for organization, QA function, design control, procurement document control, instructions, procedures, drawings, document control, control of purchased items, identification and control of items, control of special processes, inspection, test control, control of measuring and test equipment, handling, storage, and shipping, inspection, test, and operating status, nonconformances, corrective action, QA records, audits and assessments, and the applicant's provisions for continuing QA. The results of this review are documented in the MPQAP SER dated October 1, 2001 (NRC, 2001). Additional discussions on the implementation and application of the applicant's QA program description in the MPQAP, Revision 2, as supplemented by responses to requests for additional information (RAIs), were held during telephone meetings and in-office reviews, particularly on the quality level categorization commitments, including classification criteria and the categorization process. DCS incorporated all additional clarifications and commitments referred to in the MPQAP SER into Revision 3 of the MPQAP (DCS, 2002a). The staff reviewed the MPQAP, Revision 3, verified that it appropriately incorporated all commitments and clarifications, and approved it for construction application (NRC, 2003). In MPQAP, Revision 3, the applicant commits to invoking and complying with the applicable requirements of 10 CFR Part 21 for all design, procurement, fabrication, and construction activities.

15.1.2 Evaluation Findings

Based on the staff's review of revised CAR Section 15.1 and the clarifications and commitments made by the applicant in response to NRC RAIs relevant to the QA program, the staff finds that DCS has appropriately committed to invoking the applicable 10 CFR Part 21 requirements for design, procurement, and construction. Additionally, the staff finds, pursuant to 10 CFR 70.64(a)(1), that the PSSCs are being properly designed and developed, thus providing adequate assurance that IROFS will be available and reliable to perform their intended design functions when needed. However, as stated in the introduction to this chapter, the staff will perform a more detailed evaluation of management measures as part of the review of the license application, which the applicant plans to submit to the NRC if its revised CAR is approved. The staff also finds, pursuant to 10 CFR 70.64(a)(8), that the inspection and testing elements of QA are in place, so as to adequately ensure that IROFS will be available and reliable to perform their functions when needed.

Based on the above findings, together with its earlier approval of the MPQAP (NRC, 2001; DCS, 2002a), the staff concludes, pursuant to 10 CFR 70.23(b), that the QA program at the proposed facility will provide reasonable assurance of protection against natural phenomena and the consequences of potential accidents. The scope of this conclusion pertains to the construction of the facility's PSSCs and includes all related design, procurement, and fabrication activities but does not include any startup testing or operation of the proposed facility.

15.1.3 References

(ASME, 1994) American Society of Mechanical Engineers. ASME–NQA–1–1994, "Quality Assurance Requirements for Nuclear Facility Applications" (as revised by the ASME NQA-1a-1995 Addenda). New York, 1994/1995.

(DCS, 2002a) Hastings, P., Duke Cogema Stone & Webster. Letter RE: Quality Assurance Program for Construction of the Mixed Oxide Fuel Fabrication Facility. March 26, 2002.

(DCS, 2002b) Ihde, R., Duke Cogema Stone & Webster. Letter to U.S. Nuclear Regulatory Commission, RE: Mixed Oxide Fuel Fabrication Facility Construction Authorization Request Revision. October 31, 2002 (including page changes through February 9, 2005).

(NRC, 1985) U.S. Nuclear Regulatory Commission. Regulatory Guide 1.28, Rev. 3, "Quality Assurance Program Requirements (Design and Construction)." Washington, DC, August 1985.

(NRC, 2001) Persinko, A., U.S. Nuclear Regulatory Commission. Letter to P. Hastings, Duke Cogema Stone & Webster, RE: Duke Cogema Stone & Webster Quality Assurance Program for the Construction of the MFFF. October 1, 2001.

(NRC, 2003) Persinko, A., U.S. Nuclear Regulatory Commission. Letter to P. Hastings, Duke Cogema, Stone & Webster, RE: DCS Quality Assurance Program for Construction of the Mixed Oxide Fuel Fabrication Facility. January 10, 2003.

15.2 Configuration Management

15.2.1 Conduct of Review

This section of the FSER contains the staff's review of the CM system committed to by the applicant in Section 15.2 of the revised CAR (DCS, 2002). The objective of this review is to verify that the applicant has adequately planned for the implementation of an acceptable CM system which will provide reasonable assurance that the PSSCs identified by the applicant will be available and reliable to perform their safety function when needed. The review, for construction authorization, is to determine whether the applicant has adequately planned for CM to be accomplished during design and construction and whether necessary policies, personnel, procedures, and instructions will be in place to begin CM during the design and construction of the PSSCs.

The following regulations apply specifically to CM:

- 10 CFR 70.4 defines the term "configuration management" as a management measure "that provides oversight and control of design information, safety information, and records of modifications (both temporary and permanent) that might impact the ability of items relied on for safety to perform their functions when needed."

- 10 CFR 70.64(a)(1) states that, for quality standards, the MFFF design must be developed and implemented in accordance with management measures, to provide adequate assurance that items relied on for safety will be available and reliable to perform their function when needed.

Section 15.2.3, "Areas of Review," of NUREG-1718 defines the review areas and states that the description of the CM system is to be reviewed with emphasis on the processes for documenting an established baseline configuration and controlling changes to it to preclude inadvertent degradation of safety. The review is to include the applicant's descriptions of the organizational structure responsible for CM activities and the process, procedures, and documentation required for modifying SSCs, PSSCs, and IROFS and the supporting management measures. The review is to focus on the applicant's management level controls that ensure (1) the disciplined documentation of engineering, installation, and operation of modifications, (2) the training and qualification of affected staff; (3) revision and distribution of operating, test, calibration, surveillance, and maintenance procedures and drawings, (4) postmodification testing, and (5) operational readiness review. The review topics are to include CM policy, design requirements, document control, change control, and assessments.

15.2.1.1 CM Policy

The applicant described its overall CM system in revised CAR Section 15.2 and presented its CM policy in Section 15.2.1, which states that CM is provided for PSSCs throughout facility design, construction, testing, operation and deactivation, to provide the means to establish and maintain a technical baseline for the facility. The applicant's CM system during design and construction is the responsibility of the Deputy Project Manager—MFFF Engineering and Construction. The CM system controls documents in accordance with QA procedures for design control, document control, and records management. During the design and construction phases, the applicant's CM is, and will be, based on Section 3, "Design Control," of the MPQAP and associated MPQAP requirements and procedures for design and construction documents and activities that establish and maintain the technical baseline. The staff reviewed these MPQAP commitments and requirements for CM and determined that they were acceptable for construction activities, including design, procurement, and fabrication. The staff review and conclusion were documented in an SER, on October 1, 2001 (NRC, 2001). Design documents and changes undergo interdisciplinary review and verification. Proper implementation is verified and reflected in the design basis documentation. The applicant will update the CM description to include details of the operational CM program in the license application. If the facility is licensed to operate, any changes to the CM program would be governed by 10 CFR 70.72, "Facility Changes and Change Process."

Section 15.2 of the revised CAR states that CM is applied to SSCs. The applicant addressed its commitment to CM application during design and construction and specifically described (1) establishing and controlling the design bases to include all SSCs, not just PSSCs and IROFS, (2) how the CM process functions for documenting the baseline configuration and controlling all changes, and (3) how it provides for change control during construction (i.e., that the same design control procedures are to be used for all SSCs, not just PSSCs and IROFS). During any construction of the proposed facility, all field changes, as-built configurations, and nonconformances will be reviewed for impact to the design basis.

In Section 15.2 of the revised CAR, the applicant described the design documents under CM, which include calculations, safety analyses, design criteria, engineering drawings, system descriptions, technical documents, and specifications that establish design requirements. The scope of CM expands throughout the design process. During any construction, startup, and operations, the scope of documents under CM would expand to include vendor, test, and inspection data; startup, test, operating, and administrative procedures; and nonconformance reports. These documents will include those generated through functional interfaces with QA, maintenance, and training and qualifications of personnel. The applicant described how the CM system is implemented through or related to other management measures and described these interfaces and relationships.

The approved MPQAP establishes the framework for the applicant's CM system and other management measures for PSSCs (NRC, 2001; NRC, 2003). The CM system records are generated and processed in accordance with the requirements of the MPQAP. Maintenance requirements are established as part of the design basis which is controlled under CM, and records provide evidence of compliance with preventive and corrective maintenance schedules. Training and qualification of personnel are controlled in accordance with the MPQAP provisions and will be considered part of the design basis controlled under CM. Corrective actions and changes resulting from audits, assessments, and incident investigations will be evaluated and controlled in accordance with provisions of the MPQAP and QA procedures. Plant procedures will be controlled in accordance with the MPQAP and QA procedures and will be reviewed for impact on the design basis.

The applicant's description of CM includes the designation of PSSCs under the QA classification and grading provisions of Section 2 of the MPQAP. The grading approach to SSCs includes applying the most stringent QA controls to SSCs with the highest safety significance. All controls for all PSSCs are controlled under CM and documented in the same manner. The applicant's commitments for SSC classification and the grading process are discussed in the SER approving the MPQAP (NRC, 2001), and are incorporated in MPQAP, Revision 3 (NRC, 2003).

15.2.1.2 Design Requirements

The applicant stated that the organization structure and staffing interfaces for the CM system, for design and construction, will be administered by the MFFF engineering organization. The lead discipline engineers will have primary technical responsibility for the work performed by their disciplines and will be responsible for the conduct of interdisciplinary reviews. Reviews will also be conducted by construction management, operations, QA, and procurement personnel. The design control process will also interface with document control and records management processes controlled by QA procedures. The PSSCs are designated as Quality Level 1 (QL-1) using the approved MPQAP classification process, and their associated design documents will be subject to review and verification. Analyses constituting the safety assessment of the design bases are subject to these same requirements, and changes are evaluated to ensure consistency with the design bases. Design bases documented in CAR Chapters 5 through 11 will be consistent with those in, and flowed down from, the design requirements and basis of design documents, analyses, specifications, and drawings. The CM system will capture these requirements and resulting design bases in accordance with design control, document control, and records management procedures.

The SSCs will be classified based on their safety significance and role in preventing or mitigating design basis accidents in accordance with the categories of QA classification described in Section 2 of the MPQAP and approved by the staff (NRC, 2001; NRC, 2003).

15.2.1.3 Document Control

Document control will be implemented in accordance with Section 6 of the MPQAP. These provisions were reviewed by the staff and found to be adequate for design and construction of the proposed facility (NRC, 2001).

15.2.1.4 Change Control

The applicant's description of control of changes to the technical baseline identifies that these changes are controlled under Section 3, "Design Control," of the MPQAP and associated procedures. The change control process includes technical, management, and safety reviews before implementation. The review process includes reviews to ensure consistency with the approved safety assessment of the design bases of PSSCs and includes provisions for appropriate reviews at the design, construction, and operations phases.

15.2.1.5 Assessments

The applicant confirms that assessments, including initial and periodic examinations of the CM system, will be conducted to determine the system's effectiveness and to correct deficiencies. The applicant committed that such assessments will be systematically planned and conducted in accordance with an overall facility audit and assessment program as described by DCS in revised CAR Section 15.6.

The applicant also committed to updating the CM system to reflect any changes to the proposed facility made between submittal of the revised CAR and the expected submittal of its later license application.

15.2.2 Evaluation Findings

In Chapter 15.2 of the revised CAR, DCS committed to implement and update its CM system at the proposed facility. Management-level policies and procedures, including a safety review of any proposed activity involving SSCs, are described that will ensure that the relationship between design requirements, construction, and facility documentation is maintained as part of a new design or change in an existing design. The administrative control will ensure that the organizational structure, procedures, and responsibilities necessary to implement CM are in place or committed to; that the design requirements and bases are documented and supported by analyses and the documentation is maintained current; that documents, including drawings, are appropriately stored and accessible; that drawings and related documents adequately describe SSCs; that procedures adequately describe how the applicant will achieve and maintain strict consistency among the design requirements, facility construction, and facility documentation; and that methods are in place for suitable analysis, review, approval, and implementation of identified changes to SSCs. The applicant described its approach to QL categorization and grading of controls for SSCs and identified the process, criteria, and control to be applied. Pursuant to 10 CFR 70.64(a)(1), the staff finds that the applicant's proposed CM system is a management measure ensuring that the facility design is being properly developed and implemented, so as to provide adequate assurance that IROFS will be available and reliable to perform their function when needed. The applicant will describe its operation phase change process, for 10 CFR 70.72 provisions, in more detail in its license application. Accordingly, the staff concludes, pursuant to 10 CFR 70.23(b), that the CM system set forth in the MPQAP and the revised CAR is part of a QA program which will provide reasonable assurance of protection against natural phenomena and the consequences of potential accidents.

15.2.3 References

(DCS, 2001) Hastings, P., Duke Cogema Stone & Webster. Letter to U.S. Nuclear Regulatory Commission, RE: Response to Request for Additional Information (DCS-NRC-000059). August 31, 2001.

(DCS, 2002) Ihde, R., Duke Cogema Stone & Webster. Letter to U.S. Nuclear Regulatory Commission, RE: Mixed Oxide Fuel Fabrication Facility Construction Authorization Request Revision. October 31, 2002 (including page changes through February 9, 2005).

(DOE, 1993) U.S. Department of Energy. DOE–STD–1073–93–Pt.1 and –Pt.2, "DOE Standard Guide for Operational Configuration Management Program." Washington, DC, 1993.

(NRC, 2001) Persinko, A., U.S. Nuclear Regulatory Commission. Letter to P. Hastings, Duke Cogema Stone & Webster, RE: DCS Quality Assurance Program for Construction of the MFFF. October 1, 2001.

(NRC, 2003) Persinko, A., U.S. Nuclear Regulatory Commission. Letter to P. Hastings, Duke Cogema Stone & Webster, RE: DCS Quality Assurance Program for Construction of the MFFF. January 10, 2003.

15.3 Maintenance

15.3.1 Conduct of Review

This section of the FSER contains the staff's review of the Maintenance Program committed to by the applicant in Section 15.3 of the revised CAR (DCS, 2002). The objective of this review is to determine whether the proposed facility will have a maintenance program which will provide adequate assurance that IROFS (other than personnel activities) will be available and reliable to perform their safety function when needed.

The following regulations apply specifically to the staff's review of the proposed Maintenance Program:

- 10 CFR 70.4 defines the term "management measures" as including a maintenance program.

- 10 CFR Section 70.64(a)(1) states that the MFFF design must be developed and implemented in accordance with management measures to provide adequate assurance that items relied on for safety will be available and reliable to perform their function when needed.

- 10 CFR Section 70.64(a)(8) states that the design of IROFS must provide for maintenance to ensure that IROFS will be available and reliable to perform their functions when needed.

15.3.1.1 Maintenance Program

In revised CAR Section 15.3, DCS described and committed to implementation of a maintenance program, including safety controls, surveillance/monitoring, corrective maintenance, preventive maintenance, functional testing, and work control methods and describes the relationship of the maintenance elements to other management measures. The applicant will describe its maintenance program in more detail in the license application. Preventive maintenance activities, surveillance, and performance trending will be done to provide reasonable and continuing assurance that IROFS will be available and reliable to perform their safety functions.

15.3.1.2 Safety Controls

The applicant committed to providing safety controls by specifying maintenance requirements for calibration frequency, functional testing requirements, and replacement of specified components for IROFS.

15.3.1.3 Maintenance Elements

The applicant's description requires that surveillance and monitoring of IROFS, including instrument calibration and testing, be performed at specified intervals to measure the degree to which IROFS meet performance specifications. The results of surveillances would be trended and, when indicated by potential performance degradation, preventive frequencies adjusted or other corrective action taken. Incident investigations may also identify root causes of failures related to maintenance type or frequency, and lessons learned from these investigations would be factored into the maintenance program. Procedures would prescribe compensatory measures for surveillance tests that could be performed only while equipment was out of service.

Preventive maintenance measures described by the applicant include preplanned and scheduled periodic refurbishment, overhaul, or replacement of IROFS to ensure their continued safety function. Planning would include results of surveillance and monitoring and instrument calibration and testing.

Corrective maintenance would be performed for repair or replacement of equipment that has unexpectedly degraded or failed. Corrective maintenance would restore IROFS to acceptable performance through a planned, systematic, controlled, and documented approach for the activities.

Following initial installation, functional testing of IROFS would be performed as part of the applicant's periodic surveillance testing and also after corrective or preventive maintenance or calibration to ensure that the item is capable of performing its safety function when required.

The functional testing would be conducted using approved procedures, which would include compensatory measures that may be necessary while the test of equipment or systems is being conducted.

15.3.1.4 Work Control Methods

As stated in CAR Section 15.3.3, the applicant committed to maintenance-related work control methods including maintenance management and tracking, which would involve integration of maintenance activities with ongoing operations activities. Work control methods would also include appropriate interfaces with radiation protection and associated work permits, lockout/tagout requirements, and procedures.

15.3.1.5 Maintenance Relationship to Other Management Measures

The applicant's committed maintenance function will interface with the configuration management and procedure systems by obtaining the approved and controlled drawings, specifications, and procedures. Personnel would be trained in the maintenance of IROFS through the training program, and records of performance trends and maintenance history would be maintained.

15.3.2 Evaluation Findings

In Chapter 15.3 of the revised CAR, DCS described and committed to implement its Maintenance Program to be used on IROFS and associated activities at the proposed facility. Pursuant to 10 CFR 70.64(a)(1), the staff finds that the applicant's proposed Maintenance Program is a management measure providing adequate assurance that IROFS will be available and reliable to perform their function when needed. Additionally, pursuant to 10 CFR 70.64(a)(8), the staff finds that the maintenance elements of QA are in place, so as to adequately ensure that IROFS will be available and reliable to perform their functions when needed. The applicant will describe its maintenance program in more detail in its license application. Accordingly, the staff concludes, pursuant to 10 CFR 70.23(b), that the Maintenance Program set forth in the revised CAR is part of a QA program which will provide reasonable assurance of protection against natural phenomena and the consequences of potential accidents.

15.3.3 References

(DCS, 2002) Ihde, R., Duke Cogema Stone & Webster. Letter to U.S. Nuclear Regulatory Commission, RE: Mixed Oxide Fuel Fabrication Facility Construction Authorization Request Revision. October 31, 2002 (including page changes through February 9, 2005).

(NRC, 1984) U.S. Nuclear Regulatory Commission. Inspection Manual, Procedure 88025, "Maintenance and Surveillance Testing." Washington, DC, May 23, 1984.

(NRC, 1989) U.S. Nuclear Regulatory Commission. "Guidance on Management Controls/Quality Assurance, Requirements for Operation, Chemical Safety, and Fire Protection for Fuel Cycle Facilities." *Federal Register:* Vol. 54, No. 53., pp. 11590–11598. March 21, 1989.

(NRC, 1996) U.S. Nuclear Regulatory Commission. Inspection Manual, Procedure 88062, "Maintenance and Inspection. Washington, DC, January 1996.

(NRC, 1997) U.S. Nuclear Regulatory Commission. Regulatory Guide 1.160, Rev. 2, "Monitoring the Effectiveness of Maintenance at Nuclear Power Plants. Washington, DC, March 1997.

(NRC, 2003) Persinko, A., U.S. Nuclear Regulatory Commission. Letter to P. Hastings, Duke Cogema, Stone & Webster, RE: DCS Quality Assurance Program for Construction of the MOX Fuel Fabrication Facility. January 10, 2003.

15.4 Training and Qualification

15.4.1 Conduct of Review

This section of the FSER contains the staff's review of the training and qualification information provided by the applicant in Section 15.4 of the revised CAR (DCS, 2002). The objective of this review is to determine whether personnel who would perform activities relied on for safety will understand, recognize the importance of, and be qualified to perform these activities in a manner that adequately protects the public, worker health and safety, and the environment. The staff evaluated the applicant's provisions for training and qualification by reviewing the revised CAR, the applicant's QA program description, the MPQAP, responses to NRC staff RAIs, and relevant documents available at the applicant's offices but not submitted by the applicant.

The following regulations apply specifically to the staff's review of the proposed training and qualification program:

- 10 CFR 70.4 defines the term "management measures" as including a training and qualification program.

- 10 CFR 70.64(a)(1) states that the MFFF design must be developed and implemented in accordance with management measures, to provide adequate assurance that items relied on for safety will be available and reliable to perform their function when needed.

Pursuant to Section 15.4.3 of NUREG-1718, the staff's review of the proposed training and qualification program included the following areas:

- organization and management of training

- analysis and identification of functional areas requiring training

- position training requirements

- development of the basis for training, including objectives

- organization of instruction using lesson plans and other training guides

- evaluation of trainee learning

- conduct of on-the-job training

- evaluation of training effectiveness

- personnel qualification

- applicant's provisions for continuing assurance, including the needs for retraining or reevaluation of qualification

15.4.1.1 Organization and Management of Training

The applicant described its training program for the operations phase of the facility in revised CAR Section 15.4 and stated that training program requirements will apply to plant personnel who perform activities relied on for safety. The applicant's approved MPQAP addresses training and qualification requirements during the facility design and construction phases, including QA training of personnel for nondestructive examination, for inspection and test personnel, and for auditors. The staff reviewed these MPQAP commitments and requirements for training and qualification and determined that they were acceptable for construction activities, including design, procurement, and fabrication (NRC, 2001). DCS specifically committed to the establishment of an operational training program in accordance with its description in revised CAR Section 15.4 and will update the program information in any later license application it may submit.

The applicant's training program description requires that line managers are responsible for the content and effective conduct of training for their personnel. Line managers will be given the authority to implement training, and their responsibilities are included in position descriptions. The applicant's training organization will provide support with planning, directing, analyzing, developing, conducting, evaluating, and controlling a systematic performance-based training process. Plant procedures will establish the requirement for indoctrination and training of personnel performing activities relied on for safety. Lesson plans, which will be used for classroom and on-the-job training, will be included in the CM system and will be updated based on design changes or plant modifications. Auditable training records will be maintained to support management information needs for personnel training, job performance, and qualifications.

15.4.1.2 Analysis and Identification of Functional Areas Requiring Training

The applicant will perform a needs/job analysis and identify tasks to ensure that appropriate training is provided to personnel. A task list will be developed and updated as needed and will be reviewed as part of the systematic valuation of training effectiveness.

15.4.1.3 Position Training Requirements

The applicant will develop minimum training requirements for positions whose activities are relied on for safety. Entry-level criteria will be defined for the positions, including minimum educational, technical, and experience requirements. Initial identification of job-specific training requirements will be based on experience from MELOX and La Hague operations in France and from U.S. experience.

15.4.1.4 Development of the Basis for Training

The applicant will establish learning objectives that identify the training content, based on the needs/job analyses and position-specific requirements. The task list will be used to develop the desired posttraining performance objectives, including the knowledge, skills, and abilities that the trainee should demonstrate; the conditions under which required actions will take place; and the standards of performance the trainee should achieve on completion of the training activity.

15.4.1.5 Organization of Instruction Using Lesson Plans and Other Training Guides

The applicant will use the learning objectives, derived from specific job performance requirements and the needs/job analysis, to develop lesson plans and other training guides. Lesson plans are approved before use and are used for classroom and on-the-job training.

15.4.1.6 Evaluation of Trainee Learning

The applicant will evaluate trainee mastery of learning objectives through observation and demonstration or oral or written tests as appropriate. Evaluations will measure the trainee's skills and knowledge of job performance requirements.

15.4.1.7 Conduct of On-the-Job Training

The applicant's description includes requirements for on-the-job training to be performed for IROFS activities using current performance-based training materials, conducted by designated personnel who are competent in the program standards and methods of conducting the training. Completion of on-the-job training will be demonstrated by actual task performance where feasible and appropriate. When the actual task cannot be performed by the trainee, a simulation of the task is performed using the conditions encountered with task performance, including references, tools, and equipment reflecting the actual task to the extent practical.

15.4.1.8 Evaluation of Training Effectiveness

The applicant will systematically and periodically evaluate the training program's effectiveness in producing competent employees. These evaluations will include feedback from trainees and will identify program strengths and weaknesses, determine whether program content matches current job needs, and determine if corrective actions are needed to improve the program's effectiveness. Evaluation objectives will be developed, results will be documented, and changes made to procedures, practices, or training materials as necessary.

15.4.1.9 Personnel Qualification

The applicant discussed its commitments for personnel qualification in revised CAR Section 15.4.9. Qualification requirements for key management positions, addressed in revised CAR Chapter 4, are in accordance with the applicable guidance in Section 15.4.4.3 of NUREG-1718. Revised CAR Chapter 4 currently stresses the organization for design and construction of the proposed facility. The applicant will supplement the information in Chapter 4 when it submits its license application to address operations and the qualifications of key plant management positions.

15.4.1.10 Applicant's Provisions for Continuing Assurance

The applicant's description of its provisions for continuing assurance of training and qualification of plant personnel includes an evaluation of personnel performing activities relied on for safety to determine that they continue to understand, recognize the importance of, and have the appropriate qualifications needed to perform their activities. The evaluation may be by written or oral test or by performance evaluation. The evaluation results will be documented, and retraining or other appropriate action taken when indicated. Retraining will also be required for plant modifications, procedure changes, and QA program changes when needed.

15.4.2 Evaluations Findings

In Chapter 15.4 of the revised CAR, DCS described its management measures for training and qualification to be used during the operations phase of the proposed facility. Based on that information and the discussion provided in the sections above for training and qualification, the staff finds that the personnel who perform activities relied on for safety will understand, recognize the importance of, and be qualified to perform these activities in a manner that will adequately protect the public, worker health and safety, and the environment. The applicant will establish an operational training program in accordance with its description in revised CAR Section 15.4. Pursuant to 10 CFR 70.64(a)(1), the staff finds that the applicant's proposed training and qualification program is a management measure providing adequate assurance that IROFS will be available and reliable to perform their function when needed. The applicant will update the training program description for operations activities in its license application. Accordingly, the staff concludes, pursuant to 10 CFR 70.23(b), that the management measures for training and qualification set forth in the MPQAP and the revised CAR are part of a QA program which will provide reasonable assurance of protection against natural phenomena and the consequences of potential accidents.

15.4.3 References

(ASME, 1994) American Society of Mechanical Engineers. ASME–NQA–1–1994, "Quality Assurance Requirements for Nuclear Facility Applications." New York, 1994.

(DCS, 2002) Ihde, R., Duke Cogema Stone & Webster. Letter to U.S. Nuclear Regulatory Commission, RE: Mixed Oxide Fuel Fabrication Facility Construction Authorization Request Revision. October 31, 2002 (including page changes through February 9, 2005).

(NRC, 2001) Persinko, A., U.S. Nuclear Regulatory Commission. Letter to P. Hastings, Duke Cogema Stone & Webster, RE: Duke Cogema Stone & Webster Quality Assurance Program for the Construction of the MFFF. October 1, 2001.

(NRC, 1993) U.S. Nuclear Regulatory Commission. NUREG-1220, Rev. 1, "Training Review Criteria and Procedures." Washington, DC, January 1993.

(NRC, 2003) Persinko, A., U.S. Nuclear Regulatory Commission. Letter to P. Hastings, Duke Cogema Stone & Webster, RE: Duke Cogema Stone & Webster Quality Assurance Program for the Construction of the MFFF. January 10, 2003.

15.5 Plant Procedures

15.5.1 Conduct of Review

This section of the FSER contains the staff's review of procedure information submitted by the applicant in Section 15.5 of the revised CAR (DCS, 2002). The objective of this review is to determine whether the applicant can adequately control potential facility operations in areas to be identified as IROFS, by developing, reviewing, approving, and controlling the implementation of written plant procedures that would protect the workers, the public, and the environment during any potential testing, startup, and operation of the facility. The staff evaluated the applicant's provisions for plant procedures by reviewing Section 15.5 of the revised CAR, other sections of the revised CAR, responses to NRC staff RAIs, and relevant documents available at the applicant's offices but not submitted by the applicant.

The following regulations apply specifically to the staff's review of the proposed plant procedures:

- 10 CFR 70.4 defines the term "management measures" as including plant procedures.

- 10 CFR 70.64(a)(1) states that the MFFF design must be developed and implemented in accordance with management measures, to provide adequate assurance that IROFS will be available and reliable to perform their function when needed.

For the construction approval, in accordance with guidance in Section 15.5.4.3, NUREG-1718, the staff review is limited to verifying whether the applicant has adequately committed to establish a process for the production, use, and management control of written plant procedures.

15.5.1.1 Plant Procedures Commitment Description

In revised CAR Section 15.5, the applicant discussed general commitments for plant procedures applicable to the startup, testing, and operations phases of the proposed facility and stated that the MPQAP contains the applicant's description of procedures for the design and construction phases. The staff reviewed the MPQAP requirements, in particular Section 5.0, "Instructions, Drawings and Procedures," and determined that they were acceptable for construction activities, including design, procurement, and fabrication (NRC, 2001). The applicant will describe its plant procedures for the facility's proposed startup, testing, and operation in more detail in any later license application it may submit.

Revised CAR Section 15.5 provides additional description of the provisions for procedures and identifies four types of plant procedures that would be used to control activities during any facility operations—operating procedures, administrative procedures, maintenance procedures, and emergency procedures. All such procedures would be prepared, issued, used, and controlled under the CM system and the MPQAP requirements.

Operating procedures would be used to directly control process operations by workstation and control room operators. They would include directions for normal operations, including startup and some testing, operation and shutdown, as well as off-normal conditions, including alarm response. They would also include operating limits and controls, controls to ensure operational safety and hold- or checkpoints. Administrative procedures would be used to perform management control activities that support operations, including CM, safety, human-system interface, QA, design control, training and qualification, audits and assessment, incident investigations, records and document control, and reporting. Maintenance procedures would address preventive and corrective maintenance, surveillance for calibration, inspection and testing, and functional testing following maintenance. Emergency procedures would address the preplanned actions of plant personnel in the event of an emergency.

The applicant described its commitments during any facility operations to review all plant procedures at least every 5 years to ensure continued accuracy and usefulness. Emergency procedures would be initially reviewed annually and subsequently every 2 years. Additional reviews and modifications of procedures would be based on facility operating experience, incidents, and identified inadequacies.

The applicant will develop procedures for test control for the preoperational testing program. These procedures will provide testing criteria for determining when tests will be required and how the activities will be performed. Tests will simulate the most adverse design conditions feasible, and results will be documented and evaluated and acceptability determined by responsible personnel. The applicant committed to developing all required plant procedures before receiving any special nuclear material (SNM) at the site of the proposed facility and to validating the procedures during startup testing of the proposed facility (CAR Section 15.5.5).

15.5.2 Evaluation Findings

In Chapter 15.5 of the revised CAR, DCS described its management measures for establishing plant procedures to be used at the proposed facility. Based on that information and the discussion provided in the sections above for plant procedures, the staff finds that the applicant is committed to and would be capable of providing the management control necessary to support operation of the proposed facility. Pursuant to 10 CFR 70.64(a)(1), the staff finds that the applicant's proposed plant procedures are a management measure providing adequate assurance that IROFS will be available and reliable to perform their function when needed. The applicant will describe its plant procedures for startup, testing, and operation in more detail in its license application. Accordingly, the staff concludes, pursuant to 10 CFR 70.23(b), that the management measures for establishing plant procedures set forth in the MPQAP and the revised CAR are part of a QA program which will provide reasonable assurance of protection against natural phenomena and the consequences of potential accidents.

15.5.3 References

(ASME, 1994) American Society of Mechanical Engineers. NQA–1–1994, "Quality Assurance Requirements for Nuclear Facility Applications." New York, 1994.

(DCS, 2002) Ihde, R., Duke Cogema Stone & Webster. Letter to U.S. Nuclear Regulatory Commission, RE: Mixed Oxide Fuel Fabrication Facility Construction Authorization Request Revision. October 31, 2002 (including page changes through February 9, 2005).

(NRC, 1978) U.S. Nuclear Regulatory Commission. Regulatory Guide 1.33, Rev. 2, "Quality Assurance Program Requirements (Operation)." Washington, DC, February 1978.

(NRC, 1989) U.S. Nuclear Regulatory Commission. "Guidance on Management Controls/Quality Assurance, Requirements for Operation, Chemical Safety, and Fire Protection for Fuel Cycle Facilities." *Federal Register:* Vol. 54, No. 53, pp. 11590–11598. March 21, 1989.

(NRC, 2001) Persinko, A., U.S. Nuclear Regulatory Commission. Letter to P. Hastings, Duke Cogema Stone & Webster, RE: Duke Cogema Stone & Webster Quality Assurance Program for the Construction of the MFFF. October 1, 2001.

(NRC, 2003) Persinko, A., U.S. Nuclear Regulatory Commission. Letter to P. Hastings, Duke Cogema Stone & Webster, RE: Duke Cogema Stone & Webster Quality Assurance Program for the Construction of the MFFF. January 10, 2003.

15.6 Audits and Assessments

15.6.1 Conduct of Review

This section of the FSER contains the staff's review of the audits and assessment information provided by the applicant in Section 15.6 of the revised CAR (DCS, 2002). The objective of this review is to determine whether the applicant has developed and adequately described a system of audits and assessments that provides reasonable assurance that the PSSCs identified by the applicant at the MFFF construction and design stage, and the IROFS to be identified by the applicant in its license application, will be available and reliable to perform their safety function when needed. The staff evaluated the applicant's system of audits and assessments by reviewing Section 15.6 of the revised CAR, other sections of the revised CAR, responses to NRC staff RAIs, and relevant documents available at the applicant's offices but not submitted by the applicant.

The following regulations apply specifically to the staff's review of the proposed audits and assessments program:

- 10 CFR 70.4 defines the term "management measures" as including an audits and assessments program.

- 10 CFR 70.64(a)(1) states that the facility design must be developed and implemented in accordance with management measures, to provide adequate assurance that IROFS will be available and reliable to perform their function when needed.

Section 15.6.3 of NUREG-1718 defines the scope of the construction approval review pertaining to the applicant's planned system of audits and assessments. In reviewing the revised CAR, the staff has evaluated this system and the applicant's provisions for continued adherence to the system.

15.6.1.1 Audits and Assessments—General

In revised CAR Section 15.6.1, the applicant discussed general commitments to perform audits and assessments and stated that they are to be performed in accordance with the applicant's MPQAP requirements for SSCs and associated activities using a graded approach commensurate with their safety significance. The staff reviewed the commitments for audits and assessments in MPQAP Section 18.0, "Audits," and Section 2.4, "Management Assessments," and determined that they are acceptable for construction activities (NRC, 2001). The safety significance of SSCs and their associated activities will be used to determine the frequency and rigor by which they are audited and assessed. The audits and assessments are to provide DCS management with feedback on the technical adequacy of SSCs and activities by evaluating how well the QA program is being implemented and feedback on the program's effectiveness in ensuring that SSCs are properly designed and constructed.

Audits and assessments of SSCs will be scheduled to provide coverage, consistency, and coordination with ongoing work and at a frequency commensurate with the project status and importance of the work. All functional areas performing work controlled by the MPQAP will be audited at least once a year. Results of audits, assessments, surveillances, deficiencies, and corrective action reports will be used to determine the scope and frequency of functional area audits. Audits will be scheduled to begin as early in the life of the work as practical and will be continued at intervals consistent with the work schedule. External audits of PSSC suppliers will be performed before contract placement, with annual supplier evaluations and full audits required every 3 years. Annual project assessments to determine the overall effectiveness of the QA program will be conducted by the project manager, and each functional area performing work on PSSCs will perform an internal management assessment annually. Additional audits and assessments of specific functions will be conducted as directed by management to provide an adequate assessment of compliance and effectiveness.

The applicant committed to conducting its internal and external audits and assessments using procedures in accordance with the approved MPQAP requirements (NRC, 2001). These procedures will include requirements for scheduling and planning, certification of audit/assessment personnel, development of audit plans and checklists, audit/assessment performance, reporting and tracking of findings to closure, and closure of the audit/assessment. They will emphasize timely reporting and correction of findings to prevent recurrence.

The applicant identified the qualifications and responsibilities of the QA manager, who will have overall responsibility for managing the QA program including audits and assessments of quality-affecting activities. The QA verification manager is directly responsible for ensuring that audits and assessments are conducted in accordance with the MPQAP requirements, including the lead auditor/auditor certification program, audit program management, reporting findings to management, evaluating the effectiveness of QA program implementation, approving audit checklists and reports, maintaining the approved suppliers list, and providing input for continuous program improvements.

The applicant described the training and qualification requirements and responsibilities for audit and assessment personnel. These include lead auditor and auditor training and certification and assessment personnel training appropriate to their activities. Audit and assessment procedures will require that personnel be independent of the activity being audited or assessed, and that they have appropriate authority, freedom, and access to make the audit process meaningful and effective and to properly audit or assess the assigned areas or activities. Checklists will establish acceptance criteria to determine acceptable performance and audit/assessment team determinations, results, and reports will be reviewed and approved by appropriate management.

Audits and assessments will be conducted using written procedures/checklists and will include detailed walkdowns of plant areas, including out-of-the-way and limited-access areas. If findings result, the deficiencies will be accurately documented for accurate evaluation and timely corrective actions, including immediate correction, where feasible and appropriate. Audit and assessment results will be reviewed by management having responsibility in the area audited/assessed. Audit and assessment findings and recommendations will be documented and distributed to appropriate management for review, and response will be required from responsible managers. Audit and assessment results will be tracked by the applicant's QA organization. The data will be analyzed and trended and resultant reports, which indicate quality trends and the effectiveness of management measures, will go to appropriate management for review, response, corrective action, and followup.

15.6.1.2 Audits

The applicant described its requirement for audit team personnel, audit conduct, and reporting of results in revised CAR Section 15.6.2. Audit team personnel will be independent of the areas and activities being audited and have no direct responsibility for the items they audit. Technical and programmatic audits will be performed to evaluate internal project activities using applicable procedures. External suppliers will be evaluated using applicable supplier evaluation procedures. Audit reports will be issued to appropriate management on a timely basis, followup reviews will be performed to verify effective completion of corrective actions for audit findings, and status of open findings will be routinely reported to project management. During construction, internal audits as-built conditions against controlled drawings, specifications, and procedures based on committed construction codes and standards.

15.6.1.3 Assessments

The applicant's description and commitments for assessments include appropriate requirements as discussed in Section 15.6.1.1, above. In particular, annual project assessments to determine the overall effectiveness of the QA program will be conducted by the project manager, and each functional area performing work on PSSCs will perform an internal management assessment annually. Additional audits and assessments of specific functions will be conducted as directed by management to provide an adequate assessment of compliance and effectiveness.

15.6.1.4 Provisions for Continuing Assurance

The applicant's provisions for adhering to its planned system of audits and assessments are described in revised CAR Section 15.6.4 and include maintaining the applicant's QA program current through deactivation of the proposed facility. Appropriate changes to the QA program and procedures for audits and assessments will be made as a result of reorganizations, revised activities, lessons learned, changes to applicable regulations, and program improvements. The applicant also committed to update the system of audits and assessments to reflect any changes to the proposed facility made between submittal of the revised CAR and the expected submittal of its later license application.

15.6.2 Evaluation Findings

In Chapter 15.6 of the revised CAR, DCS described its planned system for conducting audits and assessments on PSSCs and IROFS at the proposed facility. Based on that information and the discussion provided in the sections above for audits and assessments, the staff finds that the applicant has adequately described its system for audits and assessments. Pursuant to 10 CFR 70.64(a)(1), the staff finds that the applicant's proposed audits and assessments system is a management measure providing adequate assurance that IROFS will be available and reliable to perform their function when needed. Accordingly, the staff concludes, pursuant to 10 CFR 70.23(b), that the system for audits and assessments set forth in the MPQAP and the CAR is part of a QA program which will provide reasonable assurance of protection against natural phenomena and the consequences of potential accidents.

15.6.3 References

(DCS, 2002) Ihde, R., Duke Cogema Stone & Webster. Letter to U.S. Nuclear Regulatory Commission, RE: Mixed Oxide Fuel Fabrication Facility Construction Authorization Request Revision. October 31, 2002 (including page changes through February 9, 2005).

(NRC, 2001) Persinko, A., U.S. Nuclear Regulatory Commission. Letter to P. Hastings, Duke Cogema Stone & Webster, RE: Duke Cogema Stone & Webster Quality Assurance Program for the Construction of the MFFF. October 1, 2001.

(NRC, 2003) Persinko, A., U.S. Nuclear Regulatory Commission. Letter to P. Hastings, Duke Cogema Stone & Webster, RE: Duke Cogema Stone & Webster Quality Assurance Program for the Construction of the MFFF. January 10, 2003.

15.7 Incident Investigations

15.7.1 Conduct of Review

This section of the FSER contains the staff's review of the system for incident investigations referenced by the applicant in Section 15.7.1 of the revised CAR (DCS, 2002). The objective of this review is to determine whether DCS has developed and adequately described a process for the systematic investigation of incidents, assignment and acceptance of corrective actions, and followup to ensure completion of the actions. The staff evaluated the applicant's proposed system by reviewing Section 15.7 of the revised CAR, MPQAP Section 16, responses to NRC staff RAIs, and relevant documents available at the applicant's offices but not submitted by the applicant.

The following regulations apply specifically to the staff's review of the proposed system for conducting incident investigations:

- 10 CFR 70.4 defines the term "management measures" as including incident investigations.

- 10 CFR 70.64(a)(1) states that the MFFF design must be developed and implemented in accordance with management measures, to provide adequate assurance that IROFS will be available and reliable to perform their function when needed.

With respect to the applicant's proposed system for conducting incident investigations, Section 15.7.4.3 of NUREG-1718 limits the staff's CAR review to verifying whether the applicant has committed to establishing a system to adequately investigate incidents. Such a system includes provisions for the assignment and acceptance of corrective actions and followup measures to ensure completion of corrective actions.

15.7.1.1 Incident Investigation and Corrective Action Process and Administration

In revised CAR Sections 15.7.1 and 15.7.2, the applicant discussed its commitments and process to perform incident investigation and corrective action and stated that the process currently in use during design and construction phases of the proposed facility is as described in the applicant's MPQAP Section 16, "Corrective Action." Section 16 of the MPQAP was previously reviewed and approved by the staff (NRC, 2001). It contains the applicant's system for identifying, classifying, following up, closing and trending conditions adverse to quality, and preparing significant event reports. The staff has verified that this system is adequate for ensuring that proper corrective action would be taken during any construction activities, including design, procurement, and fabrication (NRC, 2001). In revised CAR Section 15.7.2, the applicant committed to modifying this process before any facility startup testing to include the additional specific actions that would be needed to support an operating facility.

The applicant's description in revised CAR Section 15.7.1 of the incident investigation and corrective action process for design, construction, and operations includes management controls to promptly identify incidents/findings, evaluate the need to stop work, and assign investigation teams. It also provides for significance and root cause evaluations and corrective action planning, management approval, implementation, completion and tracking, as well as tracking and evaluation for adverse trends.

Corrective action process administration is discussed in revised CAR Section 15.7.2. The incident investigation and deficiencies corrective action process will be administered by the applicant's QA organization during the design and construction phases of the proposed facility. The applicant committed to providing in the license application a detailed description of how the incident investigations process would work during any operation of the proposed facility. Such a description will be expected to address the prompt investigation of incidents, the use of qualified investigative teams, monitoring and documenting corrective actions, investigating team plans, methodologies, personnel qualifications and independence, and appropriate documentation and records requirements.

15.7.2 Evaluation Findings

In Chapter 15.7 of the revised CAR, DCS described its planned system for performing incident investigations relevant to PSSCs and associated activities during the construction of the proposed facility, including design, procurement, and fabrication activities. Based on that information and the discussion provided in the sections above for audits and assessments, the staff finds that the applicant has adequately described its system for performing incident investigations. Pursuant to 10 CFR 70.64(a)(1), the staff finds that the applicant's proposed incident investigations system is a management measure providing adequate assurance that IROFS will be available and reliable to perform their function when needed. Accordingly, the staff concludes, pursuant to 10 CFR 70.23(b), that the system for incident investigations set forth in the MPQAP and the revised CAR is part of a QA program which will provide reasonable assurance of protection against natural phenomena and the consequences of potential accidents.

15.7.3 References

(DCS, 2002) Ihde, R., Duke Cogema Stone & Webster. Letter to U.S. Nuclear Regulatory Commission, RE: Mixed Oxide Fuel Fabrication Facility Construction Authorization Request Revision. October 31, 2002 (including page changes through February 9, 2005).

(DOE, 1992a) U.S. Department of Energy. DOE–STD–1010–92, "Guide to Good Practices for Incorporating Operating Experiences." Washington, DC, July 1992.

(DOE, 1992b) U.S. Department of Energy. DOE–NE–STD–1004–92, ""Root Cause Analysis Guidance Document." Washington, DC, February 1992.
(NRC, 1986) U.S. Nuclear Regulatory Commission. NUREG/CR–4616, "Root Causes of Component Failures Program: Methods and Applications." Washington, DC, December 1986.

(NRC, 1991) U.S. Nuclear Regulatory Commission. NUREG/CR–5665, "A Systematic Approach to Repetitive Failures." Washington, DC, February 1991.

(NRC, 1996) U.S. Nuclear Regulatory Commission. Information Notice 96–28, "Suggested Guidance Relating to Development and Implementation of Corrective Action." Washington, DC, May 1996.

(NRC, 1999) U.S. Nuclear Regulatory Commission. "Domestic Licensing of Special Nuclear Material (10 CFR Part 70)." *Federal Register:* Vol. 64, No. 146, pp. 41338–41357. July 30, 1999.

(NRC, 2001) Persinko, A., U.S. Nuclear Regulatory Commission. Letter to P. Hastings, Duke Cogema Stone & Webster, RE: Duke Cogema Stone & Webster Quality Assurance Program for the Construction of the MFFF. October 1, 2001.

(NRC, 2003) Persinko, A., U.S. Nuclear Regulatory Commission. Letter to P. Hastings, Duke Cogema Stone & Webster, RE: Duke Cogema Stone & Webster Quality Assurance Program for the Construction of the MFFF. January 10, 2003.

15.8 **Records Management**

15.8.1 Conduct of Review

This section of the FSER contains the staff's review of the facility records management system provided by the applicant in Section 15.8 of the revised CAR (DCS, 2002). The objective of this review is to verify that the applicant has developed and adequately described a facility records management system that complies with NRC requirements. The staff evaluated the applicant's facility records management system by reviewing Section 15.8 of the revised CAR, other sections of the revised CAR, responses to NRC staff RAIs and relevant documents available at the applicant's offices but not submitted by the applicant.

The following regulations apply specifically to the staff's review of the proposed records management system:

- 10 CFR 70.4 defines the term "management measures" as including a records management system.

- 10 CFR 70.64(a)(1) states that the MFFF design must be developed and implemented in accordance with management measures, to provide adequate assurance that IROFS will be available and reliable to perform their function when needed. Appropriate records of these items must be maintained by or under the control of the licensee throughout the life of the facility.

Section 15.8.3 of NUREG-1718 defines the scope of the construction approval review of the applicant's facility records management system to include (1) the process whereby records are specified, created, verified, categorized, indexed, inventoried, protected, stored, maintained, distributed, and deleted or preserved, (2) the handling and control of various kinds of records and the methods of recording media that comprise the records, including contaminated and classified records, and (3) the physical characteristics of the record storage facilities with respect to the preservation and protection of the records for their designated lifetimes.

The applicant's description of its facility records management system, in revised CAR Section 15.8, addresses commitments for the records management program, including record generation, receipt, storage, preservation, safekeeping, correction, retrieval, and disposition, for program changes and provisions for continuing adequate records management.

15.8.1.1 Records Management Program

In revised CAR Section 15.8.1, the applicant described its facility records management system, and stated that the system for controlling records management responsibilities and the generation, review, approval, classification, verification, indexing, storage, protection, maintenance, correction, retrieval and disposition of QA records will be in accordance with the MPQAP requirements. The staff reviewed the MPQAP Section 17.0, "Quality Assurance Records," commitments for audits and assessments and determined that they were acceptable for construction activities at the proposed facility (NRC, 2001). Section 17.0 of the MPQAP commits the applicant to adhere to the requirements of Criterion 17, "Quality Assurance Records," in Appendix B to 10 CFR Part 50 and Basic Requirement 17 and Supplement 17S-1 of NQA-1-1994 Part I, as revised by RG 1.28, Revision 3. The applicant did not request to be excepted from any of these requirements. The staff reviewed the applicant's commitments and the description of the QA program for records in accordance with NUREG-1718 and compared them to the applicable requirements of Appendix B and the NQA-1 provisions. The staff reviewed the MPQAP description of the facility QA records management system and verified that MPQAP Section 17.0 meets the requirements of Appendix B and NQA-1.

The applicant further described the records management system in revised CAR Section 15.8.1, including applicable project procedures, dual facility storage and fireproof backup tape storage for electronic data management system, and use of fireproof storage for other documents such as radiographs and microfilm. The procedures control the generation of records, their review and approval as records, receipt process, storage, preservation, and safekeeping. Records requiring correction or revision will be retrieved by authorized individuals in accordance with the applicable procedures. Original records will be retained and the revision processed per the applicable project procedures. Records file folders, and interfacing links to associated documents/records, will be structured to ensure timely retrievability of records. All "lifetime" QA records will be stored for the period of time in which the proposed facility may operate.

15.8.1.2 Records Management Program Changes

The applicant's discussion includes the requirements that routine audits, surveillances, and assessments of document control and records management will be performed to evaluate the implementation of the program. Findings and observations from such oversight functions and other monitoring activities may result in program improvements.

15.8.1.3 Continuing Records Management Provisions

The applicant's provisions for ensuring the continuing adequacy of its records management system include commitments to keep the program procedures current. The records management procedures will be revised based on lessons learned during implementation; corrective actions from audits, surveillances, or assessments; improvements based on trend analysis; and changes that result from regulations, commitments, reorganizations, revised schedules, or program improvements. The applicant committed to update the facility records management system to reflect any changes to the proposed facility made between submittal of the revised CAR and the expected submittal of its later license application.

15.8.2 Evaluation Findings

In Chapter 15.8 of the revised CAR, DCS described its planned records management system to be used at the proposed facility. Based on that information and the discussion provided in the sections above for the records management system, the staff finds that the applicant has adequately described its system for records management. Pursuant to 10 CFR 70.64(a)(1), the staff finds that the DCS proposed records management system is a management measure providing adequate assurance that IROFS will be available and reliable to perform their function when needed, and that appropriate records will be maintained by or under the control of DCS throughout the operating life of the proposed facility, if the NRC authorizes its operation. Accordingly, the staff concludes, pursuant to 10 CFR 70.23(b), that the records management system set forth in the MPQAP and the revised CAR is part of a QA program which will provide reasonable assurance of protection against natural phenomena and the consequences of potential accidents.

15.8.3 References

(DCS, 2002) Ihde, R., Duke Cogema Stone & Webster. Letter to U.S. Nuclear Regulatory Commission, RE: Mixed Oxide Fuel Fabrication Facility Construction Authorization Request Revision. October 31, 2002 (including page changes through February 9, 2005).

(NRC, 2001) Persinko, A., U.S. Nuclear Regulatory Commission. Letter to P. Hastings, Duke Cogema Stone & Webster, RE: Duke Cogema Stone & Webster Quality Assurance Program for the Construction of the MFFF. October 1, 2001.

(NRC, 2003) Persinko, A., U.S. Nuclear Regulatory Commission. Letter to P. Hastings, Duke Cogema Stone & Webster, RE: Duke Cogema Stone & Webster Quality Assurance Program for the Construction of the MFFF. January 10, 2003.

NRC FORM 335 (9-2004) NRCMD 3.7	U.S. NUCLEAR REGULATORY COMMISSION	1. REPORT NUMBER (Assigned by NRC, Add Vol., Supp., Rev., and Addendum Numbers, If any.)
	BIBLIOGRAPHIC DATA SHEET *(See instructions on the reverse)*	NUREG-1821

2. TITLE AND SUBTITLE	3. DATE REPORT PUBLISHED	
Final Safety Evaluation Report on the Construction Authorization Request for the Mixed Oxide Fuel Fabrication Facility at the Savannah River Site, South Carolina	MONTH	YEAR
	March	2005
	4. FIN OR GRANT NUMBER	

5. AUTHOR(S)	6. TYPE OF REPORT
	Final
	7. PERIOD COVERED *(Inclusive Dates)*

8. PERFORMING ORGANIZATION - NAME AND ADDRESS *(If NRC, provide Division, Office or Region, U.S. Nuclear Regulatory Commission, and mailing address; if contractor, provide name and mailing address.)*

Division of Fuel Cycle Safety & Safeguards

Office of Nuclear Material Safety & Safeguards

U.S. Nuclear Regulatory Commission

Washington, DC 20555-0001

9. SPONSORING ORGANIZATION - NAME AND ADDRESS *(If NRC, type "Same as above"; if contractor, provide NRC Division, Office or Region, U.S. Nuclear Regulatory Commission, and mailing address.)*

Same as above

10. SUPPLEMENTARY NOTES

11. ABSTRACT *(200 words or less)*

The U.S. Department of Energy (DOE), National Nuclear Security Administration (NNSA), has contracted with Duke Cogema Stone & Webster (DCS or applicant) to design, construct, and operate a proposed Mixed Oxide (MOX) Fuel Fabrication Facility (MFFF) that would convert depleted uranium and weapon-grade plutonium into MOX fuel. The proposed MOX facility would be located on the DOE's Savannah River Site in South Carolina. Use of the proposed facility to produce MOX fuel would be part of the DOE's surplus plutonium disposition program. The purpose of the DOE program is to ensure that plutonium produced for nuclear weapons and declared excess to national security needs is converted to proliferation-resistant forms.

Under the applicable 10 CFR Part 70 requirements, before a license to possess and use special nuclear material may be issued for the proposed MFFF, the U.S. Nuclear Regulatory Commission (NRC) must first authorize construction of the proposed MFFF. On February 28, 2001, DCS submitted to the U.S. Nuclear Regulatory Commission (NRC) a Construction Authorization Request (CAR). This Final Safety Evaluation Report (FSER) documents the NRC staff's review of the CAR, as amended, and supplemental supporting information provided by the applicant. This FSER only addresses regulatory requirements for approval of construction, and does not address operational aspects of the proposed facility.

12. KEY WORDS/DESCRIPTORS *(List words or phrases that will assist researchers in locating the report.)*	13. AVAILABILITY STATEMENT
Mixed Oxide Fuel Fabrication Facility	unlimited
MOX	14. SECURITY CLASSIFICATION
fuel cycle	*(This Page)*
MOX fuel	unclassified
plutonium	*(This Report)*
safety evaluation	unclassified
	15. NUMBER OF PAGES
	16. PRICE

NRC FORM 335 (9-2004)   PRINTED ON RECYCLED PAPER

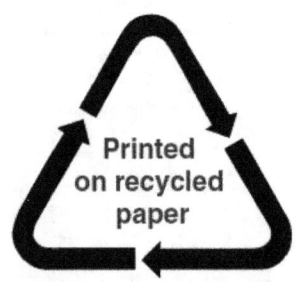

Federal Recycling Program